Radio Theory Handbook
Beginner to Advanced

Ron. Bertrand VK2DQ

Copyright © 2016 Ron. Bertrand

All rights reserved.

ISBN: 1534696121
ISBN-13: 978-1534696129

DEDICATION

For Lorraine, Amor vincit omnia.

Acknowledgments

The author extends a sincere gratitude to the Trainers, Students and Friends of the Radio & Electronics School for the assistance in producing this handbook. Without their input and gentle pushing this handbook may never have been produced.

Disclaimer of Liability

The material contain in this publication are supplied without guarantee. The author expressly disclaims all and any liability to any persons whatsoever in respect of anything done or omitted to be done by any such person in reliance either in whole or in part upon any contents of this publication. All rights reserved. © R. Bertrand.

CONTENTS

1	Basic Electricity – Part 1.	8
2	Basic Electricity – Part 2.	18
3	Basic Electricity – Part 3	28
4	Ohm's Law	40
5	Series Circuits	51
6	Parallel Circuits	57
7	Series-Parallel Circuits	66
8	Magnetism	71
9	Alternating Current – Part 1	84
10	Alternating Current – Part 2	91
11	Capacitance – Part 1	99
12	Capacitance – Part 2	112
13	Capacitive Reactance	122
14	Inductance	131
15	Meters	142
16	Electric Cells & Battery's	154
17	Microphones	166
18	Transformers	174
19	Resistive, Inductive & Capacitive Circuits	188
20	Power Supplies	208

21	Electron Tubes	231
22	Decibels	247
23	Electromagnetic Radiation	260
24	Semiconductors Part 1	272
25	Semiconductors Part 2	294
26	Amplifiers	309
27	Oscillators	325
28	Propagation	359
29	Amplitude Modulation	368
30	AM Transmitters & Receivers Part 1	384
31	AM Transmitters & Receivers Part 2	394
32	Filters in Radiocommunications	409
33	Interference	425
34	Frequency Modulation	436
35	A Complete FM Transceiver	455
36	Transmitter Faults	459
37	Antennas	469
38	Transmission Lines	499
39	Test Equipment & Measurements	530
40	Other Modes	547

1. Basic Electricity-Part 1

This book begins with the basics of electrical and radio theory and then progressively builds to a solid understanding of Advanced Amateur Radio Operator level as it applies to countries like Australia, U.K. USA and Canada. While starting at beginner level, the target standard of this book is CEPT Radio Experimenter Advanced Licence (T/R 61-01). It is imperative to have a thorough understanding of the fundamentals of electricity before any study of radio and communications. We start our study from the beginning, thus benefiting readers with a limited knowledge of the basics whilst also providing a convenient refresher for those who may have long since covered the material. Once we have the fundamentals down rock solid, we will progress to the Advanced Radio Theory.

ELECTRICITY

The complicated electronic systems involved in modern-day communication, satellites, nuclear power plants, radio and television and even up-to-date automobiles, do not require technicians to understand the functioning of electric and electronic circuits. Modern day electronics is very modular. A remove and replace, or substitution of the suspected 'faulty module' is the approach to modern electronics servicing. This, in itself, is not a bad thing, as in the real world, getting an electronic device up and going is the most important thing. However, to have a true understanding requires a strong foundation in the basics of electricity. The term "electronic" infers circuits ranging from the first electronic device, the electron tube, to the newer solid-state devices such as diodes and transistors, as well as integrated circuits (IC's). The term "electric" or "electrical" is usually applied to systems or circuits in which electrons flow through wires but which involve no vacuum tubes or solid-state devices. Many modern electrical systems are now using electronic devices to control the electric current that flows in them.

What makes such a simple thing as an electric lamp glow? It is easy to pass the problem off with the statement, "The switch connects the light to the power lines and it glows" or something to that effect. But what does connecting the light to the power lines do? How does energy travel through solid copper wires? What makes a motor turn. A radio play? What is behind the dial that allows you to pick out one radio station from thousands of others operating at the same time? How fast is electricity? There are no single simple answers to any of these questions. Each question requires the understanding of many basic principles. By adding one basic idea to another, it is possible to answer, eventually, all of the questions that may be asked about the intriguing subjects of electricity, electronics and radio.

When a light switch is turned on and elsewhere the light suddenly glows, energy has found a path through the switch to the light. The path used is usually along copper wires and the tiny particles that do the moving and carry the energy are called **electrons**. These electrons are important to anyone studying electronics and radio since they are usually the only particles that are considered to move in electric circuits.

To explain what is meant by an **electron**, it will be necessary to investigate more closely the makeup of all matter. The word "matter" means, in a general sense, anything that can be touched. It includes substances such as rubber, salt, wood, water, glass, copper and air. The whole world is made of different kinds of stuff. The ancient Greek philosophers were always trying to find the 'stuff' that the universe was made of. Even before the Greeks, the Alchemists were trying to find the basic building blocks that all matter was made from, though most of the time their driving force was not so much science but the pursuit of wealth. They figured that if they could isolate the building blocks of matter, then they would be able to 'create' matter themselves. One of their primary pursuits was the creation of the precious metal gold.

Water is one of the most common forms of 'stuff' that we call matter. If a drop of water is divided in two and then divided again and again until it can be divided no longer and still be water, then we have arrived at the smallest possible piece of water. We have a water molecule. The ancient Greeks would have called the smallest droplet of water an *atomos* (atom). The word *atomos* means *indivisible*. We know today that substances such as water can be divided into more fundamental bits.

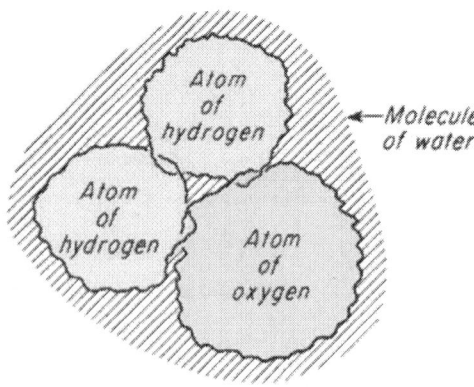

Figure 1-1

The water molecule can be broken down into still smaller particles, but these new particles will not be water. Physicists have found that three smaller particles make up a molecule of water. Two atoms of hydrogen (H) and one atom of oxygen (O) as shown in figure 1-1. The symbols 'H' and 'O' are universal symbols used to represent Hydrogen and Oxygen. Oxygen, at average temperatures, is one of the several gases that constitute the air we breathe. Hydrogen is also a gas in its natural state; it is found in everyday use as part of the gas used for heating or cooking. If a gaseous mixture containing two parts of hydrogen and one part of oxygen is ignited, a violent chemical reaction, an explosion, will occur as water is formed and excess energy released. This is not an experiment that I would recommend.

Water is made up of two atoms, hydrogen and oxygen. Water is a molecule. A molecule is a substance that is made up of groups of atoms. If you divided a droplet of water down to its smallest possible size, you would have a single molecule of water. If you had the means to split the water molecule further, you would no longer have water; you will have the atoms (hydrogen and oxygen) that make up water. The chemical name of water then is Di-Hydrogen Oxide. It has been found that atoms are also divisible. An atom is made up of at least two types of particles: **protons** and **electrons** and a third particle called a **neutron**. Don't let these names concern you too much. For our purposes, the primary particle is the electron, at this time. Electrons and protons are called electrical particles and neither one is divisible (in typical environments). All the molecules that make up all matter in the universe are composed of these electrical proton-electron pairs.

ELECTRONS AND PROTONS

Electrons are the smallest and lightest of the fundamental particles. They are said to have a negative charge, meaning that they are surrounded by some invisible field of force that will react in an electrically negative manner with other matter.

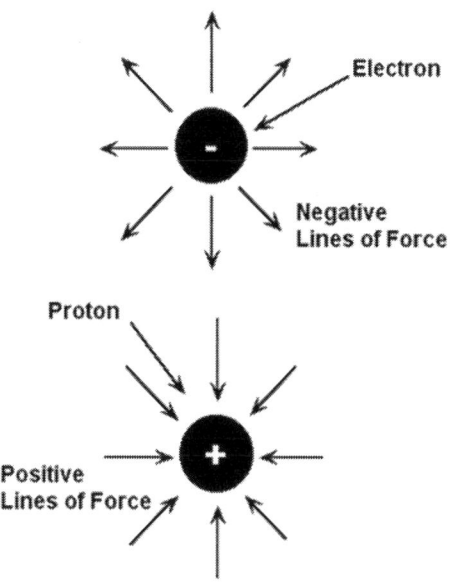

Figure 2-1

Protons are said to have a positive charge and are surrounded by an invisible force field that causes them to react in an electrically positive manner. The words negative and positive are just names to describe the so-called charge of electrons and protons and their charge describes how they interact with each other. We could just as easily call the charge of the electron the **white charge** and the charge of the proton the **black charge.** My point is, 'charge' is an electric behaviour and since there are two types of charge we need to name them so that when we talk about them we will know which behaviour we are speaking of: either the positive charge behaviour or the negative charge behaviour.

AN EXPERIMENT WITH CHARGE

You may have already done this if you have, please try it again. Tear up some tiny strips of paper and place them on the table in front of you. Make sure no one is watching! Now run a hair comb through your hair briskly several times and put the comb close to the bits of paper. Before, the comb touches the paper, a bit of paper will leap off the table and move through the air and cling to the comb. This happens because you have produced a **charge** on the comb, which will physically interact with matter around it (in our case the bits of paper).

The charge on the comb was created by friction between the comb and your hair. Protons are about *eighteen hundred* times more massive than electrons and have a positive electric field surrounding them. *The proton is exactly as positive as the electron is negative; each has a unit electric charge.*

When an electron and a proton are far apart, only a few of their lines of force (the invisible field around them) join and pull together. The attracting pull between the two charges is, therefore, small. When brought closer together, the electron and proton can link more of their lines of force and will pull together with greater force. If close enough, all the lines of force from the electron are joined to all the lines of force of the proton and there is no external field and they attract each other strongly. Together, a positive charge proton and the negative charge of an electron cancel out and they form a neutral, or uncharged, group. The neutral atomic particle, known as a **neutron**, exists in the nucleus of all atoms heavier than hydrogen. The fact that electrons repel other electrons, protons repel other protons, but electrons and protons attract each other gives us the basic law of charges:

LIKE CHARGES REPEL, UNLIKE CHARGES ATTRACT

Because the proton has about 1,800 times more mass than the electron, it seems reasonable to assume that when an electron and a proton attract each other, it will be the tiny electron that will do most of the actual moving. Such is the case. It is *the Electron which moves in electricity*. If the proton were the lower mass particle, we would probably have called what we know today as 'electricity, something like 'proton'icity. Regardless of the difference in apparent size and mass, the negative field of an electron is just as strong negatively as the positive field of a proton is positive.

Though physically small, the field near the electron is quite strong. If the field strength (field strength = the strength of the invisible field) around an electron at a distance of one-millionth of a metre is a certain amount, at two-millionths of a metre it will be one-quarter as much; at four-millionths of a metre it will be one-sixteenth as much; and so on. If the field decreases as distance increases, the field is said to vary inversely with distance. It varies inversely with the distance squared.

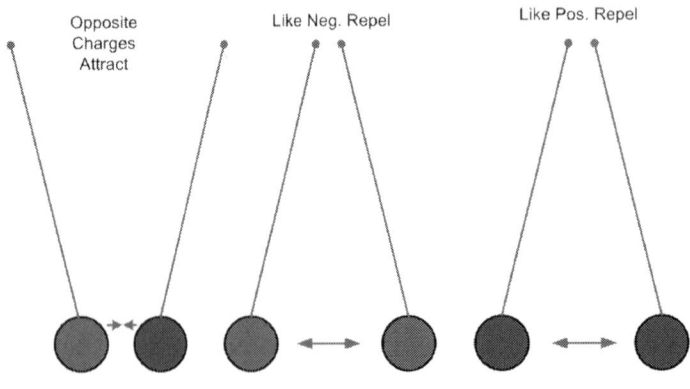

Figure 3-1

Note: a millionth of a metre has a name; it is called a 'micron'. A micron is not an SI Unit, but it is a unit which has stuck and is still used. It is too nice of a name not to be used.

When an increase in something produces an increase in something else, the two things are said to vary directly rather than inversely. Two million electrons on an object produce twice as much negative charge as one million electrons would. The charge is **directly proportional** to the number of **electrons.** The invisible fields surrounding electrons and protons are known as electrostatic fields. The word 'static' means, in this case, "stationary", or "not caused by movement". When electrons are made to move, the result is **dynamic electricity.** The word "dynamic" indicates that motion is involved. To produce a movement of an electron, it will be necessary to have either, a negatively charged field to push it, or positively charged field to pull it. Normally in an electric circuit, both a negative and a positive charge are used (a pushing and pulling pair of forces).

THE ATOM AND ITS FREE ELECTRONS

There are more than 120 different kinds of atoms, or elements, from which the millions of different forms of matter found in the universe, are composed. Let me elaborate on the last paragraph. About 100 different atoms are found in nature. Many atoms do not occur naturally. Many are only manufactured by very powerful particle colliders or inside stars.

The heavy atoms, those containing a large number of protons, electrons and neutrons, like uranium and radium are unstable. They throw off energy (they are radioactive) and decompose until they become stable non-radioactive atoms. A material, which is only made from one type of atom, is called an **element**. Water is not an element because it contains two types of atoms, hydrogen and oxygen. Water is, therefore, a molecule. Copper contains **only** copper atoms, so copper is an element. There are many other common elements.

The simplest and lightest atom (or element) is hydrogen. An atom of hydrogen consists of one electron and one proton, as shown in figure 4-1. In one respect the hydrogen atom is similar to all others: the electron whirls (orbits) around the proton, or **nucleus**, of the atom, much as planets **rotate** around the sun. Electrons whirling around the nucleus are termed planetary, or orbital, electrons! This is not an accurate model but is accurate for our purposes in radio physics. Electrons are more like clouds around the nucleus with electrons having a probability that they are more likely to be in the vicinity of one or more bands. We really can't say that they are just like planets around the Sun however we can use this analogy for our purposes.

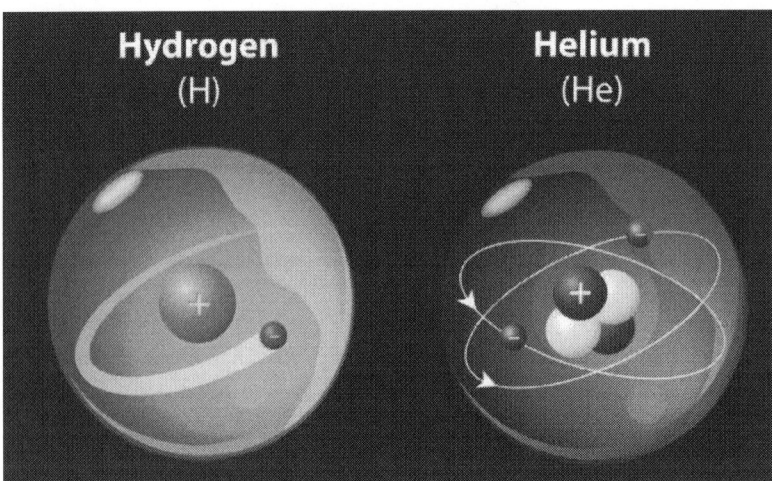

Figure 4.1

The nucleus is just the name given to the 'centre' of the atom. The next atom in terms of weight is helium, having two protons and two electrons. The third atom is lithium, with three electrons and three protons and so on. All elements are grouped in families with the atomic weights in the Periodic table of elements. Table 1-1.

Most atoms have a nucleus (centre) consisting of all the protons of the atom and also one or more neutrons. The electrons (always equal in number to the protons in the nucleus) are whirling around (orbiting) the nucleus in various **layers**. The first layer of electrons outside the nucleus can accommodate only two electrons. If the atom has three electrons, two will be in the first layer and the third will be in the next layer. The second layer is completely filled when eight electrons are whirling around in it. The third is filled when it has eighteen electrons. Some of the electrons in the outer orbit, or shell, of the atoms of many materials such as copper or silver, exist in a higher "conduction level" and can be dislodged easily. These electrons travel out into the wide-open spaces between the atoms and molecules and may be termed free **electrons**

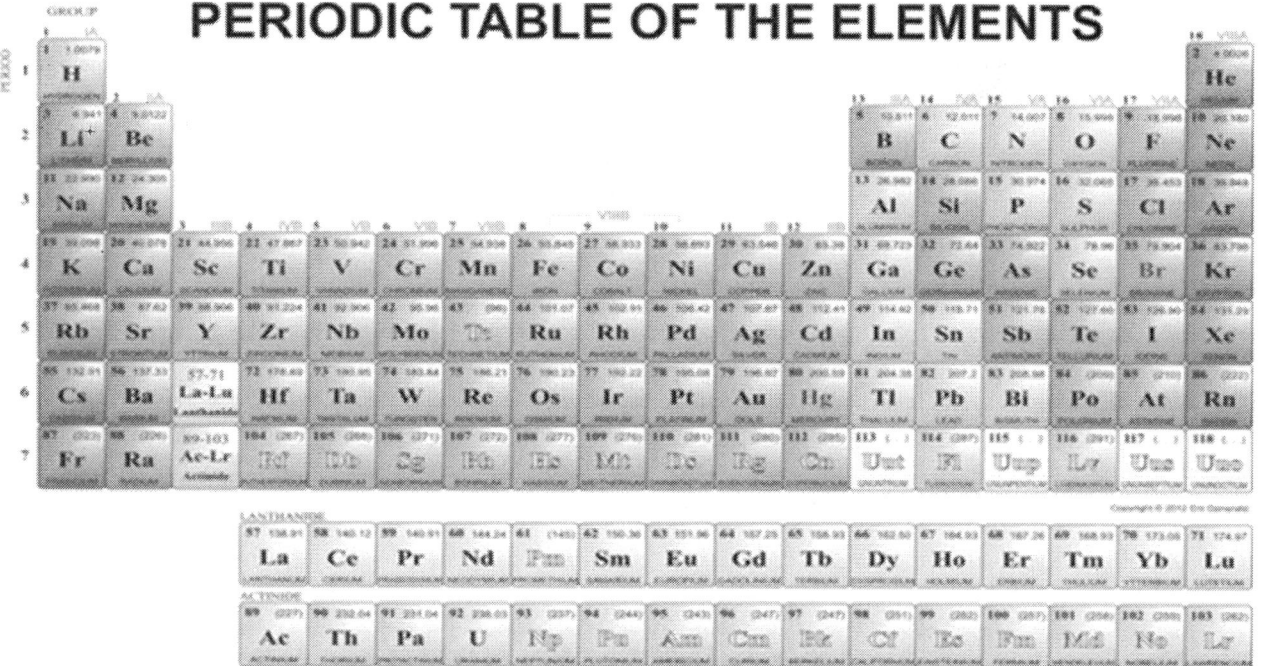

Table 1-1.

Other electrons in the outer orbit will resist dislodgment and are called <u>bound</u> or valence electrons. Materials consisting of atoms (or molecules) having many free electrons will allow an easy interchange of their outer-shell electrons while atoms with only bound electrons will hinder any electron exchange. Copper, for example, has one electron in its outer orbit or layer. This lonely little outer electron of the copper atom is very easy to 'steal' from the copper atom and made to move. The outer electron is called a free electron. It is not free but loosely bound to the atom and easy to encourage away and made to move, so we call it a free electron. Copper does not resist the movement of its outer electrons strongly, or in other words, it does not offer much resistance to us if we try to get its outer electrons to move. We will talk about how we get them to move later. A material, which does not have free electrons, is said to have a high resistance. All metals have free electrons. Most common metals when heated cause greater energy to be developed in their free electrons. The more energy electrons have the greater they resist orderly movement through the material. The material is said to have an increased resistance to the movement of electrons through it.

THE ELECTROSCOPE

An example of electrons and electric charges acting on one another is demonstrated by the action of an electroscope. An electroscope consists of two very thin gold or aluminium leaves attached to the bottom of a metal rod. The delicate metal-foil leaves and rod are encased in a glass bottle to protect them from air movement.

To understand the operation of the electroscope, it is necessary to recall these facts:

1) Normally an object has a neutral charge.
2) Like charges repel; unlike charges attract.
3) Electrons are negative.
4) Metals have free electrons.

Normally the metal rod of the electroscope has a neutral charge and the leaves hang downward parallel to each other, as shown in figure 5-1. The leaves are shown in charged position in figure 5-1. With no charge the leaves hang vertically down from the rod.

Figure 5-1.

Rubbing a piece of hard rubber with wool causes the wool to **lose electrons** to the rubber and the **excess electrons** on the rubber charge the rubber **negatively**. When such a negatively charged object is brought near the top of the rod, some of the free electrons at the top are repelled and travel down the rod, away from the negatively charged object. Some of these electrons force themselves onto one of the leaves and some onto the other. Now the two leaves are no longer neutral but **are slightly negative and repel each other**, moving outward from the vertical position as shown. When the charged object is removed, electrons return up the rod to their original areas. The leaves again have a neutral charge and hang down parallel to each other.

Since the charged object **did not touch** the electroscope, it neither placed electrons on the rod nor took electrons from it. When electrons were driven to the bottom, making the leaves negative, these same electrons leaving the top of the rod left the top positive. The overall charge of the rod remained neutral. When the charged object was withdrawn, the positive charge at the top of the rod pulled the displaced electrons up to it. All parts of the rod were then neutral again. If a positively charged object, such as a glass rod vigorously rubbed with a piece of silk, is brought near the top of the electroscope rod, some of the free electrons in the leaves and rod will be attracted upward toward the positive object. This charges the top of the rod negatively because of the excess of free electrons there. Both leaves are left with a deficiency of free electrons which means they are positively charged. Since both leaves are similarly charged again, they repel each other and move outward a second time.

A deficiency of electrons on an object leaves the object with a positive charge. An excess of electrons gives it a negative charge. If a negatively charged object is **touched** to the metal rod, some excess electrons will be deposited onto the rod and will be immediately distributed throughout the electroscope. The leaves spread apart. When the object is taken away, an excess of electrons remains on the rod and the leaves. **The leaves stay spread apart**. If the negatively charged electroscope is touched to a large body that can accept the excess free electrons, such as a person, a large metal object, or earth (the ground), the excess electrons will have a path to leave the electroscope and the leaves will collapse as the charge returns to neutral. The electroscope has been discharged. If a positively charged object is **touched** to the top of the metal rod, the rod will lose electrons to it and the leaves will separate. When the object is taken away, the rod and leaves still lack free electrons and are therefore positively charged and the leaves will remain apart.

A large neutral body touched to the rod will drain some of its free electrons to the electroscope discharging it and the leaves will hang down once more. The electroscope demonstrates the free movement of electrons that can take place through metallic objects or conductors when **electric pressures**, or charges, are exerted on the free electrons.

2. Basic Electricity-Part 2

THE THREE BIG NAMES IN ELECTRICITY

Without calling them by name, we have touched on the three elements always present in operating electric circuits:

Current: A progressive movement of free electrons along a wire or other conductor caused by electrical pressure.

Voltage: The electron-moving force in a circuit that pushes and pulls electrons (current) through the circuit. Also called electromotive force and electrical pressure.

Resistance: Any opposing effect that hinders free-electron progress through wires when an electromotive force is attempting to produce a current in the circuit.

We will be talking a lot about these three properties of an electric circuit and how they relate to each other.

A SIMPLE ELECTRIC CIRCUIT

The simplest of electric circuits consists of some electron-moving force, or source, such as that provided by a dry cell, or battery; a load, such as an electric light; connecting wires and a control device. A pictorial representation and the electric diagram of a simple circuit are shown in figure 1-2.

The diagram on the right in figure 1-2 is called a schematic diagram and is much easier to draw. The control device is a switch, to turn the bulb on and off. In effect, the switch disconnects one of the wires from the cell. In our circuit, the switch could be connected anywhere to turn the bulb on and off. In this circuit, the light bulb is the load. Although the wires connecting the source of electromotive force (the dry cell) to the load may have some resistance, it is usually tiny in comparison with the resistance of the load and is ignored in most cases.

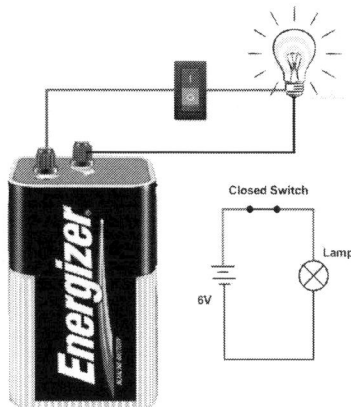

Figure 1-2.

A straight line in a schematic diagram is considered to connect components electrically and does not represent any resistance in the circuit. In the simple circuit shown, the cell produces the electromotive force (voltage) that continually pulls electrons to its positive terminal from the bulb's filament and pushes them out of the negative terminal to replace the electrons that were lost to the load by the pull of the positive terminal. The result is a continual flow of electrons through the lamp filament, connecting wires and source. The special resistance wire of the lamp filament heats when a current of electrons flows through it. If enough current flows, the wire becomes white-hot and the bulb glows and gives off light (incandescence).

CURRENT

A stream of electrons forced into motion by an electromotive force is known as a current. Here we have a definition of electric current:

Current is the ordered movement of electrons in a circuit.

In a good conducting material such as copper, one or more free electrons in the outer ring are constantly flying off at a high rate. Electrons from other nearby atoms fill in the gaps. There is a constant aimless movement of billions of electrons in all directions at all times in every part of any conductor. This random movement of electrons is not an electric current as there is not yet movement in any one direction.

Only when a voltage is applied, do we get an ordered movement of electrons.

When an electric force is applied across the conductor (from a battery), it drives some of these aimlessly moving free electrons away from the negative force toward the positive. It is unlikely that any one electron will move more than a fraction of a centimetre in a second, but an energy flow takes place along the conductor at significant fraction of the velocity of light.

Notice that I said the energy flow in the circuit is very fast (almost the speed of light - but not quite). The speed of the electrons in a circuit - or the current flow is in fact very slow. I won't bore you with calculations. However, I did once calculate how fast the electron flow was in a typical circuit and it came to be about walking speed. Electron flow or current flow is very slow. The effect of an electric current at a distance through a conductor, on the other hand, is very fast. If you want to see a calculation of the speed of electricity, see the article *How fast is electricity* in the downloads section of the RES website; www.res.net.au.

If you have trouble with this and many do, think about how fast water travels through a pipe. The dam where the water comes from may be many kilometres away from the tap. When you turn the tap on the water comes out immediately does it not? Did the water travel all the way from the dam to the tap in an instant? I am sure you would agree that it did not. If I tried to tell you that it did you would most probably laugh at me and say the water was already in the pipe, all you did by turning the tap on was to make the water move in the pipe between the dam and the tap.

Similarly, the electrons are already in the wire (conductor). When we close a switch in the circuit and apply an electromotive force all we are doing is making all the electrons move in the conductor at the same time. It may take a very long time for an electron leaving the source to reach the load if it ever does.

HOW FAST ARE MARBLES?

Let's do another analogy to make sure we have got this clear. Suppose a pipe was connected between Sydney and Melbourne. Imagine if we blocked off one end of the pipe and filled it with marbles until we could not fit any more in. We now unblock the pipe and we have a crowd of people at each end to witness the experiment and find out how fast marbles travel. The two crowds are in contact by telephone or radio and anxiously awaiting the big moment. One too many marbles are about to be inserted into the full pipe. As soon as a marble is inserted in Sydney, a marble drops out in Melbourne. The effect of pushing a marble into the pipe in Sydney caused an immediate result or effect in Melbourne. The newspaper's report "Eccentric experimenter proves that marbles travel at the speed of light". Is this right?

I hope you are shaking your head and saying 'no'. The marble, which fell out of the pipe in Melbourne, was sitting there ready to fall out as soon as the marble was pushed in at the Sydney end. So marbles definitely do not travel at the speed of light any more than do electrons in a conductor. The marbles were already in the pipe just as the electrons were already in the conductor.

The effect of an electric current at a distance is almost instantaneous; *however, the speed of the electrons is very slow.*

A source of electrical energy does not increase the number of free electrons in a circuit; it merely produces a concerted pressure on loose, aimlessly moving electrons. If the material of the circuit is made of atoms or molecules that have no freely interchanging electrons, the source cannot produce any current in the material. Such a material is known as an insulator, or a non-conductor.

The amount of current in a circuit is measured in Amperes, abbreviated 'A' or Amp. An Ampere is a certain number of electrons passing or drifting past a single point in an electric circuit in one second. Therefore, an Ampere is a rate of flow, similar to litres (or marbles) per minute in a pipe.

The quantity of electrons used in determining an Ampere (and other electrical units) is the Coulomb, abbreviated 'C'. An Ampere is one Coulomb per second. A single Coulomb is 6,250,000,000,000,000,000 electrons. This large number is more easily expressed as 6.25×10^{18}, which is read verbally as "6 point 25 times 10 to the eighteenth power". "Ten to the eighteenth power" means the decimal place in the 6.25 is moved 18 places to the right. This method of expressing numbers as a power of 10 is called Scientific Notation. The use of Scientific Notation is essential when expressing very large or small numbers.

ELECTROMOTIVE FORCE OR VOLTAGE

The electron-moving force in electricity, variously termed electromotive force (emf), electric potential, potential difference (PD), difference of potential, electric pressure and voltage (V), is responsible for the pulling and pushing of the electric current through a circuit. The force is the result of an expenditure of some form of energy to produce an electrostatic field.

An emf (I like to read this as 'electron-moving force') exists between two objects whenever one of them has an excess of free electrons and the other has a deficiency of free electrons. An object with an excess of electrons is negatively charged. Similarly, an object with a deficiency of electrons is positively charged. Should two objects with a difference in charge be connected by a conductor, a discharge current will flow from the negative body to the positive one.

Instead of writing negatively or positively charged all the time it is easier to say '-Ve' or '+Ve.' I tend not to use this notation but you shoud know what it means.

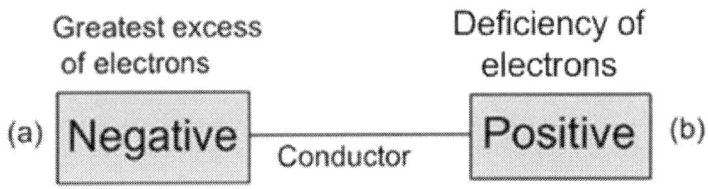

(a) is negative with respect to (b)

Current flow from most negative to less negative

Figure 2-2

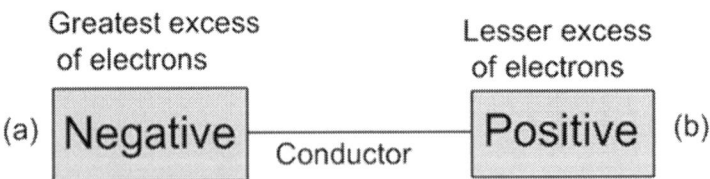

(a) is negative with respect to (b)

Current flow from most negative to less negative

Figure 3-2

An emf also exists between two objects whenever there is a difference in the number of free electrons per unit volume of the objects. In other words, both objects may have a negative charge, but one is more negative than the other. The less negative object is said to be positive with respect to the more negative object

In the electrical trades, it is common to hear of current flow from positive to negative. This is called the conventional direction of current flow. This is just what it says, a convention (popular method). Current flow is electron flow and it is from negative to positive. Conventional current flow is mostly used in Electrical Engineering. In Radio, there is a greater tendency to do what I have done in this book and that is to use electron flow; current flows from negative to positive.

The unit of measurement of electric pressure, or emf, is the Volt (V). A single torch dry cell produces about 1.5 V. A Volt can also be defined as the pressure required to force a current of one Ampere through a resistance of one Ohm.

NOTE: A battery is a collection of cells (like a battery of cannons). There is no such thing as an AA battery; it is an AA cell. On the other hand, a 9-Volt transistor battery or a car battery are samples of real batteries because they are constructed from a number of cells connected together (in series).

PRODUCING AN ELECTRON-MOVING FORCE (VOLTAGE)

1) Chemical (cells and batteries)
2) Electromagnetic (generators)
3) Thermal (heating the junction of dissimilar metals)
4) Magnetostriction (filters and special energy converters called transducers)
5) Static (laboratory static-electricity generators) remember our hair comb experiment.
6) Photoelectric (light-sensitive cells)
7) Magnetohydrodynamics (MHD, a process that converts hot gas directly to electric current)
8) Piezo-electricity – some materials produce a voltage when physical pressure is applied to them.

EFFECTS OF AN ELECTRIC CURRENT

These are the main effects:

1) Heat and light – current flowing in a conductor causes the conductors temperature to increase. If the temperature increases sufficiently, the conductor will become incandescent and radiate light.

2) Magnetic - a conductor carrying a current will produce a magnetic field around the conductor.

3) Chemical – electroplating, charging batteries. An electric current is able to cause a chemical reaction.

THE BATTERY IN A CIRCUIT

In the explanations thus far, "objects," either positively or negatively charged, have been used. A common method of producing an emf is by the chemical action in a battery. Without going into the chemical reactions that take place inside a cell, a brief outline of the operation of a Leclanche cell is given here.

Consider a torch battery. Such a battery (two or more cells form a battery) is composed of a zinc container, a carbon rod down the middle of the cell and a black, damp, paste-like electrolyte between them. The zinc container is the negative terminal. The carbon rod is the positive terminal. The active chemicals in such a cell are the zinc and the electrolyte.

The materials in the cell are selected substances that permit electrons to be pulled from the outer orbits of the molecules or atoms of the carbon terminal chemically by the electrolyte and be deposited onto the zinc can. This leaves the carbon positively charged and the zinc negatively charged.

The number of electrons that move is dependent upon the types of chemicals used and the relative areas of the zinc and carbon electrodes. If the cell is not connected to an electric circuit, the chemicals can pull a certain number of electrons from the rod over to the zinc. The massing of these electrons on the zinc produces a backwards pressure of electrons, or an electric strain, equal to the chemical energy of the cell and no more electrons can move across the electrolyte.

If a wire is connected between the positive and negative terminals of the cell, the 1.5 V of emf starts a current of electrons flowing through the wire. The electrons flowing through the wire start to fill up the deficient outer orbits of the molecules of the positive rod.

The electron movement away from the zinc into the wire begins to neutralise the charge of the cell. The electron pressure built up on the zinc, which held the chemical action in check, is decreased. The chemicals of the electrolyte can now force an electron stream from the positive rod through the cell to the zinc, maintaining a current of electrons through the wire and battery as long as the chemicals hold out.

As soon as the wire begins to carry electrons; the electrolyte also has an electric current moving through it. This motion produces an equal amount of current through the whole circuit at the same time. This is a very important concept to understand. There are no bunches of electrons moving around an electric circuit like a group of racehorses running around a track. A closed circuit is more like the racetrack with a single lane of cars, bumper to bumper. Either all must move at the same time, or none can move. In an electric circuit, when electrons start flowing in one part, all parts of the circuit can be considered to have the same value of current flowing in them instantly. Most circuits are so short that the energy flow velocity, 300,000,000 meters per second, may be disregarded.

Figure 4-2 shows the construction of a typical 9V battery from six 1.5V cells.

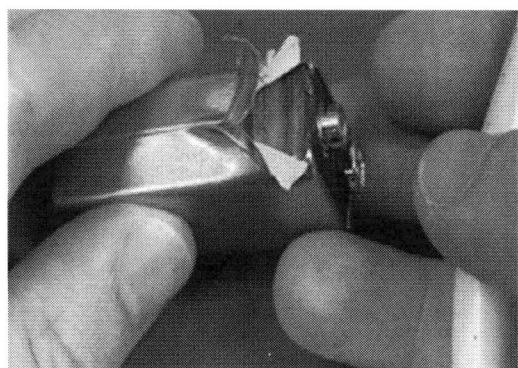

Figure 4.2

IONISATION

When an atom loses an electron, it lacks a negative charge and is, therefore, positive. An atom with a deficiency of one or more electrons is called a positive ion. On the other hand, if an atom were to gain an electron, albeit temporarily, it is a negative ion.

In most metals the atoms are constantly losing and regaining free electrons. They may be thought of as constantly undergoing ionisation. Because of this, metals are usually good electrical conductors. Atoms in a gas are not normally ionised to any great extent and therefore, a gas is not a good conductor under low electric pressures.

However, if the emf is increased across an area in which gas atoms are present, some of the outer orbiting electrons of the gas atoms will be attracted to the positive terminal of the source of emf and the remainder of the atom will be attracted toward the negative. When pressure increases enough, one or more free electrons may be torn from the atoms. The atoms are ionised. If ionisation happens to enough of the atoms in the gas, a current flows through the gas. For any particular gas at any particular pressure, there is a certain voltage value that will produce ionisation. Below this value, the number of ionised atoms is small. Above the critical value, more atoms are ionised, producing greater current flow, which tends to hold the voltage across the gas at a constant value. In an ionised condition the gas acts as an electric conductor.

Examples of ionisation of gases are lightning, neon lights and fluorescent lights. Ionisation plays an important part in electronics and radio.

TYPES OF CURRENT AND VOLTAGE

Different types of currents and voltages are dealt with in electricity:

1. Direct current (DC). There is no variation of the amplitude (strength) of the current or voltage. Obtained from batteries, dc generators and power supplies.

2. Varying direct current (VDC). The amplitude of the current or voltage varies but never falls to zero. Found in many radio and electronic circuits. A telephone is a good example of the use of varying direct current.

3. Pulsating direct current (PDC). The amplitude drops to zero periodically (such as our light bulb circuit if it was repeatedly switched on and off).

4. Alternating current (AC). Electron flow reverses (alternates) periodically and usually changes amplitude in a more or less regular manner. Produced in AC generators, oscillators, some microphones and radios in general. Household electricity is alternating current.

RESISTANCE

Resistance is that property of an electric circuit which opposes the flow of current. Resistance Is measured in Ohms. The higher the resistance in an electric circuit the lower will be the current flow. The symbol used for resistance is the Greek letter omega - Ω. If a circuit with an electric pressure of 1 Volt causes a current of 1 Ampere to flow, then the circuit has a Resistance of 1Ω.

What is resistance? It is all well and good to say it is the opposition to current flow but from where does the opposition come. Resistance causes heat. In fact, resistance is the only electrical property that does cause heat. Resistance is the effort or energy that it takes to get electrons moving. When we do get them moving some of the energy is dissipated as heat. It is lost energy. So resistance is the energy lost making electrons move in an orderly manner.

WHAT'S IN A NAME?

We have learnt quite a few new terms. Some of these terms are taken from people's names. These people were usually pioneers in the fields of physics, electricity or electronics. Read the very short biographies below and think about the person's name and what it represents in an electric circuit.

George Simon Ohm. Born March 16, 1789, Erlangen, Bavaria [Germany]. Died July 6, 1854, Munich. German physicist who discovered the law named after him, which states that the current flow through a conductor is directly proportional to the potential difference (voltage) and inversely proportional to the resistance.

Andre-Marie Ampere. Born Jan. 22, 1775, Lyon, France. Died June 10, 1836, Marseille. French physicist who founded and named the science of electrodynamics, now known as electromagnetism. Ampère was a prodigy who mastered all mathematics then existing by the time he was 12 years old. He became a professor of physics and chemistry at Bourg in 1801 and a professor of mathematics at the École Polytechnique in Paris in 1809.

Allesandro Giuseppe Antonio Anastasia Volta. Born Feb. 18, 1745, Como, Lombardy [Italy]. Died March 5, 1827, Como. Italian physicist whose invention of the electric battery provided the first source of continuous current. He became a professor of physics at the Royal School of Como in 1774 and discovered and isolated methane gas in 1778. One year later he was appointed to the chair of physics at the University of Pavia.

Charles Augustin de Coulomb. Born June 14, 1736, Angouleme, France. Died Aug. 23, 1806, Paris. French physicist best known for the formulation of Coulomb's law, which states that the force between two electrical charges is proportional to the product of the charges and inversely proportional to the square of the distance between them.

Goerges Leclanche. Born 1839, Paris. Died Sept. 14, 1882, Paris. French engineer who in about 1866 invented the battery bearing his name. In slightly modified form, the Leclanche battery, now called a dry cell, is produced in great quantities and is widely used in devices such as torches and portable radios.

3. Basic Electricity - Part 3

MORE ON RESISTANCE

As discussed briefly in Basic Electricity Part II, resistance is the opposition to current flow in any circuit. Resistance is measured in Ohms and the symbol for resistance is Ω, though for equations (we will be using them soon) the letter 'R' is often used.

Copper and silver are excellent conductors of electric current. When the same emf (voltage) is applied across an iron wire of equivalent size compared to silver, only about one-sixth as much current flows. Iron may be considered a good conductor.

When the same voltage (emf) is applied across a length of rubber or glass, no electron drift results. These materials are insulators. Insulators are used between conductors when it is desired to prevent electric current from flowing between them. To be more precise, in a normal 'electric' circuit the current flow is *negligible* through an insulator.

Silver is one of the better conductors and glass is one of the best insulators. Between these two extremes are found many materials of intermediate conducting ability. While such materials can be catalogued as to their conducting ability, it is more usual to think of them by their resisting ability. Glass (when cold) completely resists the flow of current. Iron resists much less. Silver has the least resistance to current flow.

The resistance of a wire or other conducting material is dependent on four physical factors:

1. The type of material from which it is made (silver, iron, etc.).
2. The length (the longer the conductor, the greater the resistance).
3. The cross-sectional area of the conductor (the more area, the more molecules with

free electrons and the less resistance).
4. Temperature (the higher the temperature, the greater the resistance, except for carbon and other semiconductor materials).

A piece of silver wire of given dimensions will have less resistance than an iron wire of the same dimensions. It is reasonable to assume that if a 1-metre piece of wire has a 1-Ohm resistance, then 2 metres of the same wire will have 2 Ohms of resistance.

On the other hand, if a 1-metre piece of wire has 1 Ohm of resistance then two pieces of this wire placed side by side will offer twice the cross-sectional area, will conduct current twice as well and therefore will have half as much resistance.

The basic formula for calculating the resistance of a wire is:

$R = \rho L/A$
    Where:
R = resistance in Ohms.
ρ = resistivity of the material in Ohms per metre cube.
L = length in metres.
A = cross-sectional area of the wire in square metres.

ρ, is a Greek letter spelt 'rho' and pronounced the same as in '*row*-your-boat.'

RESISTIVITY ρ

Since the resistance of a wire (or any other material) depends on its shape, we must have a standard shape to compare the conducting properties of different materials. This standard is a cube measuring 1 metre on each side. That's a pretty big cube! Smaller materials sizes are measured and then the results extrapolated to Ohm-meter. The resistance measured between opposite faces of the cube is called the resistivity.

Resistivity should not be confused with resistance. The resistivity of a material is the resistance measured for a standard size cube of that material. If you look at the table below you will see for example that the resistivity of copper is 1.76×10^{-8}. The part of this number shown as 10^{-8} is called the exponent and the 'minus 8' means that the decimal point must be moved 8 places to the left. If we take 1.76 and move the decimal point 8 places to the left, we get:

0.000 000 017 6 ohm·metre (Ω·m)

Now this is a very low resistance indeed. It is the resistance that would be measured across the opposite faces of a cubic metre of solid copper. Resistivity is not resistance. Resistivity is a measure of the resistance of a material of a standard size to allow us to compare how well that material conducts or resists current compared to other materials.

The Resistivity of Metals and Alloys at 20 degrees C.

Material	Resistivity (Ohms Per metre cube)
Silver	1.62×10^{-8}
Copper	1.76×10^{-8}
Aluminium	2.83×10^{-8}
Gold	2.44×10^{-8}
Brass	3.9×10^{-8}
Iron	9.4×10^{-8}
Nickel	7.24×10^{-8}
Tungsten	5.48×10^{-8}
Manganin	45×10^{-8}
Nichrome	108×10^{-8}

The most common conducting material used in radio is, of course, copper, for it is a good conductor and relatively cheap. You can see in the table that aluminium is not as good a conductor as copper. However, aluminium is used for conductors more than any other material because of its light weight. In overhead power line distribution, weight is a crucial consideration, so aluminium is the conductor of choice. In radio and communications, antennas are made of aluminium, again because of the light weight.

CALCULATING THE RESISTANCE OF A WIRE

A 100-metre length of copper wire is used to wind the primary of a transformer and the wire has a diameter of 0.5 millimetres. What is the resistance of the winding?

Solution:

We need to use the equation $R = \rho L/A$

Since the wire is copper, we look at the Resistivity table and get a ρ of 1.76×10^{-8} for copper. The length (L) = 100 metres. We need to calculate 'A' (the cross-sectional area) from the equation for the area of a circle.

$A = \Pi d^2/4$

Where:

Π = Another Greek letter, the mathematical constant Pi, approximated by 22/7
d = diameter of the circle (wire).

$A = 22/7 \times (0.5 \times 10^{-3})^2 / 4$
$A = 3.142 \times 0.00000025 / 4$
$A = 0.000000196375$ square metres

$R = \rho L/A$
$R = 1.76 \times 10^{-8} \times 100 / 0.000000196375$
$R = 8.96$ Ohms

You will **not** have to use this equation in and Advanced Theory Exam. However, you **must** fully understand what the equation tells us, so let's look at it again.

$R = \rho L/A$ ◄ **what does this say?**

We shall look at the numerator and the denominator on the right-hand side separately.

$R = \rho L$

This means that resistance is directly proportional to the resistivity and to the length. As either ρ or L changes so does R. If ρ or L increases by say a factor of 2 then so does R. In other words, doubling the length of a wire doubles its resistance. If we increase the length by 3.25 times the resistance is increased 3.25 times. So, from the equation, we say R is directly proportional to length and resistivity.

The cross-sectional area is in the denominator of the equation on the right-hand side. Ignoring the numerator for the moment, we can rewrite this relationship as:

$R = 1/A$

This means that resistance is inversely proportional to the cross-sectional area, A. If the cross-sectional area of a wire were to be doubled then its resistance would be halved.

This would be the same as twisting two wires together and using them as one. If the cross-sectional area of a wire was increased by a factor of say 4.5 times, then the resistance would be R/4.5 of what it originally was.

We have avoided bringing temperature into our calculations. The resistance of metals increases with temperature - we will discuss this further later.

In the real world, conductors are not round. Wire is not truly round. Copper circuit board track is definitely not round. In these cases, we work with the cross-sectional area given by the manufacturer. I would like you to know what this equation 'says' to you about the resistance of a wire.

Resistance is directly proportional to the length and resistivity and inversely proportional to the cross-sectional area.

THE METRIC SYSTEM

Scientific measurements are more and more being given in the metric system. It is a multiple-of-10. The basic prefixes of metric units of measurement are:

Atto (a)	=	quintillionth of	=	10^{-18} times
Femto (f)	=	quadrillionth of	=	10^{-15} times
Pico (p)	=	trillionth of	=	10^{-12} times
Nano (n)	=	billionth of	=	10^{-9} times
Micro (u)	=	millionth of	=	10^{-6} times
Milli (m)	=	thousandth of	=	10^{-3} times
Centi (c)	=	hundredth of	=	10^{-2} times
Deci (d)	=	tenth of	=	10^{-1} times
unity	=	1	=	1
Deka (da)	=	ten times	=	10 times
Hecto (h)	=	hundred times	=	10^{2} times
Kilo (k)	=	thousand times	=	10^{3} times
Mega (M)	=	million times	=	10^{6} times
Giga (G)	=	billion times	=	10^{9} times
Tera (T)	=	trillion times	=	10^{12} times

The common prefixes used in radio and electronics that you need to learn are underlined. You will need to memorise all of these.

Volts, Amperes, Ohms, etc. may use metric-based prefixes. Some examples of the utilisation of these prefixes are:

1kV = 1000 Volts.
1mV = one thousandth of a Volt.
10M Ω = 10 million Ohms.
56mΩ = 56 thousandths of an Ohm.
25mA = 25 thousandths of an Ampere.
65uA = 65 millionths of an Ampere.

CONVERTING FROM ONE PREFIX TO ANOTHER

Many a mark has been lost in Advanced Radio exams when a question asks you to convert from one unit to another.

Questions like; convert Megahertz to Kilohertz? – and other similar unit conversions.

Hertz is the UNIT. Mega and Kilo are prefixes of the unit.

Let's try the one above.
Megahertz is 10^6 – 10^6 means 1,000,000 or one million.
Kilohertz is 10^3 – 10^3 means 1,000 or one thousand.

Megahertz is a larger unit than kilohertz. How much larger is a megahertz than a kilohertz? How many times will one thousand divide into one million?

It takes 1,000 kilohertz to make one megahertz. A megahertz is 1000 times BIGGER than a kilohertz, so to convert megahertz to kilohertz, we must MULTIPLY megahertz by 1000.

144MHz is the same as 144,000 Kilohertz.

If you were asked to do it the other way round, that is, convert kilohertz to megahertz you would DIVIDE by 1000.

100kHz is 0.1MHz.

I think this looks all too easy to some and for this reason, it is done too quickly and very often the wrong answer is picked in the exam and 2 marks are lost all too easily.

Let's do another.

How do you convert microfarads to picofarads?

Please do it step by step, not what you immediately think it should be, unless you are very confident.

This would be my reasoning.

Micro = 1/1,000,000 or one millionth.
Pico = 1/1,000,000,000,000 or one millionth of a millionth.

This means that picofarads are one million times SMALLER than microfarads.

To convert microfarads to picofarads MULTIPLY by 1,000,000.

Example – convert 0.001 microfarads to picofarads.

If we multiply 0.001 by 1,000,000, we get 1000 picofarads.

The easiest way to multiply by 1,000,000 is to move the decimal point 6 places to the right. Can you see that moving the decimal point in 0.001 six places to the right gives you 1000? Do it on a piece of paper – it's the way I do it. Simply put your pen on the decimal point in 0.001 and draw 6 'hoops' moving to the right. Then add "zeros" to the 'hoops' that have nothing in them.
Like this:

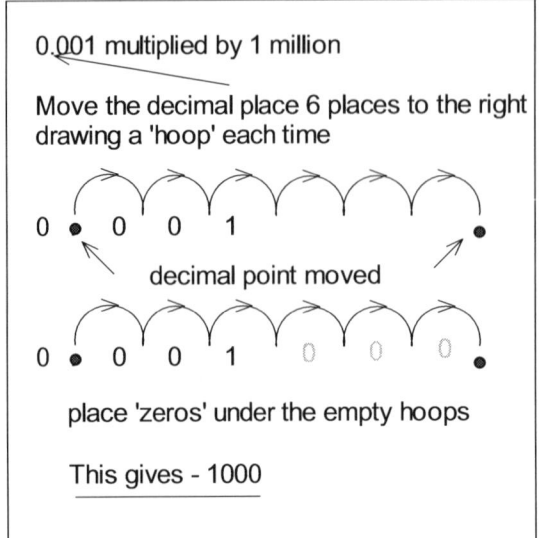

Figure 3-1

This may seem like a silly way of doing it, but it is a *safe way*. If you had to divide by 1 million (converting picofarads to microfarads), use the same method only move the decimal point 6 places to the *left*.

You can use pen and paper in the exam. Using this method, you can convert any prefix to another prefix. Do not try and memorise the conversions, but do memorise what each prefix means, then work out how to convert one to the other using the techniques described above.

SYMBOLS IN TEXT

Since some assignments (when this book is used as part a course) are done using plain text you cannot show superscripts, subscripts and symbols. Many email programs do allow this, but you should avoid using symbols as when your email is received the symbols could be stripped out.

This is the method I use (but you can use your own as long as it is clear).

For example, show:

10^6 (10 to the power 6) as 10^6
10^{-6} (10 to the power *minus* 6) as 10^-6

Use the '^' which is a 'shifted 6' to indicate that what follows is a superscript.

For symbols, you can type the symbol name in brackets. Take the formula for the resistance of a conductor that we used earlier in this chapter.

R = ρL/A
    The Greek letter ρ can be typed as (rho):
R = (rho)L/A
    If you have to type the Greek letter Π in a formula you can just type (Pi).

THE RESISTOR COLOUR CODE

Resistors can be very small electronic components. Too small to write the resistors value on, so instead, each resistor has colour coded bands which tell you its value and tolerance. The tolerance is the percentage of error, about which the resistor may vary from its coded value.

Resistors are manufactured in what are called *preferred values*. Usually, you will require a certain value of resistance in Ohms and normally you will choose a resistance with the closest preferred value. If you want a specific resistance that does not match any preferred value, then you may have to make up a resistor especially for the job, or more commonly, you will use a variable resistance and adjust it using an Ohmmeter.

There are only so many numerical values in a decade, i.e. from 0-10, or 0-100, 0-1,000, etc. Simple resistors only have 12 values in a decade e.g. 1.0, 1.2, 1.5, 1.8, 2.2, 2.7, 3.3, 3.9, 4.7, 5.6, 6.8, 8.2. This is called the E12 series

A newer series called E24 has 24 standard values and closer tolerances.

E12 Series	E24 Series	
10	10	33
12	11	36
15	12	39
18	13	43
22	15	47
27	16	51
33	18	56
39	20	62
47	22	68
56	24	72
68	27	82
82	30	91

Resistors can have either 4 or 5 coloured bands. With each type, the last band is the tolerance. Five band resistors simply have room for three significant figures even though the E24 series only really needs two bands. Most of the time these resistors are 1 or 2% tolerance (within +/- 1 or +/- 2% of the stated resistance). This will be either a brown or red band, respectively, at one end of the resistor, separated from the other bands. The value of the resistor starts at the other end.

The colour code, which you must commit to memory, is:
Black   = 0   Green = 5
Brown   = 1   Blue   = 6
Red   = 2   Violet = 7
Orange   = 3   Grey   = 8
Yellow   = 4   White = 9

The tolerance band of a resistor is coded:

Gold   = 5%   Silver = 10%
Brown = 1%   Red   = 2%

The tolerance band is always a band on the end and separated slightly from the other bands.

The second last band is the multiplier for both 4 and 5 colour banded resistors.

The first letter of each word in the following sentence may help you to remember the colour code. Big Boys Race Our Young Girls But Violet Generally Wins.

Example 1:

A resistor has 4 coloured bands. From left to right the bands are coloured:

Yellow, Violet, Yellow and Gold

The first significant number is: 4.
The second significant digit is: 7.
The third band is the multiplier, in this case, 4 and means add four zeros.
So we get 470000 Ohms or 470kΩ.
The tolerance is (gold) +/- 5%.
So the final result is 470kΩ +/-5%.

Example 2:

A resistor has 5 coloured bands. From right to left the bands are coloured:

Green, Blue, Black, Red, Brown

The first significant digit is: 5.
The second significant digit is: 6.
The third significant digit is: 0.
The multiplier in the fourth band is 2 (add two more zeros).
So we get 56000 Ohms or 56KΩ.
The tolerance band is 1, therefore, 1% tolerance.

So the final result is 56KΩ +/- 1%

The fourth band on a five-banded resistor can be Gold or Silver. Gold means the multiplier is 0.1 and Silver means 0.01. To multiply by 0.1, move the decimal place one place to the left. To multiply by 0.01, move the decimal place two places to the left.

CONDUCTANCE

There is no need for you to be concerned about the term 'conductance'. I mention it only for information and you will occasionally see the term. You will not see the term in your exam.

For some (not myself) it is more convenient to work in terms of the ease with which a current can be made to flow, rather than the opposition to current flow (resistance). Conductance is merely the reciprocal of resistance. The symbol for conductance is 'G' and G=1/R. Conductance is measured in Siemens 'S', formerly the 'mho'.

The term 1/R is simply resistance divided into 1 giving the conductance in siemens. Taking any quantity 'x' and dividing that quantity into 1 is called the reciprocal of the quantity - this you will need to remember.

RESISTORS

Pictured are some of the types and sizes that resistors are packaged in. They are all just resistors. Resistors can be made to vary; these are called rheostats, potentiometers, or just plain variable resistors.

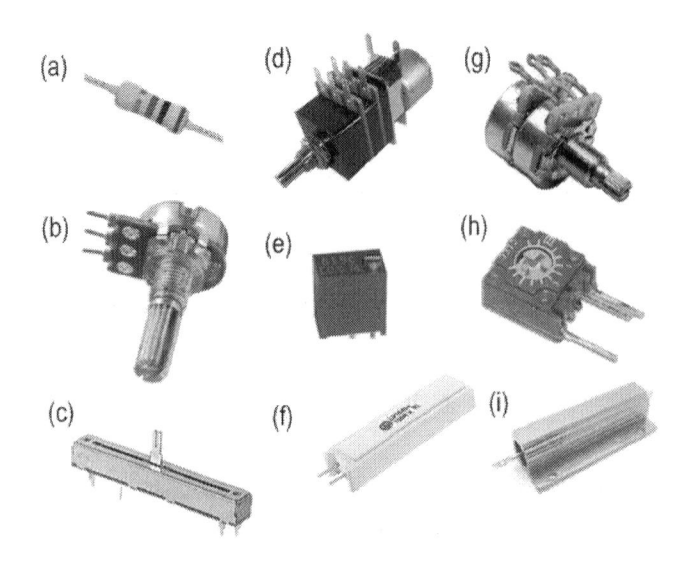

Figure 3-3.

The resistors in figure 3-3 are not shown to scale.

   (a) fixed low wattage resistor.
   (b) a variable resistor – potentiometer.
   (c) a variable 'slider' resistor.
   (d) a variable motorised resistor.
   (e) a multi-turn circuit board mounted variable resistor.
   (f) high wattage wire wound resistors.
   (g) two potentiometers on the one shaft.
   (h) a circuit board mounted variable resistor – trim pot.
   (i) A very high power resistor with heat sink.

The only reason that resistors are made large is so they can dissipate (give off to their surroundings) heat. I only add this because a physically large resistor does not mean a large resistance. The large resistor at the bottom of the picture may only be a few Ohms and the tiny one in the centre could be 1MΩ. Large physical resistors will often not use the colour code and have their value marked on them in Ohms.

SCHEMATIC SYMBOL – RESISTOR

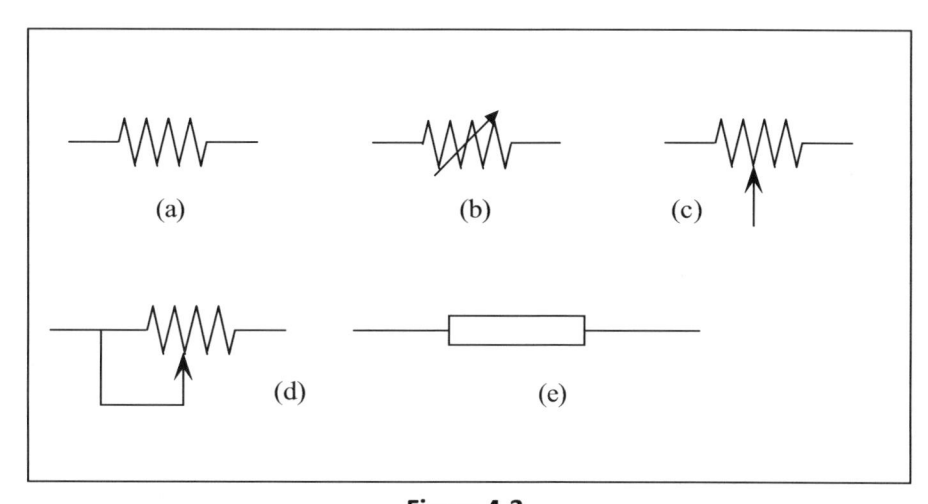

Figure 4-3.

Figure 4-3 shows the most common methods of representing a resistor in a circuit. The various types are:
   (a) Fixed resistor.
   (b) Variable resistor.
   (c) This type of variable resistor is called a potentiometer (to be discussed).
   (d) Rheostat - essentially the same as (b).
   (e) An alternative symbol for a fixed resistor.

4. Ohms Law

Ohm's Law describes the relationship between current, voltage and resistance in an electric circuit.

Ohm's Law states:

The current in a circuit is directly proportional to voltage and inversely proportional to resistance.

Let:
- I = current
- E = voltage
- R = resistance

Part of Ohms Law says: **current is directly proportional to voltage.**

Using the symbols given, we can write an equation to show a direct proportion between current and voltage.

I = E

Normally the above equation is read I 'equals' E. It can just as easily and more understandably be read as: **I is directly proportional to E.**

I know I harp on the direct proportion and inverse proportion stuff a lot. I do so because it is so important to understand this thoroughly before we come to do more complex equations.

I = E

Means that if the voltage is increased or decreased in a circuit, then the current will increase or decrease by the same amount. Double the voltage and you double the current. Halve the voltage and you halve the current. This is a direct proportion.

The other part of Ohm's Law says that **current is inversely proportional to the resistance**. This can be written as:

I = 1/R

Now 1/R is a fraction with a numerator (the top part, 1) and a denominator, R.

1/R is a fraction just like 1/4, 1/2 and 3/8 are fractions.
R is the denominator in the fraction. What happens to the whole fraction if the denominator is changed? Watch.

1/2, 1/3, 1,4, 1/5, 1/6

As the denominator increases, the fraction decreases. In fact, if the denominator doubles then the fraction is half the size. 1/4 is half the size of 1/2.

'I' is the same as 1/R. This is an inverse proportion. If I is the same as 1/R and R is increased in size by three times, then the fraction 1/R is a third the size now and since 1/R is the same as the current, then the current is a third the size also.

The complete equation for Ohm's Law then is:

I = E/R

This equation, derived from Ohms law, enables us to find the current flowing in any circuit if we know the voltage (E) and resistance (R) of the circuit.

For example, a resistor of 20 Ohms has a 10-Volt battery connected across it. How much current will flow through the resistor?

I = 10/20 = 1/2 = 0.5 Amperes

The equation I = E/R can be transposed for E or for I.

In some texts a diagram called the Ohm's Law triangle is used to help you rewrite the equation for E and R - I don't like this method, as you do really need to know how to transpose equations - not just this one. If you learn to transpose this equation, then you will be able to do it with many others. There is a memory wheel on the website for you to download that will help you remember equations for Ohm's law and power. You will also find a tutorial on transposing equations and using a calculator in the downloads area if you feel you might need some extra help. Always write to your facilitator if you require assistance as well.

We want to transpose I = E/R for E and R. The rule is: do whatever you like to the equation and it will always be correct as long as you do the same to each side of the equals sign. For example, if I multiply both sides of the equal sign by R, we get:

$$I \times R = \frac{E}{R} \times R$$

On the left-hand side (LHS) we have I x R. On the RHS we have E multiplied by R and divided by R. Can you see that the R's cancel on the RHS? R/R is 1/1.

$$I \times R = \frac{E}{1} \times 1$$

There is no need to show the 1's at all since multiplying or dividing a number by 1 does not change the number, therefore:

I x R = E

Rewriting the above with E on the LHS we get:

E = I x R or just E=IR

When there is no sign between two letters in an equation, like IR above, it is assumed the IR means I x R.

Now transpose the equation for R:

$$I = \frac{E}{R}$$

Multiply both sides by 1/E (which is the same as dividing both sides by E):

$$I \times \frac{1}{E} = \frac{E}{R} \times \frac{1}{E}$$

On the RHS the E's cancel out so we can rewrite the equation as:

$$I \times \frac{1}{E} = \frac{1}{R}$$

or

$$\frac{I}{E} = \frac{1}{R}$$

Turning both sides upside down (remember we can do anything as long as we do the same to both sides):

$$\frac{E}{I} = \frac{R}{1}$$

Remove the '1' and reverse the sides to get:

$$R = E/I$$

So the three equations are:

$I = E/R \quad E = IR \quad R = E/I$

I have probably made you bored by now - however; it is really important to be able to transpose equations for yourself. For one thing, you don't need to remember so many equations.

So if you know any two of the three in 'E', 'R' and 'I' then you can calculate the missing one.

Finding I when you know E and R:

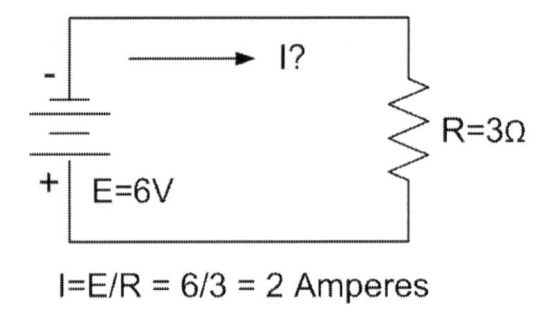

I=E/R = 6/3 = 2 Amperes

Finding E when you know I and R:

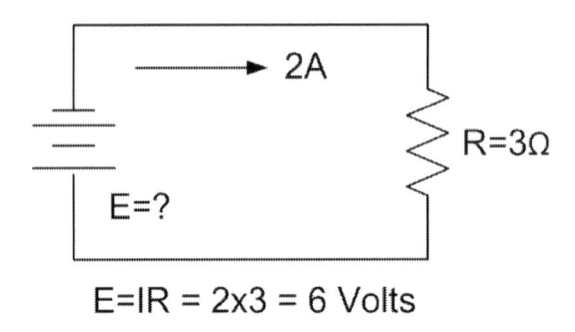

E=IR = 2x3 = 6 Volts

Finding R when you know I and E:

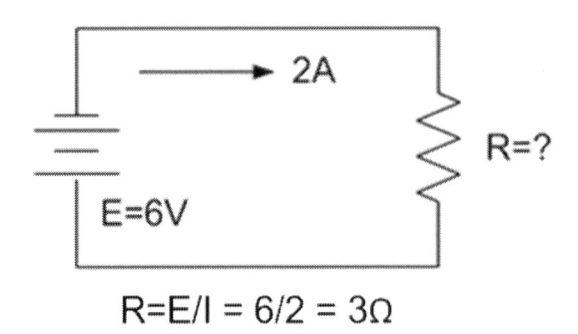

R=E/I = 6/2 = 3Ω

POWER

The unit of electrical power is the Watt (W), named after James Watt (1736-1819). One Watt of power equals the work done in one second by one Volt of potential difference in moving one Coulomb of charge.

Remember that one Coulomb per second is an Ampere. Therefore, power in Watts equals the product of Amperes times Volts.

Power in Watts = Volts x Amperes

P = E x I

Example: A toaster takes 5A from the 240V power line. How much power is used?

P = E x I = 240 V x 5 A
P = 1200 Watts

Example: How much current flows in the filament of a household 75-Watt light bulb connected to the standard 240 Volt supply?

You know P (power) and E (Volts). You need to transpose P=EI for 'I' and you get:

I = P/E

Therefore:

I = 75/240
I = 0.3125 Amperes

This amount of current is best expressed in milliAmperes. To convert Amperes to milliAmperes multiply by 1000 or think of it as moving the decimal point 3 places to the right, which is the same thing. This gives:

312.5mA

Power in Watts can also be calculated from:

$P = I^2R$, read, "power equals I squared R".
$P = E^2/R$, read, "power equals E squared divided by R".

Watts and Horsepower Units.

746 W = 1 horsepower.
This relationship can be remembered more easily as 1 horsepower equals approximately 3/4 kilowatt. One kilowatt = 1000W.

WORK

Work = Power x Time

Practical Units of Power and Work. Starting with the Watt, we can develop several other important units. The fundamental principle to remember is that power is the time rate of doing work, while work is the power used during a period. The formulas are:

$$Power = work / time$$
$$and$$
$$Work = power \times time$$

The unit of power is the Watt. One Watt used during one second equals the work of one joule. In other words, one Watt is one joule per second. Therefore, 1 W = 1 J/s. The joule is a basic practical unit of work or energy.

A unit of work that can be used with individual electrons is the *electron Volt.* Note that the electron has charge, while the Volt is potential difference. Now 1eV is the amount of work required to move an electron between two points having a potential difference of one Volt. Since 6.25×10^{18} electrons equal 1C and a joule is a Volt-Coulomb, there must be 6.25×10^{18} eV in 1J.

Kilowatt-hours. This is a unit commonly used for large amounts of electrical work or energy. The amount is calculated simply as the product of the power in kilowatts multiplied by the time in hours during which the power is used. This is the unit of energy you need to know.

Example: A light bulb uses 100 W or 0.1 kW for 4 hours (h), the amount of energy used is:

Kilowatt-hours = kilowatts x hours
= 0.1 x 4
= 0.4 kWh.

We pay for our household electricity in kilowatt-hours of energy.

POWER DISSIPATION IN RESISTANCE

When current flows through a resistance, heat is produced due to the energy consumed in moving free electrons and the atoms obstructing the path of electron flow. The heat is evidence that power is used in producing current. This is how a fuse opens, as heat resulting from excessive current melts the metal link in the fuse.

The power is generated by the source of applied voltage and consumed in the resistance in the form of heat. As much power as the resistance dissipates in heat must be supplied by the voltage source; otherwise, it cannot maintain the potential difference required to produce the current.

Any one of the three formulas can be used to calculate the power dissipated in a resistance. The one to be used is just a matter of convenience, depending on which factors are known.

In the following diagram, the power dissipated with 2A through the resistance and 6V across it is 2 x 6 = 12W. Or, calculating in terms of just the current and resistance, we get 2^2 times 3, which equals 12W. With voltage and resistance, the power can be calculated as 6^2 or 36, divided by 3, which also equals 12W.

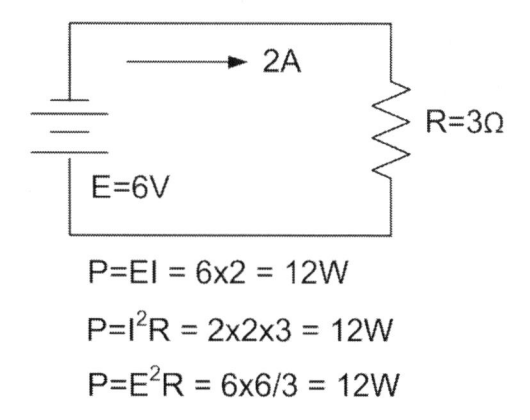

$$P = EI = 6 \times 2 = 12W$$
$$P = I^2 R = 2 \times 2 \times 3 = 12W$$
$$P = E^2 R = 6 \times 6 / 3 = 12W$$

Resistance is the "only" electrical property that dissipates heat

We have introduced a new schematic symbol here too. The schematic symbol of a battery is shown at the left. Note the small bar at the top is the negative terminal. The direction of current flow is shown correctly, from negative to positive

No matter which equation is used, 12W of power is dissipated, in the form of heat. The battery must generate this amount of power continuously in order to maintain the potential difference of 6 V that produces the 2A current against the opposition of 3Ω.

In some applications, the electrical power dissipation is desirable because the component must produce heat in order to do its job. For instance, a 600W toaster must dissipate this amount of power to produce the necessary amount of heat. Similarly, a 300W light bulb must dissipate this power to make the filament white hot so that it will have the incandescent glow that furnishes the light. In other applications, however, the heat may be just an undesirable by-product of the need to provide current through the resistance in a circuit. In any case, though, whenever there is current in a resistance, it dissipates power equal to I²R.

The term I²R is used many times to describe unwanted resistive power losses in a circuit. You will hear of the expression I²R losses as we go through this course.

ELECTRIC SHOCK

While you are working on electric circuits, there is often the possibility of receiving an electric shock by touching the "live" conductors when the power is on. The shock is a sudden involuntary contraction of the muscles, with a feeling of pain, caused by current through the body. If severe enough, the shock can be fatal. Safety first, therefore, must be the rule.

The greatest shock hazard is from high voltage circuits that can supply appreciable amounts of power. The resistance of the human body is also an important factor. If you hold a conducting wire in each hand, the resistance of the body across the conductors is about 10,000 to 50,000Ω. Holding the conductors tighter lowers the resistance. If you hold only one conductor, your resistance is much higher. It follows that the higher the body resistance, the smaller the current that can flow through you.

A safety rule, therefore, is to work with only one hand if the power is on. Also, keep yourself insulated from earth ground when working on power-line circuits since one side of the line is usually connected to earth. Also, the metal chassis of radio and television receivers is often connected to the power line ground. The final and best safety rule is to work on the circuits with the power disconnected if at all possible and make resistance tests.

Note that it is current through the body, not through the circuit, which causes the electric shock. This is why care with high-voltage circuits is more important since sufficient potential difference can produce a dangerous amount of current through the relatively high resistance of the body. For instance, 500V across a body resistance of 25,000Ω produces 0.02A, or 20mA, which can be fatal. As little as 10uA through the body can cause an electric shock. In an experiment on electric shock to determine the current at which a person could release the live conductor, this value of "let-go" current was about 9 mA for men and 6 mA for women.

In addition to high voltage, the other important consideration in how dangerous the shock can be is the amount of power the source can supply. The current of 0.02A through 25,000Ω means the body resistance dissipates 10W. If the source cannot supply 10W, its output voltage drops with the excessive current load. Then the current is reduced to the amount corresponding to how much power the source can produce.

In summary, then, the greatest danger is from a source having an output of more than about 30V with enough power to maintain the load current through the body when it is connected across the applied voltage.

RESISTANCE OF EARTH

The earth, no not the ground, I am speaking of planet earth, is not made of metal (in any great concentrated amount) so one may expect that it is not a good conductor. However, if you recall the equation R = ρL/A, where A is the cross-sectional area - well the earth indeed does have an enormous cross-sectional area. This means for many applications the earth itself can be used as a conductor to save us having to run two conductors from the source to the load. Such circuits are called earth return and they have been used for power distribution and telephone communications.

Some Revision

By now you should have a good concept of current, voltage and resistance, among other things. It should be clear in your mind that current flows in a circuit pushed and/or pulled along by voltage. Current is restricted from flowing in a circuit by resistance.

You should be aware by now that statements like; "the voltage through the circuit" are in error. Voltage is electrical pressure. voltage is never through anything. You can have voltage across the circuit or a component, but you can never have voltage through anything. Current flows through the circuit pushed along by voltage and restricted by resistance.

VOLTS PUSH AMPS THROUGH OHMS

A final point. You can have voltage without current. However, you cannot have current without voltage. A battery sitting on a bench has a voltage on its terminals – but no current is flowing. Voltage is electric pressure just like the water pressure in your tap. Current is the flow of electrons just like the flow of water from a tap. If the tap is turned off, you do not have a water flow. However, the pressure is definitely still there. Likewise, it is possible (with a disconnected battery) to have voltage (electric pressure) and no current (flow).

However, you cannot have any flow without pressure. So voltage can exist on its own, current cannot.

The unit of current is the Ampere. When 6.25×10^{18} electrons flow past a given point in a circuit in one second, the current is said to be one Ampere.

Since 6.25×10^{18} electrons is a Coulomb, this can be used in the definition of an Ampere. An Ampere of current is said to flow when one Coulomb passes a given point in one second.

5. Series Circuits

When components in a circuit are connected in serial order with the end of each joined up to the other end of the next as shown below in figure 1-5, they form a series circuit.

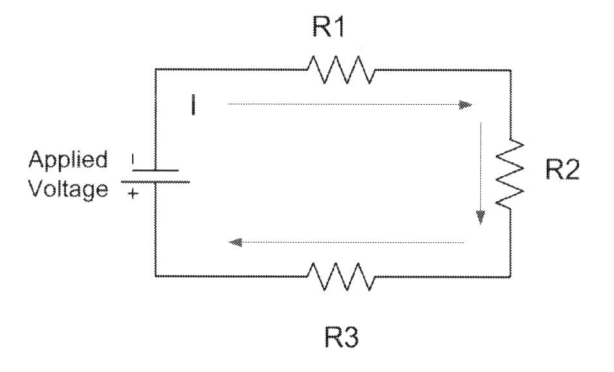

Figure 1-5.

An electric current consists of an ordered movement of electrons. In the schematic shown in figure 1-5, the current leaves the negative terminal of the battery and flows through R_1, R_2 and R_3.

It does not matter where we measure the current in a series circuit as we will always get the same value of current everywhere, as:

Current is the same in all parts of a series circuit.

The total resistance of any number of resistances in series is simply the sum of the individual resistances:

$R_t = R_1 + R_2 + R_3..n$

Suppose the resistors in figure 1-5 were 10, 20 and 30 Ohms respectively and the applied voltage was 10 Volts. What is the current flowing in the circuit?

$R_t = R_1 + R_2 + R_3$
$R_t = 10 + 20 + 30$
$R_t = 60$ Ohms

We now know the total resistance (60 Ohms) and the applied voltage is 10 Volts, so we can use Ohm's law to calculate the current flowing in the circuit.

I=E/R = 10/60 = 1/6 A or 0.1666 A or 166.6 milliamps

All of the resistances shown in the five circuits of figure 2-5 are in series. The circuits are drawn differently, but nevertheless, they are all series circuits and the current is the same in every part of the circuit. The rectangle symbol is an alternative method of drawing a resistor. In this circuit the rectangle could be any type of component.

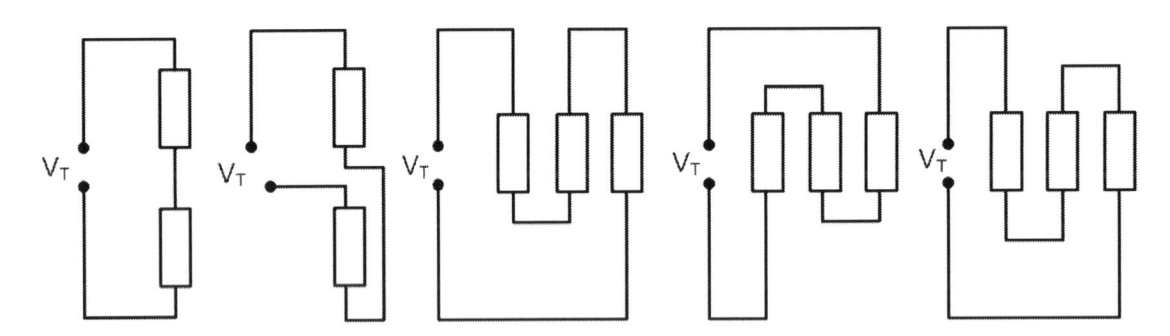

Figure 2-5.

Another rule that you must learn is:

The sum of the voltage drops in a series circuit is equal to the applied voltage.

Going back to figure 1-5, we have three resistances R_1, R_2 and R_3 in series, connected to a 10-Volt supply. We can calculate the voltage across R_1 because we know the resistance and we know the current through R_1. Let's call the voltage across R_1 'E_{R1}', then:

E_{R1} = I x R_1 = 0.1666 x 10 = 1.666 Volts

And for the other two resistances:

$E_{R2} = I \times R_2 = 0.1666 \times 20 = 3.332$ Volts

$E_{R3} = I \times R_3 = 0.1666 \times 30 = 4.998$ Volts

The sum of the voltage drops across each resistance will equal the applied voltage. There will be a small error in our example due to rounding of decimals.

$E_t = E_{R1} + E_{R2} + E_{R3} = 1.666 + 3.332 + 4.998 = 9.996$ Volts (with rounding error).

Notice how the supply voltage is distributed around the circuit - think about it for a while. The least resistance has the smallest voltage drop across it and the largest resistance has the most voltage across it. This makes sense when we realise that the current in a series circuit must be the same through all components. R_3 needs nearly half of the supply voltage to get 0.1666 amps to flow through it and R_1 requires only 1.666 Volts to produce the same current through it.

If two resistances of equal value were connected in series across a supply voltage of say 20 Volts, then each resistance would have exactly half the supply voltage across it.

You can work out the voltage drops in a series circuit by using a method known as 'proportion'.

Let's use our series circuit again and this time, we will work out the voltage drops for R_1, R_2 and R_3 without using the current.

The total resistance of the circuit is 60 Ohms (R_t) and the supply voltage is 10 Volts.

$E_{R1} = R_1/R_t \times 10 = 10/60 \times 10 = 1.6667$ Volts (rounded)
$E_{R2} = R_2/R_t \times 10 = 20/60 \times 10 = 3.3334$ Volts (rounded)
$E_{R3} = R_3/R_t \times 10 = 30/60 \times 10 = 5$ Volts

AN EXAMPLE OF A SERIES CIRCUIT

The best example I can think of for a common series circuit which will also demonstrate one of the problems with series circuits is Christmas tree lights. These lights are low voltage lights of about 10 Volts. Christmas tree lights plug into the mains voltage of 240 Volts. Suppose each bulb requires 10 Volts and you connect 24 of them in series, then each bulb will have 10 Volts across it.

At least this is the way Christmas tree lights are supposed to work. Darn dangerous things if you ask me as if you break a bulb you are exposing yourself to potentially (no pun intended) 240 Volts.

So, from the point of view of Christmas tree lights, series circuits have an advantage. The disadvantage is, should a bulb blow, no current will flow in the circuit and all the lights will go out.

What would happen if you placed a short circuit where a bulb had blown (don't do this by the way)? The remaining 23 lights would come on again as you have completed the circuit. However, they would all be a little brighter than intended as they will now have slightly more than their designated 10 Volts each. Of course running them at a higher voltage will cause them to burn out faster, so if you continued down this track and survived, you would find that the bulbs would blow faster and faster until eventually they would blow immediately.

POWER IN A SERIES CIRCUIT

The total power in a series circuit is found from P=EI where E is the applied voltage and I is the total current.

In our example: 10 x 0.1666 = 1.666 Watts

You can use any of the power equations to calculate the power dissipated in R_1, R_2 or R_3. Of course, the sum of the power dissipated in each resistance should equal the total power in the circuit (1.666 Watts).

$P_{R1} = I^2 \times R_1 = (0.1666)^2 \times 10 = 0.2775556$ Watts
$P_{R2} = I^2 \times R_2 = (0.1666)^2 \times 20 = 0.5551112$ Watts
$P_{R3} = I^2 \times R_3 = (0.1666)^2 \times 30 = 0.8326668$ Watts

$P_t = P_{R1} + P_{R2} + P_{R3}$
$P_t = P_{R1} + P_{R2} + P_{R3} = 1.6653336$ Watts (slight rounding error)

THE VOLTAGE DIVIDER

A voltage divider is a simple way to adjust the voltage in a particular part of a larger circuit. The supply voltage might be 9V and the component requires 2V on one terminal. A voltage divider can be used to produce the required 2 Volts. Two resistors can be connected to produce such a voltage divider. Such a circuit is shown in figure 3-5.

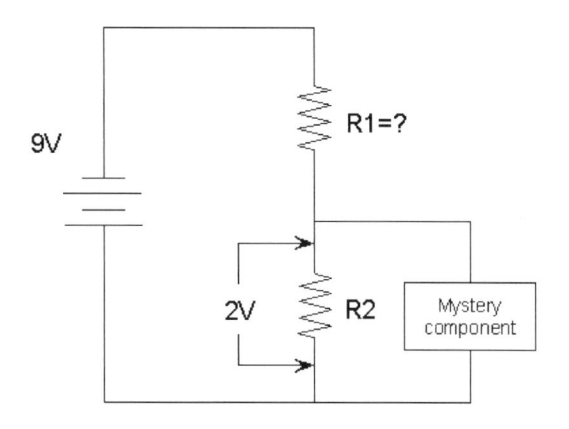

Figure 3-5.

Since we have not looked at parallel circuits yet, we will assume that the total resistance of the mystery component is 8000 Ohms and this is represented by the resistance R_2.

Our problem then is to find the value of R_1.

We know that the sum of the voltage drops across R_1 and R_2 must equal the supply voltage of 9 Volts. R_2 has 2 Volts across it so R_1 must have 7 Volts across it.

It should be evident that the resistance value of R_1 must be higher than that of R_2 since it has the higher voltage across it.

Let's try this:

E_{R2} we know is 2 Volts and R_2 is 8000Ω.

We can calculate the current through R_2:

$I_{R2} = E_{R2}/R_2 = 2/8000 = 0.25$ milliamps

We now know the current through R_1 since R_2 and R_1 are in series.

$R_2 = E_{R2}/I_{R2}$

In case the abbreviations are confusing you - E_{R2} means the voltage across R_2 and I_{R2} means the current through R_2.

R_1 = 7 Volts / 0.25 milliamps ← remember this is milliamps
R_1 = 28000 Ω

We can check this by saying:

8000/36000th of the supply voltage should be across R_1,
or 8/36 x 9 which gives 2 Volts - correct.

Also:

28000/36000th of the supply voltage should be across R_2,
28/36 x 9 = 7 Volts.

6. Parallel Circuits

When two or more components are connected directly across one voltage source, they form a parallel circuit. The two lamps in figure 1-6 are in parallel with each other and with the battery. Each parallel path is called a branch, with its own individual current. Parallel circuits have one common voltage across all the branches, but the individual branch currents can be different.

The voltage is the same across all components in a parallel circuit

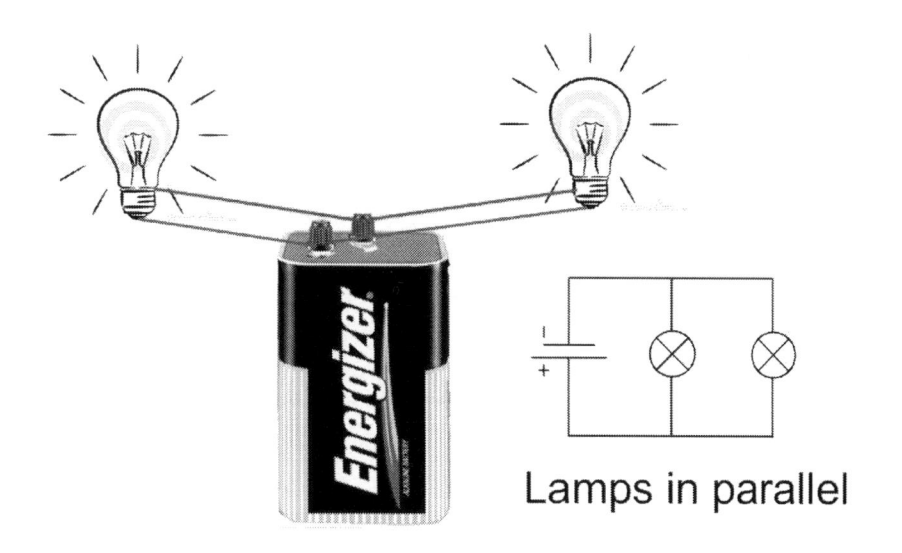

Lamps in parallel

Figure 1-6.

In figure 1-6 a pictorial diagram and the schematic equivalent circuit is shown. The two lamps are *directly* connected to the battery terminals. This is always the case with parallel circuits. If you had 10 components (they don't have to be lamps) connected in parallel, then each side of each component is connected directly to the battery (or another source).

BRANCH CURRENTS

Each resistance (or other component) in a parallel circuit is connected by a conductor directly to the source voltage. Each resistor will draw current from the source according to Ohm's law, I=E/R, for each branch. The sum of all the branch currents must then be equal to the total current drawn from the source.

In a parallel circuit the sum of the branch currents equals the total current

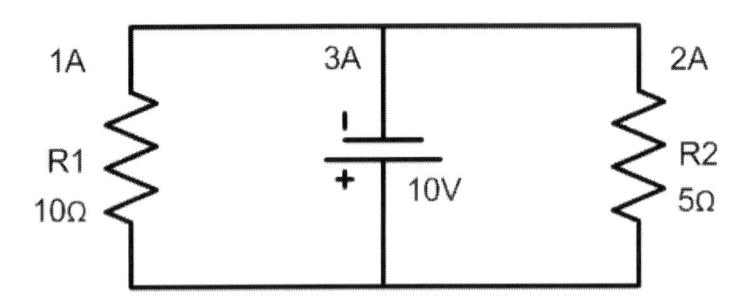

Figure 2-6.

In applying Ohm's law, it is important to note that the current equals the voltage applied across the circuit divided by the resistance of the circuit. I=E/R.

In figure 2-6, 10 V is applied across the 10Ω of R_1, resulting in a current of 1 Ampere being drawn from the battery through R_1. Similarly, the 10 Volts applied to the 5Ω of R_2 will cause 2 Amperes to be drawn from the battery.

The two branch currents in the circuit are then 1 Ampere and 2 Amperes. The total current drawn from the battery is then 3 Amperes.

Just as in a circuit with only one resistance, any branch that has less resistance will draw more current. If R_1 and R_2 were equal, however, the two branch currents would have the same value. For instance, if R_1 and R_2 were both 5Ω then each branch would draw 2 Amperes and the total current drawn from the battery would be 4 Amperes.

The current can be different in parallel circuits having different resistances because the voltage is the same across all the branches. Any voltage source generates a potential difference across its two terminals. This voltage does not move. Only current flows around the circuit.

The source voltage is available to make electrons move around any closed path connected to the generator terminals. How much current flowing in the separate paths depends on the amount of resistance in each branch.

For a parallel circuit with any number of branch currents we can then write an equation for calculating the total current (I_t):

$I_t = I_1 + I_2 + I_3 + I_4$ etc.

This rule applies to any number of parallel branches, whether the resistances are equal or unequal.

Example:

An R_1 of 20Ω and an R_2 of 40Ω and an R_3 of 60Ω are connected in parallel across a 240 Volt supply. What is the total current drawn from the supply?

Let's calculate the branch currents for R_1, R_2 and R_3:

$I_{R1} = E/R_1 = 240/20 = 12$ Amps

$I_{R2} = E/R_2 = 240/40 = 6$ Amps

$I_{R3} = E/R_3 = 240/60 = 4$ Amps

The total current drawn from the 240 Volt supply is the sum of the branch currents:

$I_t = I_{R1} + I_{R2} + I_{R3} = 12 + 6 + 4 = 22$ Amperes.

RESISTANCES IN PARALLEL

In the example, above we could have worked out the total resistance to calculate the total current being drawn from the supply.

To find the total resistance of any number of resistors in parallel we find the reciprocal of the sum of the reciprocals for each resistance. This sounds like a bit of a mouthful so I will put it in equation form and you should see what I mean.

$$R_t = \frac{1}{1/R_1 + 1/R_2 + 1/R_3}$$

Let's calculate the total resistance of our example using this equation. Firstly, find the reciprocal of each of the resistances:

Reciprocal of R_1 = $1/R_1$ = 1/20 = 0.05
Reciprocal of R_2 = $1/R_2$ = 1/40 = 0.025
Reciprocal of R_3 = $1/R_3$ = 1/60 = 0.01667 (recurring decimal).

The sum of the reciprocals above is: 0.091667

Finally, to find R_t we take the reciprocal of the sum of the reciprocals, or:

1/0.091667 which equals 10.91 rounded.

So R_t = 10.91 Ohms (with a little rounding error).

Since we know the total resistance and the applied voltage we can now calculate the total current from Ohm's law.

I=E/R = 240/10.91 = 21.998 Amps, or close enough to the 22 Amps we calculated earlier. When you get used to it, these calculations are very easy to do on a calculator. Many calculators have a reciprocal key '1/x' or 'x^{-1}'.

Example: A parallel circuit consisting of two branches, with a current of 5A through each branch, is connected across a 90V source. What is the equivalent total resistance R_t?

To find the total resistance (R_t) we need to know the applied voltage (90V) and the total current drawn from the supply. Since we are told that there are two branches and each branch draws 5 A then the total current must be 10A. We can now use Ohm's Law to calculate the total resistance:

R=E/I
R=90/10
R=9Ω

PARALLEL BANK

A combination of parallel branches is often called a bank. Radio operators often use a device called a dummy load (we will go into detail about dummy loads later). A dummy load is typically just a resistance of 50Ω which is connected to the antenna socket of a transmitter for tuning purposes. The problem is the dummy load has to dissipate all of the transmitter power. If the operator purchased and used a typical 50-Ohm carbon resistor, it would burn out, as carbon resistors cannot dissipate more than about 1W. A typical dummy load should be able to dissipate 100W.

A good dummy load can be made from twenty 1000Ω 5 Watt resistors connected in parallel. Each of the twenty resistors, being 5 Watts each are able to dissipate 5 Watts, so the bank is able to dissipate 20 x 5 = 100 Watts. The total resistance of twenty 1000Ω resistors in parallel is 50Ω with a total power rating of 100 Watts.

In practice, the dummy load is usually cooled, perhaps by immersing the resistors in oil or some sort of air cooling which enables the resistors to dissipate more power without overheating and being destroyed.

A stumbling block for many is trying to understand how adding more resistance to a parallel circuit can reduce the total resistance.

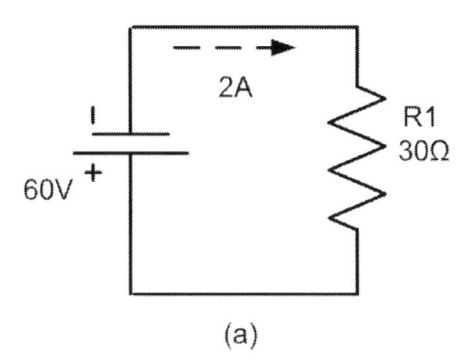

Figure 3(a)-6.

In figure 3(a)-6 you see a supply voltage of 60 Volts connected to a 30Ω load. By Ohm's law the current drawn by the 30Ω load must be 2A.

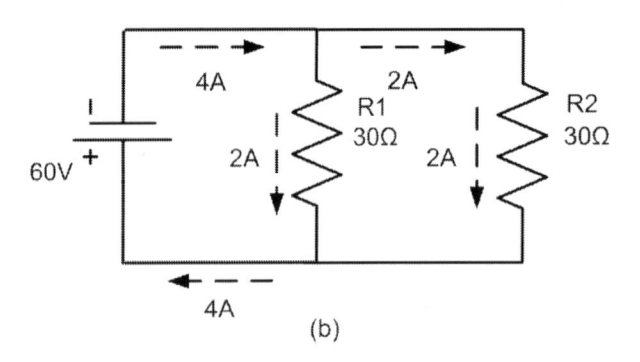

Figure 3(b)-6.

If a second 30Ω resistor is now connected as in figure 3(b)-6, then an additional 2A will flow through it. The total current drawn from the supply is now 4A.

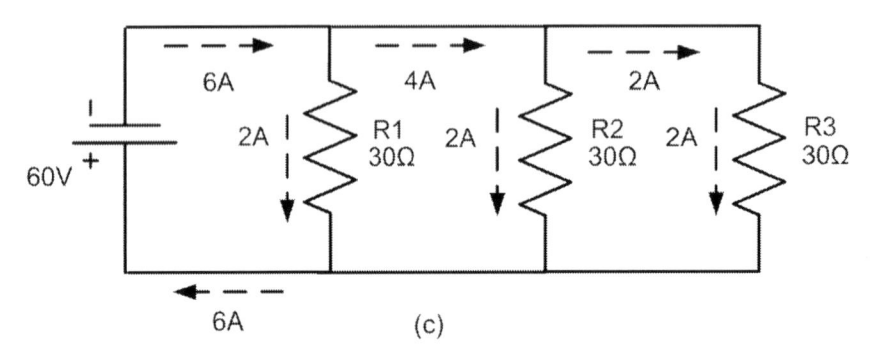

Figure 3(c)-6.

Adding a third 30Ω resistor as in figure 3(c) causes a further 2A to be drawn from the supply, bringing the total current drawn from the supply to 6A.

The total resistance of this circuit would be:
R=E/I = 60/6 = 10Ω

We can see that as we add parallel resistances to the supply, more current is drawn from the supply, so the total circuit resistance must be less. We could go on adding parallel resistances for as long as we wanted and each time we did, the total circuit resistance would become less with each added resistance.

So now you should be able to answer the question; why does adding more resistance to a parallel circuit increase the total current? Each additional parallel resistor creates an additional branch for current that would not otherwise be there.

DERIVING THE RECIPROCAL EQUATION

We discussed the reciprocal equation earlier, but we did not mention how this equation was derived.

We know the basic law: the sum of the branch currents is equal to the total current flowing in a parallel circuit.

$I_t = I_1 + I_2 + I_3$ etc.

$I_t = E/R_t$ and $I_1 = E/R_1$ and $I_2 = E/R_2$ etc.

We can substitute this into the rule for branch currents and get:

$E/R_t = E/R_1 + E/R_2 + E/R_3$ etc.

Notice how E is in the numerator on both sides of the equal sign. If we divide both sides by E, the E's cancel out and we are left with:

$1/R_t = 1/R_1 + 1/R_2 + 1/R_3$ etc.

This gives us the reciprocal of the total resistance. The reciprocal of this is the total resistance.

SPECIAL CASE OF ALL RESISTANCES THE SAME

If all of the resistances are the same in a parallel circuit, as in our dummy load example presented earlier, the total resistance can be found by dividing the number of branches into the value of one of the parallel resistances.

In the dummy load example, we had twenty, 1000Ω resistors. A quick method to find the total resistance is; 1000/20 = 50Ω. However, please do remember that this approach can only be used when all resistances are the same.

SPECIAL CASE OF ONLY TWO BRANCHES

When there are two parallel resistances and they are not equal, it is usually quicker to calculate the combined resistance by the method known as 'product over sum'.

This rule says that the combination of two parallel resistances is their product divided by their sum.

$R_t = (R_1 \times R_2) / (R_1 + R_2)$

The symbol '/' means divided by.

Example:

A resistance of 100Ω is connected in parallel with a resistance or 200Ω. What is the total resistance of the circuit using the product over sum method.

$R_t = (100 \times 200) / (100 + 200)$
$R_t = 20000 / 300$
$R_t = 66.667Ω$ (rounded).

Notice also that the total resistance of parallel resistors is always less than the lowest resistance.

TOTAL POWER IN PARALLEL CIRCUITS

Since the power dissipated in the branch resistances must come from the source voltage, the total power equals the sum of the individual values of power in each branch.

$P_t = P_{R1} + P_{R2} + P_{R3}$ etc.

Of course, if you know the total current and the applied voltage, you can find the power by P = EI.

I can recall a question that was asked in a radiocommunications exam some years ago. I am sure the question was not meant to be as difficult as it turned out and my guess is that the person writing the question must have been anxious to knock off work at the time.

The question was:
Four 10Ω, 1-Watt resistors are connected in series. What is the total power dissipation rating of the circuit?

That was it, no voltage or current mentioned. In fact, the resistance (10Ω) is totally irrelevant as is the series connection.

The answer is 4 Watts.

The question was poorly phrased in my view and could have been put:

What is the maximum power that four, 1-Watt resistors can dissipate?

ADVANTAGE OF PARALLEL CONNECTION

Household appliances, as you may know, are designed to operate at 240V. The electrical wiring to all the power points and light fittings in your home are in the form of a parallel circuit. This ensures that each appliance receives the supply voltage of 240V. Imagine the problems if your house was wired as a series circuit. Each time you turned on a light all the lights in the house would get dimmer. If you turned on something with a high resistance (low wattage), the lights would not even glow. Supplying the same voltage to various devices in a house, a radio circuit, or a car requires parallel connections.

7. Series-Parallel Circuits

In many circuits, some components are connected in series to have the same current, while others are in parallel for the same voltage. When analysing and doing calculations with series-parallel circuits, you simply apply what you have learnt from the last two chapters.

In the circuit of figure 1-7, we could work out all the voltages across all of the resistances and the current through each resistance and then total resistance. For now, I am just going to walk through the simplification of this circuit to a single resistance connected across the 100V source.

Keep in mind that any circuit (resistive) can be reduced to a single resistance. This is particularly useful when we come to do transmission lines and antennas.

For now, let's have a go at simplifying the circuit of figure 1-7. There are many ways to go about this problem. The method I prefer is to start at the right-hand side and work my way back to the source, simplifying the circuit as I go. Let's try this together: -

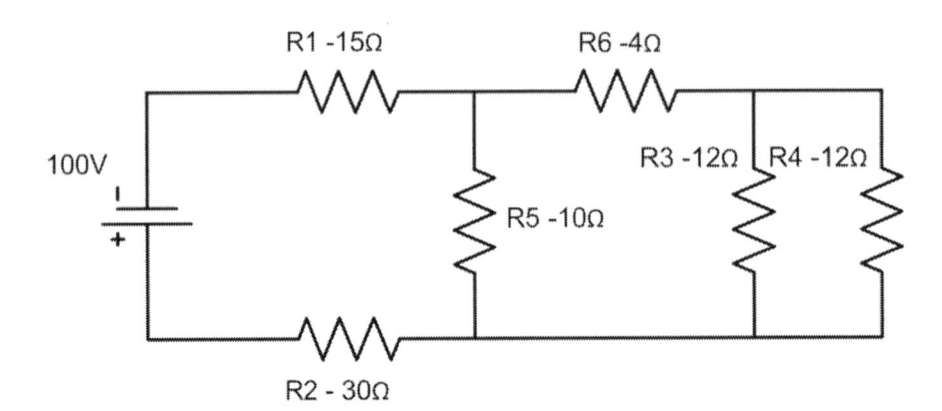

Figure 1-7.

On the right-hand side, we see R₃ and R₄ in parallel and each 12Ω. Do you remember the shortcut method when parallel resistances are all the same value? Divide the value of the branch by the number of branches: 12Ω /2 = 6Ω.

Replace the parallel pair of R₃ and R₄ with a single resistance R₇ as shown in figure 2-7.

Always redraw the circuit.

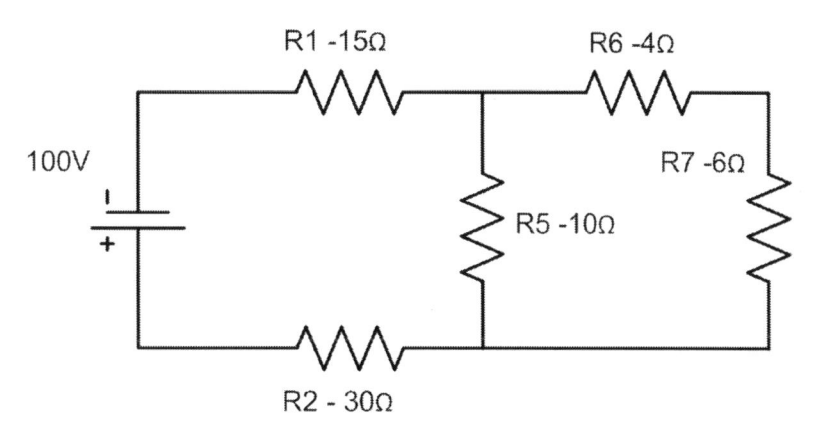

Figure 2-7.

The next logical step would be to combine the series resistance of R₆ and R₇ into a single resistance R₈ by adding them, as shown in figure 3: 4Ω + 6Ω = 10Ω.

Redraw the circuit.

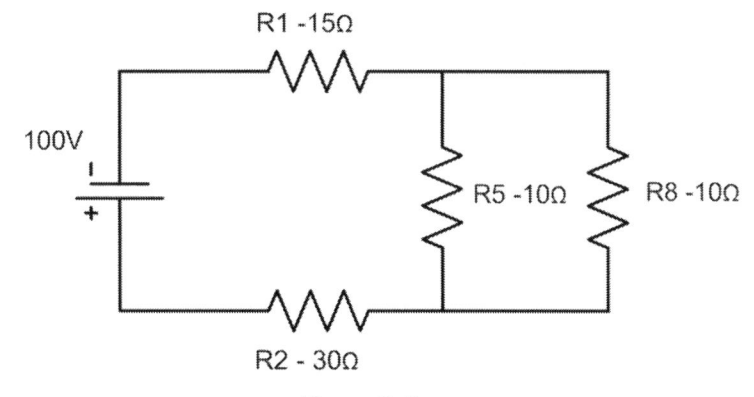

Figure 3-7.

Now combine the parallel and equal resistances of R₅ and R₈ into a single resistance R₉ of 5Ω as shown in figure 4-7.

Redraw the circuit.

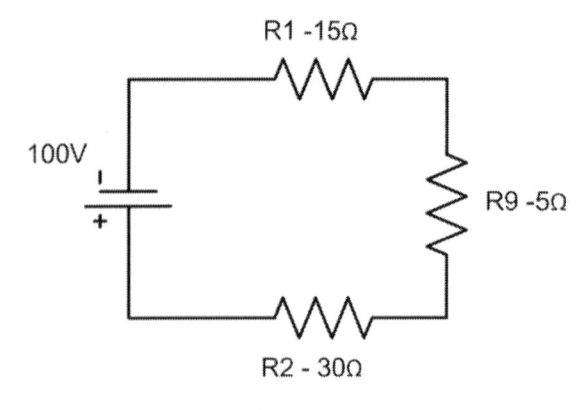

Figure 4-7.

We are now left with a series circuit consisting of R_1, R_2 and R_9. We find the total resistance of these by summing them: 15Ω + 30Ω + 5Ω = 50Ω.

You guessed it! **Redraw the circuit** *again for our final result shown in figure 5-7.*

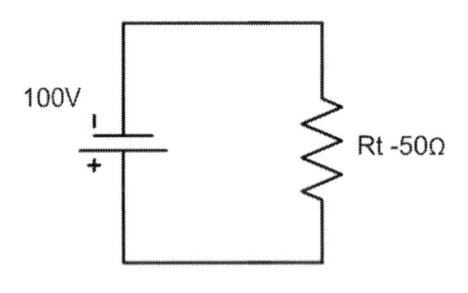

Figure 5-7.

So the total resistance of the circuit we started with is equivalent to a single resistance of 50Ω.

We could now find out how much current is being drawn by the circuit from the supply, by using Ohm's law:

I=E/R = 100/50 = 2 Amperes.

That's it! You will find that most circuits, even very complex ones, can be handled in the same manner i.e. by simplifying series and parallel branches as you work your way down to a single resistance.

I cannot emphasise enough the need to redraw the circuit as you work your way through it. The calculations for this circuit were easy arithmetically. If it were not so easy, then it is important to show all of your working out as well.

Refer to the circuit of figure 1-7. What is the voltage across R_4?

We know that the total resistance of the circuit is 50Ω and from this we worked out that the current drawn from the supply was 2A.

Therefore, if we go back to the figure 1-7 circuit, R_1 (15Ω) and R_2 (30Ω) must have 2A flowing through them since they are in series with the supply.

The voltage across R_1 will be: $E=IR_1 = 2 \times 15 = 30$ V.
Likewise, the voltage across R_2 will be: $E=IR_2 = 2 \times 30 = 60$ V.

Now if there is 30 Volts across R_1 and 60 Volts across R_2 then this leaves 100-30-60 = 10 Volts across R_5.

R_5 is in parallel with R_6 and the parallel pair of R_3 and R_4.

R_3 and R_4 simplify to 6Ω.

So we have 10 Volts across 4Ω and 6Ω in series. The voltage across the 6Ω will be the voltage across R_4 (and R_3 for that matter).

Some may see immediately without any calculation that 10 Volts across a series combination of a 4Ω and a 6Ω resistor will result in 4 Volts across the 4Ω and 6 Volts across the 6Ω. If you can't see this then don't worry, let's solve it using Ohm's law.

I = E/R = 10/10 = 1 Ampere.

So there is 1 Ampere flowing through the 6 Ω resistance that represents the combined resistance of R_3 and R_4 in parallel.

E = IR = 1 x 6 = 6 Volts.

Therefore, the voltage drop across R_4 is 6 Volts.

COMPARISON OF SERIES AND PARALLEL CIRCUITS

SERIES CIRCUIT	PARALLEL CIRCUIT
The current in all parts of the circuit is the same.	The voltage is the same across all parallel branches.
E across each series R is I x R.	I in each branch is E/R.
The sum of the voltage drops equals the applied voltage. $E_t = E_1 + E_2 + E_3$ etc.	The sum of the branch currents equals the total current. $I_t = I_1 + I_2 + I_3$ etc.
$R_t = R_1 + R_2 + R_3$ etc.	$1/R_t = 1/R_1 + 1/R_2 + 1/R_3$ etc.
R_t must be larger than any individual R. $P_t = P_1 + P_2 + P_3$ etc.	R_t must be less than the smallest branch R. $P_t = P_1 + P_2 + P_3$ etc.
Applied voltage is divided into IR drops.	Main current is divided into branch currents.
The largest voltage drop is across the largest resistance.	The largest branch current is through the smallest parallel R.
Open circuit in one component causes the entire circuit to be open.	Open circuit in one branch does not prevent current in other branches.

8. Magnetism

There are usually no questions in radio exams about magnetism. You are justified then in asking why we are doing this subject. Well, although there are not many, if any, questions in the exams, later on in this book we do have to cover topics like:

Inductance
Mutual Inductance
Alternating Current

It is impossible to gain an understanding of these subjects without a basic understanding of magnetism. I will try to keep the topic as brief as possible, bearing in mind what has to be learnt and more to the point 'understood' later on.

Magnetism is a broad subject. However, I will cut it down into what you really need to know to help with your understanding of other topics later on.

I think it is a deficiency in most Radio syllabi that there is not enough on magnetism.

MAGNETISM AND ELECTRICITY

Any wire carrying a current of electrons is surrounded by area of force called a *magnetic field*. For this reason, any study of electricity or electronics must consider magnetism.

Almost everyone has had experiences with magnets or with pocket compasses at one time or another. A magnet attracts pieces of iron but has little effect on practically everything else. Why does it single out the iron? A compass, when laid on a table, swings back and forth, finally coming to rest pointing toward the North Pole of the world.

Why does it always point in the same direction?

These and other questions about magnetism have puzzled scientists for hundreds of years. It is only relatively recently that theories seeming to answer many of the perplexing questions that arise when magnetism is investigated have been developed.

Radio and electronic apparatus such as relays, circuit breakers, earphones, loudspeakers, transformers, chokes, magnetron tubes, television tubes, phonograph pickups, tape and disk recorders, microphones, meters, motors and generators depend on magnetic effects to make them function.

Every coil (inductance) in a radio receiver or transmitter is utilising the magnetic field that surrounds it when current is flowing through it. But what is meant by the term magnetic field?

THE MAGNETIC FIELD

An electron at rest has a negative electrostatic field of force surrounding it. When energy is imparted to an electron to make it move, a new type of field develops around it, at right angles to its electrostatic field. Whereas negative electrostatic lines of force are considered as radiating outward from an electron, the electromagnetic field of force develops as a ring around a moving electron, at right angles to the path taken, around the wire.

An interesting difference between a magnetic field and an electrostatic field is that an electrostatic field can exist about a single stationary charge such as an electron or a proton. This is not the case with magnetism. A North 'monopole' does not exist any more than a South 'monopole' exists. With magnetism, you always have two poles.

Electrons orbiting around the nucleus of an atom or a molecule produce electromagnetic fields around their paths of motion. These fields (electric and magnetic) are balanced or neutralised by the effect of proton movement in the nucleus. The movement of one orbital electron is counteracted by another orbital electron whirling in an opposite direction. In almost all substances, the net result is little or no external fields.

In the case of an electric conductor carrying current, the collective movement of electrons along the wire produces a magnetic field around the conductor. The greater the current, the more intense the magnetic field.

The diagram in figure 1-8 shows a method of determining the direction of the magnetic field around a current carrying conductor.

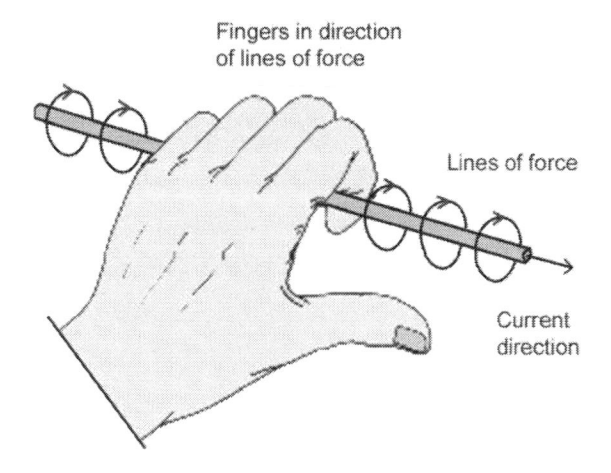

Figure 1-8.

If you place a compass next to a conductor and then pass a current through the conductor, the compass needle will move indicating the presence of a magnetic field.

Under normal circumstances, the field strength around a current-carrying conductor varies inversely as the distance from the conductor. At twice the distance from the conductor, the magnetic-field strength is one-half as much, at five times the distance the field strength is one-fifth and so on. At a relatively short distance from the conductor, the field strength may be quite weak.

When the current in a conductor is increased, more electrons flow, the magnetic-field strength increases and the whole field extends further outward.

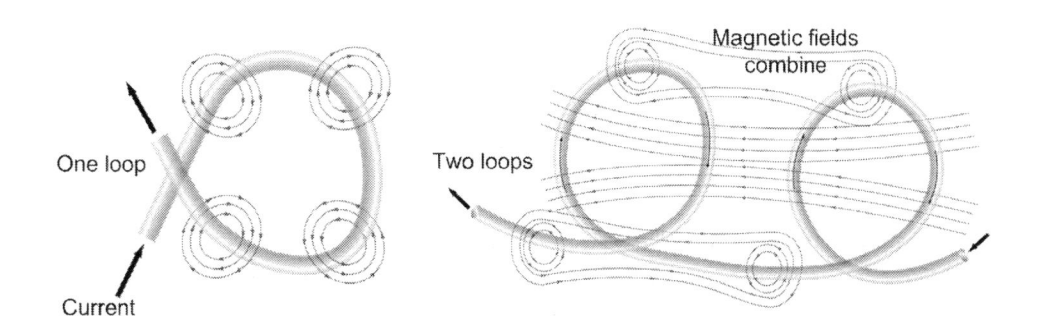

Figure2-8.

Figure 3-8.

By looping a conductor, as shown in figure 2-8, magnetic lines of force are concentrated in the central core area of the loop.

When two or more turns of wire are formed into a coil, as shown in figure 3-8, the lines of force from each turn link to the fields of the other turns and a more concentrated magnetic field is produced in the core of the coil.

By forming a coil into many loops the magnetic field is stronger and concentrated. A with enough turns and current a strong electromagnet can be created.

An experiment to try

Typically in face-to-face classes, I would demonstrate many things by experiment. This makes learning more enjoyable. We don't have that luxury. However, I will mention these experiments from time to time and perhaps you can try them yourself. If you have never made an electromagnet, I really encourage you to do so as it can be a lot of fun and fascinating. Just get an iron rod about 75mm long. A bolt with a head and a nut screwed onto one end is ideal. Wind as many turns of wire onto it as you can. I use single telephone wire. However, any small diameter wire will do the job. The more turns, the better. Connect the ends of the coil you have made to a 6-Volt lantern battery and you will have yourself an electromagnet. Which, by the way, you can use in other experiments down the track. Besides just playing with your electromagnet you can use it to magnetise some tools such as screwdrivers etc. Don't leave it connected to the battery for very long as it draws a heavy current from it. Do not use a car battery as the coil will overheat and melt. The coil (inductor) is a very low resistance. However, current is limited by using a 6-Volt lantern battery, which can only supply 3-4 amps. If you want to use a car battery you need to limit the current flow through the electromagnet. You could do this by connecting a 12 Volt automotive light bulb in series with the magnet.

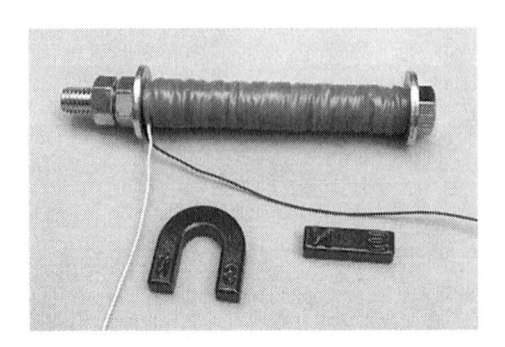

Figure 4-8.

Figure 4-8 shows an electromagnet made from a bolt and some scrap wire; a permanent bar and horseshoe magnet.

The direction of the field of force (north and south pole) can be reversed by reversing the current direction or by reversing the winding direction.

At one end of the coil the field lines are leaving and at the other end, they are entering. When a coil or piece of metal has lines of force leaving one end of it, that end is said to have a north pole. The end with the lines entering is the south pole.

The terms "north" and "south" indicate magnetic polarity, just as "negative" and "positive" indicate electrostatic polarity. They should not be used interchangeably. The negative end of a coil is the end connected to the negative terminal of the source and does not refer to the north or south magnetic polarity of the coil.

All magnetic lines of force are complete loops and may be considered somewhat similar in their action to stretched rubber bands. They will contract back into the circuit from which they came as soon as the force that produced them ceases to exist.

Magnetic lines of force never cross each other. When two lines of force have the same direction, they will oppose mechanically if brought near each other.

PERMEABILITY

When a coil of wire is wound with air as the core, a certain flux density will be developed in the core for a given value of current. If an iron core is inserted into the coil, the flux density will increase. This increase in flux density is achieved without increasing current or the number of turns.

With an air-core coil, the air surrounding the turns of the coil may be thought of as pushing against the lines of force and tending to hold them close to the turns. With an iron core, however, the lines of force find a medium in which they can exist much more easily than in air. As a result, lines that were held close to the turns in the air-core coil are free to expand into the highly receptive area afforded by the iron. This allows lines of force that would have been close to the surface of the wire to expand into the iron core. Thus, the iron core produces a greater flux density, although no more magnetising force. In other words, for the same number of turns and current, a coil will have a stronger magnetic field if an iron core is inserted. The iron core merely brings the lines of force out where they can be more readily used and concentrates them.

The ability of material to concentrate lines of force is called Permeability.

The permeability of most substances is very close to that of air, which may be considered as having a value of 1. A few materials, such as iron, nickel and cobalt, are highly permeable, with permeabilities of several hundred to several thousand times that of air. The word "Permeability" is a derivation of the word "Permeate," meaning "to Pervade or saturate" and is not related to the word "Permanent.")

Permeability is represented by the Greek letter µ (mu, pronounced mew).

Alloying iron makes it possible to produce a broad range of permeabilities. Most stainless steel exhibits practically no magnetic effect, although some may be magnetic.

Any substance that is not affected by magnetic lines of force and is reluctant to support a magnetic field is said to have the property of reluctance. Air, vacuum and most substances have unity reluctance, while iron has a very low reluctance.

In electric circuits, the reciprocal of resistance is called conductance. In magnetic circuits, the reciprocal of reluctance is called permeance. "Permeability" is used when discussing how magnetic materials behave.

THE ATOMIC THEORY OF MAGNETISM

The discussion here will be a considerably condensed version of the atomic theory of magnetism.

From atomic theory, it is known that an atom is made up of a nucleus of protons surrounded by one or more electrons encircling it. The rotation of electrons and protons in most atoms is such that the magnetic forces cancel each other. Atoms or molecules of the elements iron, nickel and cobalt, arrange themselves into magnetic entities called domains. Each domain is a complete miniature magnet.

Groups of domains form crystals of the magnetic material. The crystals may or may not be magnetic, depending on the arrangement of the domains in them. Investigation shows that while any single domain is fully magnetised, the external resultant of all the domains in a crystal may be a neutral field.

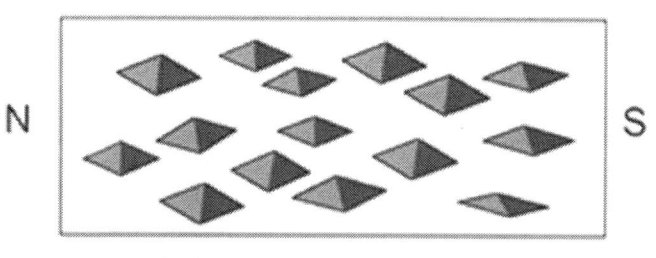

RANDOM DOMAINS   **DOMAINS LINED UP**

Figure 5-8.

Each domain has three directions of magnetisation: easy magnetisation, semi-hard magnetisation and hard magnetisation. If an iron crystal is placed in a weak field of force, the domains begin to line up in the easy direction. As the magnetising force is increased, the domains begin to roll over and start to align themselves in the semi-hard direction. Finally, as the magnetising force is increased still more, the domains are lined up in the hard direction. When all the domains have been lined up in the hard direction, the iron is said to be saturated. An increase in magnetising force no longer magnetically changes the material. The material is magnetically *saturated.*

FERROMAGNETISM

Substances that can be made to form domains are said to be ferromagnetic, which means "iron magnetic." The ferromagnetic elements are iron, nickel and cobalt, but it is possible to combine some non-magnetic elements and form a ferromagnetic substance. For example, in the proper proportions, copper, manganese and aluminium, each by itself being non-magnetic, produce an alloy which is similar to iron magnetically.

Materials made up of non-ferromagnetic atoms, when placed in a magnetic field may weakly attempt either to line up in the field or to turn at right angles to it. If they line themselves in the direction of the magnetic field, they are said to be paramagnetic. If they try to turn from the direction of the field, they are called diamagnetic. There are only a few diamagnetic materials. Some of the more common are gold, silver, copper, zinc and mercury. All materials which do not fall in the ferromagnetic or diamagnetic categories are paramagnetic. The greatest percentage of substances are paramagnetic.

Ferromagnetic substances will resist being magnetised by an external magnetic field to a certain extent. It takes some energy to rearrange even the easy-to-move domains.

Once magnetised, however, ferromagnetic substances may also tend to oppose being demagnetised. They are said to have retentivity, or remanence, the ability to retain magnetism when an external field is removed.

As soon as the magnetising force is released from a magnetised ferromagnetic substance, it tends to return at least part way back to its original non-magnetised state, but it will always retain some magnetism. This remaining magnetism is residual magnetism. Paramagnetic and diamagnetic materials always become completely non-magnetic when the external magnetising force is removed from them.

PERMANENT AND TEMPORARY MAGNETS

Ferromagnetic substances that hold magnetic-domain alignment well (have a high value of retentivity) are used to make permanent magnets. One of the strongest permanent magnets is made of a combination of iron, aluminium, nickel and cobalt called Alnico. It is used in horseshoe magnets, electric meters, headphones, loudspeakers, radar transmitting tubes and many other applications. Some magnetically hard, or permanent magnetic materials, are cobalt steel, nickel-aluminium steels and special steels.

Figure 6-8 below shows the magnetic field surrounding permanent magnets. This picture was made by placing a piece of paper over the magnets and sprinkling iron filings onto the paper.

You can see the lines of force attracting when a north and a south pole are brought near to each other and repelling when two north poles are brought near to each other. Remember:

Unlike poles attract and like poles repel.

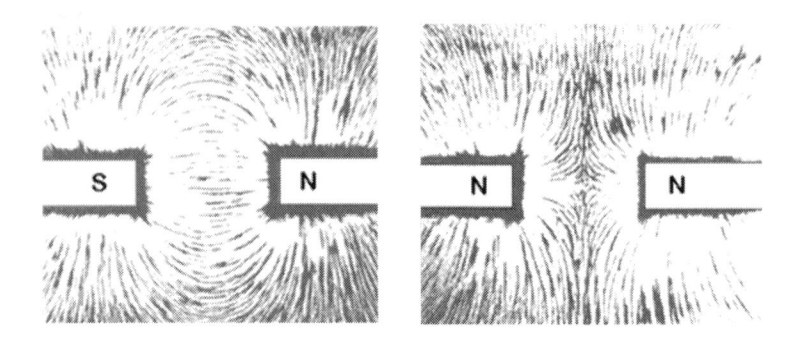

Figure 6-8.

Ferromagnetic metals that lose magnetism easily (have a low value of retentivity) make temporary magnets. They find use in transformers, chokes, relays and circuit breakers.

Pure iron and Permalloys ('Perm' derived from "Permeable," not from "Permanent") are examples of magnetically soft, temporary-magnet materials. Finely powdered iron, held together with a non-conductive binder, is used for cores in many applications. These are called ferrite cores.

MAGNETISING AND DEMAGNETISING

There are two simple methods of magnetising a ferromagnetic material. One is to wrap a coil of wire around the material and force a direct current through the coil. If the ferromagnetic material has a high value of retentivity, it will become a permanent magnet.

If the material being magnetised is heated and allowed to cool while subjected to the magnetising force, a greater number of domains will be swung into alignment, and a greater permanent flux density may result. Hammering or jarring the material while under the magnetising force also tends to increase the number of domains that will be affected.

A less efficient method of magnetising is to stroke a high-retentivity material with a permanent magnet. Have you ever magnetised a screwdriver using this approach? This will align some of the domains of the material and induce a relatively weak permanent magnetism.

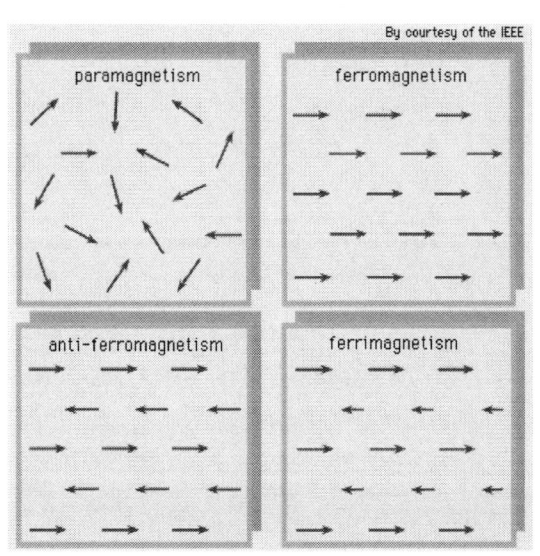

Figure 7-8.

If a permanent magnet is **hammered**, many of its domains will be jarred out of alignment and the flux density will be lessened. If heated, it will lose its magnetism because of an increase in molecular movement that upsets the domain structure. Strong opposing magnetic fields brought near a permanent magnet may also decrease its magnetism. It is important that equipment containing permanent magnets be treated with care. The magnets must be protected from physical shocks, excessive temperatures and strong alternating or other magnetic fields.

When heated, permanent magnets lose their magnetism quickly and also at a certain temperature. The temperature at which a magnet loses its magnetism is called the *Curie temperature.*

When tools or objects such as screwdrivers or watches become permanently magnetised, it is possible to demagnetise them by slowly moving them into and out of the core area of a many-turn coil in which a relatively strong alternating current (AC) is flowing. The AC produces a continually alternating magnetising force. As the object is placed into the core area, it is alternately magnetised in one direction and then the other. As it is pulled farther away, the alternating magnetising forces become weaker. When it is finally out of the field completely, the residual magnetism will usually be so low as to be of no consequence.

PERMANENT-MAGNET FIELDS

When a piece of magnetically hard material it subjected to a strong magnetising force the domains are aligned in the same direction. When the magnetising force is removed, many of the domains remain in the aligned position and a permanent magnet results. A north pole is anywhere the direction of the magnetic lines of force move outward from the magnet. A south pole is any place where the direction of the lines of force are inward.

If a magnet is completely encased in a magnetically soft iron box, all its lines of force remain on the walls of the box and there is no external field. This is known as magnetic shielding. Shielding may be used in the opposite manner. An object completely surrounded by an iron shield will have no external magnetic fields affecting it, as all such lines of force will remain in the permeable shield.

THE MAGNETISM OF THE EARTH

Sufficient quantity of the ferromagnetic materials making up the Earth have domains aligned in such a way that the earth appears to be an enormous permanent magnet. The direction taken by the lines of force surrounding the surface of the earth is inward at a point near what is commonly known to be the North Pole of the world and outward near the earth's South Pole.

The simple magnetic navigational compass consists of a small permanent magnet balanced on a pivot point. The magnetic field of the compass needle lines itself up in the earth's lines of force.

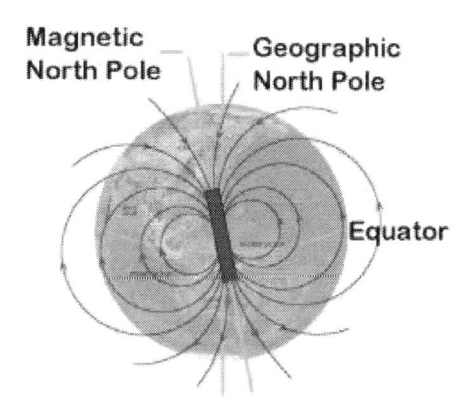

Figure 8-8.

As a result, the magnetic north end of the compass needle is pulled toward the earth's South Magnetic Pole, since unlike poles attract each other. This means that when the "north-pointing end" is pointing toward geographical north, this end (a magnetic north pole) is pointing toward a magnetic south pole

RELAYS

A relay is a relatively simple magnetic device that generally consists of a coil, a ferromagnetic core and a movable armature on which make and/or break contacts are fastened. A simple relay may be used to close a circuit when the coil is energised. This type of relay is known as single-pole single-throw (SPST), normally open (NO), or "make-contact" relay. The relay in figure 9-8 is a double pole single throw. The contacts can be used to turn something ON or OFF.

Electromechanical Relay Construction

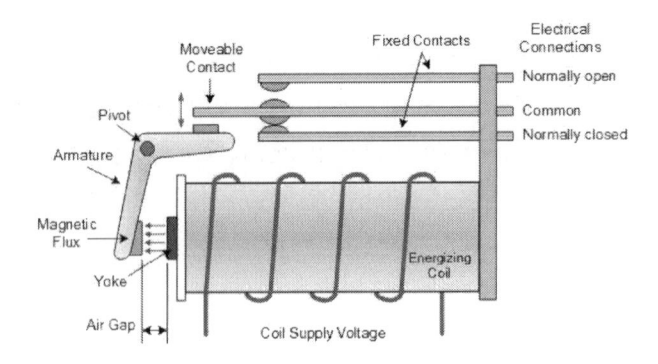

Figure 9-8.

The core, of the relay and the pivoted armature bar, are all made of magnetically soft ferromagnetic materials having high permeability and little retentivity. One of the relay contacts is attached by an insulating strip to the armature and the other to the relay body with an insulating material. The contacts are electrically separated from the operational parts of this particular relay.

When current flows through the coil, the core is magnetised and lines of force develop in the core and through the armature and the body of the relay. The gap between the core and the armature is filled with magnetic loops trying to contract. These contracting lines of force overcome the tension of the spring and pull the armature toward the core, operating the relay contacts. When the current in the coil is stopped, the magnetic circuit loses its magnetism and the spring pulls the armature up, opening the contacts.

Relays are useful in remote closing and opening of high-voltage or high-current circuits with relatively little voltage or current flow in the coil.

Relay contacts are usually made of silver or tungsten. Silver oxidises but can be cleaned by using a very fine abrasive paper, or a piece of ordinary letterhead paper rubbed between the contacts. If the contacts are pitted by heavy currents, they may be smoothed with a fine file, but the original shape of the contacts should be retained to allow a wiping action during closing to keep them clean.

When I started out in electronics, relays were pretty big components. These days' relays for switching small currents can be made very small and look no larger than a medium-sized integrated circuit.

The energy of a magnetic field.

Here is something for you to ponder. A magnetic field around a relay or a solenoid (electromagnet) has energy. The energy comes from the current flowing in the conductor coil. Suppose we switch the current OFF quickly by opening a switch. Where does the energy of the magnetic field go? It cannot just "disappear". Energy never disappears. It can change into another type of energy, though. For now, I will leave you to think about that.

Barkhausen Effect.

Earlier we explained how materials than can be magnetised contain many small tiny magnets in them called domains. For a long time, this was called the domain hypothesis of magnetism and it was not quite a fact. A scientist by the name of Barkhausen was experimenting with microphones attached to a soft iron bar while adjusting the strength of an external field. As he adjusted the strength of the external field, his microphone picked up the sound of the magnetic domains moving. So the domain theory of magnetism was proven. He also noticed that the domains did not move smoothly. They moved in discreet steps. Staying still for a while as the magnetising force was increased; then suddenly jumping to their next position. This and other experiments led to the idea of what is known today as a "quantum". A **quantum** (plural: **quanta**) is the minimum amount of any physical entity involved in an interaction. This became critical to the evolution of quantum mechanics.

9. Alternating Current Part 1

You may be thinking that when I speak of alternating current (AC), I am talking about the household electrical supply. True, the household supply is AC, however, so is the signal from a radio transmitter. The only difference is the frequency.

FARADAY'S LAW

When relative motion exists between a conductor and a magnetic field an emf is induced into the conductor.

This is a fundamental law. We will take our time and try to understand it thoroughly. What Faraday discovered was a way of producing electrical pressure (voltage or emf) from a magnetic field.

Faraday's law requires relative motion. This means there must be motion between the conductor and the magnetic field. This is not to say that the magnetic field must move. Neither does it mean that the conductor must move. Provided there is relative motion between the magnetic field and the conductor; a voltage will be induced into the conductor. In practice, this means the conductor can be stationary and the magnetic field must be moved relative to it. Alternatively, the magnetic field can be stationary and the conductor moved through it.

A Mind Experiment

Imagine a cardboard tube. I have successfully used the cardboard tube from an empty kitchen wrap. The tube is wound with at least 50 turns of insulated conductor. So, in fact, we have made a simple coil. To the ends of the coil, we attach a Voltmeter.

We have not discussed meters yet, but I think you can follow along if I just say it is an instrument with a needle, which indicates if the voltage is present. I used a special type of Voltmeter called a Galvanometer. A Galvanometer is a Voltmeter that has zero volts at centre scale. The advantage of this is that the meter's pointer can deflect left or right of centre depending on the polarity of the voltage being measured. You can see a Galvanometer in figure 1-9.

In figure 1-9 we have a galvanometer connected to our coil. Now imagine taking a permanent magnet, a bar magnet and inserting it quickly into the centre of the coil. When you do so, the pointer will deflect in one direction as you insert the magnet. When the magnet stops moving the meter will return to zero.

You still have the magnet inside the coil. Now pull the magnet out. As you do the pointer will again deflect, indicating a voltage on the coil, however, this time, the deflection will be in the opposite direction to when we inserted the magnet.

Figure 1-9

What we have done is demonstrated Faraday's law of induction. When we inserted the bar magnet inside our coil, we produced relative motion between the coiled conductor and the magnetic field of the bar magnet. Once the magnet was in and stopped moving, the relative motion stopped and the induced voltage fell to zero again. When we removed the magnet from the coil, relative motion existed again, but since the magnet was being extracted rather than inserted, the polarity of the induced voltage was in the opposite direction and the meter's pointer deflected accordingly.

If you happened to have made the electromagnet we discussed in an earlier chapter, you could place the electromagnet inside the coil. If you then energised the electromagnet by applying a voltage to it, the expanding magnetic field would induce a voltage in our coil. After a few moments, the induced voltage would fall to zero. Now, if we disconnected the voltage to the electromagnet, the magnetic field around it would collapse (relative motion) and voltage would be induced in the coil. You may not know it, but you have just demonstrated the operation of a device to be discussed shortly, called a transformer.

THE AMOUNT OF INDUCED VOLTAGE

How much voltage is induced with our coil experiment is determined by the *number of turns on the coil*, the *strength of the magnet* we use and the *speed of the relative motion*.

Of course, if we were able to continue to move our bar magnet in and out of the coil at high speed without stopping, we would get a continuously induced voltage.

Remember how the needle of the galvanometer deflected first one way and then the other. This is because when the magnet is inserted, the induced voltage is one polarity (negative or positive) and when the magnet is withdrawn the polarity reverses. This is an *alternating voltage*. An AC voltage is one which changes polarity periodically. If we connect an alternating voltage to a circuit, we will get an *alternating current*. This is a current that flows in the circuit first in one direction and then in the other direction.

CYCLES PER SECOND

Going back to our coil. Placing the magnet into the coil and then removing it is one cycle. If we could take the magnet in and out of the coil every second, the polarity of the induced voltage would go through one cycle in one second. Mechanics call this revolutions per second. In electronics, we once used cycles per second, but now use Hertz. So when you see 'Hertz' just think of 'cycles per second'.

This method of inducing an emf was the first method used for household power generation. The first household power was direct current (DC). The method used to produce this direct current was electromagnetic induction. The generator converted the AC to DC.

The alternating voltage connected to the household supply from the power grid changes polarity 100 times a second (50 cycles per second) and reaches a peak of 325 Volts in Australia. The mains voltage in some locations varies a great deal depending on consumer load.

WHERE HAVE ALL THE ELECTRONS GONE?

A bit of a strange subtitle but it's all I could think of at the time. What I want you to think about is the electrons in your household wiring! They don't move from the power station through the conductors to your house. The household supply is AC at 50 Hertz (Hz). So the electrons in the conductors in your house are probably the same electrons that were there when the wiring was installed in your home. The AC supply just causes them to move first one way and then the other. We discussed earlier how slow electrons really move. So if they (the electrons) are changing direction 50 times a second, they would not be getting anywhere. They just seesaw back and forth. For the effects of an electric current to work, we don't have to make the electrons move in one direction all the time we just have to make them move. Think about that.

Let's look at how AC voltage produced in the power station. The device which creates the household power is called a generator. It works by creating relative motion between a conductor and a magnetic field.

The figure marked '2-9(a)' is a simplified AC generator. The blue and red segments are the north and south pole of a permanent magnet. Between the magnet, in the magnetic field, we have a coil of wire. Only two turns are shown but in practice, there would be many turns. The coil in a generator is called the armature. The ends of the coil are connected to a galvanometer via slip rings and brushes. We are going to make the armature rotate and we don't want the wires to become twisted where they connect to the galvanometer, so we feed them out through slip rings. The armature is in the starting position (vertical). We are going to rotate the armature clockwise.

Look at the galvanometer. Notice how its pointer is in the centre when it has no voltage applied. There is no voltage because the armature is now stationary.

Figure 2-9(a)

Figure 2-9(b) shows what happens when we have rotated the armature clockwise by almost 1/4 of a turn or 90 degrees. The galvanometer has deflected to the right indicating an induced voltage in the armature. In the diagram, we have not quite arrived at 90 degrees. At 90 degrees the induced voltage will be at a maximum because at 90 degrees we have maximum relative motion between the armature and the magnetic field.

Figure 2-9(b)

In figure 2-9(c) the armature has gone past 90 degrees. It's at about 170 degrees, almost 1/2 of a turn. Look at the galvanometer. It has fallen and is almost back at zero (the centre). It will reach the centre (zero) when the armature reaches 180 degrees or 1/2 of a turn.

Figure 2-9(c)

In figure 2-9(d) the armature is at 180 degrees and the induced voltage on the galvanometer is zero. We have now gone through one-half of a cycle.

Figure 2-9(d)

In figure 2-9(e) the armature is at about 240 degrees. Look what the galvanometer has done. It has deflected to the opposite side of zero. This indicates that the induced voltage has changed polarity. The maximum induced voltage will occur at 270 degrees or 3/4 of a turn.

Figure 2-9(e)

After going through 270 degrees, the induced voltage will again begin to fall to zero and will become zero when the armature is back in the position shown in figure 2-9(a).

Well, there you have it. In the next chapter, we will discuss the shape of the voltage produced by the AC generator.

You may be wondering why we use AC in the household supply. Seems like a lot of trouble with all the electrons changing direction 50 times a second.

There are significant advantages of AC over DC and I am sorry, but I am not going to go into that now.

It is interesting to note that Thomas Edison advocated DC for household power distribution and claimed DC was safer than AC. Tesla, on the other hand, advocated AC as a more efficient means of power distribution. Edison installed the first DC generators for power distribution. However, Tesla's AC distribution eventually won out and for good reasons.

By the way, the difference between an AC generator and a DC generator is very little. An AC generator has slip rings like that shown above. A DC generator has a device connected to the armature called a commutator, which provides a DC output from generators. Unfortunately, or fortunately, you don't need to know how this works.

Don't forget Faraday's Law….

When relative motion exists between a conductor and a magnetic field an emf (electron-moving force) is induced into the conductor.

10. ALTERNATING CURRENT - PART 2

In the last chapter, we discussed how an alternating current, or voltage, can be produced using a generator.

We are now going to look more closely at the shape (waveform) of this AC voltage. Recall how we discussed that the voltage starts out at zero, builds up to a maximum, then falls to zero. The polarity of the voltage then reverses. It builds up to a maximum again and then falls back to zero. This is one cycle. The number of cycles in 1 second is called the *frequency* and is measured in *Hertz*.

Now the shape (waveform) of the wave from a generator is called a sine wave or sinusoidal wave. Some textbooks just flash this shape in front of you and say here it is. When I was learning I found the concept of a sine wave a little difficult, so I am going to assume the same for you.

A sine wave is the graphical representation of the alternating current or voltage plotted against time.

AN EXPERIMENT

Get yourself a piece of paper, about A4. On the left-hand side of the paper place a pen down and draw a vertical line about 50mm long. Now continue to trace over the line going up and down, over and over. Consider the centre of this line to represent zero Volts. Why aren't you doing it?

Continue to move your pen up and down, redrawing the line. I want you to think that the middle of the line represents zero Volts. Your pen moving upwards toward the top of the line represents current or voltage increasing in one direction. When you get to the top move down again and you eventually pass through the zero point. Once through zero your pen movement is now representing a change in the direction of the current and it begins to increase towards the bottom of the line, which represents maximum voltage in this direction. When you reach maximum at the bottom your pen now moves back toward the zero point and so on.

So you are sitting there copying over this vertical line. Do it reasonably quickly and get used to it.

Now without stopping your up and down motion, move your hand to the right without even thinking about the pen.

This is what I got, shown below in figure 1-10, though I have added a couple of notations.

The graph is very rough, but it does illustrate clearly the waveshape of an alternating voltage plotted against time. The waveshape is called a sine wave.

An AC voltage can be fed into an oscilloscope and its screen will display the waveform precisely.

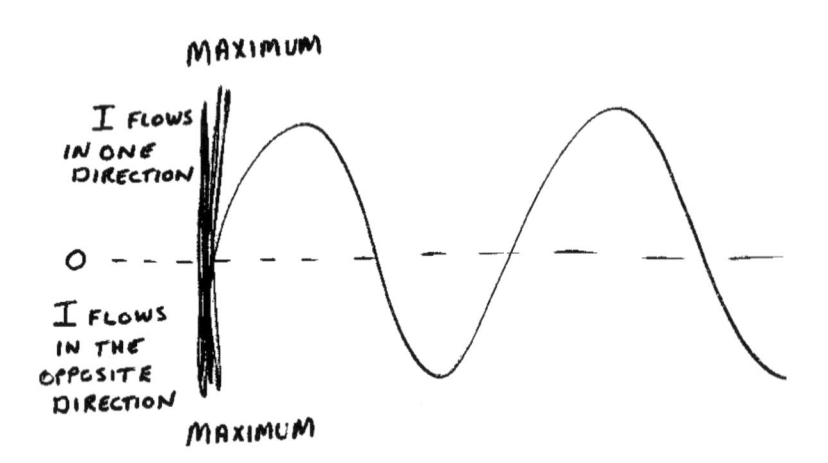

Figure 1-10

Figure 2-10 is how you will see the sine wave represented in most textbooks. Sometimes positive and negative symbols are drawn on the vertical axis. Just think of positive as being current flow in one direction and negative as current flow in the opposite direction. How many cycles you see will depend on the time scale on the horizontal axis. On an oscilloscope, the horizontal and vertical axes are calibrated accurately in time and voltage.

So we can read things like the maximum voltage and we can work out the frequency in Hertz (number of cycles per second).

The shape of the AC voltage delivered to your home is a sine wave. A signal from a radio transmitter, without modulation, is also a sine wave.

One cycle, or revolution, is 360 degrees. Table 1-10 shows the values of the voltage or current for any sine wave at different parts of the cycle. 'Sin' is a trigonometric function that can be used to calculate points on a sine wave. You will find the 'Sin' key on most calculators and if you enter 30 (degrees) into the calculator and press 'Sin' you will get 0.5. At 30° the instantaneous value of a sine wave is 0.5; with 1 being its peak value.

This is why the waveform is called a **sine wave.**

Degrees	Sin (degrees)	voltage
0	0	Zero
30	0.500	50% of maximum
45	0.707	70.7% of maximum
60	0.866	86.6% of maximum
90	1.000	Positive maximum
180	0	Zero
270	-1.000	Negative maximum
360	0	Zero

Table 1-10

VOLTAGE AND CURRENT VALUES FOR A SINE WAVE

Suppose you wanted to buy electricity from Edison or Tesla. They both have the same monetary charge per kilowatt-hour. However, Edison says he will supply you with 100 Volts direct current (DC), none of this sine wave stuff from him. He offers to provide 100 Volts just like you would get from a battery. No current reversals. Tesla offers you 100 Volts AC. From whom would you buy electricity?

Well, Edison's offer is very straightforward. Tesla, on the other hand, is offering a sinusoidal wave and you would have to ask more questions. Tesla is offering you 100 Volts but what and when is it 100 Volts? Is it 100 Volts at the peak value of each half cycle? Which would mean that you would only get 100 Volts for a brief moment twice during each cycle. Does this matter? You bet it does. If you were in the middle of winter in front of the electric bar heater, you would want 100 Volts all the time not just for a brief period, each half cycle.

100 Volts peak AC will not produce the same heating effect as 100 Volts DC. So there is a need to be able to compare AC with DC, in terms of their heating effect. That is the same as comparing how much work each can produce.

It can be found mathematically or by experiment that the equivalent heating effect of a sinusoidal waveform compared to DC is 0.707 of its peak value. This value is known as the root mean square or effective value and is written RMS.

So if Tesla was to supply 100 Volts peak, the actual equivalent heating value would be:
0.707 x 100 = 70.7 Volts RMS, in which case you would go for Edison's deal.

The average value of a sine wave is:
Peak x 0.637
I have never found a use for this value.

Sometimes an AC voltage is given as Peak to Peak. So if the maximum peak voltage was 100 Volts, then the peak to peak value would be 200 Volts - again a need to know, however, I have never found any use for the peak to peak value.

Your household electricity supply is 240 Volts RMS. This means it provides the same heating value as 240 Volts DC. What is the peak value of the household power supply in Australia? The Standard for AC mains in Australia is now 230V AC RMS . However, it is nearly always closer to 240V AC RMS.

RMS = 0.707 x Peak

Transpose for Peak by dividing both sides by 0.707

Peak = RMS/0.707 = 240/0.707 = 339.46 Volts

So twice during each cycle, the household mains voltage actually reaches 339 Volts. The frequency of the household supply is 50 Hertz. How many times a second does the mains voltage reach 339 Volts?

Since there are 50 cycles per second and two peaks during each cycle, the supply voltage will reach 339 Volts 100 times each and every second.

Summary of equations

RMS = 0.707 x Peak
Peak = RMS / 0.707
Average = Peak x 0.637
Peak-to-Peak = Peak x 2

Note: 1/0.707 is 1.414 and 1.414 is the square root of 2

The duration of one cycle in seconds is called the Period.

What is the period of the household mains in Australia?
Since the frequency is 50Hz, there are 50 cycles in 1 second. Therefore 1 cycle has a Period of 1/50th of a second or 20 milliseconds.
Period = 1 / frequency.

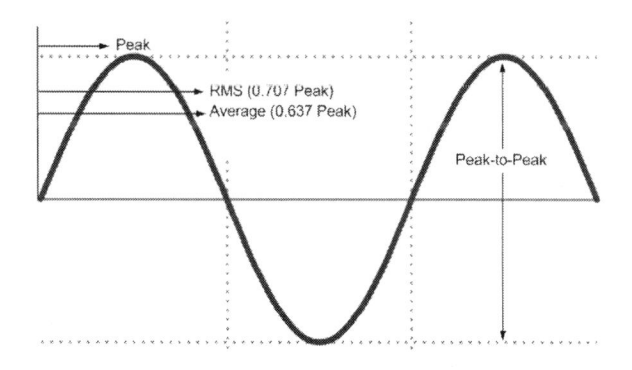

Figure 2-10.

WHAT DOES ROOT MEAN SQUARE MEAN?

We have learnt what the RMS value of an AC voltage means. We also know the RMS value is 0.707 of the peak. However, what is this term 'root mean square'?

The factor 0.707 for RMS is derived as the square root of the average (mean) of all the squares of a sine wave. Now, if that still sounds like gobbledygook, let me show you how it is done and at least we will take away the mystery of the term RMS.

Interval	Angles	Sin(angle)	Sin(angle)^2
1	15	0.26	0.07
2	30	0.50	0.25
3	45	0.71	0.50
4	60	0.87	0.75
5	75	0.97	0.93
6	90	1.00	1.00
7	105	0.97	0.93
8	120	0.87	0.75
9	135	0.71	0.50
10	150	0.50	0.25
11	165	0.26	0.07
12	180	0	0.00
	Total	7.62	6.00
	Average	7.62/12 = **0.635**	SQRT(6/12) = **0.707**

Table 2-10.

WHAT'S IN A NAME?

Michael Faraday, who became one of the greatest scientists of the 19th century, began his career as a chemist. He wrote a manual of practical chemistry that reveals his mastery of the technical aspects of his art. He discovered a number of new organic compounds, among them benzene and was the first to liquefy a "permanent" gas (i.e., one that was believed to be incapable of liquefaction). His major contribution, however, was in the field of electricity and magnetism. He was the first to produce an electric current from a magnetic field; invented the first electric motor and dynamo; demonstrated the relationship between electricity and chemical bonding; discovered the effects of magnetism on light; and discovered and named diamagnetism, the peculiar behaviour of certain substances in strong magnetic fields. He provided the experimental and a good deal of the theoretical, foundation upon which James Clerk Maxwell worked out the classical electromagnetic field theory.

Nikola Tesla (b. July 9/10, 1856, Smilijan, Croatia. d. Jan. 7, 1943, New York City), Serbian-American inventor and researcher who discovered the rotating magnetic field, the basis of most alternating-current machinery. He emigrated to the United States in 1884 and sold the patent rights to his system of alternating-current dynamos, transformers and motors to George Westinghouse the following year. In 1891 he invented the **Tesla** coil, an induction coil widely used in radio technology. **Tesla** was from a family of Serbian origin. His father was an Orthodox priest; his mother was unschooled but highly intelligent. A dreamer with a poetic touch, as he matured **Tesla** added to these earlier qualities those of self-discipline and a desire for precision.

Thomas Alva Edison was the quintessential American inventor in the era of Yankee ingenuity. He began his career in 1863, in the adolescence of the telegraph industry, when virtually the only source of electricity was primitive batteries putting out a low-voltage current. Before he died in 1931, he had played a critical role in introducing the modern age of electricity. From his laboratories and workshops emanated the phonograph, the carbon-button transmitter for the telephone speaker and microphone, the incandescent lamp, a revolutionary generator of unprecedented efficiency, the first commercial electric light and power system, an experimental electric railroad and key elements of motion-picture apparatus, as well as a host of other inventions. Singly or jointly he held a world-record 1,093 patents. In addition, he created the world's first industrial research laboratory. Born in Milan, Ohio, on Feb. 11, 1847, **Edison** was the seventh and last of four surviving children of Samuel **Edison** Jr. and Nancy Elliott **Edison**. At an early age, he developed hearing problems, which have been variously attributed but were most likely due to a family tendency to mastoiditis. Whatever the cause, **Edison**'s deafness strongly influenced his behaviour and career, providing the motivation for many of his inventions.

11. CAPACITANCE Part 1

In this chapter, we are going to talk about capacitance. We need to make a distinction between the capacitor and capacitance. A capacitor is a device, whereas capacitance is an electrical property. First, we will discuss the capacitor and then the property of capacitance. We will avoid mathematics where possible.

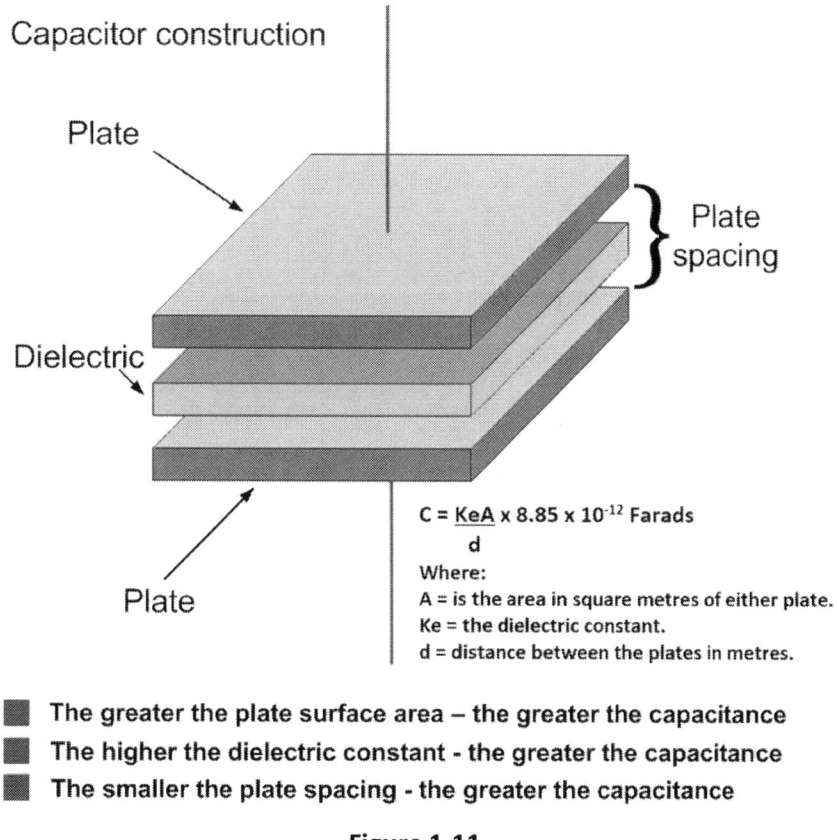

$$C = \frac{K_e A}{d} \times 8.85 \times 10^{-12} \text{ Farads}$$

Where:
A = is the area in square metres of either plate.
Ke = the dielectric constant.
d = distance between the plates in metres.

- The greater the plate surface area – the greater the capacitance
- The higher the dielectric constant - the greater the capacitance
- The smaller the plate spacing - the greater the capacitance

Figure 1-11.

As you can see a capacitor is a two-terminal device. Two conductive plates separated by an insulator. This should suggest to you that current never flows through a capacitor.

Basic schematic symbols for capacitors:

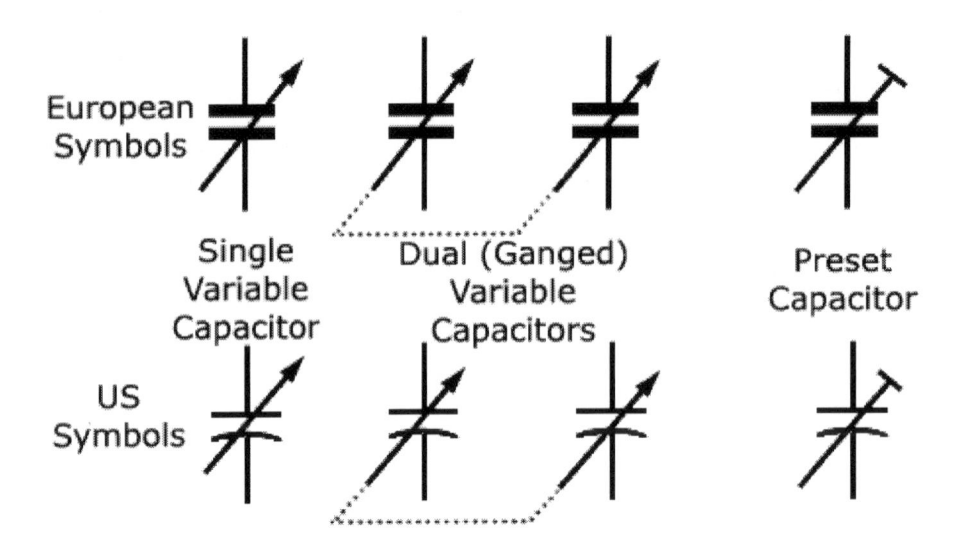

Figure 2-11.

Figure 2-11 shows the standard schematic symbols for a capacitor. Capacitors with a polarity marked on them are called polarised capacitors and have to be connected to that polarity.

Some capacitor types and packages

Figure 3-11.

Figures 3-11 – shows a variety of capacitor types. The yellow/brown are ceramic capacitors (because they use ceramic as the dielectric). The green capacitor in the foreground is the polyester type. The blue and black cylindrical capacitors are called electrolytic. The pale blue bottom left is a tantalum capacitor. Electrolytic and tantalum capacitors have their dielectrics formed electrically, for this reason, they are polarised capacitors that have a negative and positive leg. Connected incorrectly they can create internal gas and explode.

Larger 200-400pF air variable capacitors are used in transmitter output stages and antenna tuners. Air dielectric capacitors can withstand higher voltages. It takes about 10kV to flash over 1cm of dry air.

Figure 4-11.

Charging and discharging a capacitor

The interesting stuff starts to happen when we connect an emf to the plates of a capacitor. Have a look at the test circuit of figure 5-11.

Capacitor charge/ discharge circuit

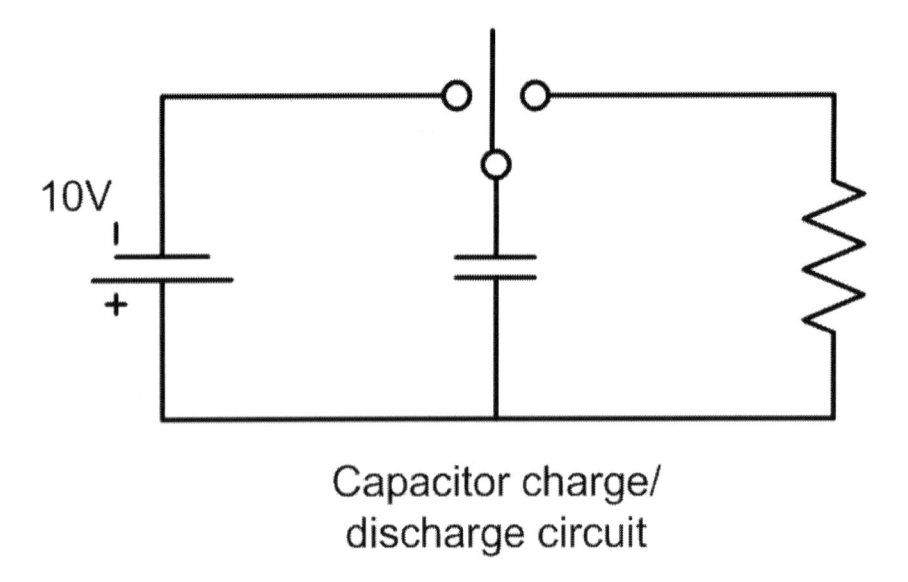

Figure 5-11.

The capacitor can be switched so that it will be connected to the 10-Volt battery when the switch is thrown to the left, or when the switch is thrown to the right, it will be connected to the resistor. The capacitor can be connected by the switch to the battery or the resistor, but not both at the same time.

The negative terminal of the battery has an excess of electrons on it, created by the chemical action within the battery. The positive terminal has a deficiency of electrons. Now recall that unlike charges attract. If there were any way for the electrons to get from the negative terminal of the battery to the positive, they would. for now I just want you to imagine a battery by itself with two terminals. There is an electrostatic field between the two terminals of any battery created by the unlike charges on each terminal. In other words, there is a very slight tugging from the positive terminal and a very slight pushing from the negative terminal in a vain attempt to move electrons from the negative to the positive terminal. No current flows between the unconnected terminals of a battery because the electrostatic fields are very small due to the spacing of the battery terminals and the very high air resistance between them.

If you find this hard to imagine look at figure 6-11 of two charges separated by air.

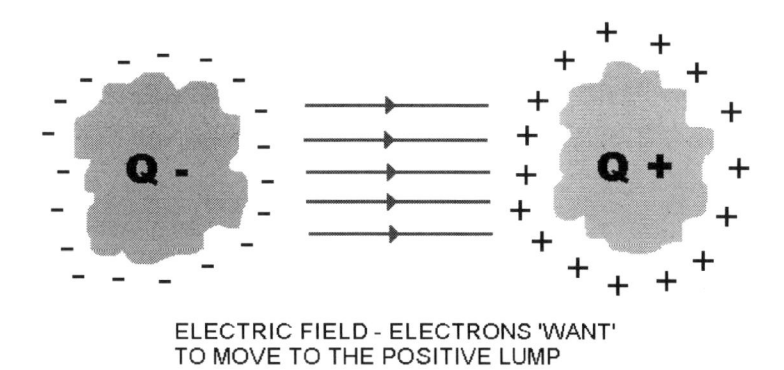

ELECTRIC FIELD - ELECTRONS 'WANT'
TO MOVE TO THE POSITIVE LUMP

Figure 6-11.

Here we have two lumped charges Q1 and Q2. There is an electrostatic field between them. This field is trying to pull electrons from the negative charge over to the positive charge. There is an electrical strain here. No current flows because the two charges are too far apart. Let's say they are 100mm to start with. Now move the two lumped charges so that they are only 50mm apart. The electrostatic field will now be stronger. The strain will be stronger. Still no current flows. Let's push the issue. Move the two lumped charges so that they are only 5mm apart. Now (depending on the charges) the electrostatic field will be much stronger. The tugging of electrons from the negative charge will be much greater. Still no current flows, as the air insulation resistance between the charges is too high.

The electrons on the negative lumped-charge want to traverse the gap to the positive charge. Do you think the electrons would be evenly distributed on the negative lump? On the side of the negative lump closest to the positive lump (the inside) the electrons will be crowding up trying to jump the gap.

Can you figure out what will happen if we continue to move them closer, say, to 1 mm? Well, I think you will agree that a point will be reached where the electrostatic field is so strong that electrons will jump off the negative lump and flow through the air to the positive lump. There will be an arcing of electric current. This is what happens in a lightning storm.

Returning to our capacitor that, at the moment, is not connected to anything. The switch is thrown to the left. The plates of the capacitor have a very large surface area. The dielectric between the plates is extremely thin but still a very good insulator. When we throw the switch to the left as in figure 7-11, we are extending the charges on the battery terminals to the plates of the capacitor and there will be a strong electrostatic field across the plates of the capacitor and through the dielectric.

Electrons are going to move from the negative terminal of the battery and bunch up on the top plate of the capacitor. Similarly, electrons are going to move from the bottom plate and travel to the positive terminal of the battery. No electrons will be able to flow through the dielectric. Its insulating properties are too good. So, if you like, there is going to be a redistribution of charge. This movement of charge will continue until the electrons flowing into and out of the capacitor create a potential difference on the plates of the capacitor equal to the battery voltage, namely 10 Volts. The capacitor is now said to be charged.

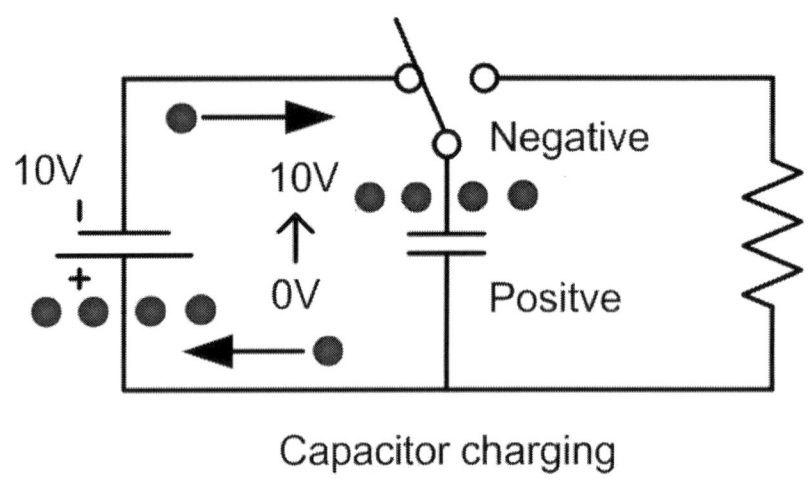

Capacitor charging

Figure 7-11

This charging of the capacitor does not happen instantaneously - it takes a little time. Suppose now we move the switch so the capacitor is disconnected from the circuit, as shown in figure 8-11.

Capacitor charged

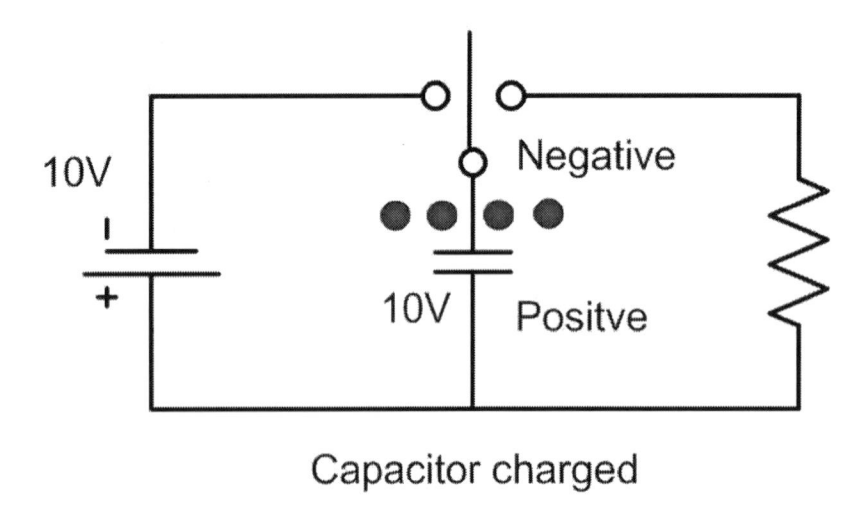

Figure 8-11

The capacitor is now left charged, even though it has been disconnected from the battery. The capacitor has 10 Volts across it. Older capacitors would not hold this charge for very long as a current would slowly leak through the dielectric and the capacitor would eventually self-discharge. Some modern capacitors can hold their charge for many days or even weeks.

The capacitor has stored energy in it in the form of a charge on the plates. If we connect a circuit to the capacitor, it will discharge as shown in figure 9-11.

Capacitor discharging

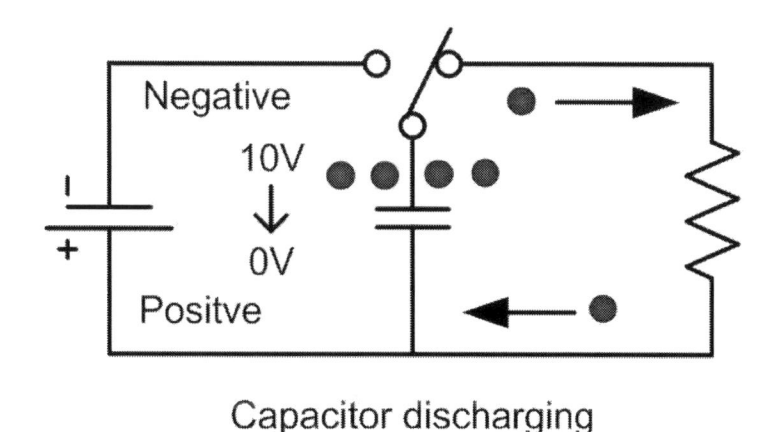

Figure 9-11

Electrons will now flow from the top plate of the capacitor through the resistance until the capacitor becomes discharged. If the resistor was a small light bulb, it would flash brightly at first and then slowly dim as the capacitor discharges.

At no time did current flow through the dielectric of the capacitor. Current flows into and out of a capacitor giving the illusion that current flows through a capacitor.

Suppose the resistor was a lamp. Also, suppose we continue to move the switch rapidly back and forth between the left and right position. The lamp would perhaps flicker a bit, but be continuously lit.

So you should now see that a capacitor can be used to store charge and we can use that charge to do something.

Where is the energy stored?

Though we have said that energy has been stored by the charge on the plates, it is more correct to say that **the energy is stored in the electric field**. It is the charge on the plates that forms the electric field between the plates. When current flows into a capacitor, charging it, the electric field becomes stronger (stores more energy). When the current flows out of the capacitor, the voltage across the plates decreases and hence the strength of the electric field decreases (energy moves out of the electric field).

UNIT OF CAPACITANCE

The unit of capacitance is called the Farad. The Farad is the measure of a capacitors ability to store a charge. If one Volt is applied to the plates of a capacitor and this causes a charge of 1 Coulomb to be stored on the plates, the capacitance is 1 Farad.

In practice 1 Farad is an enormous capacitance. More practical sub-units of the Farad are used. Microfarad and picofarad are the most common sub-units.

1 microfarad = 1×10^{-6} Farads
1 nanofarad = 1×10^{-9} Farads
1 picofarads = 1×10^{-12} Farads

PERMITTIVITY OR DIELECTRIC CONSTANT

The insulating material between the plates (dielectric) determines the concentration of electric lines of force. Just like different materials will concentrate magnetic lines of force to a greater or lesser extent, materials also vary in their ability to concentrate electric lines of force.

If the dielectric was air, then a certain number of lines of force will be set up. Some papers have a dielectric constant twice that of air, which would cause the *density* of the electric lines of force to be double and the capacitance would be double. The ability of a dielectric to concentrate electric lines of force is called the dielectric constant or permittivity. The higher the dielectric constant, the greater the capacitance for a given plate area.

Suppose an air dielectric capacitor (dielectric constant close enough to 1) of 8 microfarads had its air dielectric replaced with mica, without changing the distance between the plates. The capacitance would increase in direct proportion to the dielectric constant.

In other words, the capacitance would increase from 8 microfarads to 5-7 times that value, or 40 to 56 microfarads proportional to the new dielectric constant.

DIELECTRIC CONSTANTS

Material	Dielectric Constant
Vacuum	1
Air	1.0006
Rubber	2-3
Paper	2-3
Ceramics	3-7
Glass	4-7
Quartz	4
Mica	5-7
Water	80
Barium titanate	7,500

Table 1-11

FACTORS DETERMINING CAPACITANCE

A formula to determine the capacitance of a two-plate capacitor is:

$$C = \frac{K_e A}{d} \times 8.85 \times 10^{-12} \text{ F}$$

Where:
A = is the area in square metres of either plate.
Ke = the dielectric constant.
d = distance between the plates in metres.

The constant 8.85×10^{-12} is the absolute permittivity of free space.

Example. Calculate C for two plates, each with an area of 2 square metres, separated by 1 centimetre (1×10^{-2} metres), with a dielectric of air.

We will take the dielectric constant of air as 1. Even though it is more accurately 1.0006, 1 is close enough, so that:

$$C = \frac{1 \times 2 \times 8.85 \times 10^{-12}}{1 \times 10^{-2}}$$

$$= 200 \times 8.85 \times 10^{-12}$$
$$= 1770 \times 10^{-12}$$
$$= 1770 \text{ pF (picofarads)}$$

For examination purposes, you do not have to use this equation. However, you most definitely do need to know what the equation says about the factors determining capacitance.

$$C = \frac{KeA}{d} \times 8.85 \times 10^{-12} \text{ F}$$

Capacitance is directly proportional to the dielectric constant (Ke).
Capacitance is directly proportional to the area of one of the plates (A).
Capacitance is inversely proportional to the distance between the plates (d).

CAPACITORS IN SERIES AND PARALLEL

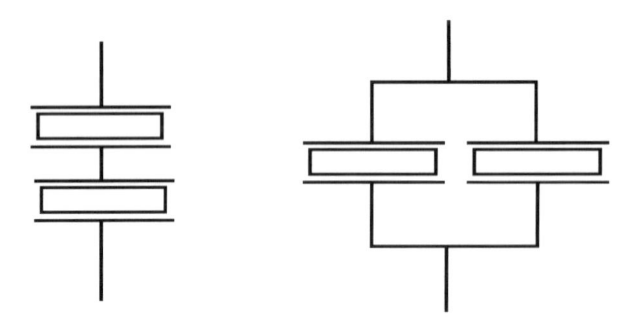

Figure 10-11 - (a) Series and (b) Parallel capacitors.

In figure 10-11 all the capacitors are the same. Let's say they are 10 microfarads. Think about what happens when you connect two identical capacitors in series. Think about it in terms of the factors that we just discussed; that affect capacitance. Can you visualise by looking at the series capacitors (Figure 10-11(a)) that we have actually doubled the thickness of the dielectric? Doubling the thickness of the dielectric is exactly the same thing as doubling the distance between the plates.

If the distance between the plates is doubled and capacitance is inversely proportional to the distance between the plates, then the capacitance must be half that of a single capacitor on its own. The total capacitance of two 10 microfarad capacitors in series must then be 5µF.

The equation for any number of capacitors of any value in series is:

$$1/C_t = 1/C_1 + 1/C_2 + 1/C_3 + ..n$$

OR

$$C_t = \frac{1}{1/C_1 + 1/C_2 + 1/C_3 + ..n}$$

The easiest way to use this equation on a calculator is:

1. Find the reciprocal of each capacitance.
2. Add all of the reciprocals together (sum them).
3. Find the reciprocal of this sum.

In other words, the reciprocal of the sum of the reciprocals.

Worked example

Take three capacitors of 7, 8 and 12 microfarads in series.

Find the reciprocal of 7:   1/7 = 0.143 (rounded to 3 decimal places)
The reciprocal of   8:   1/8 = 0.125
The reciprocal of   12:   1/12 = 0.083

The sum of the above is: 0.351

Find the reciprocal of this sum: 1/0.351 = 2.85 microfarads.

Three capacitors of 7, 8 and 12 microfarads in series has a total capacitance of 2.85µF. Notice how the total capacitance is always less than the lowest value capacitor.

Note: If just two capacitors are in series then you can use the simplified product over sum formula.

For two capacitors (only) in series:

$C_t = (C_1 \times C_2) / (C_1 + C_2)$

When the same two capacitors are connected in parallel, the distance between the plates and all other factors remain the same, except we have doubled the effective area of the plates. So the capacitance has doubled.

The equation for any number of capacitors of any value in parallel is:

$C_t = C_1 + C_2 + C_3 + ...n$

These equations are opposite to that of resistances in parallel and series, so be careful not to confuse the two. The shortcuts we took with resistors in parallel work the same for capacitors in series. For example, if we have just two capacitors of any value in series we can use the product over sum method to find the total capacitance.

VOLTAGE ACROSS SERIES CAPACITORS

If two equal capacitors were connected in series across a 100-Volt DC supply and we were to measure the voltage across each capacitor, we would get 50 Volts across each. Since the capacitors are equal, we get an equal voltage drop.

For unequal capacitances in series the voltage across each C is inversely proportional to its capacitance. In other words, the smallest capacitance would have the largest voltage drop. The largest capacitance would have the smallest voltage drop.

The amount of charge on each capacitor in series capacitor is given by:

Q = CE

If a 10µF capacitor was charged to 10 Volts, then the charge in Coulombs on the capacitors would be:

10µF x 10 Volts = (10×10^{-6})F x (10)V = 10^{-4} Coulombs. (100µC)

The equation can be transposed for voltage across a capacitor and we get:

E = Q/C

In a series circuit, each capacitor regardless of its capacitance will have the same charge. Q is the same for all capacitances in series. For a smaller capacitance to have the same charge as larger capacitors in a series circuit, it must have a higher E.

We could do numerical examples. However, you do not need this in practice or for examination purposes. You do need to understand and be able to visualise the voltage drops across capacitances in a series circuit.

Example. Two capacitances of 1µF and 2µF are connected in series across a 900-Volt DC supply. What is the voltage drop across each capacitor?

Now, the voltage drops have to be unequal because each capacitor will have the same charge (Q).

$E = Q/C$

E is inversely proportional to C. Therefore, the smallest C (1µF) must have the greatest voltage drop. But how much greater? Since the 1µF capacitor is half the value of the 2µF capacitor, it must have twice the voltage to achieve the same charge. The 1µF capacitor must then have 600 Volts across it, leaving 300 Volts on the 2µF capacitor.

VOLTAGE RATING OF CAPACITORS

All capacitors are given a maximum voltage rating. This is necessary as the dielectric of capacitors can breakdown and conduct causing the capacitor to fail and in most cases be destroyed. Some capacitors, if placed across a voltage which is too high, will create gas within them and explode somewhat violently.

12. CAPACITANCE – Part 2

TYPES OF CAPACITORS

AIR DIELECTRIC CAPACITORS

Capacitors using an air dielectric are used mostly as variable capacitors. Air dielectric variable capacitors are large but can withstand high voltages; this is just what is needed for transmitter output stages and antenna tuning units.

It has been seen that capacitance is proportional to the plate area and inversely proportional to the spacing. Even with close spacing, the area of the plate must be large, in order to obtain the required capacitance. Instead of using two large plates, which would be inconvenient, a number of smaller plates can be interleaved.

The capacitance can easily be varied by sliding the plates in and out of mesh since the capacitance is proportional to the cross-section of the dielectric between the plates connected to opposite terminals.

A practical capacitor is constructed with a set of fixed plates and a set of moving plates that rotate on a spindle. As the moving plates are rotated through 180° the meshing and therefore the capacitance varies from a minimum to a maximum value. The maximum value of the capacitance is generally about 500pF. It is common practice to use two or three of these capacitors ganged together on a common spindle to form the main tuning control of a transmitter.

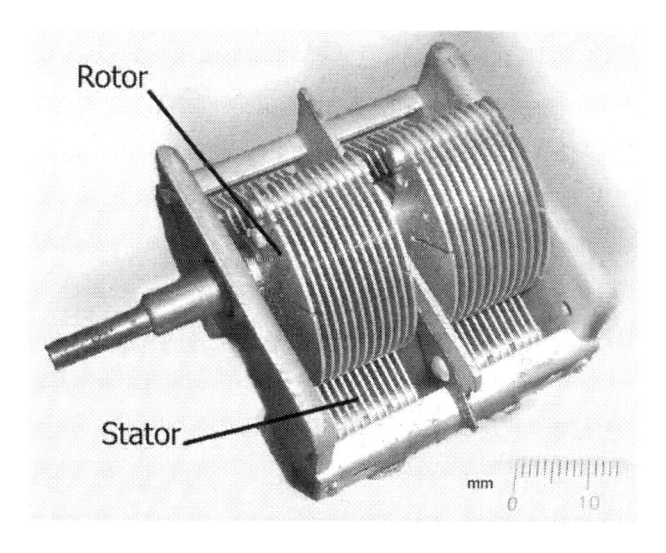

Figure 1-12.

Over the years, the physical size of variable capacitors has decreased considerably due to the use of smaller spacing between the moving and fixed plates. However, with the introduction of small communication receivers; air spaced capacitors are too large. In receivers, a synthetic dielectric is now used. In these capacitors, a solid dielectric in the form of thin plastic sheets is placed between the plates. In this way, the capacitance is increased approximately by the value of the relative permittivity of the dielectric. The sheets are a fairly loose fit between the plates so that the moving plates can still slide over the plastic sheets (which are fixed).

Figure 2-12.

In radio transmitters of significant power, air-dielectric variable capacitors are the only option, as the air dielectric can withstand high voltage radio frequencies.

SILVER MICA CAPACITORS

This capacitor consists of thin mica sheets as the dielectric. The mica sheets are coated with a thin layer of silver, which forms the plates. A number of plates may be stacked together to obtain the required capacitance. The capacitor is protected by a wax, lacquer or plastic coating. These capacitors are available up to 10,000 pF and have a small loss, small tolerance (e.g. ± 1 %), are stable in capacitance value and have a low temperature coefficient. They are mainly used where an accurate, *highly stable*, good quality capacitor (high Q) is required. Typical applications of a mica capacitor include the crystal *oscillators* and other *critical RF circuits*.

There was a time not that long ago that you could identify most capacitors from the colour of the package and their general appearance. That is not always the case now. You may need to look up a manufacturers code to identify a capacitor type. Mica capacitors are generally dark or light brown. Do watch out as these conventions may no longer apply. The package shape could be anything; such as shown in 2-12, disc shape and many others.

CERAMIC CAPACITORS

These capacitors use a ceramic for the dielectric and are very widely used as they are physically small and inexpensive. The term 'ceramic' covers a very large range of materials and the properties of the capacitor depend on the type of ceramic used. The permittivity of some ceramic materials is very high, as much as 16,000; this results in a physically small capacitor. Capacitors can be made using relatively low permittivity ceramics, which have different temperature coefficients and can be used for temperature compensation. When a high permittivity ceramic is used the temperature coefficient may be very large and not linear, i.e. varies greatly with temperature. These also have a capacitance that changes with the applied voltage and because of this, the tolerance may be large, as much as -20% to +80%. Obviously, such capacitors must only be used where the capacitance value is not important.

The maximum working voltage may range from, say, 100V to 10kV or more.

Ceramic capacitors are widely used owing to their small size, low cost and because so many types are available.
There are many *bypass* and *coupling capacitors* (discussed later) in a transceiver. There could be a hundred or more. These *bypass* and *coupling capacitors* then need to be small, inexpensive and stable. This is an ideal application for the ceramic capacitor. Remember that tip.

1. Leads
2. Dipped coating
3. Soldered connection
4. Silver electrode
5. Ceramic Dielectric

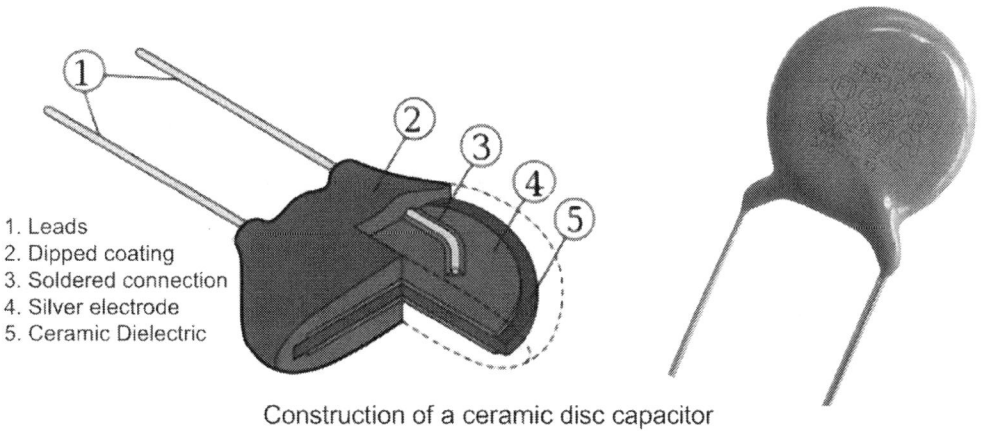

Construction of a ceramic disc capacitor
Figure 3-12.

PAPER CAPACITORS

At one-time paper capacitors were very common. They consisted of a metal foil for the electrodes and paper (impregnated with oil or wax) for the dielectric. Instead of a number of plates, two plates only are used, say 2 to 5 cm wide and of a length corresponding to the capacitance required. The plates and dielectric are now wound into a roll to form a tubular capacitor. Paper capacitors are rarely seen today.

PLASTIC FILM CAPACITORS

These are constructed similar to a paper capacitor, but with a plastic film instead of paper. There are several types: polystyrene, polycarbonate, polyester, polyethylene terephthalate or polypropylene. The plates may be foil (e.g. with polystyrene) or metallised plastic film. Polystyrene ones do not normally have a case, but the others are usually enclosed in a plastic case that may be cylindrical or rectangular. The various films have different detailed characteristics but will be considered as one class.

ELECTROLYTIC CAPACITORS

This type is used where a high value is required in a small space. It depends on the principle of depositing, by electrolytic action, an extremely thin insulating film on one plate, the film then acting as the dielectric. With such a thin dielectric, the capacitance for a given plate area is large. There are two basic types: aluminium and tantalum. These capacitors are polarised. If connected to the wrong polarity they will be destroyed.

Aluminium Capacitors

These are made in a similar way to a paper capacitor, but use two aluminium plates and an absorbent paper. The paper is impregnated with a suitable electrolyte, which produces an aluminium oxide coating on the positive plate. The paper acts as a conductor and container of the electrolyte, the only dielectric being the oxide coating. The capacitance may be increased by etching the positive plate (the one with the oxide coating), thus increasing the effective surface area. In normal capacitors, the aluminium coating must be maintained by the application of a direct voltage across the plates in the correct direction, i.e. the capacitor is polarised. If the voltage is reversed the coating will be removed and the capacitor ruined. Working voltages vary from 3 to 500 Volts and it is important that the working voltage is not exceeded as they may explode. The capacitor passes a small leakage current, particularly when first switched on, the current reforming the oxide layer. This current rises rapidly if the working voltage is exceeded and damage then results. The capacitors are usually fitted into aluminium cases, the case often being one of the connections. These capacitors are commonly used in transistor equipment. Values are available from 1µF to 100,000µF. The tolerance is high, -25% to +100% and the losses are high, particularly at high frequencies.

There are some special electrolytic capacitors made that are not polarised, in which both plates have oxide coatings. They are used in a few special applications where a large capacitor is required and there is no polarising voltage, e.g. crossover networks in loudspeakers.

Figure 4-12.

Packages differ a great deal; from the small ones shown in 4-12 to large cylindrical packages. A typical application is *coupling* and *bypassing* in low-frequency circuits to large filter capacitors found in *low voltage DC power supplies*.

When electrolytic capacitors fail they often bulge due to the internal generation of gas. When looking for a fault check for this bulging. If you find two or three electrolytic capacitors like this on the one circuit board it is sound engineering practice to replace all of the electrolytics on that board.

Tantalum Capacitors

The basic principles are the same as aluminium, with tantalum in place of aluminium. They have a better shelf life and lower leakage current but are more expensive. There are three types: wet electrolyte sintered anode, foil electrode wet electrolyte and solid electrolyte sintered anode. The last type is the cheapest and the more commonly used in domestic equipment, so no further details will be given of the first two types. The solid type uses a sintered tantalum anode with a solid electrolyte of manganese dioxide. It comes in various forms, in particular, tubular and "tear drop" types. Capacitance values vary from 0.1 to at least 100μF with working voltages up to 50V. The orange and brown capacitors in figure 4-12 are tantalum.

CHARGE AND DISCHARGE OF A CAPACITOR

If a capacitor 'C' is connected to a DC supply through a resistor 'R', as shown in figure 5-12, a current will flow through R until the capacitor is charged to the supply voltage. As the capacitor charges, the voltage across it will rise to 63.2 percent of its final value, E, in a time equal to CR seconds (where C is in Farads and R is in Ohms). This time is known as the *time constant* of the circuit. The voltage across R is the difference between the supply voltage and the voltage across the capacitor and therefore decreases as the capacitor charges. As the capacitor charges the current will decrease and fall to zero at full charge.

If the supply is now disconnected the capacitor will remain charged, apart from the small leakage through the dielectric of the capacitor. It may remain nearly fully charged for hours or even days if it is a high-Q capacitor. A large capacitor (say 1F) charged to a high voltage (say 500 Volts) may store considerable energy and it is unwise to touch the terminals or a nasty shock will result. If the resistor is connected across the charged capacitor, it will discharge by passing a current through R, the value of the current being determined by Ohm's law. The voltage will now decrease by 63.2% in a time equal to the time constant 'CR'.

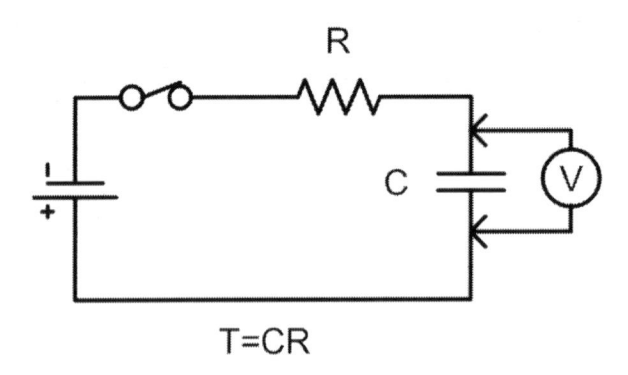

Figure 5-12

Example. In figure 5-12, say the resistance is 1MΩ and the capacitor 10μF. What is the time constant of the circuit?

T=CR = 10 x 10^{-6} x 1 x 10^6 = 10 seconds.

When the switch is closed, what will be the voltage on the charging capacitor after 10 seconds?

10 seconds is 1 time constant period. The capacitor will charge to approximately 63.2% of the applied voltage after 1 time constant period has elapsed. Hence, after 10 seconds, the voltage on the capacitor will be about 63% (rounded) of 10 Volts or 6.3 Volts.

I would like to pause here for a moment to emphasise something. Consider a single resistor connected across a battery with a switch and the switch was closed. How long would it take for the supply voltage to appear across the resistor? Answer - immediately. With a capacitive and resistive circuit, however, the voltage is delayed from building up on the capacitor, determined by the RC time constant. This gives us a better definition of capacitance.

Capacitance is that property of an electric circuit, which opposes changes of voltage.

Going back to our circuit in figure 5-12. The capacitor has charged to 6.3 Volts after 1 time constant period. How much further does it have to charge before it equals the supply voltage? The answer is: 10 - 6.3 = 3.7 Volts.

During the next time constant period (10 seconds) the capacitor will charge a further 63% of the remaining voltage. That is, 63% of 3.7 Volts, which is 2.331 Volts. So after 20 seconds the capacitor will have charged to about 6.3 + 2.331 = 8.631 Volts.

In other words, the capacitor charges to 6.3 Volts in one time constant period (10 Sec) and it charges a further 2.331 Volts in the second time constant period. So, after 20 seconds, the total voltage on the capacitor will be 8.631 Volts. During the third time constant period the capacitor will again charge a further 63% of the *remaining* voltage and so on. A capacitor can be considered to be fully charged after 5 time constants.

THE CONCEPT OF LEADING CURRENT

If we had a resistor, a switch and battery in a series circuit and we operated the switch, how long would it take for the voltage across and the current through the resistor, to reach the values determined by Ohm's law? The answer is of course immediately.

However, in a resistive and capacitive circuit, we have learnt about 'time constant' and the fact that a capacitor takes time to charge.

When we first throw the switch in our capacitive circuit the current into the capacitor is at first maximum and the capacitor charges most during the first time constant. The instant we throw the switch the voltage on the capacitor is 0V and rises to 63% of the applied voltage in one time constant.

Now let's get this clear. When we throw the switch, the current into the capacitor is at its maximum and the voltage across the capacitor is at a minimum. This is in high contrast to a resistive circuit.

The current starts off at a maximum and decreases as the capacitor charges and the voltage on the capacitor is zero to begin with and rises to a maximum (the supply voltage) when it is fully charged.
So current and voltage in a capacitive circuit are not in sync. Current is said to lead the voltage in a capacitive circuit.

Current leads (voltage) in a capacitive circuit.

Analogy of a charging capacitor

Suppose you were to blow up a balloon. Some balloons have an initial resistance to being inflated so let take one that has already been blown up and deflated. The flow of your breath going in the balloon is current. The pressure you create with your breath is voltage. When you first start to inflate the balloon it is easy and there is a high inrush of air (current).

So air flow or current is high at first. After just one breath the balloon has pressure pushing back against you. After one breath the balloon does not have much back pressure (voltage).

As you continue to inflate the balloon, the backward pressure from it becomes greater. It becomes harder to inflate further. The amount of flow (current) into the balloon is less due to the amount of pressure (voltage) that the balloon is pushing back.
When the back pressure of the balloon matches your maximum inflating forward pressure – that is it – you cannot inflate the balloon further and the flow (current) comes to a stop. The balloon is now at maximum inflation, assuming it does not burst.

Can you see this is exactly how a capacitor behaves when it is being charged?

STRAY CAPACITANCE

When we build a capacitor, we want capacitance. However, all we need for capacitance is *any* two conductors separated by an *insulator* (dielectric). In any circuit, there are literally hundreds of stray and sometimes unwanted capacitances. More often these are not a problem, however, sometimes they can be. We will discuss this in more detail later.

A lifetime ago I was teaching PMG (telephone) linesmen about fixing telephone lines. A typical telephone line is just two insulated conductors twisted together and they can go for many kilometres.

A telephone line then by its very construction is a capacitor. Suppose a telephone line is only a kilometre long. Also, that nothing is connected to the telephone line - no telephone and no exchange equipment. The line is said to be open circuit. When a linesman connects an Ohmmeter to the line, they expect to see infinite resistance. This is what an Ohmmeter does on a good telephone line, but not immediately. The Ohmmeter will swing across and to the right and slowly fall back to infinite Ohms after one to two seconds. The Ohmmeter shows the capacitance of the line charging. It was difficult for me to explain why this happened (to linesmen anyway).

An Ohmmeter is a battery, resistor and a current meter connected in series. When the Ohmmeter is attached to the line, the line acting as a capacitor charging and after five time constants it is fully charged and the current meter reads zero.

Linesmen, without understanding the mechanism involved, use this method to get a rough idea of how long the telephone pair is. They may be trying to find an open circuit.

There is stray capacitance everywhere. One place that is not often thought about is between and antenna and ground. Remember the ground is a conductor. When a radio experimenter adjusts the length of an antenna; or its height above ground. They are adjusting the capacitance (among other things) between the antenna and ground.

13. CAPACITIVE REACTANCE

CAPACITOR IN A DC CIRCUIT

We have already discussed the operation of a capacitor in a DC circuit, however, let's just go over the main principles again.

If a capacitor is connected to a battery (or other DC source), it will charge according to its time constant (T=CR), to the battery voltage.

If a lamp is connected in series with the capacitor while it is charging, the lamp would give off light indicating that current is flowing. The lamp would be bright at first and then slowly dim and extinguish as the capacitor becomes charged. Current into the charging capacitor is high at first and then tapers off to zero when it is fully charged. The voltage on the capacitor is low at first and increases to the supply voltage when the capacitor is fully charged. We can see that the voltage across and the current into a capacitor, are not in sync (phase). Current in a purely capacitive circuit leads the voltage (by 90 degrees).

Imagine a capacitor and a lamp in series. Connect the circuit to a DC source and the lamp will light momentarily. The capacitor is now charged and the lamp is extinguished. Suppose now we reverse the battery (or supply terminals). The capacitor will now discharge through the lamp and the battery and then recharge with the opposite polarity.

Imagine now continuing to do this very quickly, that is, reversing the battery every time the capacitor is charged or near to fully charged. Because the capacitor will be continually charging and discharging, there will be current flow in the circuit all the time and the lamp will be continuously lit.

Instead of reversing the battery, it would be much easier to supply the circuit with AC voltage, which by definition, automatically changes polarity each half cycle.

CAPACITOR IN AN AC CIRCUIT

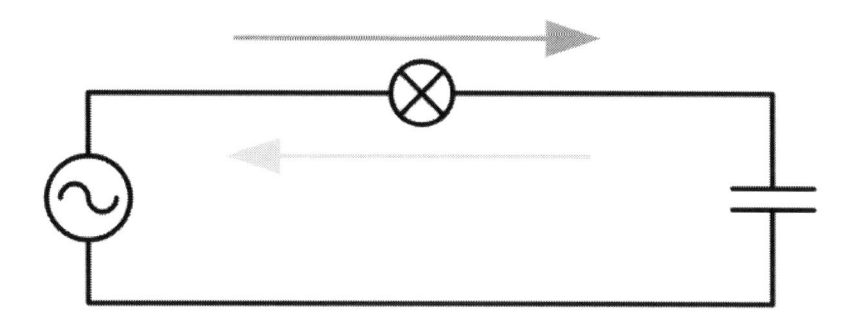

Figure 1-13

In the circuit of figure 1-13, a capacitor and a lamp are connected via a switch to a source of AC. The capacitor will continually charge and discharge as the AC supply changes polarity. AC current will flow in the circuit continuously and the lamp will remain lit. The coloured arrows indicate the *charge* and *discharge* currents.

So a capacitor 'appears' to allow an AC current to pass through it. I say 'appears' because while there is current flowing in the circuit at all times, at no point does current actually flow through the capacitor. Remember it has a dielectric which will not pass current. The current flows continuously in the circuit because the capacitor is constantly charging and discharging. This type of current is referred to as a *displacement current*.

Capacitance is the property of a circuit that opposes *changes* of voltage. A capacitor, therefore, has an opposition to current flow. The opposition to current flow produced by a capacitor is called *Capacitive Reactance* and is measured in Ohms. The shorthand for capacitive reactance is X_C.

FACTORS DETERMINING X_C

What do you think we could do in the above circuit to increase the brightness of the lamp without changing the lamp or the supply voltage?

Firstly, let's reflect on the frequency of the AC supply. It helps to go to extremes and imagine a very low frequency supply. As a capacitor charges, a voltage builds up on its plates, which opposes the supply voltage. This is why the charge current of the capacitor is at first very high and then tapers off to zero as the capacitor becomes charged.

If the frequency of the AC supply is high enough (polarity reversal is fast enough) then the capacitor will be in the early part of its charge cycle, where the current is greatest when a polarity reversal takes place.

A high frequency will then cause the capacitor to charge and discharge in the early part of its *first time constant period* and this will cause a greater current to flow in the circuit for the same supply voltage.

We have deduced that capacitive reactance (X_C) is dependent on the frequency of the AC supply. In fact, we have deduced that the higher the frequency of the AC supply the more current flows and therefore the lower the capacitive reactance.

A larger capacitance would allow higher charge currents to flow. Therefore, changing the capacitance to one of a higher value would increase the current and therefore decrease the capacitive reactance.

The two factors determining capacitive reactance are *frequency* and *capacitance*.

EQUATION FOR CAPACITIVE REACTANCE

$X_C = 1/2\pi fC$ **Equation 1-13.**

Where:

X_C = capacitive reactance in Ohms.
2π = a numeric constant.
f = frequency in Hertz.
C = capacitance in Farads.

Look at the equation. $2\pi fC$ is in the denominator. 2π is a constant (it does not change).

We can say then, *that capacitive reactance is inversely proportional to both frequency and capacitance.*

In other words, if either capacitance or frequency were to double then capacitive reactance would halve. If capacitance or frequency were to halve then, the capacitive reactance would double and so on.

EXAMPLE CALCULATION

A capacitor of 0.05µF is connected to a 10 Volt AC supply that has a frequency of 500kHz. What is the capacitive reactance in Ohms and how much current will flow in the circuit?

$X_C = 1 / 2\pi f C$
$X_C = 1 / (6.2831 \times 500 \times 10^3 \times 0.05 \times 10^{-6})$
$X_C = 6.366$ Ohms

The current is found from Ohm's law by substituting X_C for R in the equation.

$I = E / X_C = 10 / 6.366 = 1.57$ Amps

If we are given the capacitive reactances of a circuit and you need to find the net total capacitive reactance, use the same rules as for finding the total resistance of a circuit. For example, two capacitive reactances of 100 Ohms in series are equal to 200 Ohms and in parallel, the same combination would be 50 Ohms.

To reinforce what we have learnt have a look at the circuit shown in figure 2-13.

We could deduce the voltage across the 1µF and 2µF capacitors using the proportion method. Instead, let's work out how much current is flowing in the circuit, then using Ohm's law work out the voltage across each of the capacitors.

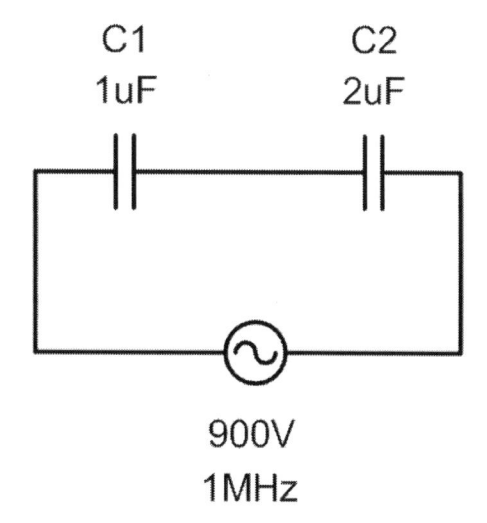

Figure 2-13.

Reactance of the 1µF capacitor:

$X_{C1} = 1 / 2\Pi fC$
$X_{C1} = 1 / (6.2831 \times 1 \times 10^6 \times 1 \times 10^{-6})$
$X_{C1} = 0.159154943$ Ohms

Reactance of the 2µF capacitor (it should be half as much because the capacitance is double):

$X_{C2} = 1 / 2\Pi fC$
$X_{C2} = 1 / (6.2831 \times 1 \times 10^6 \times 2 \times 10^{-6})$
$X_{C2} = 0.079577471$ Ohms

The total capacitive reactance is the sum of the reactances:

$X_C(total) = X_{C1} + X_{C2} = 0.159154943 + 0.079577471$ Ohms
$X_C(total) = 0.238732414$ Ohms

The current flowing in the circuit is:

$I = E / X_C(total) = 900 / 0.238732414 = 3769.911194$ amps

The voltage across the 1µF (it should be 600 Volts):

$E = IX_{C1} = 3769.911194 \times 0.159154943 = 599.9998392$ Volts

The voltage across the 2µF (it should be 300 Volts):

$E = IX_{C2} = 3769.911194 \times 0.079577471 = 299.9999987$ Volts

So C1 the smallest capacitor (1µF) has 600VAC RMS across it and C2 (2µF) has 300VAC RMS across it.

What if the circuit were DC? What would be the voltage across C1 and C2 in figure 3-13?

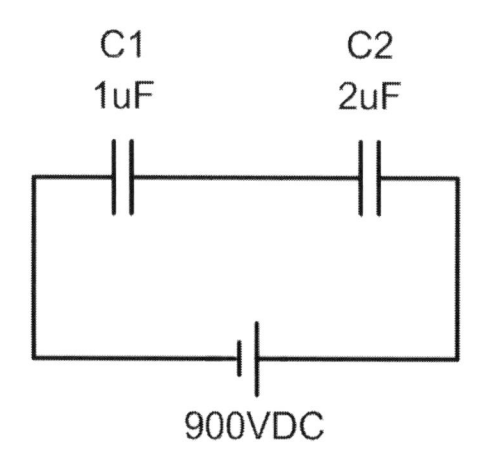

Figure 3-13.

Well, we no longer have a frequency as it is DC; however, we do know that the highest voltage will be across the smallest capacitor and the lower voltage across the larger and that the sum of these voltages must equal 900V. Why?

Figure 3-13 is a series circuit. The current is the same in all parts of a series circuit. This means the amount of charge (Q) on C1 and C2 must be the same. For the smaller capacitor C1 to have the same charge on it as the larger C2, there must be more voltage across C1. (double the voltage in this case).

Q=CE and Q is the same for both capacitors. We also know the C1 is half the value of C2.

For C1 to have the same charge as C2, C1 must have twice the voltage of C2.

Logically then; C1=600V and C2=300V and the sum, of course, the applied voltage of 900V

As you can see from the graph in figure 4-13, as frequency increases capacitive reactance decreases.

A GRAPH OF X$_C$ VERSUS FREQUENCY

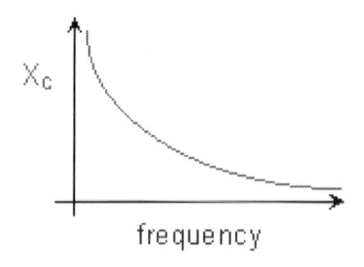

Figure 4-13.

Similarly, figure 5-13 shows that as capacitance increases X$_C$ decreases.

A GRAPH OF X$_C$ VERSUS CAPACITANCE

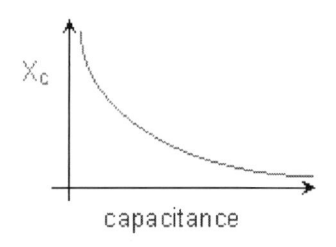

Figure 5-13.

Capacitive reactance (X$_c$) is inversely proportional to both frequency and capacitance. Now if you cannot see that, without graphs, you have forgotten to "read" equations that we did back in the early chapters.

$$Xc = \frac{1}{2\pi fC}$$

This equation should be speaking to you. The $2\pi fC$ is in the denominator. The right hand side of the equal sign in this equation is a fraction. It has a numerator (1) and a denominator ($2\pi fC$). When the denominator becomes larger in any fraction the entire fraction (which is Xc) becomes smaller.

So at a glance, even if you do not have a clue to what the symbols mean you should recognise that Xc is inversely proportional to f and C.

POWER IN A CAPACITIVE CIRCUIT

Capacitance does not dissipate power. The only circuit property that can dissipate power is resistance.

All of the power taken from a purely capacitive circuit during the charge cycle is returned during the discharge cycle. So a pure capacitance does not dissipate any power. In practice, every capacitor has some resistance in its leads and plates. Dielectric leakage and dielectric loss also dissipates some power.

In practical capacitors, a small amount of power is lost in the dielectric, called *dielectric loss*. The atoms within the dielectric of a capacitor are placed under stress and do move slightly due to the electric field. Particularly at higher frequencies, the dielectric loss can become significant.

The losses in a real capacitor are due to the small amount of resistance in the leads and the plates – this is called resistive losses. Dielectric loss accounts for the greatest amount of loss. All losses increase with increasing frequency. Dielectric loss is caused by the alternating stress placed on the dielectric's atomic structure. In industry, there are many processes that utilise dielectric heating. Food is lossy dielectric and can be cooked through dielectric heating. Plastics can be welded. Laminated wooden furniture is cured through dielectric heating.

Note: "Real" means – a real world or actual capacitor as opposed to a pure capacitor which on exists only on an engineering drawing.

Summary

The unit of capacitance is the Farad.

If a charge of 1 Coulomb produces a potential difference of 1 Volt across the plates of a capacitor, then the capacitance is 1 Farad.

Current flows into and out of, but never through, a capacitor.

The capacitance of a capacitor is directly proportional to the area of the plates and the dielectric constant and inversely proportional to the distance between the plates.

Capacitance is that property of a circuit which opposes changes in voltage.

The opposition to current flow in a capacitive circuit is called Capacitive Reactance and is measured in Ohms.

Capacitive reactance is inversely proportional to both frequency and capacitance.

A purely capacitive circuit does not dissipate any power.

Current leads the voltage in a capacitive circuit.

Capacitors in parallel are treated like resistors in series.

Capacitors in series are treated like resistors in parallel.

Capacitors are given a voltage rating, which if exceeded, could cause the dielectric to conduct, destroying the capacitor.

14. INDUCTANCE

Coils of wire were mentioned in the chapter on magnetism when we discussed the magnetic field about a coil carrying a current. An equally important aspect of the operation of a coil is its property to *oppose any change* in current through it. This property is called *inductance*.

Inductance is that property of an electric circuit that opposes changes of current

Not to be confused with capacitance, which is, the property of an electric circuit to oppose changes of voltage.

When a current of electrons starts to flow along any conductor, a magnetic field starts to expand from the centre of the conductor outward. These lines of force move outward, through the conducting material itself and then continue into the air. As the lines of force sweep outward through the conductor, they induce an emf in the conductor. This induced voltage is always opposing the current that produced it. Because of its opposing polarity, it is called a counter emf, or a back emf. This opposing affect is called Lenz's Law and it is much like one of Newton's law's; *every action has an equal and opposite reaction.*

The effect of this backward pressure built up in the conductor is to oppose the immediate establishment of maximum current. It must be understood that this is a temporary condition. When the current eventually reaches a steady value in the conductor, the lines of force will no longer be in the process of expanding or moving and a counter emf will cease to be produced. In other words, there will be no relative motion between the conductor and the magnetic field. At the instant when current begins to flow, the lines of force are expanding at their greatest rate and the greatest value of counter emf will be developed. At the starting instant, the counter emf value almost equals the applied source voltage.

Current is minimum at the start of current flow. As the lines of force move outward, however, the number of lines of force cutting the conductor per second becomes progressively smaller and the counter emf becomes progressively less. After a period of time the lines of force expand to their greatest extent, the counter emf ceases to be generated and the only emf in the circuit is that of the source. Maximum current can now flow in the wire or circuit since the inductance is no longer opposing the source voltage.

This property of a coil or more correctly and inductor to oppose changes of current by self-inducing an opposing (counter) emf is called inductance. The unit of inductance is the Henry and the symbol for inductance is 'L'.

SELF-INDUCTION

When the switch in a current-carrying circuit is suddenly opened, an action of considerable importance takes place. At the instant the switch breaks the circuit, the current due to the applied voltage would be expected to cease abruptly. With no current to support it, the magnetic field surrounding the wire will collapse back into the conductor at a fast rate, inducing a high-amplitude emf in the conductor. Originally, when the field built outward, a counter emf was generated. Now, with the field collapsing inward, a voltage in the opposite direction is produced. This might be termed a counter-counter emf, but is usually known as a self-induced emf. This self-induced emf is in the direction of the applied source voltage. Therefore, as the applied voltage is disconnected, the voltage due to self-induction tries to establish current flow through the circuit in the same direction, aiding the source voltage. The inductance induces a voltage to try and prevent the circuit current from decreasing. With the switch open it might be assumed that there is no path for the current, but the induced emf immediately becomes great enough to ionise the air at the opened switch contacts and a spark of current appears between them. Arcing lasts as long as energy stored in the magnetic field exists. This energy is dissipated as heat in any circuit resistance and the arc radiates energy as electromagnetic waves.

With circuits involving low current and short wires, the energy stored in the magnetic field will not be great and the switching spark may be insignificant. With long lines and heavy currents, inductive arcs many centimetres long may form between opened switch contacts on some power lines. The heat developed by arcs tends to melt the switch contacts and is a source of difficulty in high-voltage high-current switching circuits.

In a previous chapter, I gave you an example of the capacitance of a telephone line. A telephone line also has inductance. The normal operating voltage of a telephone line is about 50 Volts DC. When a telephone line is suddenly open circuited by a technician or a linesman, the self-induced voltage on the line can be in the order of 2000 Volts and this will produce a harmless but significant electric shock. The shock is harmless, as the amount of current is so small.

Remember, regardless of how current changes in a circuit containing inductance, the induced emf created by the inductance will oppose the *change of current*.

The unit of measurement of inductance is the Henry, defined as the amount of inductance required to produce an average counter emf of one Volt when an average current change of one Ampere per second is under way in the circuit. Inductance is represented by the symbol L in electrical problems and Henrys is indicated by 'H'.

COILING A CONDUCTOR

It has been indicated that a piece of wire has the ability to produce a counter emf and therefore has a value of inductance. A small length of wire will have an insignificant value of inductance by general electrical standards. One Henry represents a relatively large inductance in many circuit applications, where milli, micro or nanohenries are typically more practical. A straight piece of No. 22 wire one meter long has about 1.66 µH. The same wire wound onto an iron nail, or other high-permeability core may produce 50 or more times that inductance.

A given length of wire will have much greater inductance if wound into a coil. If two loops are wound on the same conductor but a distance apart then each loop will have a small inductance. If the two loops are wound together side by side, they will collectively have twice the inductance of one loop.

When ten loops are wound next to each other as shown in figure 1-14, with the same current flowing ten times the number of magnetic lines of force cutting each turn. Compared to 1 turn, this coiled inductor would produce a counter emf, ten times greater.

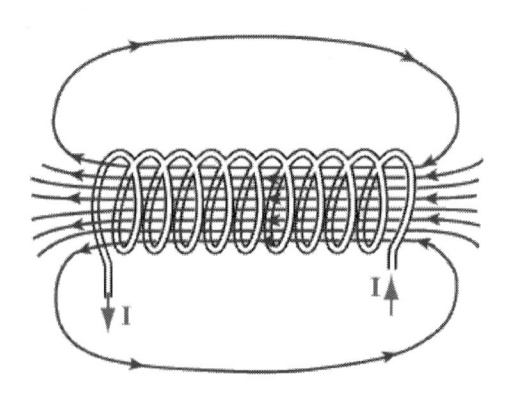

Figure 1-14.

If the turns are stretched out, the field intensity will be less and the inductance will be less. Stretching the coil but keeping the same number of turns increases the length of the magnetic circuit and the inductance decreases.

The larger the radius or diameter of the coil, the longer the wire used and the greater the inductance. In single-layer air-core coils with a length approximately equal to the diameter, a formula that will give the approximate inductance in microhenries is:

$$L = \text{Inductance}(\mu H) = \frac{r^2 N^2}{24r + 25l}$$

Where   L = is the inductance in µH.
         N = the number of turns.
         r = radius in centimetres.
         l = length in centimetres.

The inductance of straight wires is found in antennas, in power lines and in ultra-high frequency equipment. In most electronic and radio applications where inductance is required, space is limited and the wire is wound into either single layer or multilayer coils with air, powdered-iron-compound (ferrite), or laminated (many thin sheets) cores. The advantage of multilayer coil construction for high values of inductance becomes obvious when it is considered that, while 2 closely wound turns produce 4 times the inductance of 1 turn, the addition of 2 more turns closely wound on top of the first 2 will provide almost 16 times the inductance. The direction of winding has no effect on the inductance value of a given coil.

In many applications, coils are constructed with ferrite cylinders (slugs) that can be screwed into or out of the core space of the coil. This results in a controlled variation of inductance, maximum when the iron-core "slug" is in the coil and minimum with it out.

A particular type of coil is the toroid. It consists of a doughnut-shaped ferrite core, either single layer wound as shown in figure 3-14. Its advantages are high values of inductance with little wire and therefore little resistance in the coil and the fact that all the lines of force are in the core and none outside (provided there is no break in the core). As a result, it requires no shielding to prevent its field from interfering with external circuits and to protect it from effects of fields from outside sources. Two toroids can be mounted so close that they nearly touch and there will be almost no interaction (magnetic coupling) between them. Inductors come in a huge number of packages from tonnes to a few grams in weight. With some packages, it is not evident that the device is an inductor as they are small and encapsulated.

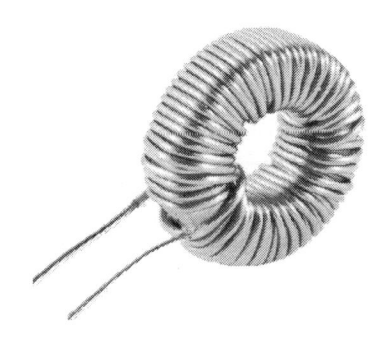

Figure 3-14.

There are so many different packages for inductors it is too much to show them all. Some look very much like resistors. Others are copper tracks etched into a circuit board. Others may be formed with a length of transmission line.

SCHEMATIC SYMBOLS OF INDUCTORS

Air core inductor

Ferrite core

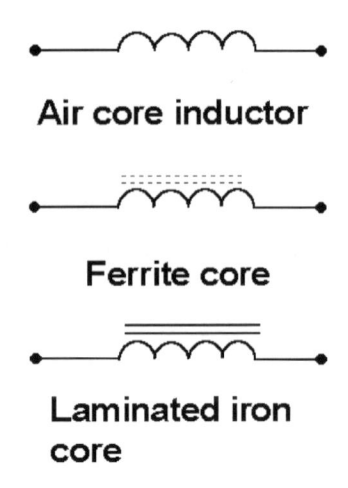

Laminated iron core

Figure 4-14.

THE TIME CONSTANT OF AN INDUCTANCE

The time required for the current to rise to its maximum value in an inductive circuit after the voltage has been applied will depend on both the inductance and the resistance in the circuit. The greater the inductance, the greater the counter emf produced and the longer the time required for the current to rise to maximum. With a constant value of inductance in a circuit and more resistance, then less current can flow.

As with capacitive and resistive circuits, the time required for the current to rise to about 63% (more precise 63.2%) of the maximum value (called the time constant) can be determined by:

T = L/R

Where T = time in seconds.
L = inductance in Henrys (H).
R = resistance in Ohms.

EXAMPLE

Calculate the time constant of a 10 Henry inductance with 10 Ohms of resistance.

T=L/R=10/10 = 1 second

Suppose the emf applied to this inductor was 10 Volts. What would be the current flowing after 1 second?

The final current after 5 time constants will be:

I = E/R = 10/10 = 1A

After 1 time constant the circuit current will have reached 63% of what will be its final value, or 63% of 1A = 630mA.

CURRENT LAGS IN AN INDUCTIVE CIRCUIT

In a resistive circuit, the current and voltage obey Ohms law. If you increase the voltage across a resistance, the current increase straight away to the new Ohms law value determined by I=E/R. In an inductive circuit, it takes time for the current to build up to its final value. The current lags behind. The current will eventually reach its Ohms law value after 5-time constants. However, if the voltage is continually changing as in an AC circuit, the current will always lag the voltage across the inductor by 90 degrees.

Think of 'L' for inductance and 'L' for lag – current lags voltage in an inductive circuit.

THE ENERGY IN A MAGNETIC FIELD

The current flowing in a wire or coil produces a magnetic field around itself. If the current suddenly stops, the magnetic field held out in space by the current will collapse back into the wire or coil. Unless the moving field has induced a voltage and current into some external load circuit, all the energy taken to build up the magnetic field will be returned to the circuit in the form of electric energy as the field collapses.

CHOKE COILS

The ability of a coil to oppose any change of current can be used to smooth out varying or pulsating types of current. In this application, an inductor is known as a choke coil since it chokes out variations of amplitude of the current. For radio frequency (RF) AC or varying DC an air-core coil may be used. For lower frequency circuits greater inductance is required. For this reason, iron core choke coils are found in audio and power frequency applications.

A choke coil will hold a nearly constant inductance value until the core material becomes saturated. When enough current is flowing through the coil to saturate the core magnetically, variations of current above this value can produce no appreciable counter emf and the coil no longer acts as a high value of inductance to these variations. To prevent the core from becoming magnetically saturated, a small air gap may be left in the iron core. The air gap introduces so much reluctance in the magnetic circuit that it becomes difficult to make the core carry the number of lines of force necessary to produce saturation. The gap also decreases the inductance of the coil. An air coil cannot be saturated.

DEFINITION OF A HENRY

If a coil has a rate of change of current of one Ampere per second and this produces a counter emf of 1 Volt, then the coil is said to have an inductance of one Henry.

MUTAL INDUCTANCE

If one coil is placed near another so that the magnetic fields interact with each other, then the moving magnetic field in one will induce a voltage into the other. This ability of one coil to effect another is called *mutual inductance*.

The farther apart the two coils are, the fewer the number of lines of force that interlink the two coils and the lower the voltage induced in the second coil.

The mutual inductance can be increased by moving the two coils closer together or by increasing the number of turns of either coil.

When all the lines of force from one inductor are linked to another, unity coupling is said to exist and the mutual inductance is:

$$M = \sqrt{L1 \times L2}$$

Where M is the mutual inductance in Henries.

The above formula assumes 100% coupling between the two inductors L_1 and L_2. This equation assumes that all the magnetic lines of force from L_1 cut all the turns of L_2. If this is not the case then M is determined by:

$$M = k\sqrt{L1 \times L2}$$

Where k is the percentage of coupling.

For example: suppose a 5 Henry coil has 75% of its lines of force cutting a 3 Henry coil. What is the mutual inductance?

$$M = 0.75\sqrt{5 \times 3}$$
$$M = 2.9H$$

COEFFICIENT OF COUPLING

The degree of closeness of coupling of two coils can also be expressed as a number between 0 and 1 rather than as a percent. A percentage of 100 is equal to a coefficient of coupling of 1 or unity. 75% coupling is 0.75 as above. No coupling is 0.

The formula for mutual inductance can be transposed for coefficient of coupling:

$$k = \frac{M}{\sqrt{L1 \times L2}}$$

INDUCTANCES IN SERIES

When you add inductances in series, you are in effect simply increasing the number of turns on the inductor. Therefore, to find the total inductance, sum the individual inductances.

$L_t = L_1 + L_2 + L_3 + \ldots n$

INDUCTANCES IN PARALLEL

If the inductances are connected in parallel, the total inductance is calculated by using a formula similar to the parallel resistance formula.

$1/L_t = 1/L_1 + 1/L_2 + 1/L_3 + \ldots n$

Both of these equations assume that the magnetic lines of force from all the inductors are not coupled (linked) to the others i.e. the mutual inductance is zero. Another way of describing how inductances are linked is called coefficient of coupling (k). If k=0 then there is no coupling. If k=1, the two inductances are completely coupled.

INDUCTIVE REACTANCE

It has been explained that DC flowing through an inductance produces no counter emf to oppose the current. With varying DC, as the current increases, the counter emf opposes the increase. As the current decreases, the counter emf opposes the decrease. Alternating current is in a constant state of change and the effect of the magnetic fields is a continually induced voltage opposition to the current. This reacting ability of the inductance to oppose a changing current is called inductive reactance. Inductive reactance is the opposition to current flow presented by an inductance in a circuit with changing current and is measured in Ohms.

$$X_L = 2\pi f L$$

So inductive reactance is directly proportional to frequency and inductance.

Just as with resistance and capacitive reactance, the total inductive reactance in Ohms is found using the same form of the equation.

In series:

$X_L(total) = X_{L1} + X_{L2} + X_{L3} + ...n$

And parallel:

$1/X_L(total) = 1/X_{L1} + 1/X_{L2} + 1/X_{L3} + ...n$

The voltage across and the current through an inductive reactance can be determined using Ohms law.

EXAMPLE

An inductance of 100µH is in series with an inductance of 200µH and connected to a 10 Volt AC supply with a frequency of 500kHz. How much current will flow in this circuit?

We need to find the total inductive reactance of the circuit. We could find the individual reactances of each inductor and add them, or we could find the total inductance and then the total inductive reactance. I will do it using the latter method.

$L_t = L_1 + L_2 = 100\mu H + 200\mu H = 300\mu H$
$X_L = 2\pi f L$
$X_L = 2\pi \times 500 \times 10^3 \times 300 \times 10^{-6}$
$X_L = 942\Omega$

I=E/R = 10/942 (x 1000 for milliamps) = 10.6mA

How much power is dissipated in the above example?

There is no resistance in the circuit, so no power is dissipated. You cannot/must not substitute X_L or X_c for R in the power equations. For example, $P=I^2 R$ is okay but $P=I^2 X_L$ is not. Reactance, either inductive or capacitive does not dissipate power – only resistance does.

Take a pure inductance reactance of 240 Ohms and imagine it being connected to a power point. The current that flows would be I=E/ X_L = 240/240 = 1 Ampere. Even though 1 Ampere would flow in the circuit, no power is dissipated as there is no resistance. All of the power supplied charging the inductor during one cycle is returned to the power point (the power supply) on the alternate half cycle. A pure inductor has no losses.

Only resistance dissipates power

Power authorities dislike reactive loads like this as current drawn does not register on the Energy Meter. The power lines to your house carry the charge and discharge 'reactive' currents *do* have resistance – the lines *do* dissipate power – and this power loss is *not* metered.

LOSSES IN A REAL INDUCTOR

A real world inductor does have losses. These losses include resistive losses of the windings and core losses. Hysteresis loss is a core loss caused by the continuous realignment of the magnetic domains. Eddy currents are small circulatory currents (eddies) induced into a conductive core. Core losses increase substantially with increasing frequency. Hysteresis and Eddy current loss will be discussed in more detail in the chapter on transformers.

15. METERS

Though most Radio Theory Exams will not ask you to describe the operation of a moving coil meter. I do believe it is a good idea to include them as it will reinforce much of the material already covered. We do need to know about using moving coil to build Ammeters, Voltmeters and Ohmmeters.

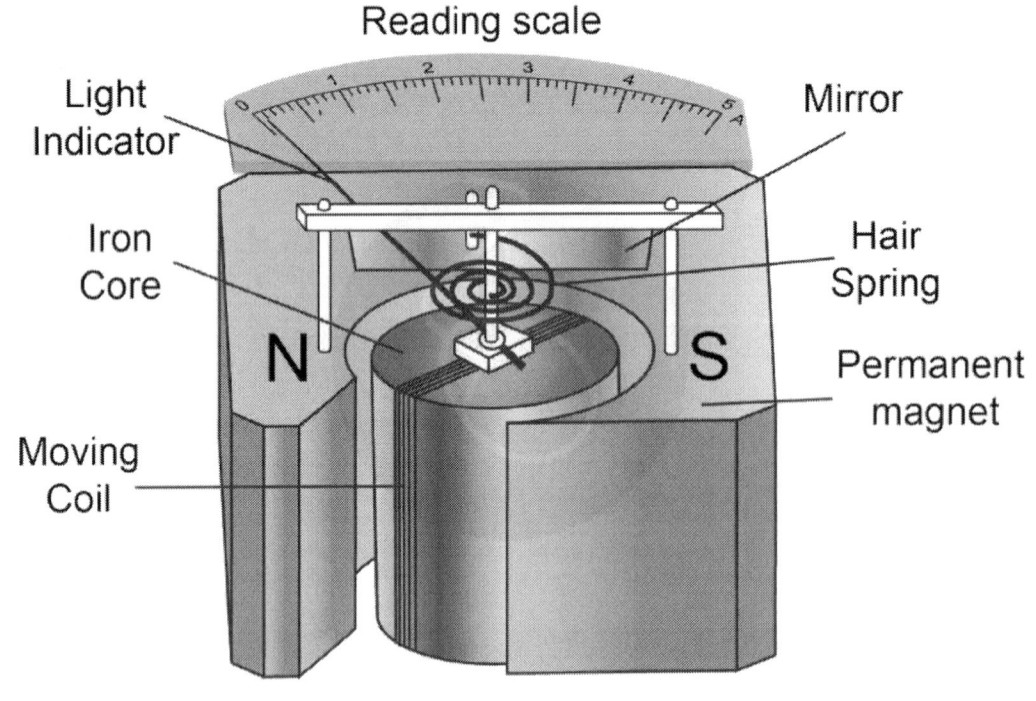

Figure 1-15.

A pictorial diagram of the moving coil meter is shown in figure 1-15. The moving coil is a coil wound on an aluminium former. The coil is free to move clockwise and anticlockwise through about 130 degrees, that in turn moves the pointer on the scale from left to right. Without any current flowing through the moving coil, the pointer will be towards the left off the scale, pushed there by a small coil return spring not shown. The pointer will not go past the left end of the scale as there is a small pointer stop; not shown. When current is passed through the moving coil, a small magnetic field is created around it which interacts with the field created by the permanent magnet in such a way as to cause the moving coil and the pointer along with it to rotate clockwise. So we have a simple but very effective method of detecting current flow with this instrument. Importantly, the degree of deflection from left to right by the pointer is *directly proportional to the current*. If a certain current causes the pointer to move to 1/4 scale, then twice that current will cause the pointers to move to half scale - this is a direct proportion.

The other important feature of the moving coil meter is that it is extremely sensitive to very small currents, typically in the order of microamperes and therefore does not consume much power from the circuit in which it is placed. The moving coil meter is a microammeter. Typical current to enable full-scale deflection ranges from 1uA to 30mA. The moving coil does have some resistance of course. A movement with a smaller full-scale deflection current has a higher coil resistance, as more turns of fine wire are needed to obtain that deflection.

We can make use of this meter by calibrating it and using it to measure current, voltage and resistance. From now on I will draw the moving coil meter as a circle, but it does help to remember what is inside.

The other point I would like to make is that many modern meters are digital. They do not use a moving coil. The digital circuitry performs an analogue to digital conversion of the applied current or voltage and displays the amount as a numerical readout. These meters are very accurate. However, the moving coil meter (called an analogue meter) has advantages that will see it in use for a long time to come. The greatest advantage of the analogue meter is its ability to show *changes* in voltage or current. For example, if an analogue meter is placed across a charging capacitor, the pointer will slowly rise up the scale as the capacitor charges. I have an all bells and whistles digital meter. When I place it across a charging capacitor, because the voltage is changing, the digital display just "blinks" as it is unable to show a varying voltage adequately.

MEASUREMENT OF CURRENT

Whether we are measuring Amperes, milliamperes or microamperes, two important facts to remember are:

1. The current meter (ammeter) must be in series with the circuit in which the current is to be measured. The amount of deflection depends on the current through the meter. In a series circuit, the current is the same in all parts of the circuit. Therefore, the

current to be measured must be made to flow through the meter as a series component in the circuit.

2. A DC meter must be connected with the correct polarity for the meter to read up-scale. Reversed polarity would make the meter read down-scale, forcing the meter pointer hard against the left-hand stop.

An ammeter should have a very low resistance when compared to the circuit in which it is placed. An arbitrary figure is 1/100th of the resistance of the circuit in which the ammeter is placed.

AMMETER SHUNTS

A meter shunt is a precision resistor connected across the meter movement for the purposes of shunting or bypassing, a specific fraction of the circuit current around the meter movement. Shunts are usually inside the meter case. The schematic symbol for an ammeter does not usually show the shunt.

While shunts are called resistors, they are extremely low-value resistors. A high current shunt may be a brass bar and measure in the in the order of milli-Ohms or less. Figure 2-15 shows a typical high current range shunt.

Figure 2-15.

One way of adjusting precisely the resistance of a high current shunt is to make small cuts in it using a very fine metal saw.

For example, we may have a 25mA meter movement with a moving coil resistance of 1.2 Ohms. We want the meter to be able to read 50mA full scale. In other words, to double the range of the meter from 0-25mA to 0-50mA.

To achieve this, a shunt resistance of 1.2 Ohms (equal to the resistance of the moving coil) would do the job.

Rm (Meter Resistance) = 1.2Ω

[25mA meter with shunt Rs below]

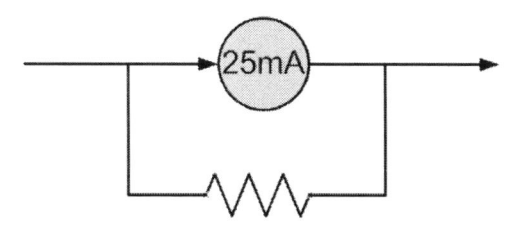

Rs (Shunt Resistance) = 1.2Ω

Figure 3-15.

With a 1.2-Ohm shunt, half of the total current will flow through the meter and half through the shunt since both form a parallel circuit consisting of two 1.2 Ohms resistances. The use of the shunt in this instance has extended the full-scale deflection (FSD) of the meter from 0-25mA to 0-50mA.

With a correct shunt, a moving coil meter can be used to measure any amount of current (FSD). The shunt is a low value resistor. In typical ammeters, the shunt is often a solid copper bar with a resistance in milli-Ohms, particularly for high current ranges on the meter.

It is common to hear a current meter called an ampmeter - this is wrong; the correct name for a current meter is an ammeter with no 'p.'

CALCULATING ANOTHER SHUNT

Figure 4-15 shows an ammeter. The moving coil has a resistance of 2000 Ohms and is deflected full scale with 50 microamperes. We wish to modify the meter to measure 1-Ampere full-scale deflection. Calculate the value of the shunt resistor that would extend the range of the ammeter to 0-1 Amperes.

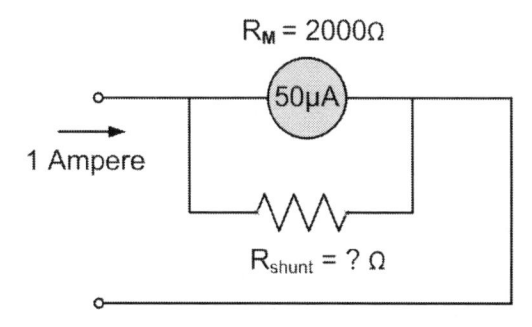

$R_M = 2000Ω$

[50µA meter with shunt below, 1 Ampere input]

$R_{shunt} = ?\ Ω$

Figure 4-15

This is just an Ohms law problem. Assume that one Ampere is flowing in the circuit. You know there is 50uA flowing through the meter movement. The current through the shunt is then:

I_{shunt} = 1amp - 50uA = 0.99995 Amperes.

We now know the current through the shunt. What is the voltage across it? We know the resistance of the meter and the current through it. Since the meter and the shunt are in parallel, they will have the same voltage. So, if we calculate the voltage across the meter we will know the voltage across the shunt:

$E_{shunt} = I_m \times R_m$ = 50uA x 2000 Ohms = 0.1 Volts.

Now calculate the resistance of the shunt:

$R_{shunt} = E_{shunt} / I_{shunt}$ = 0.1 / 0.99995 = 0.100005 Ohms.

This shunt, which is just over a 1/10th of an Ohm, would be a solid bar inside the meter. The manufacturer (or the builder) would use a reference current and adjust the resistance of the shunt accurately.

VOLTMETER

Although a meter movement responds only to current moving in the coil, it is commonly used for measuring voltage by the addition of a high resistance in series with the movement. Such a high series resistance is called a *multiplier*.

The multiplier must be higher than the coil resistance in order to limit the current flow through the coil. The combination of a meter movement with its added multiplier then forms a Voltmeter.

Since a Voltmeter has a high resistance, it must be connected in parallel to measure the potential difference between any two points in a circuit. Otherwise, the high resistance multiplier would add so much series resistance to the circuit the current would be reduced to a very low value.

The circuit is not opened to connect a Voltmeter in parallel. Because of this convenience (not having to break the circuit), it is common to make Voltmeter tests when troubleshooting rather than ammeter tests. If you need to know the current through a resistor, it is far easier to measure the voltage across it and then work out the current using Ohm's law.

Figure 5-15 shows how a 1mA meter movement is connected with a multiplier to enable it to be used as a Voltmeter with an FSD of 10 Volts. In other words, a Voltmeter with a 10V range.

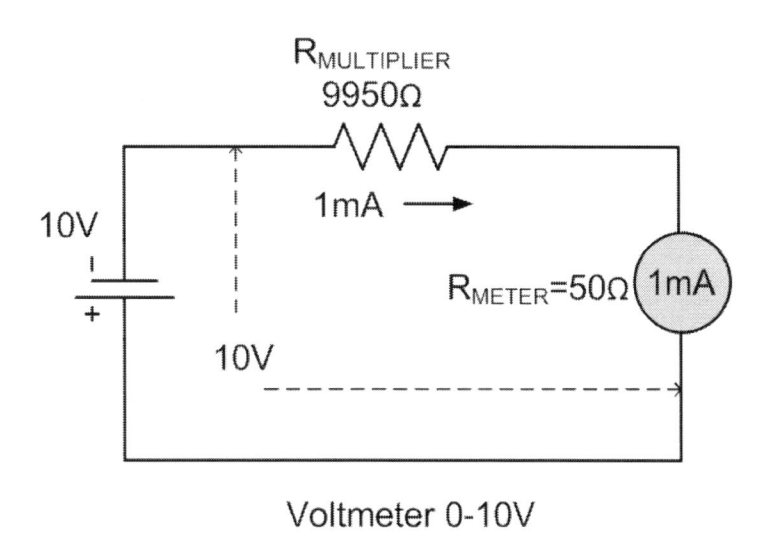

Figure 5-15.

With the 10V applied by the battery, there must be a total of 10,000 Ohms of resistance to limit the current to 1mA for full-scale deflection (FSD) of the meter movement. Since the movement has a resistance of 50 Ohms, 9950 is added in series, resulting in a total resistance of 10KΩ. Current through the meter is then: I = 10 / 10,000 = 1 mA.

With 1mA in the movement, the FSD can be calibrated as 10V on the meter scale. Of course, since deflection is directly proportional to current the meter can be calibrated proportionately from 0-10 Volts. For example, half scale would be 5 Volts, 1/4 scale 2.5 Volts, etc.

With the battery removed the circuit now consists of the meter movement and the multiplier. This is our 0-10V Voltmeter. Different multipliers can be switched in for other voltage ranges.

Let's calculate another shunt to make sure we understand this.

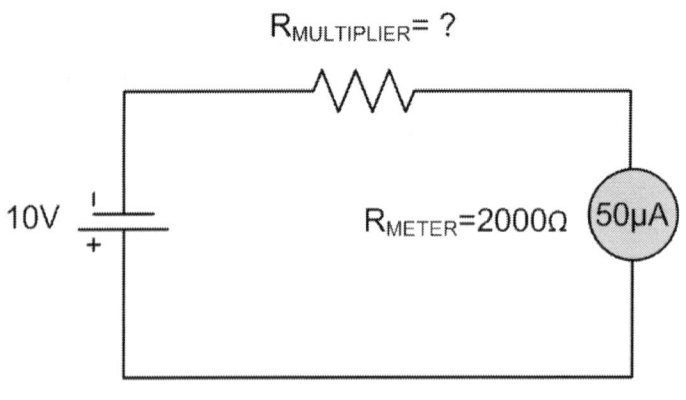

Figure 6-15.

The circuit of figure 6-15 is a Voltmeter. It has a very sensitive meter movement (50µA). What is the value of the multiplier resistor required to allow this meter to measure from 0-10 Volts?

We know that at full-scale deflection (FSD) the current in the circuit and through the multiplier will be 50µA. We can work out the voltage across the meter movement and subtract this from the applied voltage (10V) to get the voltage across the multiplier.

voltage across meter = I_{meter} x R_{meter} = 50uA x 2000 Ohms = 0.1 Volts

$R_{multiplier}$ = $E_{multiplier}$ / $I_{multiplier}$

$R_{multiplier}$ = (10-0.1 Volts) / 50uA = 9.9 Volts / 50uA = 198,000Ω

VOLTMETER RESISTANCE

The high resistance of a Voltmeter with a multiplier is essentially the value of the multiplier resistance. Since the multiplier is changed for each range, the Voltmeter resistance changes. Most moving coil voltmeters are rated in Ohms of resistance needed for 1V of deflection. This value is the **Ohms-Per-Volt** rating of the Voltmeter.

Figure 7-15 is a circuit of a multi-range Voltmeter. Why don't you check this circuit yourself by calculating two or three of the multipliers?

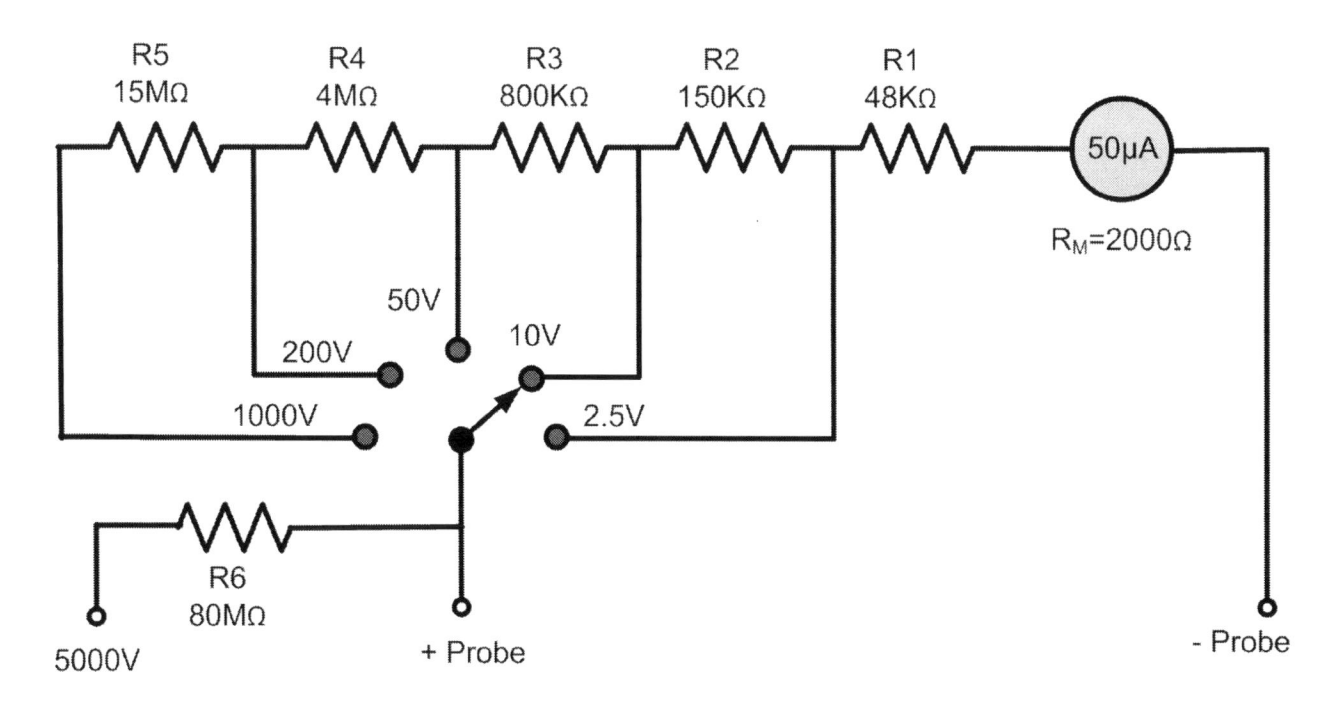

```
     R5           R4          R3          R2          R1
    15MΩ         4MΩ        800KΩ       150KΩ        48KΩ
   ──VVV────────VVV────────VVV─────────VVV─────────VVV──────(50μA)──────┐
   │        │          │          │                                      │
   │        │          │ 50V      │                                      │
   │        │  200V    │    10V   │                         R_M=2000Ω    │
   │        │      ●   ●   ●      │                                      │
   │     1000V          ↑    2.5V │                                      │
   │         ●          │   ●     │                                      │
   │                    │         │                                      │
   │    ──VVV───────────┘         │                                      │
   │       R6                                                             │
   │      80MΩ                                                            │
   ○                    ○                                                ○
 5000V                + Probe                                         - Probe
```

Figure 7-15

The leads on the Voltmeter are shown by the + and − signs. When this Voltmeter is switched to the 2.5-Volt scale, can you see that the only multiplier resistance used is the 48KΩ. So the total resistance of the Voltmeter is 48KΩ + 2000 Ohms = 50KΩ.

The *total resistance* of this Voltmeter on the 2.5-Volt scale is 50KΩ.

If we divide the total resistance of the Voltmeter by the scale to which it is switched, we get the sensitivity of the Voltmeter in Ohms-Per-Volt:

Ohms-Per-Volt = Total resistance / scale = 50KΩ / 2.5 = 20,000 Ohms per Volt.

This sensitivity works for all other scales as well. If we want to know the total resistance of the Voltmeter on any scale, just multiply the sensitivity by that scale.

Example: What is the total resistance of this Voltmeter if it is switched to the 10-Volt scale?

Total resistance of Voltmeter = Ohms-Per-Volt x scale.
Total resistance of Voltmeter = 20,000 x 10 = 200,000 Ohms.

Is this right? Check it in the circuit. On the 10 Volts scale the Voltmeters resistance consists of:

150KΩ + 48KΩ + 2000Ω = 200,000 Ohms.

Using this method, you can work out quickly the total resistance of any Voltmeter on any scale – provided you know the Ohms-Per-Volt.

A QUICK WAY TO WORK OUT SENSITIVITY

A quick way of working out the Ohms-Per-Volt or sensitivity of *any* Voltmeter is to take the reciprocal of the full-scale deflection current of the movement. In the last example, current FSD is 50uA and so the reciprocal of 50uA is $1/50 \times 10^{-6}$ which gives 20,000 Ohms Per Volt.

We need to know the sensitivity of a Voltmeter so that we can determine if it is going to affect significantly the circuit in which it is placed.

Suppose we were using a 20,000 Ohms-Per-Volt Voltmeter on a 10-Volt range. The Voltmeters resistance would be 20,000 x 10 = 200KΩ. So our Voltmeter has a resistance of 200,000 Ohms when switched to the 10-Volt range. When we place our Voltmeter in parallel with a component to measure the voltage across that component, the Voltmeter is adding 200,000 Ohms of resistance across the circuit.

Now, if the component we are measuring the voltage across is a 1000 Ohm resistor, placing the Voltmeter resistance of 200KΩ in parallel with 1000Ω is not going to disturb the resistance of the circuit much at all. However, if we were to use the same Voltmeter across a 200KΩ resistor, we would be reducing the resistance of the 200KΩ resistor to 100KΩ. 200K in parallel with 200K is 100KΩ.

This does not always matter as long as you know what affect the Voltmeter is having on the circuit. In some cases, though, the interaction of the Voltmeter with the circuit may cause the circuit to stop operating, particularly in high resistance radio frequency circuits.

Very good Voltmeters have a very high resistance in the vicinity of 10MΩ-per-Volt. Whether or not you have to be concerned about the interaction of the Voltmeter and the circuit still depends on the circuit in which you are taking the measurement.

Most digital meters have an input resistance which is not scale dependent. So you may see a digital Voltmeter with 10MΩ resistance irrespective of the voltage scale used.

OHMMETERS

An Ohmmeter consists of an internal battery, the meter movement and a current limiting resistance. When measuring resistance, the Ohmmeter leads are connected across the resistance to be measured. The circuit power is turned off. Only with the power off can you be sure that it is only the Ohmmeters battery that is producing current and deflecting the meter movement. Since the amount of current through the meter depends on the external resistance, the scale can be calibrated in Ohms.

The amount of deflection on the 'Ohms' scale indicates the measured resistance directly. The Ohmmeter reads up-scale regardless of the polarity of the leads because the polarity of the internal battery determines the direction of current through the meter movement.

The leads of a meter are normally coloured 'black' for negative and 'red' for positive. It is important to remember that a multi-function meter, when switched to Ohms, may, because of the internal battery or cell, supply a positive voltage to the negative lead and vice-versa for the other lead. This can be of particular importance when testing semiconductors.

The Ohmmeter circuit shown in figure 8-15 has 1500 Ohms of resistance. The 1.5-Volt cell will then produce 1mA if the leads are shorted and the pointer will go to full scale. In a practical Ohmmeter, a small variable resistor would be in series to 'zero' the meter exactly.

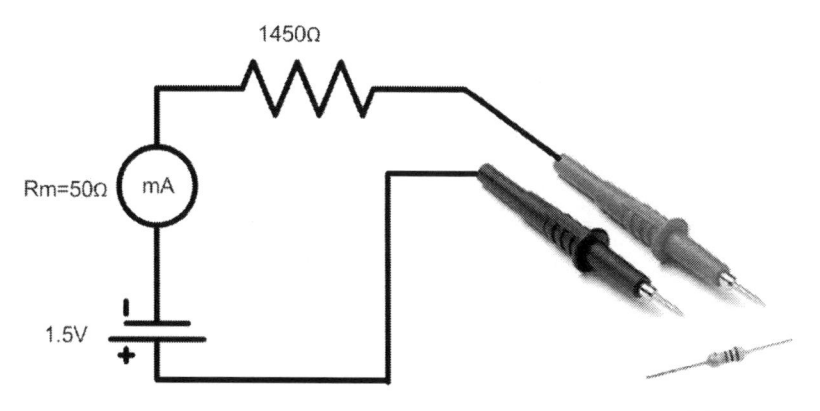

Figure 8-15.

When the Ohmmeter leads are short-circuited the meter will show full scale or zero resistance. With the Ohmmeter leads open (not touching), the current is zero and the Ohmmeter indicates infinite resistance.

Therefore, the meter face can be marked zero Ohms at the right for full-scale deflection and infinite Ohms on the left for no deflection. The in between values can be marked in Ohms by calibrating the meter against known resistances.

Notice with an Ohmmeter the red lead is negative and the black positive. When this meter is switched to Volts or amps, the lead colours are correct as voltage or current is coming from outside the meter and going in. With an Ohmmeter, the current and voltage is coming out of the meter. This is handy to know when testing semiconductors in the circuit as the polarity of the leads can be used to reverse or forward bias PN-junctions.

MULTIMETERS

All of the above meters can be built into the one box. Shunts, multipliers and the battery can be switched in and out of circuit as required to perform the various functions. Multimeters can be digital or analogue. There are advantages and disadvantages to each. Digital meters make static (non-changing) readings clear. Analogue meters make changing voltages and currents clear. Any good digital meter will have auto ranging and over volt and current protection. For an investment of a few dollars, you can buy a cheap multimeter and just do some measurements around the workshop or the house – you will learn a lot.

A word of caution, though; cheap meters like the one shown in 9-15 are great to learn on and you really do not care if you destroy if for the price – BUT – be very careful using such meters on high voltages. They tend to have cheap probes and a brittle plastic case. As long as you are careful, you can learn a lot from the meter and when you graduate to a better one it can go in the car.

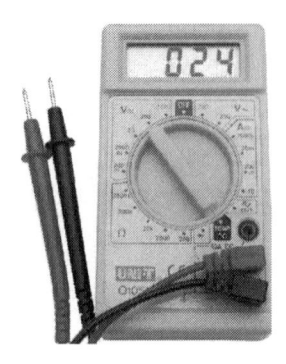

Figure 9-15.

Great for learning with but be careful of high voltages; these cheap meters are not safe on mains electricity regardless of what the manufacturer says.

When a moving coil meter is not in use it is good practice to short circuit the terminals and put the meter on a low current scale. This has the affect of physically dampening the meter from being jostled around. The meter is a coil on a pivot in a magnetic field. In effect a simple motor. If the meter is bumped the needle will the coil will attempt to deflect. However, in doing so there will be a counter emf induced into the coil that opposes this motion. A backward torque is electrically produced. The meter movement is damped.

Moving coil meters often have a mirrored scale. When reading the meter look over the top of it straight down so that you line up the pointer with its reflection in the mirrored scale. This helps you to avoid parallax error.

16. ELECTRIC CELLS & BATTERIES

There are many different types of batteries. The basic principle for all of them is the same as the ones we will cover here. There is much material here that you do not need to know, however, without the preliminary overview I would just have to give you the bare facts and say "remember this," not a pleasant way of learning, at least I think so. It might seem a bit strange, but let's start off our discussion about batteries by starting with corrosion. The two topics are related chemically and knowledge of corrosion is useful particularly when it comes to antennas.

CORROSION

Corrosion is a chemical reaction. Corrosion involves removal of metallic electrons from metals and the formation of more stable compounds such as iron oxide (rust), in which the free electrons are usually less numerous. In nature, only rather chemically inactive metals such as gold and platinum are found in pure or nearly pure form; most others are mined as ores that must be refined to obtain the metal. Corrosion simply reverses the refining process, returning the metal to its natural state. Corrosion compounds form on the surface of a solid material. If the compounds are hard and impenetrable and if they adhere well to the parent material, the progress of corrosion is arrested. If the compound is loose and porous, however, corrosion may proceed swiftly and continuously.

Aluminium is a good example. When aluminium corrodes, it becomes covered with aluminium oxide. Aluminium oxide adheres well to aluminium and prevents further corrosion. This makes aluminium an excellent choice for antennas. An added bonus is its light weight and strength.

HOW THINGS CORRODE

If two different metals are placed together in a solution (electrolyte), one metal will give up ions to the solution more readily than the other. This difference in behavior will bring about a difference in electrical voltage between the two metals. If the metals are in electrical contact with each other, electrons will flow between them and they will corrode; this is the principle of the galvanic cell or battery. Though useful in a battery, this reaction causes problems in a structure. For example, steel bolts in an aluminum framework may, in the presence of rain or fog, form multiple galvanic cells at the point of contact between the two metals, corroding one or both them.

An electrolyte is any liquid that conducts electricity. With corrosion, the electric cell effect of dissimilar metals in the presence of an electrolyte is undesirable. However, it is what we desire in an electric cell. Although the term *battery*, in strict usage, designates an assembly of two or more Voltaic cells capable of such energy conversion, it is commonly applied to a single cell of this kind.

THE VOLTAIC CELL

When two different electrodes are immersed in an electrolyte, the chemical reaction which takes place results in a separation of charges. The arrangement required to convert chemical energy into electrical energy is called the Voltaic cell.

The charged conductors are the electrodes, serving as the connection of the cell to an external circuit. The potential difference resulting from the separated charges enables the cell to function as the source of applied voltage.
Electrons from the negative terminal of the cell flow through the external circuit and return to the positive terminal. The chemical action in the cell continuously separates charges to maintain the terminal voltage that produces the current in the circuit.

AN EXPERIMENT

If you have a Voltmeter with a low DC Volts scale, you can easily demonstrate the action of a Voltaic cell. All you need are two different conductors and some type of electrolyte. I have had great success using a lemon or an orange. Just take the lemon and push two electrodes into it. I have used different nails, paper clips; almost anything will work. Place your Voltmeter across the conductors and you will measure a potential difference, usually a significant fraction of a Volt. You could find out if two coins are made of the same alloys or not! By the amount of voltage developed by two coins you can tell (roughly) how dissimilar they are. No doubt this technique has an application in Metallurgy.

SEPARATION OF CHARGES

When metals dissolve in an electrolyte, the chemical action causes separation or dissociation of the molecules, which results in charged ions. Figure 1-16 shows a generic cell with two electrodes immersed in an electrolyte which could be liquid or paste. A chemical reaction with the electrodes causes a flow of ions through the electrolyte. This flow of ions to the two electrodes causes one to become negatively charged and the other to be positively charged.

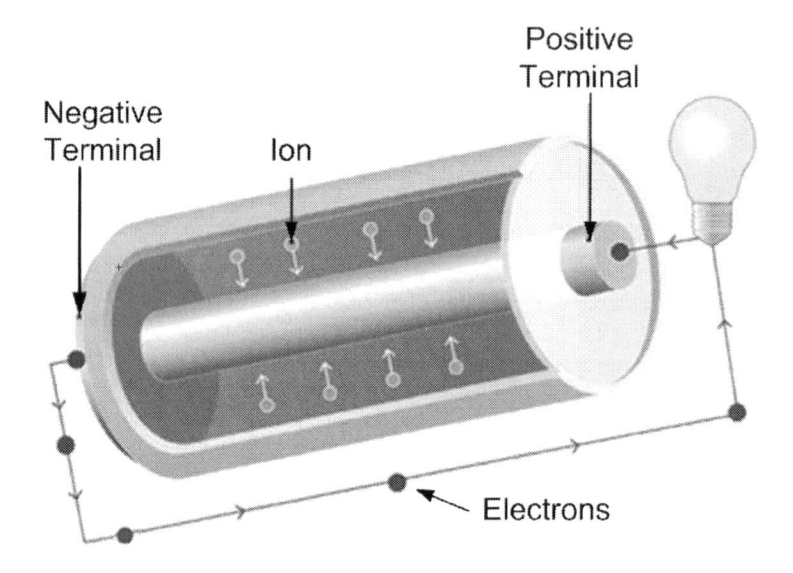

Figure 1-16

As long as the reaction continues and the external circuit can be connected between the negative and positive terminals and electron flow will occur between them and do work for us. In this case, the electron flow is lighting (by current causing heat) a light bulb. The actual chemistry is not important to us as cells use many different types of electrodes and electrolytes. It is the principle of the disassociation of charge due to ionic movement in the electrolyte that makes one electrode negative and the other positive. By the way, the ionic current flow in the electrolyte is an "electric current". Electric current is not "just" the flow of electrons. Electric current is the ordered flow of any charge. One of the first practical batteries developed used zinc and copper electrodes immersed in ammonium chloride. These early cells were assembled using glass jars. To obtain the required output voltage, the cells (jars), were connected in series. These banks of cells connected together were called a Voltaic pile; we call it a battery today. This something I am pleased about as whenever I hear or read the term Voltaic pile, I think of a hemorrhoid with a potential difference!

PRIMARY CELLS

In a primary cell, the chemical action of forming the solution is not reversible. For instance, zinc can dissolve in ammonium chloride, but the process cannot be reversed to form the zinc electrode from the solution.

SECONDARY CELLS

In secondary cells, the chemical action occurring in the electrolyte can be reversed. The electrodes can dissolve in the solution when the cell provides current to the external circuit. In this case, the cell is discharging. When an external voltage is applied to the cell to make current flow in the reverse direction, the metal comes out of the solution and is deposited back onto the electrodes, recharging the cell. Since a secondary cell can be recharged, it is also called a storage cell.

The carbon-zinc dry cell is a modern day version of the Leclanche cell – see figure 2-16.

We will look at this cell with a view to understanding how most cells work in general. You do need to know a couple of things specific to other cells and these will be covered in this chapter.

The Leclanche cell is a zinc-carbon cell. The carbon rod in the center is the positive electrode and the zinc case is negative electrode but also the housing for the entire cell. The electrolyte in the cell is ammonium chloride. The cell is a primary cell in that it cannot be recharged. The Ever Ready Red is a common zinc-carbon cell.

As the cell discharges the zinc electrode becomes dissolved in the ammonium chloride electrolyte. The electrolyte is in the form of a paste. The cell is completely discharged when the electrolyte is converted to zinc ammonium chloride, which usually corresponds with the zinc, in the zinc case almost being depleted. Early carbon-zinc cells would often leak electrolyte into equipment when they were discharged, as the zinc case would get holes in it.

Construction of Zinc-Carbon cell (Leclanche)

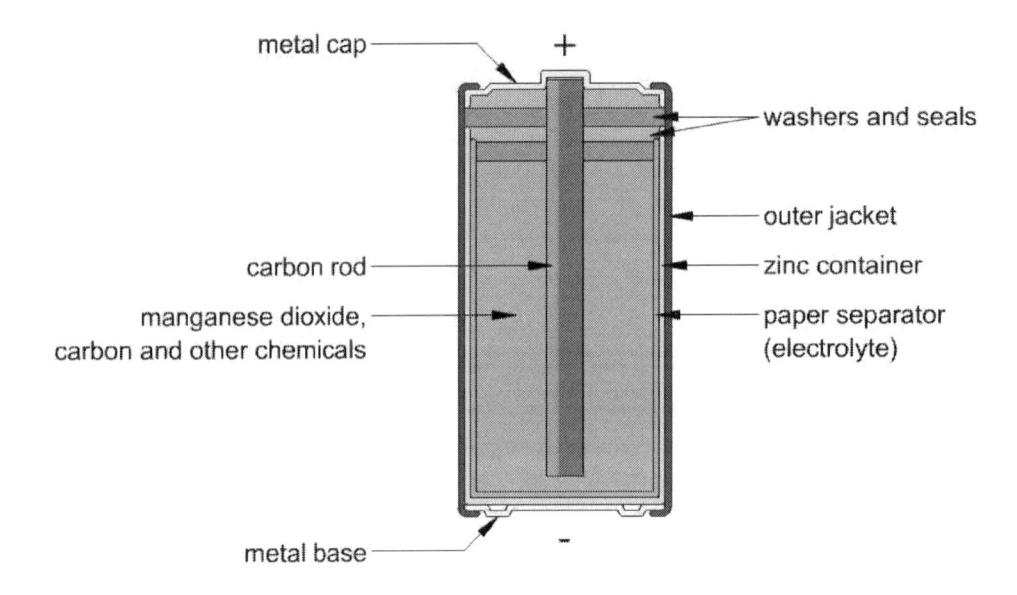

Figure 2-16.

POLARISATION

Now, if you're old enough, do you remember when the zinc-carbon battery would no longer work, it could be heated by placing it in or near oven for a few minutes and it would come to life again, for a while at least.

Well, one of the problems with the zinc-carbon cell is that as the chemical reaction takes place to charge both of the electrodes, hydrogen gas forms around the carbon rod in the form of small bubbles. This buildup of hydrogen gas around the carbon rod will begin to insulate the carbon rod from the electrolyte. Since the surface area of the carbon rod in contact with the electrolyte is reduced, the terminal voltage of the cell will drop dramatically and the cell(s) will no longer operate the equipment. This unwanted buildup of hydrogen gas (and other gases in other cells) is called *polarisation*.

Placing the dry cell in the oven would drive the hydrogen gas from the carbon rod bringing it back to life for a while.

A chemical called manganese dioxide is mixed in with the electrolyte. Manganese dioxide absorbs the hydrogen gas and prevents the gas from building up on the carbon rod. Manganese dioxide is rich in oxygen and this oxygen combines with the hydrogen to form water. Manganese dioxide in the zinc-carbon cell is called the depolariser, as removing hydrogen from the carbon rod is called depolarisation.

LOCAL ACTION

If the zinc electrode (the case) contains impurities, small Voltaic cells are formed which do not contribute to the output voltage of the cell. Also, these unwanted cells caused by impurities in the zinc, consume chemicals and dissolve the zinc. This is called 'local action.' To minimise local action, the zinc electrode is coated with mercury by a process called amalgamation.

ALKALINE CELL

The carbon-zinc cell is now largely being superseded by the alkaline cell, which has better discharge characteristics and will retain more capacity at low temperatures. The alkaline cell typically has an Ampere-hour capacity of about twice that of the carbon-zinc cell. Its nominal voltage is also 1.5 V.

The negative electrode is manganese dioxide and the positive electrode is zinc. The electrolyte is potassium hydroxide or sodium hydroxide. High conductivity (low resistance) of the electrolyte results in higher current ratings than the carbon-zinc cell. The alkaline cell is supposed to be a primary cell. However, I have been recharging them for years after finding out by experiment that it could be done. I charge these cells at 50mA constant current overnight. I first placed them inside a plastic garbage bin in case they did explode, but I have never had a misadventure with one yet. I have found that over the last few years' chargers have become available to recharge them! I guess they just did not want the public to know about it! These are the Copper Top and Energizer type batteries.

All chemical reactions are slowed by reduced temperature. Even at zero degrees Centigrade, some cells give up the ghost. Alkaline cells have superior low-temperature characteristics compared to the older Zinc-Carbon Cell.

LEAD-ACID STORAGE BATTERY

The most widely used high-capacity rechargeable battery is the lead-acid type. In automotive service, the battery is usually expected to discharge partially at a very high rate and then to be recharged promptly while the alternator is also carrying the electrical load. If the conventional car battery is allowed to discharge fully from its nominal 2.2 V per cell to 1.75 V per cell, fewer than 50 charge-discharge cycles may be expected, with

reduced storage capacity.
The typical lead-acid car battery consists of 6 cells connected in series.

Many older car batteries could be physically sliced into six individual 2.2V cells.
The plates in a lead-acid battery are lead peroxide for the positive and spongy lead for the negative. The electrolyte is dilute sulphuric acid.

SULPHATION

As a lead-acid cell discharges, lead sulphate is deposited onto the plates. This is commonly referred to as *sulphation*. Lead sulphate is a white powder often seen on the outside of old batteries or on the terminals. If lead-acid batteries are allowed to remain discharged for very long the plates will be covered in lead sulphate and the capacity of the battery greatly reduced as the lead sulphate insulates the plates from having contact with the electrolyte.

For radio use, lead-acid batteries make a very cheap alternative to an expensive power supply. They are cheap to buy and can be trickle charged with an inexpensive charger.

Lead-acid batteries are also available with gelled electrolyte. Commonly called gel cells, these may be mounted in any position if sealed, but some vented types are position sensitive.

Lead-acid batteries with liquid electrolyte usually fall into one of three classes:

1. Conventional, with filling holes and vents to permit the addition of distilled water lost from evaporation or during high-rate charge or discharge;
2. Maintenance-free, from which gas may escape, but water cannot be added;
3. Sealed.

SPECIFIC GRAVITY

Specific gravity is the weight of a substance compared to the weight of water. A hydrometer can be used to draw some electrolyte from a lead-acid cell into a reservoir tube; a weighted float can be used to measure the specific gravity of the electrolyte. This reading gives a very good indication of the state of charge of a cell. In a fully charged cell, the specific gravity is in the vicinity of 1.280 (which means the density of the electrolyte is 1.280 times that of water). When the specific gravity is down to about 1.150, the cell is fully discharged. Water has a specific gravity of 1.0 so at an electrolyte level of 1.15, means we almost have water for an electrolyte.

As you can imagine, when a lead-acid cell discharges, its electrolyte becomes weaker in sulphuric acid and approaches water.
As a lead-acid cell discharges, the specific gravity or density of its electrolyte decreases (it turns more towards water).

WARNING

Lead-acid cells produce hydrogen and oxygen when being charged, these are highly flammable gases in combination with air. The bubbles in a liquid electrolyte are hydrogen. The oxygen remains mostly locked in the cell due to other chemical reactions. It is the combination of liberated hydrogen with air that is hazardous.

I have seen a telephone exchange leveled to the ground by the ignition of hydrogen gas from the charging cells. Neither hydrogen nor oxygen is flammable independently, but they are flammable when found in combination with air. Pure hydrogen will not burn by itself nor will oxygen. A spark in a container of hydrogen won't cause an explosion. Likewise, a spark in a container of pure oxygen may only form a bit of ozone. But when oxygen and hydrogen are combined, a vigorous reaction takes place when the mixture is ignited.

It is dangerous to allow salt water to get into lead acid cells; though it is a bit hard to avoid this in a sinking boat. Salt water in combination with sulfuric acid releases huge amounts of chlorine gas, a very heavy green gas which if inhaled causes severe burning of the respiratory system.
If a lead-acid cell is short circuited, enormous current will be drawn from the battery. It will overheat in seconds and explode spraying sulphuric acid everywhere. This sometimes unfortunately happens in car accidents.

NICKEL-CADMIUM BATTERY (NiCd)

The most common type of smaller rechargeable battery is the nickel-cadmium (NiCd), with a nominal voltage of 1.2 - 1.25 V per cell.

The NiCd cell has a positive electrode made of Nickel and a Cadmium negative electrode. The electrolyte is potassium hydroxide.

Carefully used, these are capable of 500 or more charge and discharge cycles. To ensure longer life, the NiCd battery should not be fully discharged. Where there is more than one cell in the battery, the most discharged cell may suffer polarity reversal, resulting in a short circuit, or seal rupture. All storage batteries have discharge limits and NiCd types should not be discharged to less than 1.0 V per cell.

Nickel cadmium cells are not limited to "D" cells and smaller sizes. They also are available in larger varieties ranging to mammoth 1000 Ah units having carrying handles on the sides and caps on the top for adding water, similar to lead-acid types. These large cells are sold to the aircraft industry for jet engine starting and to the railroads for starting locomotive diesel engines. They also are used extensively for uninterruptible power supplies. Although expensive, they have very long life. Surplus cells are often available through surplus electronics dealers and these cells often have close to their full rated capacity.

Advantages for radio use is that these vented-cell batteries are the availability of high discharge current to the point of full discharge. Also, cell reversal is not the problem that it is in the sealed cell since water lost through gas evolution can easily be replaced. Simply remove the cap and add distilled water. By the way, tap water should never be added to either nickel-cadmium or lead-acid cells, since dissolved minerals in the water can hasten self-discharge and interfere with the electrochemical process.

NiCd batteries must be charged with a *constant current* source. A constant current source is one where the current through the cell does not change as the battery is charged. The rule of thumb is $1/10^{th}$ the milliamp hour rating for 15 hours. So if a NiCd battery is rated at 500mA/hours, then it should be charged for about 15 hours at 50mA constant current if it is fully discharged. If the NiCd is only, 1/3 discharged then it should be charged for 5 hours.

LITHIUM CELLS

Like any other battery, a rechargeable lithium-ion battery is made of one or more power-generating compartments called cells. Each cell has essentially three components as all other chemical cells: a positive electrode, a negative electrode and an electrolyte in between them.

The positive electrode is typically made from a chemical compound called *lithium-cobalt oxide* ($LiCoO_2$) or, in newer batteries, from *lithium iron phosphate* ($LiFePO_4$). The negative electrode is generally made from *carbon* (graphite).

The electrolyte varies quite a lot with many options available. Electrolytes comprise of lithium salts or organic solvents, such as ethylene carbonate, dimethyl carbonate and diethyl carbonate.

All lithium-ion batteries work in broadly the same way. When the cell is charging up, the lithium-cobalt oxide, the positive electrode gives up some of its lithium ions, which move through the electrolyte to the negative graphite electrode and remain there. The battery takes in and stores energy during this process. When the battery is discharging, the lithium ions move back across the electrolyte to the positive electrode, producing the energy that powers the battery. In both cases, electrons flow in the opposite direction to the ions around the outer circuit. Electrons do not flow through the electrolyte: it's effectively an insulating barrier, so far as electrons are concerned.

The movement of ions (through the electrolyte) and electrons (around the external circuit, in the opposite direction) are interconnected processes and if either stops so does the other. If ions stop moving through the electrolyte because the battery completely discharges, electrons can't move through the outer circuit either—so you lose your power.

Lithium cells if overcharge or short circuited can cause an explosion. Large batteries usually have some electronic control and safety protection. Generally, lithium ion batteries are more reliable than older technologies such as nickel-cadmium (NiCd) and don't suffer from a problem known as the "memory effect" (where NiCd batteries *appear* to become harder to charge unless they're discharged fully first). Since lithium-ion batteries don't contain cadmium (a toxic, heavy metal), they are also (in theory, at least) better for the environment—although dumping any batteries (full of metals, plastics and other assorted chemicals) into landfills is never a good thing. Compared to heavy-duty rechargeable batteries (such as the lead-acid ones used to start cars), lithium-ion batteries are relatively light for the amount of energy they store.

CELLS IN SERIES AND PARALLEL

Cells can be connected in series, positive to negative to increase the output voltage. This method is common to all batteries. The small 9-Volt transistor battery is 6 cells connected in series. The 6V lantern battery has 4 completely separate cells inside the case. If you get the opportunity, I suggest you break an old one open and have a look. It isn't messy and you will find what looks like 4 oversized 'D' cells inside.

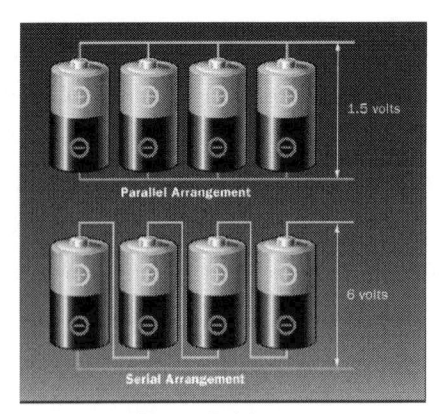

Figure 3-16

It is not typical for batteries or cells to be connected in parallel though it can be done to increase the current capacity. The batteries must be identical. A more desirable way to get greater capacity is to use a bigger battery in the first place.

It is very rare indeed to have batteries connected in a combination of series-parallel. This would only be done in specialist applications and is not recommended.

TESTING BATTERIES

The best way to test a lead-acid cell is by measuring the specific gravity of the electrolyte as already discussed. Smaller batteries and cells must be tested on load. This means the equipment in which the battery is installed should be turned on and then the terminal voltage of the cell or battery measured. A fully discharged battery will still show its correct terminal voltage if it is not under load. It is important you understand this and the reasons why. As a battery ages its internal resistance increases.

A new battery will have an *internal resistance*, which is a small fraction of an Ohm. The electrolyte between the plates forms part of the total circuit. If the battery is supplying 1A and the internal resistance of the battery is 0.1 Ohms then a mere 0.1 Volts will be dropped across the internal resistance, which is insignificant. As the battery discharges, the internal resistance will increase. However, no voltage is lost across the internal resistance unless the battery or cell is placed on load. The load should be the nominal load that the battery is expected to deliver.

ELECTROLYTIC CORROSION

The use of dissimilar metals in engineering (say when building an antenna) is likely to cause considerable trouble due to electrolytic corrosion. Every metal has its own electro-potential and unless metals of similar potential are used the difference will cause corrosion at the point of contact even when dry. When moisture is present, this effect will be even more severe. If for any reason, dissimilar metals must be used then considerable care should be taken to exclude moisture, the corrosive effects of which will vary with atmospheric pollution.

Metals can be arranged in what is called the electrochemical series, as follows:

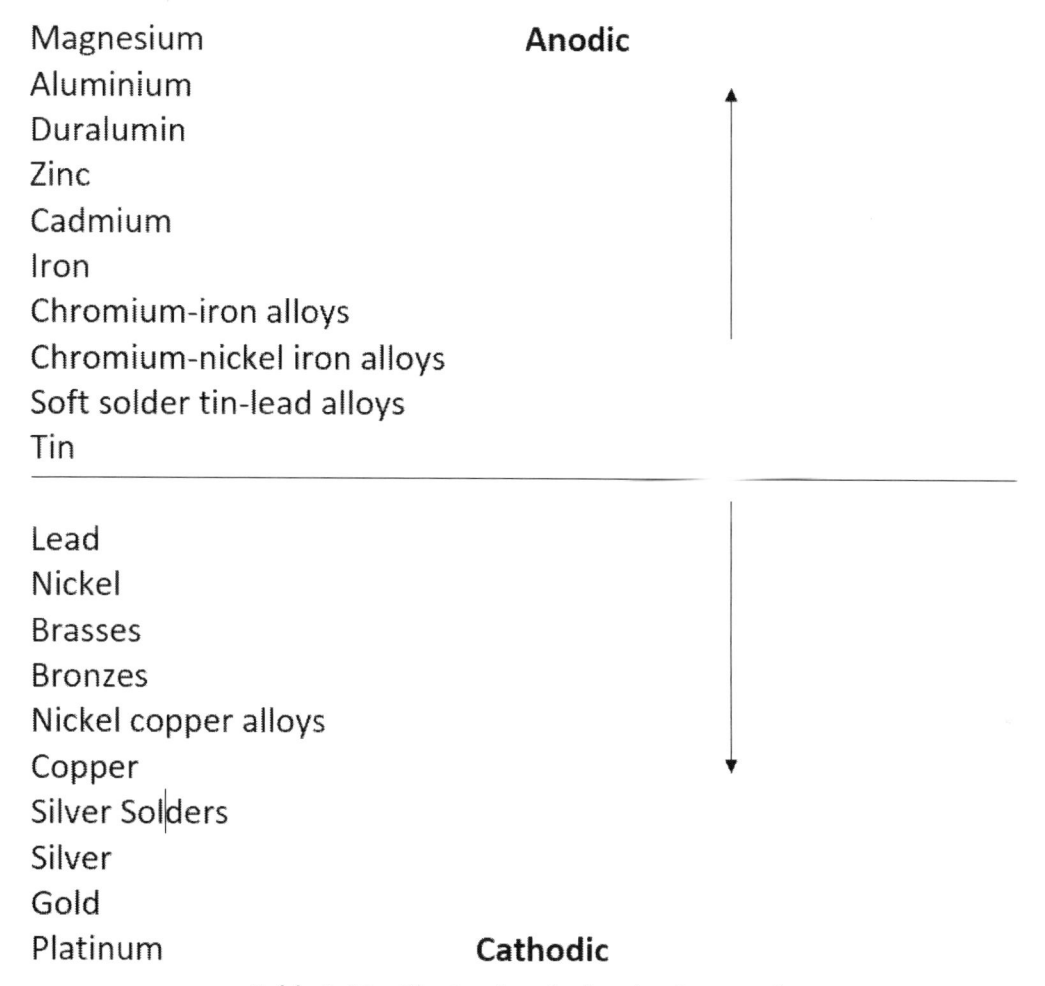

Magnesium	**Anodic**
Aluminium	↑
Duralumin	
Zinc	
Cadmium	
Iron	
Chromium-iron alloys	
Chromium-nickel iron alloys	
Soft solder tin-lead alloys	
Tin	
Lead	
Nickel	
Brasses	
Bronzes	
Nickel copper alloys	
Copper	↓
Silver Solders	
Silver	
Gold	
Platinum	**Cathodic**

Table 1-16 – Electrochemical series for metals.

Metals from within the same group of Table 1-16 may be used together with little corrosion, but metals used from across the groups will suffer greatly from the voltaic corrosion effects. Also, since the above groups are arranged in order, the greater the spacing in the list, the greater will be the effect.

The metals of the lower group in Table 1-16 will corrode those in the upper portion. For example, brass or copper screws in aluminum will corrode the aluminum considerably, whereas with cadmium plated brass or copper screws there will be much less corrosion of the aluminum.

17. MICROPHONES

SOUND

Sound consists of small fluctuations in air pressure. We hear sound because these changes in air pressure produce fluctuating forces on our eardrum. You only have to place your hand in the front of your mouth while you are talking, or in front of a sound system, to feel the pressure waves.

Similarly, microphones respond to the changing forces on their components and produce electric currents that are effectively proportional to those pressure waves.

The best way of detecting sound is with a diaphragm ('dia'-'fram'). Our eardrums have a diaphragm. Just like the skin or membrane covering of a musical drum. When sound waves enter the ear, the force of the sound waves causes the eardrum to vibrate. Vibrations from the eardrum are transmitted by various structures in the ear to our brain for processing.

All microphones have a diaphragm, which usually consists of some type of taut material like plastic, paper, or even very thin aluminium. With all microphones, sound waves strike a diaphragm and cause it to vibrate. This vibration is then, by the mechanism of the microphone, made to make a voltage or current vary in sympathy with the sound waves. The voltage or current thus produced can be made to travel long distances (if necessary) along conductors and then we can recover the sound by converting current variations back into sound wave pressure variations.

THE CARBON MICROPHONE

Carbon microphones are no longer used, as they are low quality. However, a short discussion of the carbon microphone is a good introduction to all microphones. The carbon microphone was for many decades the only type of microphone used in telephones and two-way radios.

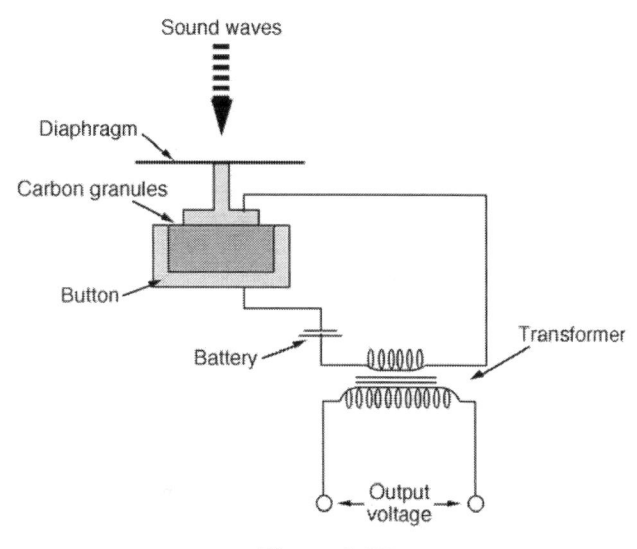

Figure 1-17

The principle of operation was both simple and ingenious. Sound waves strike the diaphragm and move it. The movement of the diaphragm causes a plunger or piston to move with it.

The plunger compresses and decompresses a chamber filled with carbon granules. Now, carbon granules will conduct electric current. If a battery is connected to the microphone terminals, a current will flow.

How much current flows will depend on the degree of compression of the carbon granules. If the sound waves forces the diaphragm inwards, then the granules are compressed and their resistance decreases and the current increases. Similarly, if the sound amplitude (loudness) decreases, the tension of the diaphragm allows it to move outwards, the carbon granules decompress, the resistance increases and so the current decreases.

In other words, the resistance of the carbon microphone varies in sympathy with the sound waves that strike the diaphragm and this, in turn, causes a varying current in the circuit which is a good representation of the original sound wave.

Carbon microphones were used for many years. You may have seen people bang an older type telephone handset against something to get rid of the noise. This is because the carbon granules would cling together in older microphones and make a noise called 'frying' because the noise sounded like something frying in a pan.

THE DYNAMIC OR MOVING COIL MICROPHONE

In the dynamic microphone, the diaphragm is connected to a lightweight coil. Inside the coil is a permanent magnet. When sound strikes the diaphragm, the coil will move back and forth in the magnetic field of the permanent magnet. We have relative motion between a conductor and a magnetic field (Faradays Law), so an emf or voltage is induced into the coil. This induced emf is a very good electrical representation of the original sound wave. A dynamic microphone produces its own output voltage and does not need a battery like the carbon microphone.

Dynamic microphones can be made in various impedances. (Impedance is the total opposition to current flow in an AC circuit). The dynamic microphone or some variation of it is widely used in telephones, radiocommunications and quality sound recording.

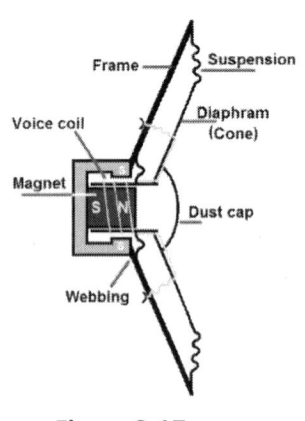

Figure 2-17

The speaker is essentially the same as the dynamic microphone, except the current carrying the sound to be recovered is fed through the coil, which creates a moving magnetic field around the coil. The interaction between the magnetic field of the coil and the field of the permanent magnetic cause the diaphragm to move back and forth. The motion of the diaphragm recreates the original sound wave.

A SIMPLE TELEPHONE (FAR SOUND)

You do not need to know how a telephone works, however, looking at a simple telephone circuit will help to consolidate what you do need to know. In figure 3-17, we have a carbon microphone and a receiver in each handset. A speaker is called a 'receiver' in telephony. A source of DC is applied to each microphone. The carbon microphones are coupled to the receiver and telephone line via a transformer. Sound on either carbon microphone will be converted to varying DC in the handset. The transformer will convert varying DC to AC and sound will be heard in both receivers. The current contains all of the information present in the original sound wave (amplitude and frequency). Since the current flows through the receiver (most likely a moving coil), the diaphragm (cone) of both receivers will move back and forth with the same frequency and relative amplitude of the original sound wave and therefore recreate it.

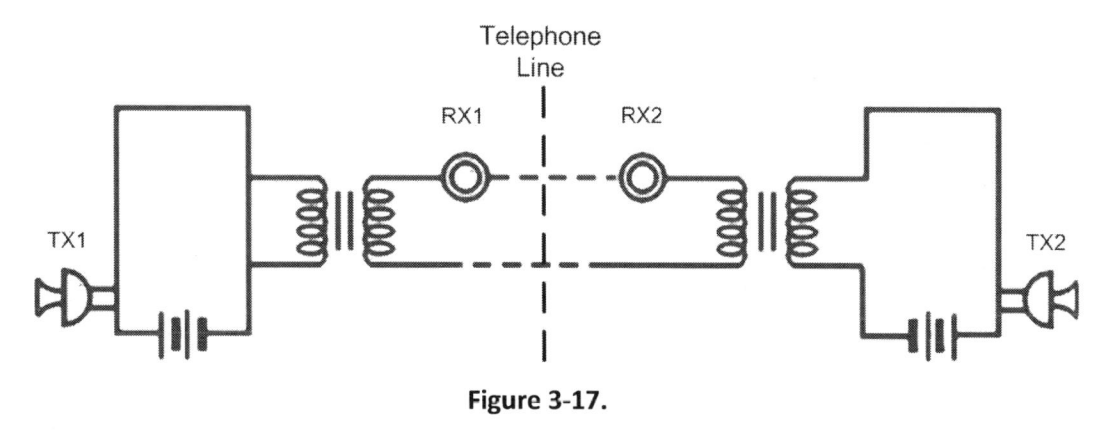

Figure 3-17.

The distance between the microphone and the speaker can be a long way apart enabling voice communications over a long distance through wires. The modern landline telephone is really not much more than this in principle.

The carbon microphone that was used in telephones for many decades has been replaced by a variant of the moving coil type called a 'rocking-armature'. The rocking armature microphone is just a moving coil microphone with an inbuilt mechanical advantage between the diaphragm and the coil.

It seem simple does it not? Yet the impact of the telephone when it was first invented had a profound effect on the world. Of course, the telegraph came first. However, imagine how mysterious and wonderful it must have been for the first telephone users to hear and talk to someone thousands of miles away.

The word telephone comes from the Greek roots 'tele' ("far") and 'phone' ("sound"). Alexander Graham Bell invented the telephone *(disputed)*, though many others improved on the invention. The U.S. patent granted to Bell in March 1876 (No. 174,465) for the development of a device to transmit speech sounds over electric wires, is often said to be the most valuable patent ever issued.

CONDENSER (CAPACITOR) MICROPHONE

When we learned about capacitance, we found that the capacitance of a capacitor was determined by the area of the plates, the type of dielectric and the distance between the plates.

Now, suppose a capacitor is connected to a source of emf and allowed to charge fully (at least 5-time constants). What would happen if we left the capacitor in the circuit and somehow altered its capacitance?

If we increased the capacitance of the capacitor in the circuit, it would be able to hold more charge. So increasing the capacitance would cause the capacitor to charge further and current would flow while the charging process was taking place. If we then decreased the capacitance, the capacitor would no longer be able to hold as much charge and it would discharge back into the source of the applied emf. While it is discharging, current would again flow in the circuit.

In the capacitor microphone, that is what happens. One of the plates of the capacitor is made of very light material and is the diaphragm of the microphone. Sound waves striking the diaphragm (one plate) will cause it to vibrate in sympathy with the sound waves. The moving diaphragm (plate) will cause the capacitance to change. As the diaphragm moves in, the capacitance will increase and a charge current will occur. As the diaphragm moves out, the capacitance decreases and a discharge current occurs. Since the motion of the diaphragm and the capacitance are in sympathy with the sound waves, the charge and discharge currents are once again an electrical representation of the original sound wave.

The diaphragm of the microphone can be one of the plates of the capacitor microphone, or a diaphragm can be connected mechanically to make a capacitor plate move.

Like a carbon microphone, the capacitor microphone needs a source of emf, as it does not generate any of its own.

The term 'condenser' is an old term for 'capacitor'. When you see these microphones referred to, the words 'condenser' and 'capacitor' are often used interchangeably. I prefer 'capacitor microphone' but it seems for many the term 'condenser' has stuck from tradition. Capacitor microphones have the advantage of being very small, which makes them attractive for use in small equipment, such as handheld two-way radios (transceivers), small voice recorders and the like.

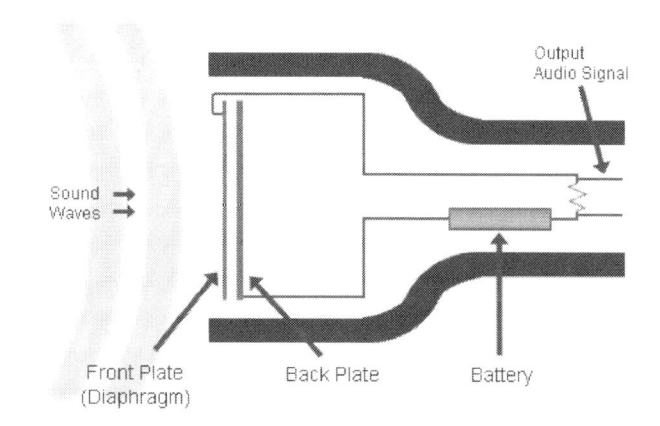

Figure 4-17

There is an another type of capacitor microphone called the 'electret microphone.' The principle of operation of these microphones is identical to that of a capacitor microphone. However, a special electret material is used as the dielectric in the capacitor. The electret microphone does not require a bias battery. An electret material is one that holds a charge for many years similar to the way a magnetic material can hold a magnetic field for a long time.

For exam purposes, you need only remember that an electret is the same as an ordinary capacitor microphone. However, it can be used without a bias voltage.

The dynamic and capacitor microphones are the most popular microphones in use today for radiocommunications.

CRYSTAL MICROPHONE

The Piezoelectric Effect (pee-zo-electric)

If a piece of quartz crystal is held between two flat metal plates and the plates are pressed together, a small emf will be developed between the plates as if the crystal became a small battery for an instant. How much emf is produced is proportional to the pressure applied. When the pressure on the plates is released the crystal springs back and an opposite polarity emf is produced on the plates. In this way, mechanical energy is converted into electrical energy by the crystal. Piezoelectricity means "pressure-electricity.

If an emf is applied to the plates of a crystal, the physical shape of the crystal will distort. If an opposite polarity emf is applied, the crystal will reverse its physical distortion. In this way, a quartz crystal converts electrical energy into mechanical energy and vice versa.

Some ceramic materials also exhibit a piezoelectric effect.

These two reciprocity qualities of a quartz crystal and ceramics are known as the *piezoelectric effect*.

Well, if you are thinking ahead of me, here we have another way to make a microphone. After all, the function of a microphone is to convert mechanical energy (sound waves) into electrical energy and the piezoelectric effect does just this.

Sound waves striking the diaphragm cause varying pressure to be applied to the crystal, which in turn causes the microphone to produce an output voltage in sympathy with the sound waves. A crystal microphone does not require a battery. Like the dynamic microphone, it directly converts mechanical energy into electrical energy.

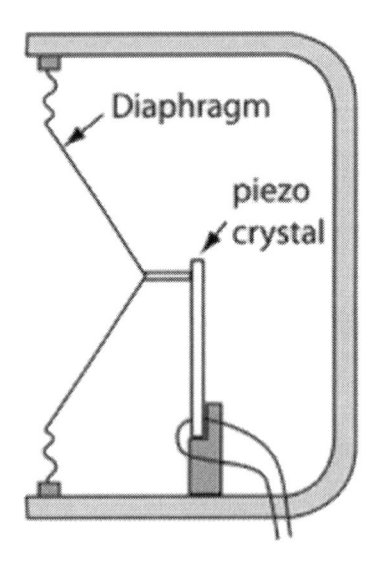

Figure 5-17

The crystal earpiece is the same principle used in reverse. Crystal microphones are not high quality, but they have the advantage of being small and inexpensive. The quartz crystal is not used in the form you may have seen them in nature. They are cut into thin slices and ground, like a gemstone, is ground, to the exact size.

Well, that's about it for microphones. Microphones used for radio communications are not high fidelity (high quality audio reproduction). Talk to a recording or sound engineer about microphones and they can go on about all of the different types used in their field. For example, special microphones that pick up sound from some directions and not others, etc. However, for us radio communication people, a simple dynamic or capacitor microphone does the job nicely.

MICROPHONE AMPLIFICATION

Because the output of a microphone is very low in electrical terms, in almost all applications, an amplifier called a microphone amplifier or pre-amplifier, is connected to the microphone to increase the level of the output signal. Many microphones come with a built in amplifier in the microphone housing.

Too much microphone amplification is a mistake

Many radio operators add microphone amplification to their transceiver. This is fine, however adding too much amplification may cause distortion to the transmitted signal that does not sound good (though many think it does). This can cause interference to other radiocommunications users. We will be discussing this in depth later; it is called 'overmodulation.'

WHAT'S IN A NAME

Alexander Graham Bell (born March 3, 1847, Edinburgh. Died, August 2, 1922, Beinn Bhreagh, Cape Breton Island, Nova Scotia, Canada). Scottish-born American audiologist best known as the inventor of the telephone (1876). For two generations his family had been recognised as leading authorities in elocution and speech correction, with Alexander Melville Bell's Standard Elocutionist passing through nearly 200 editions in English. Young Bell and his two brothers were trained to continue the family profession. His early achievements on behalf of the deaf and his invention of the telephone before his 30th birthday, bear testimony to the thoroughness of his training. Biography source - Encyclopaedia Britannica.

Guess where the term decibel i.e. tenths of a Bel originated.

There is much controversy over who really invented the telephone. There is strong evidence that an Italian immigrant to the United States by the name of Antonio Meucci invented the telephone before Bell and that is apparatus was misplaced or stolen. The evidence is so strong the US Congress passed a Bill naming Meucci as the inventor of the Telephone. There is a copy of this Bill on the RES website.

18. TRANSFORMERS

One of the most common devices used in electricity, electronics and radio, is the transformer. The name itself indicates that the device is used to transform, or change, something. The transformer may be used to step up or step down AC voltages, to change low-voltage high-current AC to high-voltage low current AC, or vice versa, or to change the impedance of a circuit to some other impedance in order to transfer energy more efficiently from a source to a load.

A BUZZER

A buzzer has nothing to do with transformers. However, I am going to describe how a buzzer works as this will nicely lead into transformers.

Take a close look at the pictorial diagram of a buzzer in figure 1-18. It is so easy to make one of these on a piece of board. No need for you to make one, though, but it will be good if you can picture how it is made and how it works. The electromagnet is just a soft iron bolt (any bolt would do) about 75mm long. Wind as many turns of thin single strand insulated wire on the bolt as you can get. The armature is just a strip of tin about 5mm wide and 75mm long. A nail holds it on the board on the left-hand side. The contact on the right hand side is just another nail.

The strip of tin (the armature) forms part of the circuit for current flow. The circuit as it is now in figure 1-18 can conduct current. However, after the battery has been connected for a fraction of a second the magnetic field builds up around the electromagnet and the electromagnet pulls the armature towards it. When the armature moves towards the electromagnet, the circuit current is broken.

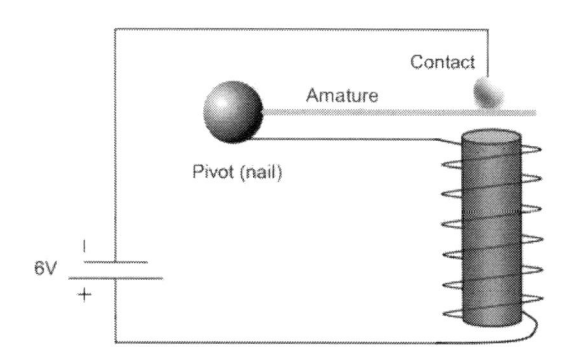

Figure 1-18.

The magnetic field about the electromagnet collapses and the armature springs back to the contact. Current can now flow again. This cycle is repeated over and over. The result is that the motion of the armature hitting the electromagnet makes a buzzing sound. We have a buzzer! If the armature was extended with a rod and on the end of that rod we had a small weight (hammer) that could strike a gong; we would have an electric bell.

What type of current flows in this circuit?

I hope you thought pulsating DC, or DC which is being turned on and off rapidly. Now think about the magnetic field around the electromagnet. It is a pulsating magnetic field, or perhaps more importantly, it is a magnetic field which is continually varying. It never stays still as it is either expanding when the current is on or collapsing when the current is off.

Let's make a modification to our simple buzzer circuit to demonstrate the action of a transformer. Refer to figure 2-18.

All we have done is wind a second coil of wire, either on top of or next to, the electromagnet using insulated wire and connected it to a Voltmeter. The two coils (inductors) are insulated from each other.

Now, do you remember when we discussed inductance and alternating current? We talked about Faraday's law of magnetic induction.

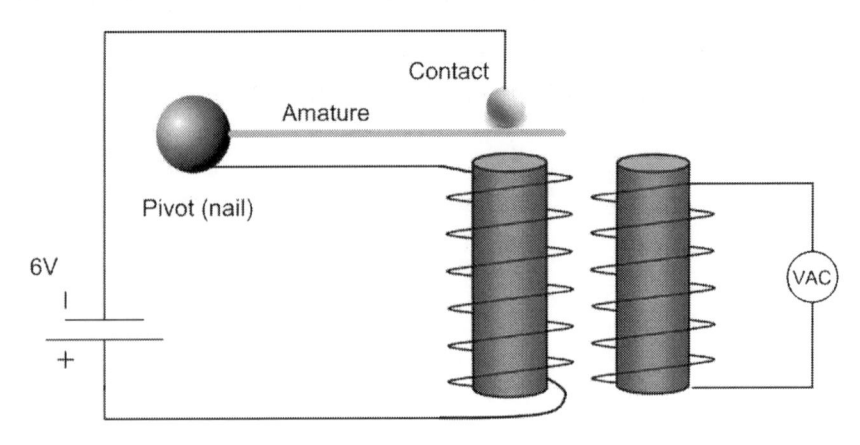

Figure 2-18

Faraday's law states: *When relative motion exists between a conductor and a magnetic field an emf is induced into the conductor.*

Is there relative motion between the conductors on the second coil and a magnetic field? Yes, there is. An emf (voltage) is induced into the second conductor and this will be shown by the Voltmeter. We have created a simple, though very inefficient transformer. Generally, a transformer has two windings. The primary winding is supplied with some type of current that will cause the magnetic field around it to vary. In our experiment, the primary winding is the coil of wire around the electromagnet. The other winding of a transformer is called the secondary winding.

So a transformer is basically two or more windings (coils, inductors) wound on a common core or former. The primary winding is fed with some type of varying current so that a moving magnetic field is created around it. The moving magnetic field created by the primary winding causes an induced emf into the secondary winding.

For a transformer to work the primary winding must have a current through it that produces a varying magnetic field. This is virtually any current other than DC, such as AC, pulsating DC, varying DC and the like. *The secondary voltage (type) will always be AC.*

What we need to learn is the uses for a transformer and how we can determine the amount of induced voltage into the secondary.

Obviously, we could reason that the amount of induced voltage into the secondary would have something to do with the voltage on the primary and the ratio of the number of turns on the primary and secondary. We will discuss this relationship in more detail shortly.

Transformers do not have to have a magnetic material such as iron for a core. The core can be air. The basic function of a transformer is to step voltage up or to step voltage down, which is also a way of matching unequal impedances.

Schematic symbol of a transformer

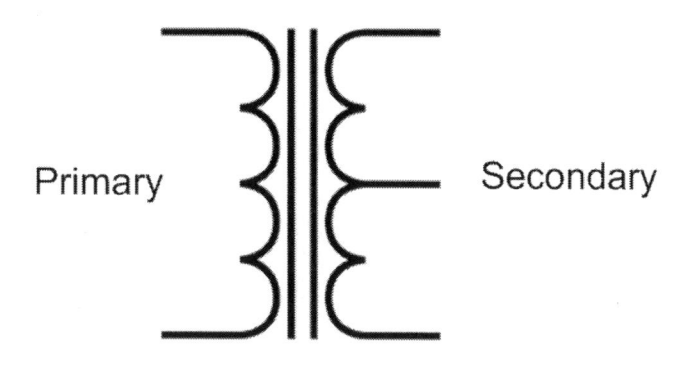

Figure 3-18.

The schematic symbol for a transformer is shown in figure 3-18 (excluding the writing which I have added). The two vertical lines between the two windings indicate that this transformer has a laminated iron core. If no lines were shown, it would be an air-core transformer. This transformer has a centre tapped secondary. You do not know what the number of conductor turns are on the primary or secondary from the symbol. Of more importance is the *ratio* of the number of turns on the primary to the secondary. This ratio is called the *turns ratio*. If there are more turns on the secondary, then voltages are stepped up. Conversely, fewer turns on the secondary and voltage is stepped down. The symbol may show you if the transformer is step up or step down but it will not give the turns ratio. The *voltage ratio* is the same as the *turns ratio*. If voltages for primary and secondary are given, then you know the turns ratio since voltage and turns ratio are the same.

Transformer construction

Figure 4-18 shows one construction method for a laminated iron core transformer. Here the primary and secondary are wound separately around a laminated iron core. An alternative is to wind one over the top of the other.

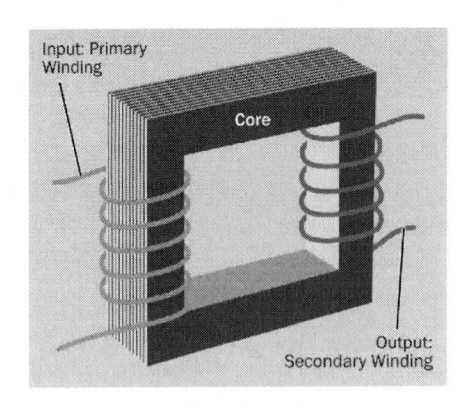

Figure 4-18

Laminated iron core

Winding the primary and secondary on an iron core improves the efficiency of the transformer since the iron core concentrates the magnetic lines of force.

The problem with iron cores is that the core is also a conductor. Go back to Faraday's law again. The core of a transformer (iron) is in a moving magnetic field. This means that small currents will be induced into the core. These unwanted core currents are called *Eddy Currents*.

These currents are given this name because they flow in circles in the iron core much like eddies of water around a propeller.

Eddy currents are undesirable because of $P=I^2R$ losses. What this means is that current flowing through any resistance produces heat. This is fine if you want to make heat. However, the function of a transformer is not to make heat but to transform from one voltage to another. By the way, you may hear the term "I squared R losses" used. This means the same thing, though "I squared R losses" can occur other ways besides eddy currents.

What is the product of the current and voltage in a circuit? Power = E x I. A transformer should ideally have no losses. This means the power in the primary circuit should be equal to the power in the secondary circuit. The power in each circuit is the product of the voltage and current in the respective circuit.

In practical transformers, particularly iron core ones, the major form of loss of power is due to eddy currents, otherwise also known as "I squared R losses."

If we could increase the electrical resistance of the core without changing its magnetic properties, we would reduce the magnitude of the eddy currents. When we discussed the properties that determine the resistance of a conductor, one of these properties was the cross-sectional area, as expressed in $\rho L/A$. This is what we are doing when we laminate the iron core.

Imagine taking a solid iron core and slicing it into lots of thin sheets. Each sheet (lamination) will have a higher resistance to current flow than the solid iron core. We now spray each sheet with an insulating lacquer and bolt or glue the entire iron core back together again. We now have a laminated iron core, which will have less power lost in it due to eddy currents.

Eddy currents are not a problem if the core of a transformer is made of air or some other insulating material.

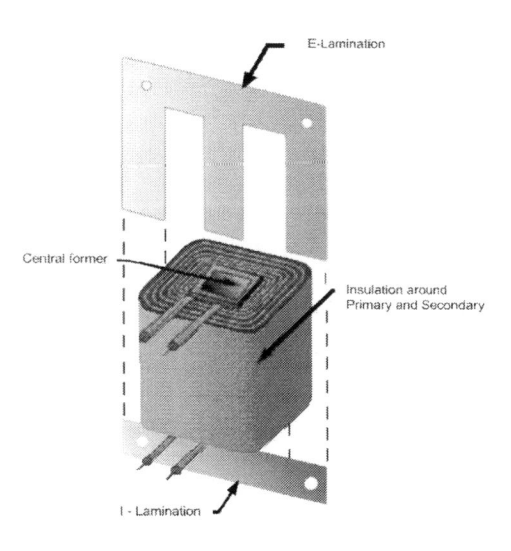

Figure 5-18

Another construction method for a transformer is shown in the diagram of figure 5-18. Eddy currents are reduced in amplitude by laminating the core. I have shown one E-lamination and one I-lamination. Imagine 50 or more of such laminations. Insulated from each other, then bolted together to form a solid laminated iron core.

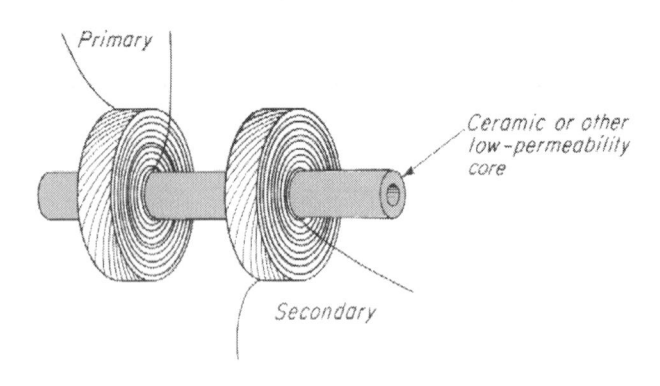

Figure 6-18

The type of transformer shown in figure 6-18 is used at high frequencies. Iron core transformers are not used at radio frequencies, as the eddy current losses become too great.

HYSTERESIS - LOSSES

If iron is in an unmagnetised state, its magnetic domains are not arranged in any particular manner. The domains are randomly orientated. When a magnetising force is applied to them, the domains rotate into a position in line with the magnetising force. If the magnetising force is reversed, the domains must rotate to the opposite position. In rotating from one alignment to the opposite, the domains must overcome a frictional hysteresis, or a resisting effect, in the core. In some materials the resisting effect is small, in others it is appreciable. Due to this effect, we lose some energy as heat. Heat is nearly always lost energy. This loss is called *hysteresis loss*.

Hysteresis occurs in iron cores of transformers. As frequency is increased, the alternating magnetising force will no longer be able to magnetise the core completely in either direction. Before the core becomes fully magnetised in one direction, the opposite magnetising force will begin to be applied and start to reverse the rotation of the domains. The higher the frequency, the less fully the core magnetises.

Transformers operated on low-frequency AC may not have much hysteresis loss, but the same cores used with a higher frequency have more hysteresis and be less efficient.

COPPER LOSS

Iron-core transformers are subject not only to eddy-current and hysteresis losses in the core but also to a *copper loss* which occurs as heat lost in the resistance of the copper wire making up the windings. The current flowing through whatever resistance exists in these windings produces heat. The heat in either winding, in Watts, can be found by the power formula P = I²R. For this reason, the copper loss is also known as the I²R loss. The heavier the load on the transformer (the more current that is made to flow through the primary and secondary), the greater the copper loss.

With one layer of wire wound over another in a transformer, there is a greater tendency for the heat to remain in the wires than if the wires were separated and air-cooled. Increased temperature causes increased resistance of a copper wire. As a result, it becomes necessary to use heavier wire to reduce resistance and heat loss in transformers than would be required for an equivalent current value if the wire were exposed to air during operation.

EXTERNAL-INDUCTION LOSS

Another loss in a transformer is due to external induction. Lines of force expanding outward from the transformer core may induce voltages and therefore currents into outside conducting bodies. These currents flowing through any resistance in an outside body will produce a heating of the external resistance. The power lost in heating these outside circuits represents a power loss to the transformer, since the power is not delivered to the transformer secondary circuit. Actually, in a well-designed transformer, the amount of power lost in this fashion is usually small. The efficiency of a well-designed transformer is around 98%.

THE VOLTAGE RATIO OF TRANSFORMERS

One of the main uses of transformers is to step up a low voltage AC to a higher voltage. This can be accomplished by having more turns on the secondary than on the primary.

The turns ratio of a transformer is the ratio of the number of turns on the primary to the number of turns on the secondary.

If a transformer has an equal number of turns on the primary and secondary (it happens), then the turns ratio is 1:1.

If a transformer has 100 turns on the primary and 10 turns on the secondary, then the turns ratio is 100:10 which is the same as 10:1.

If a transformer has 300 turns on the primary and 900 turns on the secondary, then the turns ratio is 300:900 which is the same as 1:3.

The shorthand for primary turns is usually N_p and for secondary turns N_s.
The voltage of the primary is usually designated E_p and for the secondary E_s.

The turns ratio of a transformer is the same as the voltage ratio. In other words:

$N_p/N_s = E_p/E_s$

A Worked Example.

A transformer has 100 turns on the primary and 10 turns on the secondary. If 500 Volts AC is applied to the primary what will be the secondary voltage?

$N_p/N_s = E_p/E_s$

We could transpose this equation for E_s but for exam purposes an easier approach is to see that the turns ratio of the transformer Np/Ns is 100/10 or 10:1. For every 10 Volts on the primary, there will be 1 Volt on the secondary. Therefore, the secondary voltage will be 500/10 or 50 Volts.

This is a step-down transformer because the secondary voltage is lower than the primary voltage. A step up transformer is where the secondary voltage is higher than the primary voltage

The turns ratio is equal to the voltage ratio

A lot of learners get confused, especially with the step up transformers. You can make a transformer to convert 10 Volts AC to 10,000 Volts AC. At first sight, it may appear that you are getting something for nothing - no such luck.

The power in the primary circuit is equal to the power in the secondary circuit (disregarding losses).

No power is gained in a transformer. If a transformer steps the voltage up, there is a corresponding decrease in secondary current. So we may get more secondary voltage, but at the sacrifice of the secondary current.

Take an Arc Welder. This is basically a huge step-down transformer. 240 Volts AC to the primary and only a few Volts on the secondary, but a huge amount of current (60-100 Amps or more) for melting metals is available in the secondary circuit.

TRANSFORMER ISOLATION

One major advantage of transformers is the safety isolation they can provide. Small battery eliminators are really just small transformers. The primary connects to 240 VAC and the secondary may be 12 VAC (or converted to DC).

Because there is no metallic contact between the secondary and the primary, it means the user of the low voltage equipment is isolated from the mains power, providing a significant safety advantage.

If you build an interface between your radio transceiver and your computer for using digital modes, you will most likely use some form of isolation. There are several methods, but one method is to use an audio isolation transformer. This will prevent DC levels in your transceiver affecting DC levels in the computer sound card.

AUTO TRANSFORMERS

An autotransformer consists of a single winding with one or more taps on it, as shown in figure 7-18.

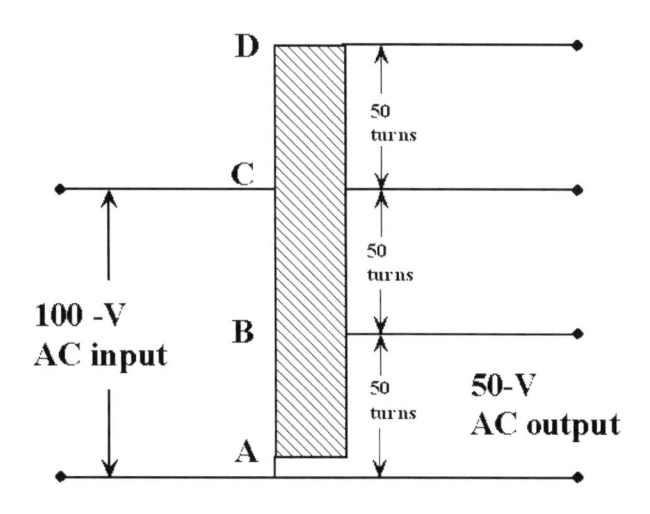

Figure 7-18

Auto transformers are not "automatic". The name "auto" comes from their use as a car (auto) ignition coil.

Everything we have discussed about other transformers (mutually coupled transformers) applies to autotransformers.

Autotransformers used on high voltages are dangerous, as they provide no electrical isolation between the user and the supply. These transformers are not commonly used in radio and communications anymore. Auto transformers have the advantage requiring less copper, therefore less weight and cost. However, in nearly all applications today better options are available than using autotransformers. An exception in radio is where RF auto transformers are used as an antenna matching device in vehicle installations.

IMPEDANCE MATCHING

We have not discussed the purpose of impedance matching yet and I am not going to now. Suffice to say, in electronics; it is important to connect components together which are of the same impedance (impedance is the total opposition to current flow in an AC circuit).

For example, your sound system may specify that you use 8-Ohm speakers. If you use speakers other than 8 Ohms on your sound system, you may damage the sound system or at least get reduced quality sound.

A transformer could be used to connect any impedance speaker to a sound system. The impedance ratio of a transformer is related to the turns ratio. All I will give you, for now, is the equation:

$$\frac{N_p}{N_s} = \sqrt{\frac{Z_p}{Z_s}}$$

Z_p = impedance looking into the primary terminals from the power source.
Z_s = impedance of load connected to secondary and
N_p / N_s = turns ratio, primary to secondary.

The basic equation tells us that the turns ratio is equal to the squareroot of the impedance ratio. Now this is something you really need remember for an Advanced exam.

Example.

A 100Ω source is connected 25Ω load using a transformer. What is the turns ratio of the transformer? Well, the impedance ratio is given as 100:25 which is 4:1 – this is the impedance ratio. Look at the equation above – the turns ratio is on the left and the impedance ratio on the right so one equals the other except there is a squareroot sign on the right-hand side. So turns ratio N_p/N_s equals the square root of the impedance ratio. The square root of 4:1 (just take the square root of each number 4 and 1) is 2:1.

So to match a 100Ω source to a 25Ω load we would use a transformer with a 2:1 turns ratio.

How is a transformer able to match impedance?

Suppose we had a transformer with 2400 turns on its primary and 150 turns on its secondary. The turns ratio of this transformer is 2400 / 150 or 16:1.

The turns ratio is 16:1. Let's see if we can work out what the impedance ratio would be.

For the sake of this exercise let's apply a primary voltage of 240 Volts. Since the voltage ratio is equal to the turns ratio, the secondary voltage must be 15 Volts.

The turns ratio is 16:1 and this gives us the same voltage ratio, that is, 240 / 15 = 16:1.

Now for calculation purposes, I have to nominate a maximum secondary current that the transformer can handle. This value is arbitrary and any value can be chosen for our purposes. The maximum secondary current I will choose is 1 Ampere.

This transformer is shown in figure 8-18.

Primary Power = E_p/I_p Secondary Power = E_s/I_s

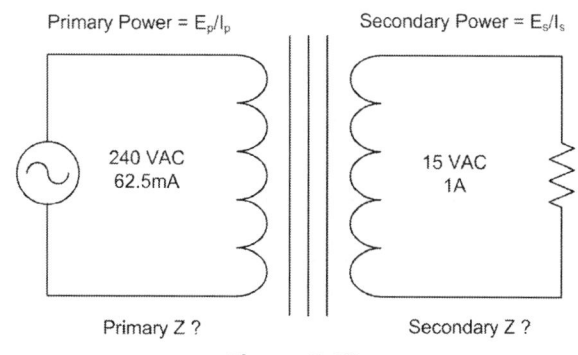

240 VAC
62.5mA

15 VAC
1A

Primary Z ?   Secondary Z ?

Figure 8-18.

Since we know the secondary current and voltage (for full load) we can work out the secondary power:

$P_{secondary} = E_{secondary} \times I_{secondary} = 15 \times 1 = 15$ Watts.

Since the power in the primary is equal to the power in the secondary (neglecting losses) we can, knowing the primary power is 15 Watts, work out the primary current:

$I_{primary} = P_{primary} / E_{primary} = 15 / 240 = 62.5$ mA.

Where is all this taking us? Well, impedance is the ratio of voltage to current:

$Z = E / I$.

We know what the current and voltage is in the primary and secondary circuit, what is the primary and secondary impedance?

Impedance of the primary = $E_{primary} / I_{primary}$ = 240Volts / 62.5mA = 3840Ω.
Impedance of the secondary = $E_{secondary} / I_{secondary}$ = 15 / 1 = 15Ω.

In other words, the impedance ratio of this transformer is 3840:15 or 256:1. I have gone about it the long way just to show you how it works.

The rules you need to remember:

(a) The turns ratio is equal to the square root of the impedance ratio

or

(b) The impedance ratio is equal to the square of the turns ratio.

N_p / N_s = Square root of (Z_p/Z_s)

USING THE FORMULA

Let's work out the impedance ratio from the formula. From the last example, we know the voltage ratio is 240:15 which is the same as the turns ratio Np/Ns.

N_p / N_s = square root of (Z_p/Z_s) (remember Z_p/Z_s is the impedance ratio)

$240 / 15$ = square root of (Z_p/Z_s)
16 = square root of (Z_p/Z_s)

We do need to get rid of the "square root" on the right-hand side. To do this we "square" both sides and we are left with:

$16 \times 16 = Z_p/Z_s$
or
Impedance ratio = Z_p/Z_s = 16 x 16 or 256 : 1

This agrees with our earlier calculation.

If you have trouble using this equation, ask your trainer for help. Many people do, so don't be afraid to ask for help.
The type of question you need to answer in Advanced exams is something like this.
What is the turns ratio of a transformer that is required to match a 600Ω source to a 50 Ohm load?

$Z_p = 600Ω$ $Z_s = 50Ω$

Turns ratio = N_p / N_s = square root of (600/50) = 3.464:1.
For every turn on the secondary, there will be 3.464 turns on the primary.

A last word about transformers.

The types of transformer that we have been discussing here is the *mutually (magnetic) coupled transformer*. They can be used for impedance matching; however, their bandwidth is very narrow. A good example might be an RF auto-transformer in the boot of a car to match a mobile vertical whip. This will work fine but because of such a transformer will have a narrow bandwidth the usual thing is to have many tappings on it and when changing frequency even slightly on a short whip you might have to readjust the impedance match.

A better broad impedance match is the transmission line balun (to be discussed in transmission lines). You should know that while these may look similar to mutually coupled transformers, they have nothing in common at all. Apply nothing you have learned in the section to transmission line baluns.

19. RESISTIVE, INDUCTIVE & CAPACITIVE CIRCUITS

The purpose of this chapter is to understand the combination of resistive, inductive and capacitive circuits and the concepts of impedance, quality factor or 'Q' and resonant circuits.

I am going to avoid the maths where possible, though I will be giving you some equations. The ones you need to remember I will mention. However, you will not have to use many of the equations in an Advanced Radio exam.

Impedance (symbol [Z])

Impedance is the opposition to current flow in an AC circuit, measured in Ohms. There is no inductive or capacitive reactance in a DC circuit. However, in an AC circuit, there is resistance, inductive and capacitive reactance. All of these oppose current flow and is this combined opposition that we call impedance.

Since impedance includes capacitive and inductive reactance it is frequency dependent. Resistance is not frequency dependent.

Impedance cannot, except in the case were reactances are equal and opposite, be expressed as a single number. There are two standard methods of expressing impedance.

1. The *polar form* and;
2. The *rectangular form*

The rectangular form is more often used in radio. Modern impedance meters display both forms. The rectangular form is favoured by radio experimenters as they are usually concerned about the amount of unwanted inductance or capacitance in antennas

POLAR FORM

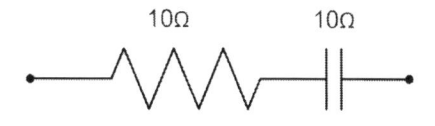

Figure 1-19.

The circuit of figure 1-19 consists of 10Ω of resistance in series with 10Ω of capacitive reactance. In your assessment, you will have to use the rectangular form of expressing impedance, but we will do both here.

When impedance is represented in Polar Form, the two numbers are called **Magnitude and Phase Angle**. The magnitude is found by using the equation in equation 1-19(a) and the phase angle 'θ' (the amount of lead or lag of current) is found from: -

Tan θ = X/R or θ = Arctan X/R. (Arctan means Inverse Tangent) Eq. 1-19(b)

Let's work out the impedance of the circuit in figure 1-19 in "polar form". Note there is no X_L, so we just use X_C.

$$Z_{mag} = \sqrt{R^2 + (X_l - X_c)^2}$$
$$Z_{mag} = \sqrt{10^2 + 10^2}$$
$$Z_{mag} = 14.14 \Omega$$

Equation 1-19(a)

and now the phase angle:

$$Tan\theta = \frac{Xc}{R}$$
$$Tan\theta = \frac{10}{10}$$
$$Tan\theta = 1$$
$$ArcTan = 1 = 45°$$

Equation 1-19(b)

So the impedance in figure 1-19 in polar form is said to be 14.14Ω / Phase angle 45⁰. The current leads (since it's a capacitive circuit) by 45⁰.

When we discussed capacitive reactance, we talked about how the current leads in a capacitive circuit. In a capacitive only circuit, the current leads the voltage by 90⁰. Another way of saying this is, in a capacitive circuit the phase angle is 90⁰ (leading). In a circuit containing both resistance and capacitance, the phase angle will be between 0 and 90⁰. There is no requirement for us to calculate impedance in polar form in the current Advanced radio syllabus.

RECTANGULAR FORM

For our purposes, the impedance above is best described just as it is in the circuit, i.e. 10Ω of resistance in series with 10Ω of capacitive reactance. Radio experimenters and most others use this sometimes more convenient and easier rectangular form of describing and impedance. This is especially so when working with antennas.

In rectangular form impedance is expressed and two numbers and those numbers are called **Real and Imaginary**. So an impedance shown in rectangular form has a Real part and an Imaginary Part. The Real part is the resistance and the Imaginary part is the reactance. Mathematicians use the symbol "i" preceding an imaginary number. Engineers use "j". Imaginary numbers cannot be manipulated like Real numbers. Numbers in polar or rectangular form fall into a family of numbers called *Complex Numbers*. With complex numbers, we can't do calculations the way we do regular arithmetic calculations using ordinary numbers. The good news is you do not have to do complex number calculations. Though I encourage you to learn more about them and how to use them as these are vital in AC circuits, antennas and transmission lines to name only three.

The second number in a complex number in rectangular form is called an imaginary number because it does not fall on the standard number line. It is used to do real calculations and get actual results. For us now and imaginary number expresses the reactive part of an impedance.

A rectangular form method of writing the impedance in figure 1-19 is - 10-j10Ω (said as ten minus Jay ten Ohms). The 'j' indicates that this part of the impedance is reactive. The minus sign indicates that the reactance is capacitive. The "j" alone shows the impedance is inductive and "j0" means there is no net reactance at all

Another Example:

50-j4Ω is a rectangular form for 50Ω of resistance in series with 4Ω of capacitive reactance. If the minus sign is not present, then the reactance is inductive.

70j30Ω is shorthand for 70Ω of resistance in series with 30Ω of inductive reactance.

If a radio experimenter measures their antennas impedance either with a noise bridge or and impedance meter and gets $_{anything}j0\Omega$, then that antenna is resonant because it has no reactance.

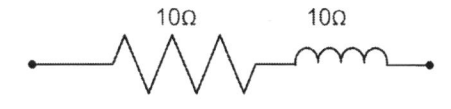

Figure 2-19.

In the circuit of figure 2-19, it is sufficient to describe this impedance as 10j10Ω.

It is important to remember that any circuit can be simplified to a resistance and a single reactance (X_L or X_C). If there are two reactances in series the larger cancels, the smaller. If there are two reactances in parallel then smaller cancels out, the larger.

A circuit cannot be inductive and capacitive at the same time. The effect of inductive reactance on current is to cause the current to lag the voltage. The effect of capacitive reactance is to cause the current to lead the voltage. Both of these effects are opposite. Therefore, the circuit will either be capacitive or inductive, but not both.

If you have trouble remembering which leads and lags, remember this:

L is for inductance - L is for Lag.

That is, current lags the voltage in an inductive circuit.

Complicated circuits with lots of resistors, capacitors and inductors can be reduced to a single resistance and a single reactance, either capacitive or inductive.

The circuit of 3-19 is a series RLC circuit. Since the inductive reactance is 100Ω and the capacitive reactance is 50Ω, the net reactance of the circuit is 50Ω of inductive reactance.

Figure 3-19.

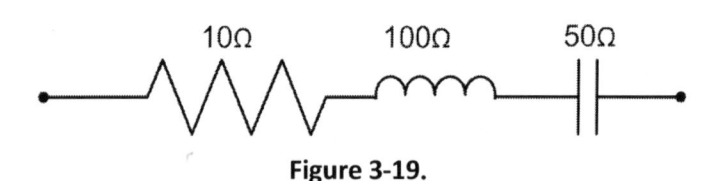

The impedance of the figure 3-19 circuit is 10j50Ω.

WHEN $X_L = X_C$

When a circuit contains the same amount of inductive and capacitive reactance, the net reactance is zero and the circuit is resistive.

So if a circuit contains 100Ω of resistance in series and 200Ω of each reactance, then the impedance of the circuit is 100j0Ω or simply 100Ω.

TUNED CIRCUITS OR RESONANT CIRCUITS

When inductance and capacitance are combined in a circuit so that their reactances are equal (no net reactance), the circuit is said to be resonant.

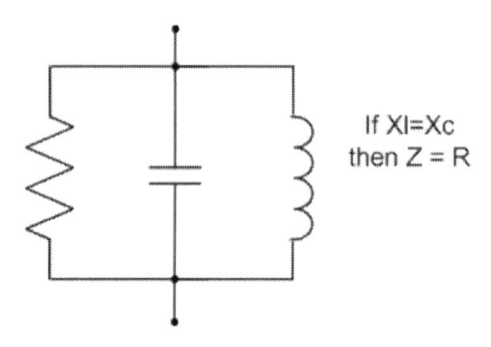

Figure 4-19.

In the parallel circuit of figure 4-19, the inductive and capacitive reactances are equal, so the net impedance is the value of the resistance. This circuit is also resonant if the two reactances are equal.

RESONANCE.

Resonant circuits have some very useful properties in electronics and radiocommunications, however, let's talk a bit about what resonance is.

Electrical and mechanical resonance are very similar. A taut string, like that on a guitar, if plucked, will resonate at a particular frequency. The length of the string determines the frequency of resonance.

A metal tuning fork is designed to resonate at a particular frequency. When you strike a tuning fork, it will vibrate at the same single frequency every time, because the tuning fork is resonant at that frequency.

If you hit the rim of a wineglass with a hard object it will "ring" and it will always ring at the same frequency because it is resonant at that frequency.

In all the cases above, we had to deliver some energy to the object (guitar string, tuning fork or wineglass) to get it to resonate.

Have you ever been listening to loud music and at certain times the window may rattle, or an object on a table may start to move (vibrate with the music)? The window will start to vibrate if sound waves at its resonant frequency strike it. Since music contains many frequencies, the window will only vibrate when its resonant frequency is present in the music.

HOW DOES A SINGER BREAK A GLASS?

It is true that a singer provided they can reach the right note can break a wine glass. A wine glass has a frequency at which it will resonate. Making a loud noise will not break a wine glass. However, if a singer reaches and sustains a note which is equal to the resonant frequency of the wine glass, then the wine glass will absorb energy from the sound wave and it will begin to resonate (vibrate) and if the amplitude (loudness) is enough, it will shatter.

The wineglass absorbs energy on its resonant frequency. All objects, even electrical circuits, will absorb energy at their resonant frequency.

TUNING FORK EXAMPLE

Suppose we had two tuning forks designed to create the same note (frequency) and we struck one fork to make it resonate. Then place the two tuning forks say 50 mm apart, the tuning fork that you did not strike would begin to resonate. The second tuning fork has absorbed sound wave energy from the first tuning fork and starts to vibrate in sympathy.

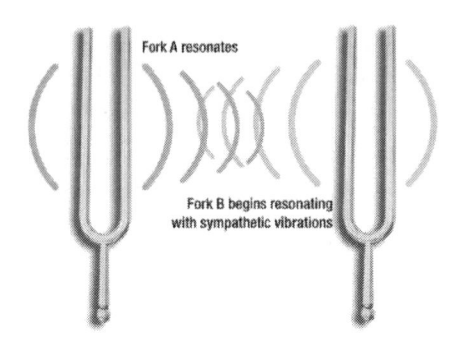

In electronics, we don't say something vibrates. We say it oscillates. A mechanic might say the guitar string when plucked would vibrate at 800 times per second. A musician would say it makes such and such a note. An electronics person will say it oscillates at 800 Hertz.

DAMPED OSCILLATIONS

If a guitar string or tuning fork is given energy, we have learnt that it will oscillate at its resonant frequency. All things being perfect, the wave produced will be the shape of a sine wave. Now what happens to the sound wave emitted from a guitar string or whatever, after the initial pulse of energy is given to it?

The guitar string will oscillate vigorously at first and then slowly decrease in amplitude until it no longer makes any sound. Importantly - the frequency stays the same, but the amplitude dies down to nothing. This is a damped oscillation.

CONTINUOUS OSCILLATION

A swing (perhaps with a child on it though the child is not an essential ingredient) will oscillate as it is a pendulum. We know that a pendulum will move back and forth at the same frequency (or period) every time. That is a fact because we use pendulums to create clocks. The amplitude of the swing of the pendulum will slowly dampen, but the frequency will remain the same.

Back to the child on the swing. How do we keep the rhythm going (oscillating)? Well, we have to keep giving it energy. When do you give it energy? Well if you are a good swing pusher, you know to give the extra push to sustain the motion of the swing when it is at the top of its cycle (about to change direction). You do not have to push hard to sustain the swing's oscillations. Once the swing is going, you only need to keep supplying enough energy to overcome the losses in the swing system. It is the losses in the swing, which slow it down. The losses in a swing are from friction.

Everything we have just discussed about resonance in mechanical systems (swings, guitars and the like) are equally applicable to resonant electronic circuits.

LC RESONANT CIRCUITS

Not every LC (inductive and capacitive) circuit is resonant. A resonant circuit is only resonant if we use it as a resonant circuit. Though every LC circuit will have a resonant frequency, we don't have to use it as a resonant circuit. We are going to talk about LC circuits that we want to resonate.

If we connect a capacitor and an inductor in parallel as shown in figure 5-19, nothing happens, as there is no energy in the circuit. This is just like the child's swing before we push it.

Now suppose we give the parallel LC circuit some energy. There are many ways we can give this circuit a pulse of energy. One way would be to expose the inductor to a moving magnetic field (Faradays Law of Induction). We could disconnect the capacitor, charge it then connect it back to the inductor.

The latter is what I have done in figure 5-19. I have charged the capacitor and then connected it back in parallel with the inductor. What happens?

LC - Oscillations

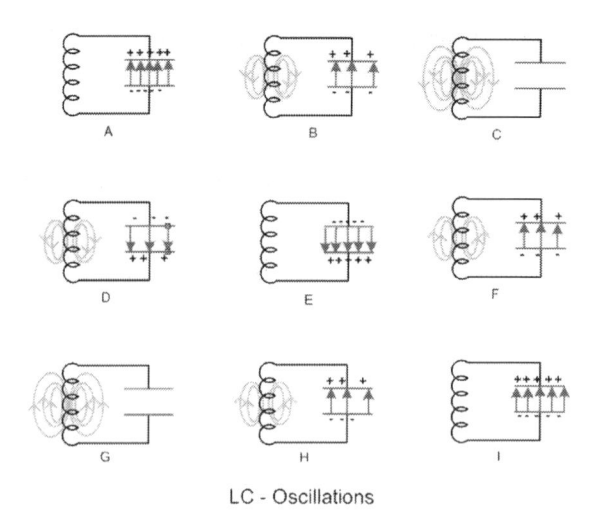

Figure 5-19.

Look at 5A-19. We have given the capacitor some energy and all of that energy is stored in the electric field of the capacitor. The capacitor will begin to discharge through the inductor. As it does so, a magnetic field will be created around the inductor; 5B-19. Eventually, the capacitor will be discharged; 5C-19.

Now the energy has moved from the electric field of the capacitor into the magnetic field of the inductor. When the capacitor stops creating current in the circuit the magnetic field around the inductor cannot stay there, as it has no current to support it.

The magnetic field about the inductor will now begin to collapse into the inductor. In doing so, there will be relative motion between the inductor and the magnetic field and an emf will be induced into the inductor.

This emf will cause a current to flow in the opposite direction and will cause the capacitor to charge; 5D-19. Eventually, all the energy from the magnetic field will have returned to the circuit and the capacitor will be charged (storing energy in its electric field) 5E-19.

The capacitor will begin to discharge through the inductor again and so on.

This circuit is oscillating. The shape of the voltage produced is a sinewave. These oscillations would go on forever except, just like the swing, there are losses in the circuit, which dissipate energy. The losses are the resistance in the wire of the inductor, the circuit and the plates of the capacitor. There is also a little energy lost in the dielectric of the capacitor. The oscillation of this circuit then is a damped oscillation.

If we wanted a continuous oscillation, we would have to supply the circuit above with energy to overcome the losses. Such a circuit is called an oscillator.

We know what moves with mechanical resonance as we can see or feel it. The guitar string vibrates, the tuning fork vibrates back and forth. What does the moving with electrical resonance? It is the electric and magnetic fields that move. The capacitor dumps its electric field energy into the magnetic field of the inductor; then the inductor dumps its magnetic field energy back into the electric field of the capacitor. This happens at a certain rate and that rate is known as the resonant frequency of the circuit.

It should be intuitively obvious that if we had small L's and C's that could not hold much energy that the rate of dumping to each other would be faster – a higher resonant frequency. If we had large L's and C's then they could hold a significant amount of energy; then the rate of dumping would be much slower – a lower resonant frequency.

RESONANT FREQUENCY

If you thought about the circuit we have just described, you could probably guess what determines the resonant frequency or frequency of oscillation. Imagine if the capacitor was huge and took ages to discharge through the inductor. Likewise, if the inductor is large. You would expect the oscillations or resonant frequency to be low. You would be right. Large values of L and C resonate at a low frequency whereas small values of L and C resonate at a higher frequency.

To work out the exact resonant frequency (fr) of a tuned circuit we can use the following important equation:

$$fr = \frac{1}{2\Pi\sqrt{LC}}$$

Equation 1-19.

The resonant frequency of a tuned circuit is inversely proportional to the square root of the product of the inductance and capacitance. The equation gives the resonant frequency in Hertz and L and C must be entered in Henrys and Farads.

You may have to use this equation in an exam. You may be asked to identify the equation from among others. You must know how to increase or decrease the resonant frequency of a tuned circuit.

The same equation is also used to find the resonant frequency of a series LC circuit.

Equation 1-19 is probably the most important fundamental equation in radio. Now you might be wondering just how this equation is derived. Recall we said resonance occurs when $X_L=X_C$? This is equivalent to saying resonance happens when: -

$$2\pi fL = \frac{1}{2\pi fC}$$ Equation 2-19

We know this equation is right we can verify it by experiment. If you are game, try for yourself transposing the equation for f if you succeed you will get the resonance equation of 1-19.

Using the resonance equation of 1-19.

A tuned circuit (series or parallel) consists of an inductor of 100 microhenries and a capacitor of 250 picofarads. On what frequency will it resonate?

100 microhenries = 100 x 10^{-6} Henries
250 picofarads = 250 x 10^{-12} Farads

Inserting these values into the equation we get:

Resonant frequency (fr) = 1 / (2 x $\prod$ x square root (100 x 10^{-6} X 250 x 10^{-12}))

Resonant frequency = 1.007 megahertz.

It was not by accident that I chose the values of L and C used in this example. The resonant frequency is close to 1MHz, which is roughly the centre of the AM radio broadcast band. You would need approximately these values of L and C to make a broadcast band radio receiver or a 'crystal set'.

THE 'Q' OF A RESONANT CIRCUIT

The 'Q' is a term used for the 'quality' of a tuned circuit. We want as high a Q as possible. The more losses a tuned circuit has, the lower the Q. The fewer losses a tuned circuit has, the higher the Q. There are equations for calculating the Q of a tuned circuit and for that matter for capacitors and inductors. However, you just really need to know what Q is.

If we did want to lower the Q of a circuit (to increase bandwidth) we could add some resistance to it. Drawing energy from a resonant circuit always reduces its Q.

IMPEDANCE OF TUNED CIRCUITS

When you read 'tuned circuits,' you can also think of *resonant circuits* as we are discussing one and the same thing. Parallel and series tuned circuits have different properties which allow us to use them in electronics for different purposes.

Try to remember this:

If there were no losses in tuned circuits (never possible) the following would be true:

Parallel resonant circuits have infinite impedance.
Series resonant circuits have zero impedance.

In the real world though resonant circuits have some losses and so we have to say:

Parallel resonant circuits have a maximum impedance at resonance.
Series resonant circuits have a minimum impedance at resonance.

The impedance type for both series and parallel LC circuits at resonance is resistive.
The impedance of a series resonant LC circuit is minimum and resistive. The impedance of a parallel LC circuit at resonance is maximum and resistive.

There is no net reactance at resonance since $X_L=X_C$. That is why we say that resonant circuits are resistive.

IMPEDANCE TYPE ABOVE AND BELOW RESONANCE

Now this topic is not an exam requirement. However, it is very useful for all manner of things to understand what the impedance *type* is a resonant circuit above and below resonance. This is very helpful when it comes to tuning antennas.

First, what do we mean by impedance type?

Well, impedance is made up of three oppositions:

1. *Resistance* (not affected by frequency).
2. *Inductive reactance* – increases with frequency.
3. *Capacitive reactance* – decreases with frequency.

All of these oppositions are measured in Ohms. All oppose current flow. The total combined effect on a circuit of all of these oppositions is called impedance and it too is measured in Ohms.

Now think of a circuit at resonance. At resonance $X_L=X_C$. So at resonance the only opposition is resistance, or a better way of putting it, the impedance type of a resonant circuit is resistive. The impedance type of both series and parallel circuits at resonance is resistive because $X_L=X_C$. Anything which is resonant has no reactance, only resistance.

Impedance type of a series LC circuit – off resonance

A simple way to work out what the impedance type of a series circuit used above or below its resonant frequency is to think, which reactance (L or C), has the largest voltage drop.

Above resonance X_C will decrease and X_L will increase. Therefore, X_L has the largest voltage drop and the impedance type is inductive. Below resonance the X_C is greater than X_L. The capacitance will have the greater voltage drop and the circuits impedance type is capacitive.

Impedance type of a parallel LC circuit – off resonance

With parallel LC circuits, the largest branch current determines the impedance type. Say to yourself, which branch will have the least reactance and therefore draw the most current. Above resonance X_L increases and X_C decreases and the capacitive branch will draw the most current. Above resonance a parallel circuit is capacitive. Below resonance then, it has to be inductive.

Summary – Impedance Type (i.e. L or C)

Parallel LC circuits:
1. Above their resonant frequency are capacitive.
2. Below their resonant frequency are inductive.

Series LC circuits:
1. Above resonance are inductive.
2. Below resonance are capacitive.

Knowing this and the logic behind it is not required for any Advanced exam. This knowledge is very useful when you experiment with tuned circuits and antenna tuning.

IMPEDANCE OF LC CIRCUITS VERSUS FREQUENCY

A parallel LC circuit has a maximum impedance at resonance. At resonance, the impedance is maximum and resistive. Either side of resonance the impedance will be less. An LC circuit with no losses has an infinite impedance at resonance. If there were a resistance in parallel with a parallel resonant circuit, then the impedance of the circuit would be equal to the parallel 'R'. For example, an infinite impedance (which is the same as an open circuit) in parallel with a 100kΩ resistor would give a total impedance at resonance of 100kΩ.

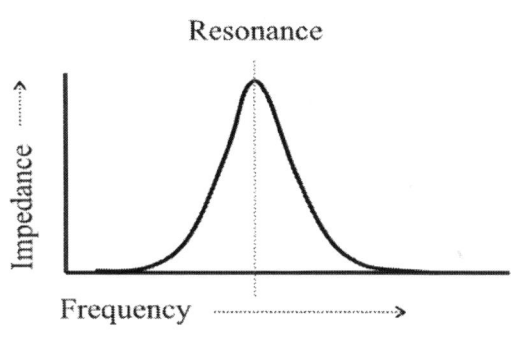

Parallel LC circuit - At resonance impedance is Maximum

Figure 6-19

A series LC circuit has a minimum impedance at resonance. At resonance, the impedance is minimum and resistive. Either side of resonance the impedance will be greater. A series LC circuit with no losses has zero impedance at resonance. If there were a resistance in series with a series resonant circuit, then the impedance of the circuit would be equal to the series 'R'. For example, a zero impedance (which is the same as a short circuit) in series with a 100kΩ resistor would give a total impedance at resonance of 100kΩ.

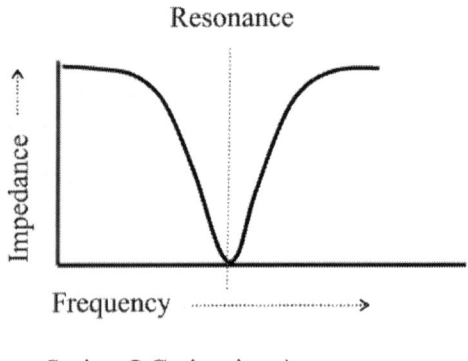

Series LC circuit - At resonance impedance is Minimum

Figure 7-19

THE EFFECT OF LOSSES ON THE 'Q' OF TUNED CIRCUITS

We know that, *in the real world,* we cannot have parallel or series tuned circuits with no losses at all. The effect of losses on a tuned circuit is to flatten out the impedance curves shown in figure 8-19. The higher the Q the sharper the curve.

Tuned circuits can be used to make filters and select radio stations. Sometimes we want the curve (often referred to as the selectivity curve) to be sharp (a high 'Q'). Other times we want it to be flat and not as selective so we may add losses in the form of resistance to flatten the curve out.

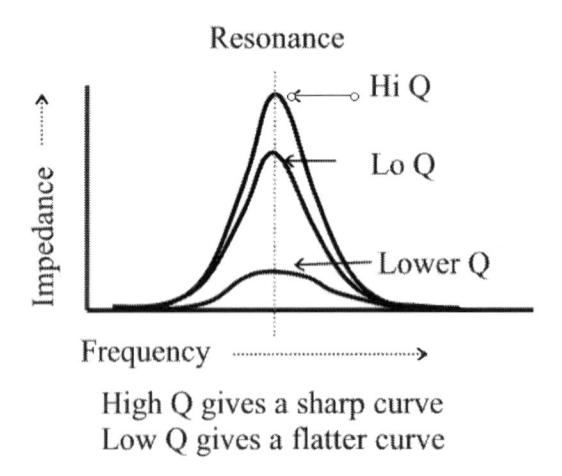

High Q gives a sharp curve
Low Q gives a flatter curve

Figure 8-19

Figure 8-19 shows the curves of a parallel tuned circuit with three values of 'Q'. For a series resonant circuit, the graph would be inverted.

'Q' OF A LONE INDUCTOR OR CAPACITOR

Just a reminder that tuned circuits are not the only thing that has a 'Q' (quality factor). A perfect (also called pure) inductor has no losses and therefore does not dissipate any power. Such an inductor would have an infinite 'Q'. The same applies to a capacitor with no losses; its 'Q' would be infinite and it also would not dissipate any power.

In reality, an inductor has losses due to the resistance of its winding and some core losses, hysteresis loss and perhaps some eddy currents if it is a choke. A capacitor has mostly dielectric loss and some loss due to the resistance in its plates. Losses in an inductor or capacitor can be represented by a resistor in series with them. The equation for finding the 'Q' of a lone inductor or capacitor is simply to divide the reactance by the resistance. You must remember $Q=X_L/R$.

'Q' of a lone inductor or capacitor:

$$Q = X_L / R \quad \text{or} \quad Q = X_C / R \qquad \text{Equation 3-19}$$

Adding resistance to a capacitor, inductor or any resonant circuit, either in series or parallel reduces its Q. This is because resistance takes energy from the circuit. Recall that resistance is the only electrical property that dissipates heat.

The equation used to find the Q of RLC circuits is $Q=X_L/R$. Why? The losses in an inductor compared to anything else is relatively so high that other losses can be ignored. The losses of the L swamp or mask all other losses. So we just use $Q=X_L/R$ for the Q of any RLC circuit.

Flywheel Effect

A steamroller has a Flywheel shown here as the smaller wheel.

What is a flywheel? In mechanics, a flywheel is a wheel with a large mass (weight). If you get the flywheel turning, it will continue to turn for a long time because of its mass. A flywheel is a means of storing energy and releasing it later.

The science fiction writer Isaac Asimov predicted that one-day electric power stations would use gigantic flywheels weighing hundreds of thousands of tonnes. Perhaps floating on mercury or using magnetic levitation. Power stations have a demand on the generators that varies enormously. They cannot, particularly with fossil fuel plants, turn the furnaces up and down at a whim. Imagine though if you had a gigantic flywheel, perhaps floating on mercury or on a magnetic field to avoid friction. All of the energy created by the power station is used to make the enormous flywheel move. You would be storing a huge amount of energy in this flywheel. It is so massive that it would continue to rotate for months left untouched. When we wanted energy from the flywheel, we could make contact with it and have it drive a generator for us. So, even though the output of the power station may be irregular, the power available from the flywheel would be regular because of the massive amount of stored energy.

Now a resonant circuit exhibits the same properties of a flywheel. This effect of resonant circuits is called the flywheel effect. How it works.

Suppose we want a sinewave to be produced. We know that a resonant circuit will generate a sine wave if it is given energy. We know that the sine wave output from a resonant circuit will dampen if we don't keep on providing it with energy.

Suppose we have pulses of energy, very short pulses and assume we have 1 million of them per second. Let's use these pulses to start a resonant circuit oscillating. Let's make the resonant frequency of the tuned circuit 3 megahertz. This means that the resonant circuit would get a pulse of energy and produce 3 cycles of sine wave output. Now a tuned circuit of reasonable Q will be able to run fine for 3 cycles. Then our next pulse comes along and away the tuned circuit goes again for another three cycles. The output of the resonant circuit is a sine wave at 3MHz. The input is some 'cruddy' pulses at 1MHz.

So, the resonant circuit is producing a 'flywheel effect' for us.

There are many situations in electronics where we want to convert pulses to sine waves or double or triple the frequency of a signal. One way of doing this is the flywheel effect of a resonant circuit.

APPLICATIONS RESONANT CIRCUITS

There are hundreds of applications for tuned circuits and it will all come together as we continue the course. However, I feel compelled to give you a preview.

Application using a parallel resonant circuit:

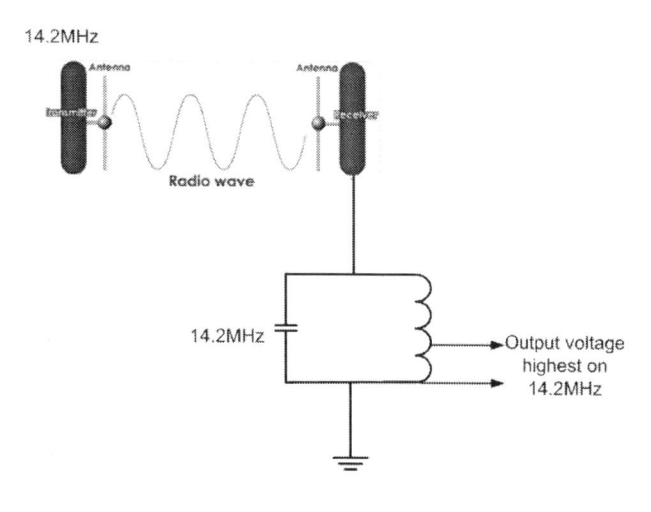

Figure 9-19

The diagram of figure 9-19 shows an antenna connected to a parallel tuned circuit that is in turn attached to the ground.

Now, remember a parallel tuned circuit has a very high impedance at one frequency only, its resonant frequency.

All of the radio waves passing by the antenna will induce a minuscule voltage into the antenna. Let's say our resonant circuit is tuned to 14.2 MHz. That is the radio signal that we want to select and reject all others.

A radio wave on 15MHz passing by the antenna will induce a voltage into it. This voltage will cause a current to flow down the antenna cable and through the tuned circuit to earth. At 15MHz, the inductor and capacitor are not resonant and the impedance of the parallel circuit will be very low. E=IR or if you like E=IZ (Z=impedance). The voltage created across the LC parallel circuit will be very low as its impedance low. The same story will go for all other radio signals that induce a voltage into the antenna except at 10.2MHz. At 10.2MHz, the LC circuit is parallel resonant and will be very high impedance. The small current through the parallel resonant circuit will produce a significant voltage across it compared to all the other radio signals.

A voltage will appear at the output terminals of the signal that the parallel tuned circuit is tuned to. So here we have the basic method of selecting the desired radio signal from the many that are present at the antenna

An application for a series resonant circuit:

Recall that at resonance a series tuned circuit has extremely low impedance. If you like, you can think of it as zero impedance or even a short circuit. Zero impedance is like a piece of conductor. If you place a conductor across the back of your TV on the terminals where the signal comes into the set, do you think you would get much of a television picture?

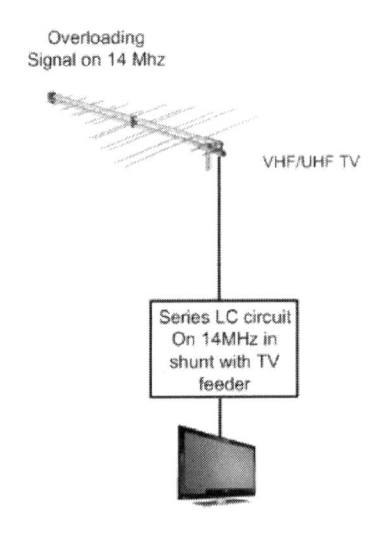

Figure 10-19

I hope you said 'no'. A short circuit on a TV antenna would stop all signals from entering the TV set and you would get no signal.

A series resonant circuit has zero (extremely low) impedance at one frequency only, its resonant frequency.

In figure 10-19, we have interference from a 14MHz transmitter getting into a nearby television set. This type of interference is called receiver overload. The problem is that the TV can't reject the 14MHz signal, as it is so strong and so close.

We have placed a filter in the box shown on the TV cable. In that box, there is a series LC circuit which is resonant on 14MHz.

A 14MHz signal trying to get to the TV will 'see' a short circuit and will be stopped. However, all other wanted TV signals will be unaffected.

A final word on Q and something to remember because it seems to puzzle many; in radio circuits:

Low Q means broad bandwidth or poorer selectivity
High Q means narrow bandwidth or better selectivity

Adding losses to any antenna, or tuned circuit causes its Q to decrease. These losses are always through 'resistance' of some sort. The resistance of the wire used in a helically wound vertical, ground losses under an antenna and the resistance of the wire used for an inductor in a tuned circuit are good examples. If the wrong core material is used for and RF inductor or a balun, then the losses can be very high and the Q low. Resistance is sometimes deliberately added to tuned circuits to lower their Q and increase their bandwidth.

Describing a selectivity curve

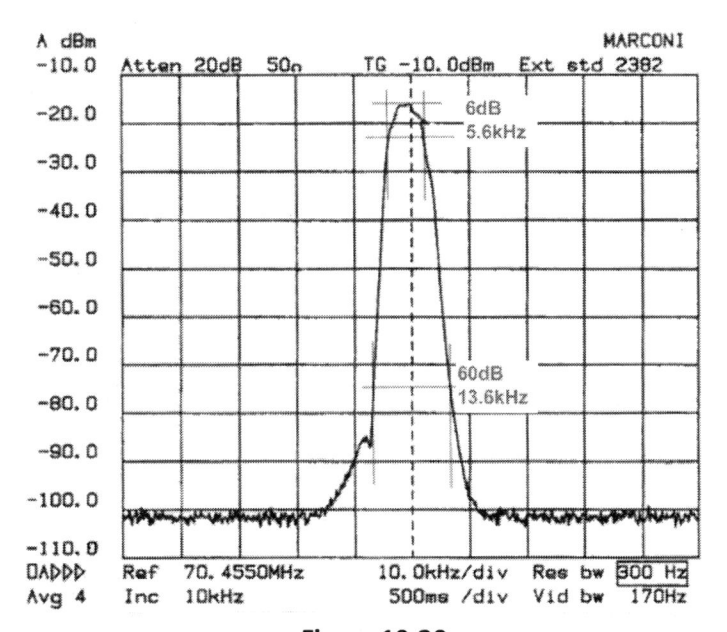

Figure 10-20

It is somewhat difficult to describe an entire selectivity curve using words. One way that is used is to state the curves bandwidth at the 6dB and 60dB points. This method is used to describe the selectivity of a receiver or a filter. Figure 10-20 shows a bandpass on a spectrum analyser. The bandwidth at 6dB is 5.6kHz and at 60dB it is 13.6kHz. This looks like a good double sideband filter or receiver front end selectivity to me.

20. POWER SUPPLIES

THE RECTIFIER DIODE

A rectifier is another name for a diode. I am not going to explain the internal operation of a rectifier now as we will be covering that in full later. I will explain what a rectifier diode does and ask you to accept that internal operation for now.

The schematic symbol of a rectifier is shown in figure 1-20. A rectifier (diode) has two terminals, a cathode and an anode as shown. If a negative voltage is applied to the cathode and a positive to the anode, the diode is said to be forward biased and it will conduct. If it is reverse biased, it will not conduct. So a diode will only conduct current in one direction i.e. from the cathode to anode.

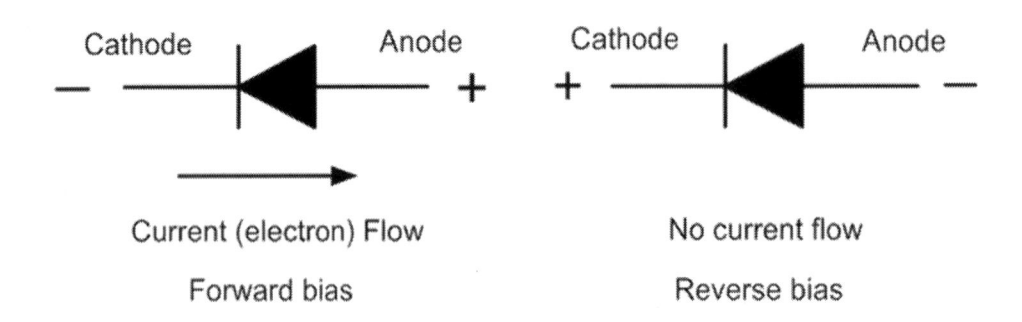

FIGURE 1-20.

PEAK INVERSE VOLTAGE (PIV)

Though we say a diode will not conduct in the reverse direction, there are limits to the reverse electrical pressure that you can apply. Manufacturers of diodes specify a 'peak inverse voltage' (PIV) for each particular diode. The PIV is the maximum reverse bias voltage that the manufacturer guarantees that the diode will withstand. In practice, the PIV is well below the pressure that will cause the diode to breakdown and conduct. However, except in special circumstances, the PIV should never be exceeded.

ELECTRON FLOW VERSUS CONVENTIONAL CURRENT FLOW

Some textbooks show current flow from anode to the cathode. This is called the conventional direction of current flow and is a hangover from the days when the current was thought to flow from positive to negative. Blame Benjamin Franklin, he arbitrarily set the flow of an electric current from positive to negative. That was before J.J. Thompson discovered the electron. Electrical engineers and electricians seem to hang on to this conventional direction of current flow. Electrons flow from negative to positive, from the cathode to the anode. In radio, electronics and communications, 'electron flow' is almost universally used. Electron flow is of course from negative to positive.

The vast majority of 'modern' references agree on the direction of current flow. Even textbooks that use conventional flow acknowledge that this direction (positive to negative) is just a convention. There is no harm in using conventional flow if this is what you prefer. However, for circuit descriptions in this book, electron flow is used.

To help remember which way current flows through a diode look at the symbol - it has an arrow and current flows against the arrow. This will help you not only with the diode but other semiconductor devices to be discussed later.

Let's have a look at the operation of a single diode rectifier if we apply an alternating voltage source to it. Since the diode will only conduct when the potential on the cathode is negative with respect to the anode, only half of the sinewave of alternating voltage will cause current to flow in the circuit of figure 2-20.

Half wave rectification

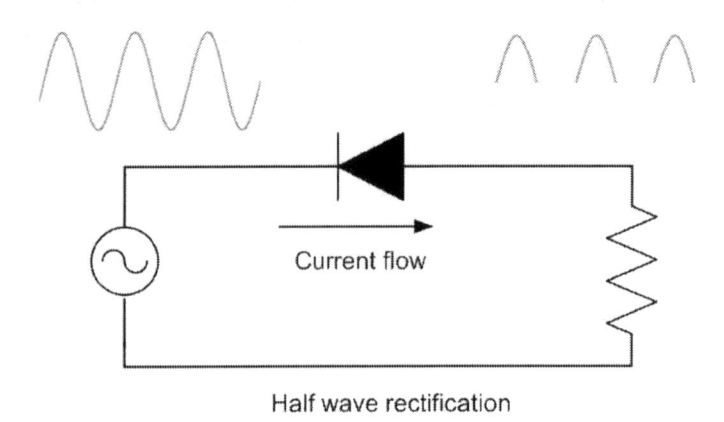

Figure 2-20.

The current flow in the circuit of figure 2-20 is DC, but it is pulsating DC. The diode refuses to conduct for a complete half cycle of input voltage. The diode conducts (is forward biased) during the half cycle where the cathode is negative and the anode is positive.

This is called half-wave rectification. This terminology *rectification* has always intrigued me a little. To rectify something means to 'make it right,' to 'fix it up!' This is where the term comes from. The radio pioneers considered converting AC to DC as fixing up the AC voltage, hence the term rectification. I often wonder if Edison had anything to do with the terminology. Edison was strongly opposed to AC power distribution. During his feud with Tesla. Tesla and Edison were bitter rivals. It was Tesla who invented and implemented the AC power and distribution system we use today. Edison even electrocuted animals in public performances to demonstrate the dangers of AC. As it turned out, Tesla's AC system was more efficient and adopted. The first power generating plant was constructed at Niagara Falls. Let me make the point that AC does not need 'fixing up', but we are left with this term 'rectification', which I find rather amusing. In many cases, we actually need to convert DC to AC, so I suppose we should call this un-rectification!

So, in trying to convert AC to DC, all we have managed to do so far is convert AC to half-wave pulsating DC (PDC).

Diodes make it very easy to produce full wave rectification. I am going to start with the most common method, the bridge rectifier. A bridge rectifier has four diodes as shown in figure 3-20. Rather than apply an AC voltage source, I have shown how the bridge network will operate when two different polarities are applied to it, which in essence is simulating AC.

As you can see from figure 3-20, two of the bridge diodes will always conduct because they have a negative voltage applied to their cathodes with respect to their anodes.

Notice how current flows against the arrow. More importantly, notice how current always flows through the load resistor in the *same direction*, regardless of the polarity applied to the bridge.

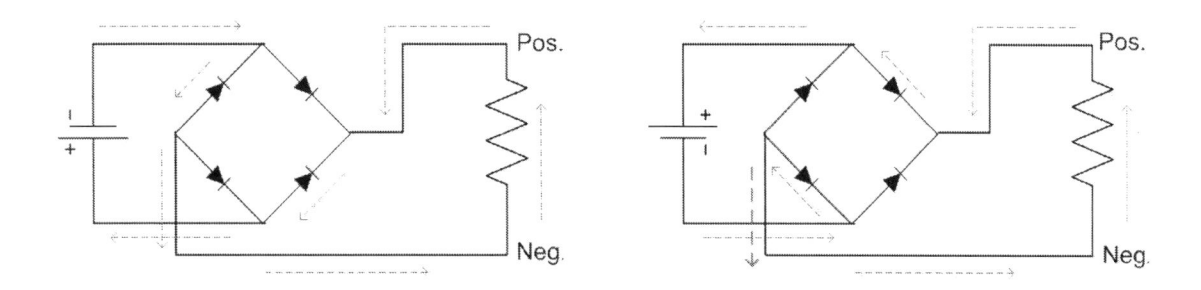

Figure 3-20.

We have created full wave rectification.

You simply must commit the diagram of a bridge rectifier to memory and be careful, because the diodes can all be reversed and you still have a working bridge. I like to think of the four diodes as positions on a clock face. So, we have diodes at 2, 4, 8 and 10 o'clock. The diodes at 2 and 8 o'clock can be reversed, provided the diodes at 10 and 4 o'clock are reversed. You also need to be able to determine the polarity of the output, or the direction of the current through the load. A way to do this easily is to pretend you are an electron out for a stroll. You can go wherever you want provided you go *into* a cathode and you will seek to get to the positive terminal of the battery or supply. You can then trace out the current flow mentally or on paper and determine if the bridge is drawn correctly.

The bridge I have drawn is a diamond or kite shape, but it can also be drawn as a square. Tracing the current to determine if current will flow through the load in the same direction regardless of polarity, as I have done, is your surest way of determining if the bridge circuit is drawn correctly.

Using a centre tapped transformer for full wave rectification.

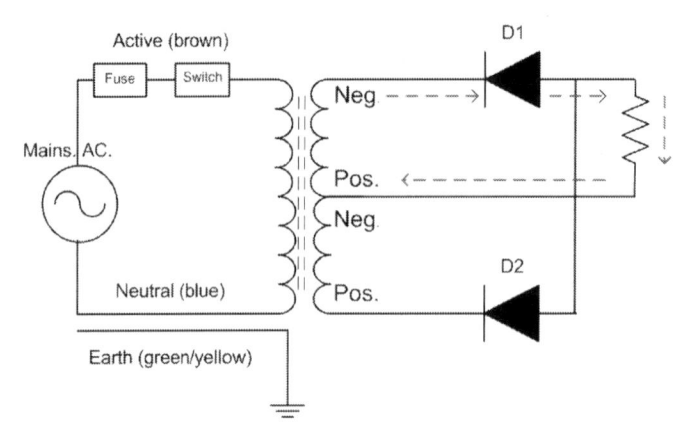

Figure 4-20

A less conventional method (today) of obtaining full wave rectification is to use a centre tapped transformer and only two diodes. I have added some extra detail to the schematic in figure 4-20, which I will fully explain as we go along. A bridge rectifier would *not* use a centre tapped transformer as shown.

I have shown the polarities on the transformer for one-half cycle of AC input from the mains supply. As you can see, the top diode has a negative potential on its cathode so it will conduct with the path of the current as shown. During the next half cycle, all of the polarities will be the opposite of that shown and the bottom diode will conduct. In either case, the current will be the same through the load. So, full wave rectification is achieved again.

Can you see that this circuit is two half wave rectifiers? Each diode rectifies one-half of the AC sinewave giving full wave rectification. For this reason, the secondary winding has twice the turns compared to transformer used for a full wave bridge rectifier. Twice as many turns means twice the copper on the secondary, more cost and more weight.

This circuit was once very popular, as it only required two rectifier diodes. This circuit is just as effective as the bridge rectifier. Which one to use comes down to cost and convenience. A centre-tapped transformer is more expensive than one without a centre tap. Electron tubes (to be discussed) can also be used as diodes. Electron tubes are relatively expensive compared to semiconductor diodes. So if the power supply were using vacuum tubes, this approach would be the most cost effective, as it requires only two diodes and that outweighs the cost of the centre tapped transformer.

Except in very high power applications, the modern trend is to use semiconductor diodes, so this circuit (figure 4-20) is not often used today. Still, you are expected to know it and it does have a useful purpose when using junk box parts. Note also that the total secondary voltage is double that required for the bridge rectifier circuit, as only one-half of the secondary is being used on alternate half cycles.

CAPACITIVE FILTERING - WITH A HALF WAVE RECTIFIER

So far the output of the power supplies has not been smooth DC, which, perhaps with the exception of a simple car battery charger, is our requirement. Adding a capacitor to the output of the power supply will smooth the pulsating DC a great deal. The circuit of figure 5-20 shows half wave rectified DC fed to a capacitor.

Capacitive Filtering

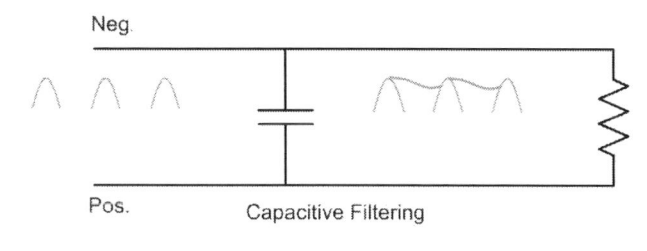

Figure 5-20

If pulsating DC is fed to a capacitor as shown, the capacitor will charge to the peak value of the pulses. Without any load, the capacitor will charge to the peak value and stay there. I have shown the effects of a load. Between the pulses, the capacitor will discharge through the load (not shown) and the voltage will drop a little. I have left the pulses under the DC output for illustration purposes only. The capacitor is a high value electrolytic – more on this shortly.

The DC output, in this case, is not pure. Pure DC would be a flat line indicating no voltage variation. The DC rises and falls a little, the rate of rise and fall being related to the frequency of the DC pulses. The frequency of the AC from the mains is 50Hz. For every cycle of AC rectified by a half wave rectifier we get one pulse of DC, so the pulses have a frequency of 50Hz as well. We can, therefore, conclude that the *ripple frequency* of a half wave rectifier is the same as the mains frequency i.e. 50Hz.

Just to emphasise that a little, the DC output is not flat, it has a ripple on it, a ripple is a periodic rise and fall. The rate of rise and fall of the DC output is called the ripple frequency and is equal to the mains frequency, which in Australia is 50Hz.

CAPACITIVE FILTERING - WITH A FULL WAVE RECTIFIER

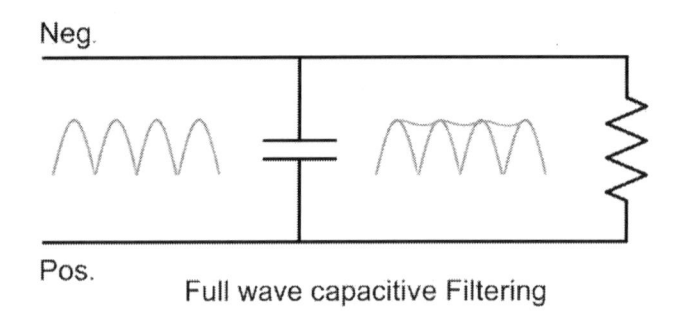

Figure 6-20.

The ripple frequency of a full wave rectifier is 100Hz. Full wave rectification with filtering is much better than a half-wave. However, with capacitive filtering alone there is still some ripple. The frequency of the ripple will always be the same, equal to the mains frequency for half wave rectification and twice the mains frequency for full wave rectification.

A COMPLETE CIRCUIT OF A PRACTICAL POWER SUPPLY.

Figure 7-20 is the circuit of a full wave bridge rectifier showing the transformer, the mains connection, a switch, fuse(s) and a capacitor for a filter.

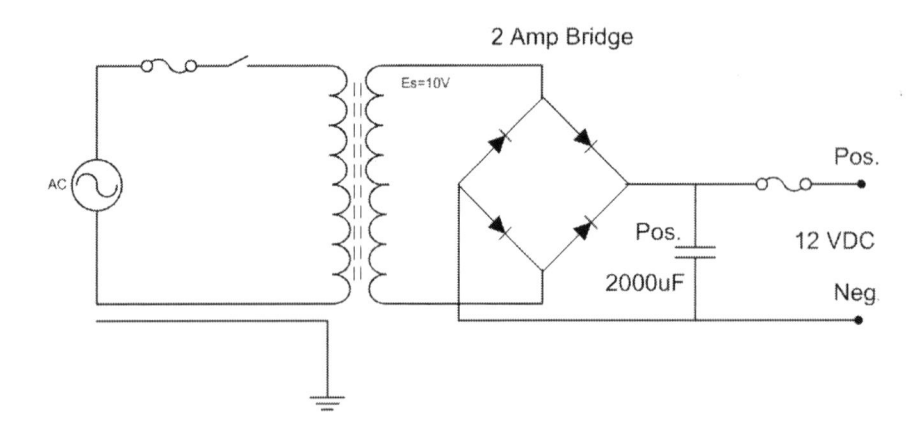

Figure 7-20.

Most of this circuit has been explained. I will point out some important features, as such knowledge applies to subjects other than power supplies.

Notice how the secondary voltage E_s is shown as 10 Volts. This is the RMS value. The output of the power supply is the peak value, RMS x 1.414, or near enough to 12 Volts. There will be a small voltage drop across the two conducting diodes. A silicon rectifier diode will have a forward voltage drop of 600-700mV. The negative side (also called the negative rail) of the power supply shows a chassis connection. This just means that if the power supply is in a metal box, the negative terminal is connected to the box. This is not done for safety purposes. It's for ease of construction, just in the same way that the negative terminal of a car battery is connected to the chassis (body) of the car. Of course, the positive output could be connected to the chassis. This is a construction issue and will depend on the type of semiconductor devices used. However, it is more common for the negative to be connected to the chassis.

I have shown a switch in the active lead from the mains. The switch must always be in the active lead. The earth lead from the mains or power point is connected to the chassis of the power supply. We will have much more to do with safety issues later.

I have shown two fuses. There really only needs to be one. The important thing about fuses is to remember that a fuse operates on current. Should something go wrong in the power supply (or outside the power supply) excessive current will be drawn and the fuse will melt, opening the circuit.

What may not be obvious to you is the difference in choice of fuses - in the primary of the transformer and the output of the power supply. Remember the power in a transformer is the same in the primary and secondary, disregarding losses. However, the voltage and current are not the same. Their product (the power) is the same. Fuses work on current. Say this power supply is 12 Volts at 2 amps' maximum. The fuse in the secondary circuit will be around 2 amps. It will melt if more than two amps are drawn from the supply and stop the supply or the load from being damaged further by excessive current.

Think about the fuse in the primary circuit. It would not be anything like 2 amps. If it were 2 amps, then 480 Watts would have to be drawn from the mains before it would melt. I will leave the calculation of the primary fuse to you. As a hint how much power can be drawn from a 12 Volt 2-amp power supply?

Let's get a bit more serious about power supply filtering. What you are about to read may in some areas go beyond what is required for the examination on power supplies. Though, where concepts are needed to be known, perhaps not in power supplies but in other subjects, I have included the information pertinent to power supplies knowing that this is required knowledge elsewhere.

CAPACITIVE INPUT FILTERING

Capacitance is that property of an AC circuit which opposes changes of voltage. If a large capacitor is connected across the output of the rectifier circuit, i.e. in parallel with the load, the capacitor will smooth out the pulsating DC to DC with a small ripple voltage. The filter capacitor does this by taking energy from the circuit and storing this energy in its electric field. If the voltage across the load tends to fall, the capacitor will discharge some of its energy back into the circuit, which smooths out the voltage. Because a high value of capacitance is required (1000μF to 10,000μF or more), the capacitor has to be an electrolytic type, as this is the only type of capacitor which can provide such a high value of capacitance. Capacitive filtering alone can be used for power supplies which need to deliver only a few Amperes of current.

One disadvantage of capacitive input filtering is that of surge current. When the power supply is turned off the capacitor is discharged. When the power supply is first turned on, the capacitor will charge *rapidly* and draw a large current from the transformer and rectifier circuit. This surge current can be so high that it may exceed the current rating of the rectifiers and damage it, or blow the fuse, whichever comes first. This is not so much an operational problem of the power supply, but a design problem.

To ensure the filter capacitor(s) in a power supply discharges when a power supply is turned off, a high-value resistor is often connected across the filter capacitor. A high resistance, say 10KΩ, in parallel with the filter capacitor will have no effect on the operation of the power supply and provide a path for the capacitor to discharge through. This resistance is known as a *bleeder resistor*. You may have noticed on some power supplies which have a light to indicate that the supply is turned on, that this light will slowly dim and eventually go out some time after the supply is turned off. This is the filter capacitor discharging. Sometimes the "on indicator lamp" acts as *a bleeder resistor* and discharges the filter capacitor.

Best is a high-value *bleeder resistor* connected across the output of the power supply to discharge the filter capacitor(s) when the supply is switched OFF.

INDUCTIVE INPUT FILTERING

Inductance is that property of an AC circuit which opposes changes of current. If we are trying to reduce ripple from a power supply, it should be no surprise then, that an inductance can be used to filter the DC output.

When an inductance is connected in series with a rectifier, a filtering or smoothing action results. An inductor used in this manner is always iron core to obtain the high level of inductance required. Inductors used in power supplies for filtering (and some other applications) are referred to as chokes.

Pulses of current through the choke build up a magnetic field around it, taking energy from the circuit to produce the field. As a pulse of current tries to decrease in amplitude, the magnetic field collapses and returns energy to the circuit, thereby tending to hold the current constant. If the inductance is the first component in the circuit after the rectifiers, then this type of filtering circuit is called *inductive input filtering*. It is not practical to use inductive input filtering alone in a power supply. Inductive filtering is always used in conjunction with capacitive filtering. A disadvantage of inductive input filtering is that there is voltage dropped across the internal resistance of the choke (the resistance of the wire making up the choke). The voltage dropped across the choke reduces the output voltage of the power supply. In practical terms, this means that the secondary voltage of the transformer has to be a little higher in voltage output to compensate for this.

SWINGING CHOKES

You will recall from our study of inductance that the purpose of the iron core is to increase the value of the inductance. The magnetic iron core does this by concentrating the magnetic lines of force. Usually, the magnetic iron core of a choke has no air gap to prevent the core from magnetically saturating. In the swinging choke, the magnetic iron core has little or no air gap. This means that the magnetic iron core will begin to saturate at average current values for the power supply. When low current flows through a swinging choke, it has a high inductance and filters effectively. With high current the magnetic core saturates and it has less choking effect. Thus, with a light load and little current, the swinging choke has a high reactance and develops a significant voltage drop across it. When the load increases, the choke saturates and has less reactance and consequently less voltage drop across it. This means the voltage output from the power supply tends to remain constant under varying loads, improving its regulation. Typically, a swinging choke will change (swing) from 5H for a heavy load (high current) to 20H for a light load.

COMBINATIONS OF CAPACITANCE AND INDUCTANCE

For better regulation, combinations of chokes and filter capacitors are used. It is not necessary for you to remember these combinations (for exam purposes). It is important to bear in mind that whatever filter type is used, it will get its name from the first component seen, looking from the rectifier towards the filter.

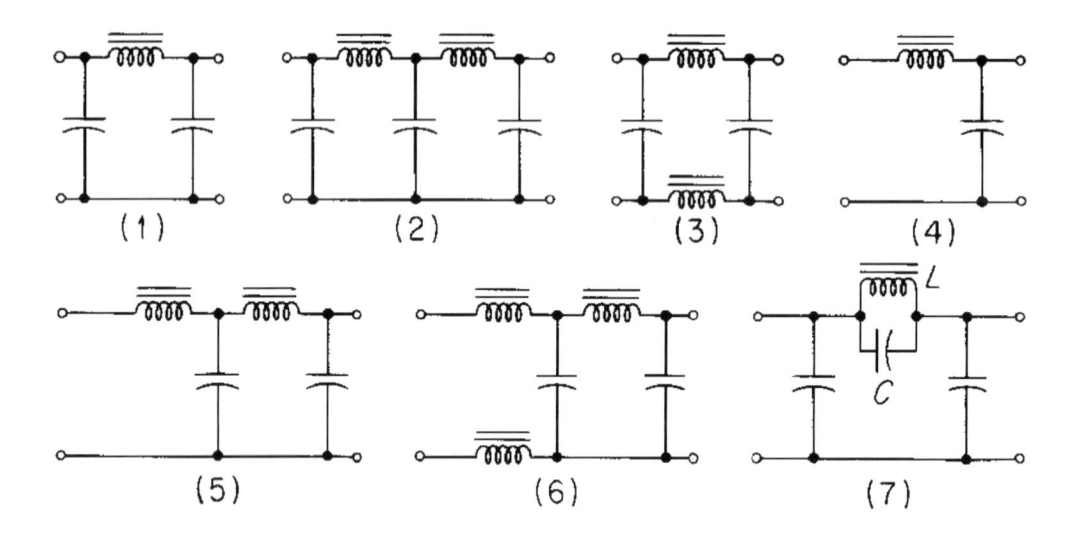

Figure 8-20.

If the first component is an inductor(s), then it is an *inductive input filter*. Likewise, if the first component is a capacitor, it is called a *capacitive input filter*. You may be asked the advantages and disadvantages of each type.

(1) Pi - Capacitive input filter.
(2) Capacitive input filter.
(3) Capacitive input filter.
(4) Inductive input filter.
(5) Inductive input filter.
(6) Inductive input filter.
(7) Capacitive input filter.

All of the power supply filter circuits shown in figure 8-20 use the principles we have discussed (except for circuit 8-20(7)). The purpose of a power supply is to provide smooth DC output, that is, a constant voltage with varying load conditions. In all circuits, each parallel capacitor has a capacitance that opposes changes in voltage and the inductance of the inductors oppose changes in current.

Figure 8-20(7) uses another principle, which you don't often see in power supplies. However I will discuss it as the principle is one we need to know for other purposes. In circuit 8-20(7) the parallel inductor and capacitor are resonant at the ripple frequency. A parallel tuned circuit has very high impedance only at its resonant frequency. The parallel LC is resonant at the ripple frequency. Any 100Hz ripple (assuming full wave rectification) will be blocked significantly from passing through this LC combination. An ingenious technique.

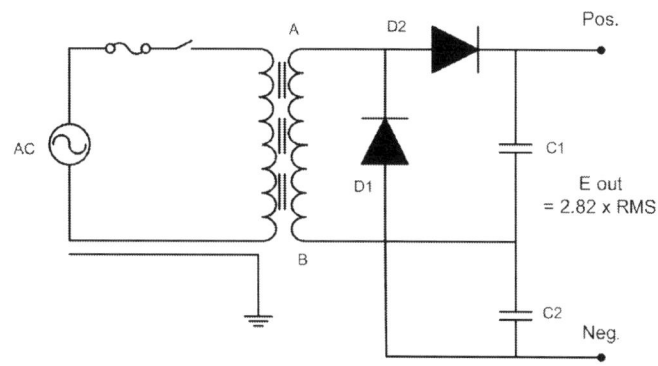

Figure 9-20

THE VOLTAGE DOUBLER (FULL WAVE)

The diagram in figure 9-20 is that of a full-wave voltage doubler. With a standard full-wave power supply (e.g. bridge) the output voltage is approximately the peak value of the RMS secondary voltage, or RMS x 1.414.

When lines are crossed as shown, they are not connected.

When the AC voltage makes point 'A' negative and 'B' positive, D1 will conduct and charge C2. When the AC voltage makes 'B' negative and 'A' positive, then D2 will conduct and charge C1. The capacitors C1 and C2 are in series and so the voltage at the output is the sum of the voltage on C1 and C2, or 2.82 times the RMS secondary voltage.

The voltage doubler can't deliver very high currents. However, it is an excellent method of obtaining the higher voltages required for an electron tube.

EQUALISING RESISTORS AND TRANSIENT PROTECTION CAPACITORS

In very high voltage power supply circuits, many rectifier diodes are often connected in series to increase the peak inverse voltage (PIV) rating. If a rectifier has a PIV of 400 Volts, then 'theoretically' two such rectifiers connected in series should have a PIV of 800 Volts. More diodes can be added in series for even higher PIV.

However, for proper series rectifier operation it is important that the PIV be divided equally amongst the individual diodes. If it is not done, one or more diodes in the string may be subjected to a greater PIV than its maximum rating and as a result may be destroyed. As most failures of this type lead to the diode junction going short circuit, the PIV of the remaining diodes in the string is raised, making each diode subject to a greater value of PIV. Failure of a single diode in the series (stack) can, therefore, lead to a 'domino effect', which will destroy the remaining diodes.

Forced voltage distribution in the stack is necessary when the diodes vary appreciably in reverse resistance.

The resistors in figure 10-20 provide PIV equalisation across the two diodes.

The reverse resistance of the diodes will be very high, in the order of megohms. Now, 470kΩ in parallel with megohms is roughly still 470kΩ. So irrespective of the diodes reverse resistance the PIV is distributed roughly equally across each diode.

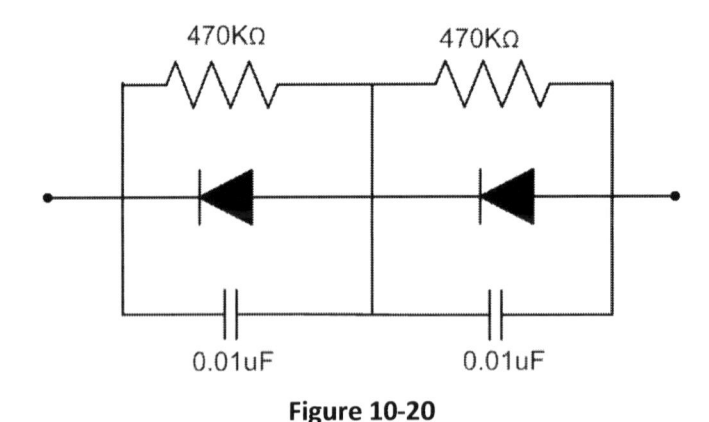

Figure 10-20

TRANSIENT PROTECTION CAPACITORS

Power supplies, high voltage ones, in particular, are prone to producing high voltage spikes called *transients*. These transients are caused primarily because of inductive effects of the circuit (the secondary of the transformer). These very short duration high voltage transients can destroy the rectifier diodes. The capacitors shown across the diode in figure 10-20 above will absorb or smooth out any transients. Recall that capacitance is that property that opposes voltage change.
You need to be able to identify the circuit of figure 10-20 and it purpose.

In summary, equalising resistors distribute the applied voltage almost equally across a string of diodes when they are reverse biased. This ensures that each diode, irrespective of its reverse resistance, will share the same amount of reverse voltage drop. Transient protection capacitors absorb short duration high voltage transients caused by inductive effects in the circuit.

ELECTRONIC POWER SUPPLY REGULATION

For power supplies that require excellent regulation, electronic regulation can be added to the filter network. This consists of either an electronic circuit comprised of discrete components or, far more commonly these days, an integrated circuit that contains all the necessary electronics.

There are three basic types of regulation:

Series Regulation - where the regulator is in series with the current path.
Shunt Regulation - where the regulator is placed in parallel or 'shunted' across the power supply output.
Switching regulators – These are the most efficient – that is they dissipate the least amount of wasted power.

These days' series regulation is almost always used (and the type you need to know for the exam). The disadvantage with the shunt regulator is that the power dissipated by the regulator circuit is much higher. This translates to lower power supply efficiency and more heat to get rid of in the regulator circuit.

Figure 11-20 is a schematic diagram of a power supply that uses a three-terminal integrated voltage regulator; this is a series regulator. More efficient than a shunt regulator.

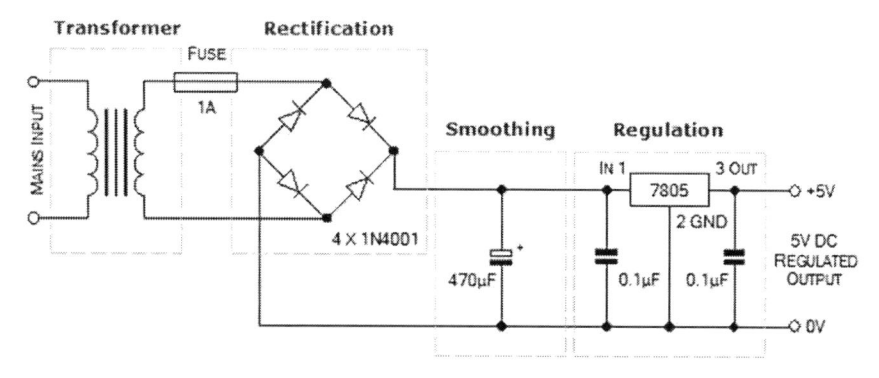

Figure 11-20.

The 7805 voltage Regulator is just one of many off the shelf integrated voltage regulators.

The IC 7805 is a 5V voltage Regulator that regulates the voltage output to 5V. It comes with a provision to add a heatsink. The maximum value for input to the voltage regulator is 35V. It can provide a constant steady voltage flow of 5V for higher voltage input till the threshold limit of 35V. If the voltage is near to 7.5V, then it does not produce any heat and hence no need for a heatsink. If the voltage to high, then excessive power is dissipated as heat from the 7805.

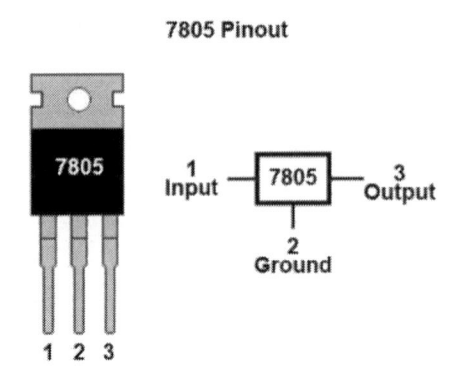

Figure 12-20.

IC 7805 is a series of 78XX voltage regulators. It's a standard, from the name the last two digits 05 denotes the amount of voltage that it regulates. Hence, a 7805 would regulate 5v and 7806 would regulate 6V and so on.

The shunt regulator method shown in figure 13-20 is only used for very low power circuits as shunt regulation is the least efficient of all types of regulation.

A good practical example of a shunt regulator and the only one you need to know is that of a Zener diode along with its current limiting resistance.

We have not discussed the operation of a Zener diode yet – we will do so in the semiconductors chapter. For now, a Zener diode is one that is *meant to be operated with reverse bias and* beyond its PIV or breakdown voltage. Usually, a diode would be destroyed if operated beyond breakdown. The series resistance R1 (refer to figure 13-20) limits the reverse current so that the Zener diode is not destroyed.

Under such conditions, the Zener diode will have a constant voltage drop across it. Zener's come in a range of voltages sometimes a bit odd like 9.1 Volts. Zener regulators are used a lot inside transceivers to regulate the voltage to the active devices. A typical example is an oscillator which is low power consumption and must have a stable voltage.

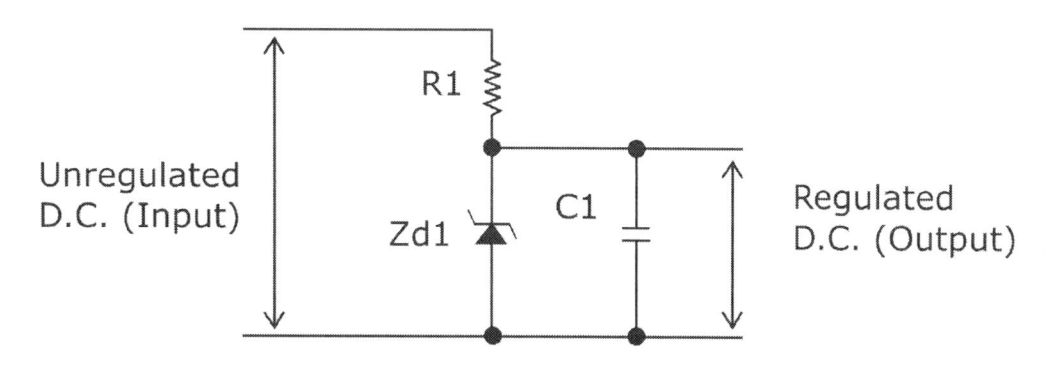

A simple shunt regualtor using a zener diode

Figure 13-20.

Assuming this Zener is 9.1V then unregulated as input voltage varies from say 10 to 15 Volts, the voltage across the Zener will remain (almost) constant at 9.1 Volts. However, the voltage across R1 will vary. The sum of the Zener voltage (9.1V) and the voltage across R_s will equal the unregulated input voltage as this is a series circuit.

The Zener circuit draws maximum current when there is no load as shown now in 14-20. This is why shunt regulators are so inefficient. They are dissipating maximum power when doing no work. When a load is applied some of the current going through the Zener will now go through the load reducing the heat dissipated by the Zener. The sum of the currents through the load and the Zener is constant. A Zener shunt regulator circuit can be converted into a series regulator circuit by the addition of *pass transistors* (to be covered).

A CONSTANT CURRENT SOURCE

This simple circuit (figure 14-20) is a favourite of mine, as I have found so many uses for it. First, allow me to explain what constant current means. The purpose of most power supplies is to maintain a constant regulated DC voltage output. For example, a radio transceiver may require around 13 Volts DC. The transceiver when in use will draw whatever current it requires from the power supply, so the current varies. The purpose of the power supply is to maintain the constant voltage output of 13 Volts, regardless of the changing current demands of the transceiver.

Now, there are situations where you *don't care about the voltage output of a power supply*. What you care about is current. This may seem a little strange because usually, we apply a voltage to a load and the load determines the current. Suppose we want to force a particular current through the load irrespective of what the load wants. A classic example is charging many NiCd batteries, which are very common in portable radio communications equipment. NiCd batteries should be charged at one tenth of their amp-hour rating for 15 hours (if fully discharged). So a 500 mA hour cell or battery should be charged at 500/10 = 50mA for 15 hours. This cannot be done with a constant voltage output power supply. We need a constant current supply. A three terminal regulator can be used to do this for a cost of less than $5. Commercial devices for charging NiCd'S (including cell phone batteries) are frankly a rip-off as most of the time they are just a plastic case with the schematic of figure 14-20 in it.

I have made many of these chargers operate from the car cigarette lighter. You need an input voltage of at least 2 Volts (preferably 2.5V) higher than the battery you are charging.

The resistor is in series with the load and the three terminal regulator is producing a constant voltage across it. The resistor will, therefore, have a constant current through it. Since it is in series with the load, the load will have the same current as the resistor.

All you need to do is buy a regulator, which will handle the current you want and calculate the value of the resistor and its minimum wattage.

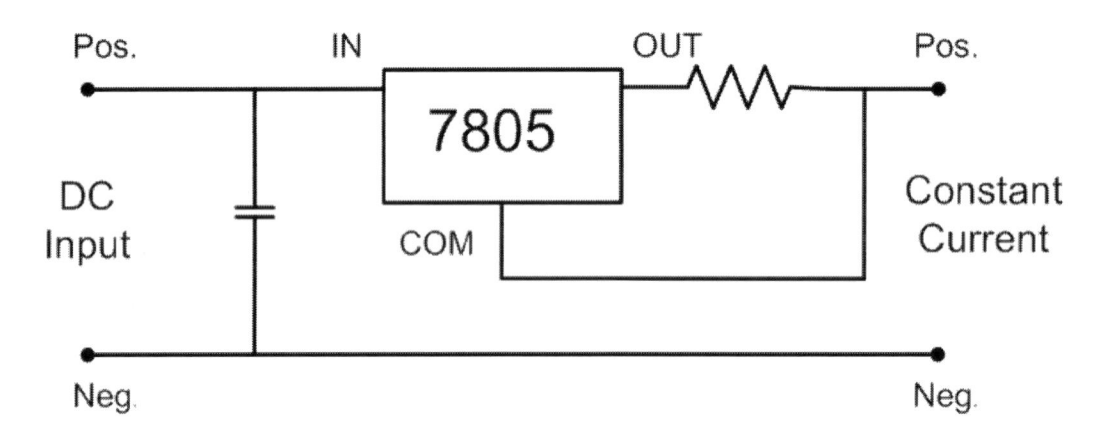

Figure 14-20.

Suppose we had a 7.5 V NiCd battery that has a capacity of 500 mA/hours. We need a minimum DC input voltage to the circuit in figure 14-20 of 7.5 + 2 = 9.5V. Now we need to work out the value of the resistance in Ohms. Let's say we are using a 5 Volt three terminal regulator. Our battery requires 500/10 = 50mA. R = E/I = 5/0.05 = 100Ω. That's it!

Of course, you need to work out the wattage of the resistor; you can do that can't you?

POWER SUPPLIES – PASS TRANSISTORS

The Zener diode cannot handle a significant amount of current, usually no more than a few hundred milliamps. Also, a Zener diode is not a perfect regulator. There is a way of improving the current handling capabilities of a Zener regulator circuit and improve regulation at the same time. This is done by adding a pass transistor.

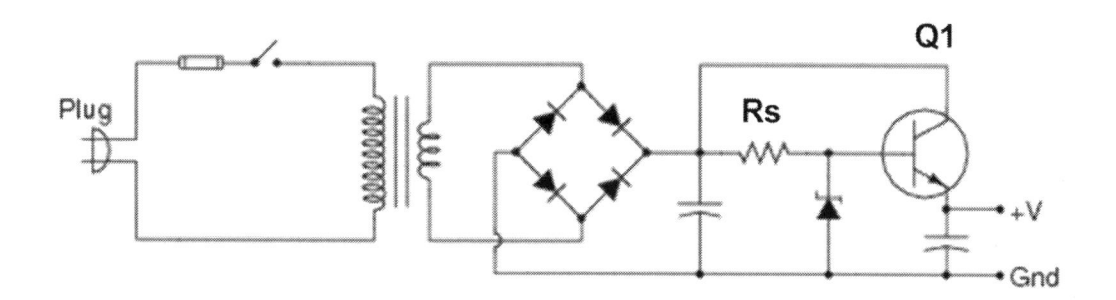

Figure 15-20.

Figure 15-20 is the schematic diagram of a Zener regulator with a pass transistor added. You will recognise R_s as the usual current limiting resistance that prevents the Zener from being destroyed. The Zener is reverse biased, with a positive voltage applied to its cathode.

The Zener regulated voltage drives the base of a transistor connected as an emitter follower (common collector). The output of the emitter follower goes to the load. Because of the current gain (β) of the emitter follower, any change in load current is reduced by a factor of β. This means voltage regulation is improved by a factor of β, which is a significant improvement.

Use of the pass transistor increases the maximum load current the regulator can handle. You need to be able to identify this circuit as a Zener regulator with a pass transistor. The purpose again – to improve regulation and the current handling capabilities of the Zener.

As a matter of interest, a pass transistor can be added to a three terminal voltage regulator in much the same way and for the same reasons. Will be discussing β and transistors more deeply in the chapter on semiconductors and amplifiers.

THE DARLINGTON PAIR

Two transistors can be connected in a configuration known as the Darlington pair. As far as i am aware, you do not even have to know what a Darlington pair is. We are jumping ahead of ourselves, but it is best, like the series pass transistor to mention them now. The Darlington pair is an improvement on the single pass transistor.

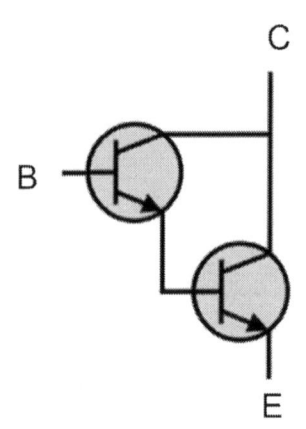

Figure 16-20.

The higher the β; the higher the input impedance of an emitter follower. One way to increase current gain or β is with a Darlington pair, two transistors of the one type connected as shown in figure 16-20.

The overall current gain is equal to the product of the two individual betas.
Transistor manufacturers can put a Darlington pair inside a single transistor package. Notice how the external connections to the two transistors are a base, emitter and collector.

A traditional application of a Darlington pair is 'pass transistor' in a power supply. A Darlington pair can also be created with two PNP transistors.

A POWER SUPPLY WITH PASS TRANSISTORS (Darlington pair)

All of the regulator circuits we have looked at are series regulators. The pass transistors are in series with the load. Series regulators are more efficient than the shunt type. The shunt type regulator (which is in shunt or parallel with the load) is rarely used, except in low power circuits, as they dissipate too much power.

Figure 17-20 is a full wave power supply. It uses two diodes. Therefore, it must have a centre tapped transformer. C1 and C2 are both filter capacitors, probably around 1000 microfarads and therefore electrolytic. R1 is the current limiting resistor for the Zener regulator D3.

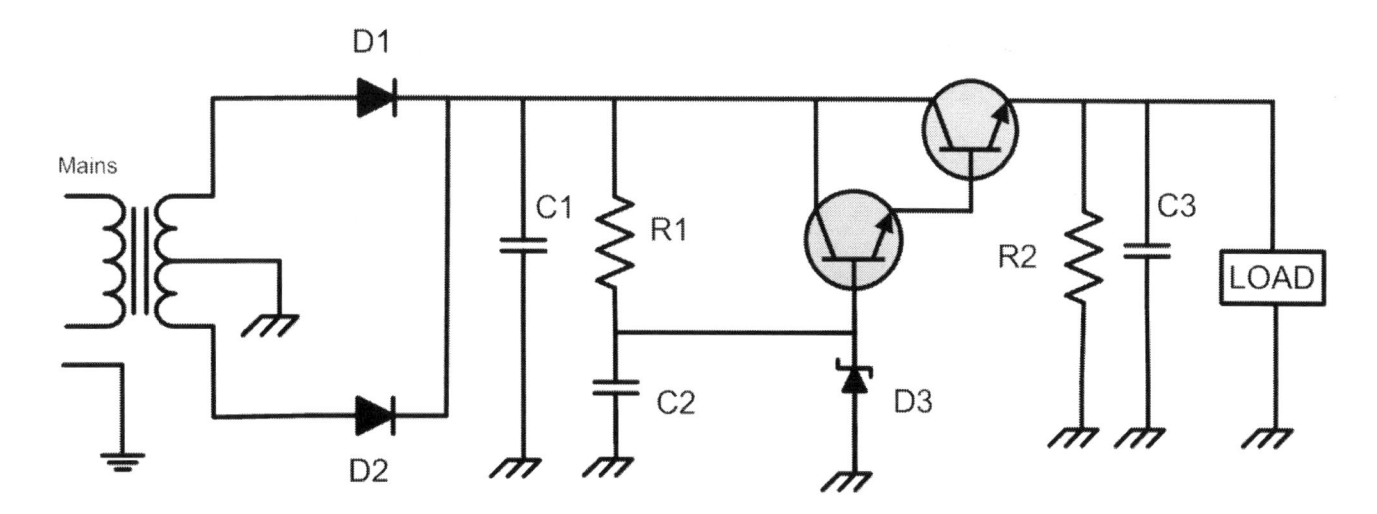

Power supply – Darlington Pair

Figure 17-20.

The Zener regulator has a Darlington pair as the pass transistors. The Darlington pair does a superior job compared to a single transistor, improving the regulation and the current handling abilities of this power supply significantly. Read the section on pass transistors and the Darlington pair again if you need to.

R2 supplies a light load on the power supply should the 'real' load not be connected. Its value would be about 1kΩ. C3 is an RF bypass capacitor about 0.01µF.

Is the Zener diode reverse biased? Yes, the positive rail (line) is at the top of the circuit. See if you can deduce this from the power supply diodes and how they would conduct and charge C1.

What is the purpose of R1? Current limiting.

Well, we have had a pretty good look at power supplies. There is one area that we have neglected a bit - the mains supply. This is an assessment favourite as there are safety issues involved.

THE MAINS SUPPLY

We have covered in earlier chapters the frequency, voltage relationships, RMS, peak, etc. and the wave shape of the mains. What we have not covered is the wiring of the mains.

Household wiring is a parallel connection. If it were not, when we switch an extra light on, all other lights would dim a bit (series circuit).

The three wires in the (Australian) mains circuit are:

The Active - Colour **Brown** (old red).
The Neutral - Colour **Blue** (old black).
The Earth - Colour **Green/Yellow** (old green).

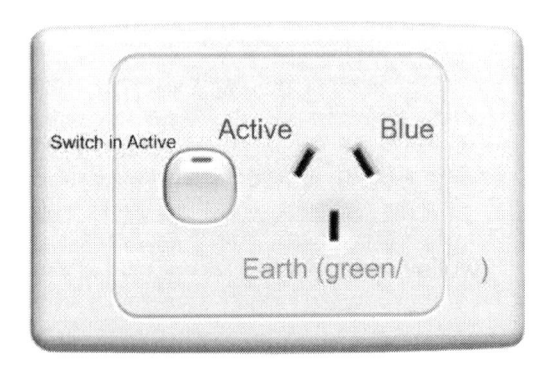

Figure 18-20.

These names are a little confusing. First of all, active and neutral deliver the 240 Volts AC throughout your home. There is 240 Volts AC RMS between active and neutral. The Standard for the mains voltage in Australia is really 230V AC RMS but this varies a lot with load and still referred to as the 240V AC Mains.

The earth is not ground in the sense that radio operators call ground. Earth is just that, EARTH, dirt if you like. The earth connection is usually somewhere near your power box and is a copper rod going into the ground.

Also, one side of the mains is connected to the earth as well. This earthed side of the mains is called neutral. The neutral is connected to the 'neutral bar' in the switchboard and the neutral bar to the electrical ground. This means that there must be 240 Volts AC RMS between active and earth. The potential difference between neutral and earth should be zero Volts, but they may be a few Volts (up to 5-6) due to capacitive coupling.

The locations of these wires are shown for a power point (socket) in figure 18-20. The active is on the left on a power point (socket). The active is on the right looking at a plug.

The photo in figure 19-20 shows the connections for both a plug and a socket (extension lead).

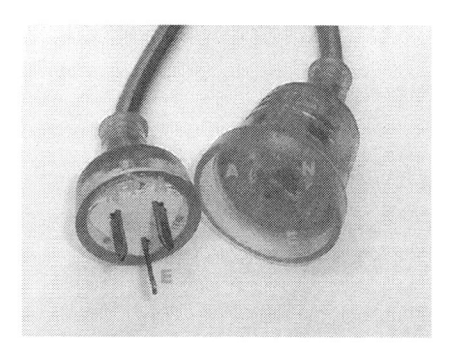

Figure 19-20.

Every power switch in your house should be in the active wire.

Should the active wire come adrift and contact the metal chassis of any appliance a fuse will blow. The metal (conductive) parts of appliances are connected to earth, as is the neutral. Should an active touch a metal part and therefore potentially 'you', the active and neutral are effectively short circuited and a fuse will blow.

All being well since there is no potential difference between neutral and earth you should be able to touch the neutral wire in your house and nothing will happen – *never put this to the test*, though by all means, use a Voltmeter to check.

It is extremely dangerous to assume that the wiring in your home is correct. If someone has wired a switch into the neutral wire instead of the active, then the appliance will still work. However, the active will still be going to it when it is switched off. For example, if someone has placed the switch in the neutral wire of light, the light will still work as it should (as will all other appliances). If you turn the light switch OFF and it is wired incorrectly like this, the active is still present at the light socket. If you touch the active terminal in the light socket, you will receive an electric shock that could be fatal.

How easy it is for someone to reverse the active and neutral accidentally or carelessly on the plug or socket of an extension lead. Doing so could be fatal. In passing, electrocution is death by electric shock. You cannot be electrocuted and survive.

You need to know the potential difference between each wire in your household mains. This is not so you can work on your house mains. It is necessary for you to know this if you build equipment that you connect to the mains. You need to know how to test a power point with a Voltmeter. This is safe provided you do not contact the metal parts of your multimeter leads. If you measure less than 240V between active to earth and active to neutral, it may mean you have a problem. Any voltage between earth and neutral indicates a problem.

An electrician should do the testing for an adequate earth. The cheap plug-in testers do not do this adequately.

If you are in a new home, you will have an 'earth leakage detector' fitted. Older homes did not have them, so it is your choice to have one fitted by an electrician.

Only qualified electricians should work on the household wiring. Being a licensed radio operator or a professor of physics does not qualify you to tamper with house electrical wiring. This part of the chapter is not a definitive explanation of safety procedures. The best procedure is to remove equipment from the mains before working on it. However, as a licensed radio operator, you are qualified to build equipment which you can connect to the mains. Be aware when purchasing home brew equipment which is to be connected to the mains. Perform an electrical safety check if you are competent to do so. If you are not competent then have someone who is competent test the equipment for you.

Different countries will have specific Federal, State, or Local government laws. You need to know what your local requirements regarding electrical safety are.

21. ELECTRON TUBES

One of the most significant developments of the early twentieth century was the invention of the electron tube. The British call it a valve. In the USA it is referred to as the "tube". Thomas Edison, the inventor of the electric light bulb (among other things), came extremely close to inventing the electron tube. When Edison was trying to develop a light bulb his most significant problem was to find a filament that did not burn out. He concluded early in his experiments that a filament would last longer if it were placed in a vacuum. Edison, being the entrepreneur that he was, would be kicking himself today if he knew just how close he had come to the discovery of the electron tube.

Edison tried hundreds of different types of filaments, without using true scientific method (Edison did not like scientists), more like a laborious trial and error approach. One of the things that frustrated Edison was that some of his filaments were giving off material which was deposited on the inside of the glass bulb causing the light to dim and overheat. He even inserted a metal electrode into some of his light bulbs and found this prevented these deposits from occurring. We now call that additional electrode an anode. It worked, but he dismissed the idea as impractical for a light bulb.

In radiocommunications, electron tubes are not used as much as they once were. One exception is very high power radio transmitters and amplifiers. Electron tubes are extensively used in industry and commercial radio communications.

Also, many of the principles of amplification, oscillation, rectification and the like, are the *same* with modern semiconductor devices as they are with electron tubes. It is somewhat beneficial to understand these principles with electron tubes first.

THERMIONIC EMISSION

When metals are heated to very high temperatures electrons are boiled off. These boiled off electrons have left the metal and form a cloud called a space charge. Since electrons have left the metal, it becomes positive. The space charge, consisting of electrons, is negative. As long as the metal remains heated the space charge will exist, as it requires energy to keep the negative electrons away from the positive metal. Metals that are designed to enhance their ability to create a space charge are called cathodes. Since electrons have left the heated cathode, the cathode will be positive with respect to the space charge. The cathode will continually attract electrons back. However, electrons will be continuously ejected by the thermionic emission.

Thermionic Emission

Figure 1-21.

CATHODES

A cathode is something capable of emitting electrons. For example, in a light bulb, a piece of wire carrying an electric current becomes hot. If sufficient current flows, the wire becomes white-hot. Heat is 'really' the measure of the internal kinetic (moving) energy of atoms or molecules. Electrons in the outer orbits of the atoms move so rapidly due to increased temperature that some of the less tightly bound ones fly outward and away from the wire into the surrounding space. This creates a cloud of free electrons around the wire, as long as it is hot enough. The wire has become a cathode. When the atoms in the filament wire lose electrons, the filament is left with a positive charge. The electrons expelled outwards are then attracted back toward these positively charged atoms in the filament (cathode); some do make it back, only to be expelled again. This results in a constant in and out, back and forth movement of electrons, in the area directly surrounding the cathode. This creation of the space charge around a hot filament is called the Edison effect or thermionic emission.

The space charge has no practical use in a light bulb. Filaments designed for creating a space charge are called cathodes. There are a number of materials used for cathodes, however, by far the most popular is thoriated tungsten. Thorium is an excellent electron emitter, but this material is not strong enough for practical use. Mixing thorium with the more rugged tungsten produces a thoriated-tungsten cathode that will permit many electrons to be emitted at a relatively low temperature (1600 degrees centigrade). Cathodes of this type are used as radio transmitting tubes.

To enhance the space charge creating properties of the cathode, it is placed in a glass envelope and the air is pumped out. To improve the vacuum further, the entire assembly is heated to force out any additional gas molecules in the metal electrodes. Then a small amount of magnesium (called the getter) placed inside the envelope during manufacture, is ignited. The resultant chemical action of the vapourising magnesium with the gases released from the metal removes the final traces of gas in the tube. The by-products of the vapourising magnesium condense on the inside of the electron tube forming the silvery coating commonly seen on electron tubes.

THE DIODE

The cathode of an electron tube can be heated by a separate filament, in which case the filament is called a heater. Such an arrangement is known as an indirectly heated cathode. Alternatively, the heater itself can be treated with thorium to act as a cathode. This method is referred to as a directly heated cathode.

Figure 2-21 shows an indirectly heated cathode, the heater is a third element connected to its own power supply.

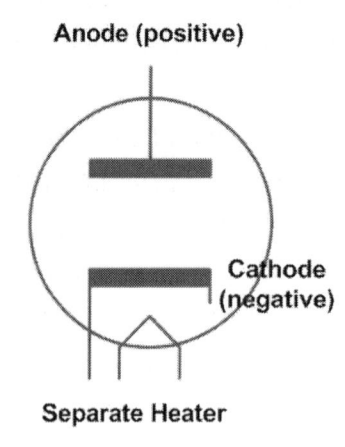

Figure 2-21

It should now be obvious that electric current does not have to occur in a conductor. In the last chapter, we discussed power supplies and rectification. The diode electron tube just described can be used in any of those power supply circuits. A diode electron tube will always conduct if its anode (also called a plate) is positive with respect to the cathode and a space charge exists. Electron flow from anode to cathode is not possible.

THE TRIODE

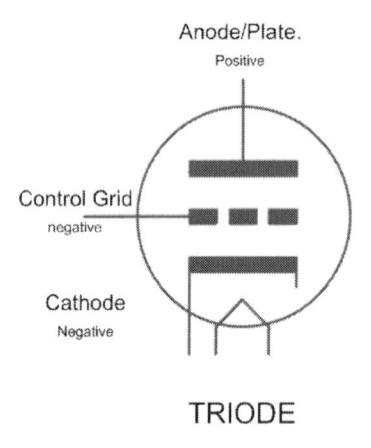

Figure 3.21

A circuit schematic of a Triode (tri=3) appears in figure 3-21. In a diode, a cathode and anode are required to enable electrons to flow. Also, a triode has a third electrode called the *control grid*. The grid is a fine metal wire, usually nickel, molybdenum, or iron. The anode and cathode are both solid electrodes. The control grid of the triode is a *fine spiral wire* placed between the cathode and the anode. All electrons attracted to the plate from the cathode go through the openings in the control grid. The grid is connected to the base of the electrode so that voltage can be applied to it. Imagine if you will a Triode operating without voltage applied to its grid, it will act just like a diode.

If we make the grid slightly negative with respect to the cathode, the negative grid will inhibit (restrict) the flow of electrons from the cathode to anode (like charges repel). The voltage between cathode and anode may be hundreds of Volts. On the other hand, a *small negative voltage* on the control grid will have a substantial effect on the flow of current from the cathode to the anode. The significance of this is that we can control a large cathode-anode current with a very small negative grid voltage. Making the grid more negative will decrease the anode current; making the grid less negative will increase the anode current.

A pictorial diagram of the construction of an electron tube is shown in figure 4-21. The electrodes and grids are in cylinders surrounding each other. All of the electrodes are usually connected to pins out of the bottom of the electron tube.

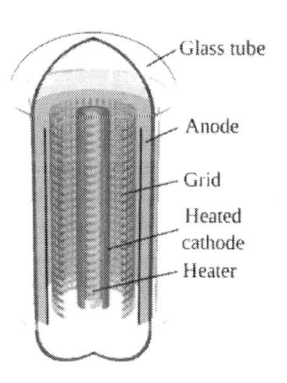

Figure 4-21

In passing, this is why the electron tube got its name - the valve. Think about how a water valve works. A minor input of energy from your hand can control a huge pressure of water. Likewise, a small voltage on the control grid can control a large anode/plate current.

Under normal circumstances the control grid is negative. Because of this, it does not draw any electrons from the cathode. It is the negative voltage alone on the control grid; that can control what can be a substantially large current from the cathode to anode (voltage without current is no power P=EI).

The control grid is placed closer to the negatively charged cathode and further from the highly positively charged anode. This placement allows it to function more effectively than if it were simply midway between the cathode and anode.

The cathode-anode circuit of the electron tube is the output circuit. The control grid-cathode circuit of the electron tube is the input circuit. No current flows in the control-grid cathode of the electron tube. There are special cases where current does flow, but it is still very small; see *grid-leak bias*. On the other hand, the voltage and current in the output circuit of the electron tube are substantial. And as we learnt, the product of current and voltage is power (Watts). Therefore, with the triode, we have the means to control significant amounts of power by simply applying a small negative voltage to the control grid.

We could, for example, connect a microphone to the control grid of an electron tube and use the very small voltage from this microphone to change the negative potential on the control grid, that in turn would control the anode-cathode current. The microphone produces small voltages corresponding to sound waves striking its diaphragm. These small voltages are then applied to the control grid, which in turn controls the larger anode-cathode current. The anode-cathode current will contain all of the information (intelligence) that was initially fed into the microphone.

What we have just described is amplification. The Triode was the first electronic device ever to be able to produce amplification.

The Triode as an amplifier

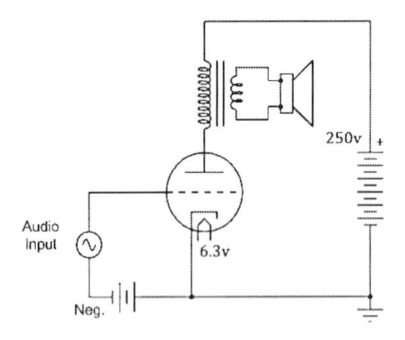

Figure 5-21

The triode amplifier in figure 5-21 has an audio input that could be from the microphone. It has an output transformer to match the output impedance of the triode to the impedance of the speaker. The small battery on the left makes the control-grid negative at all times – this is called negative bias. The audio signal will cause the control grid to become more or less negative. In practice, the battery or power supply voltage sources would be bypassed with a capacitor. In a practical amplifier, batteries would not be used.

Negative variations on the control grid cause the much larger anode-cathode current to vary and an amplified sound comes out of the speaker.

DISADVANTAGE OF THE TRIODE

When we discussed capacitance earlier in this book, we found that any two conductors separated by an insulator have capacitance. The problem with the triode is the capacitance between the control grid and anode. (more on why this capacitance is bad shortly). We also learned in our study of capacitance that capacitors in series decrease the overall circuit capacitance. Inserting another grid called the screen grid between the control grid and the anode, reduces the capacitance between the control grid and the anode.

Why is the control-grid to anode capacitance bad?

The anode is in the output circuit of the triode. The control grid is in the input circuit of the triode. The unwanted capacitance we're talking about is between the control grid and anode. This unwanted capacitance is inside the electron tube. However, it may as well be a physical capacitor connected externally from the anode to the control grid. This means that there is a connection through this capacitance from the output of the triode to the input of the triode, this is undesirable. This unwanted capacitance can cause positive feedback.

You will be well aware of the problems with feedback from a public address system. If the sound from the speakers in a public address system get into the microphone of the same system, you will have feedback and it will oscillate (squeal).

If we wanted the triode to oscillate that would be just fine. However, if we want it to amplify, then this oscillation (because of feedback) is a problem. The inclusion of a screen grid between the control-grid and anode reduces the control-grid to anode capacitance and improves the stability of the triode as an amplifier. A triode with a screen grid is no longer a triode; it is called a Tetrode.

TETRODES

The screen grid of tetrodes (tetra=four) is not used to control the plate current but has a steady positive DC voltage on it to help accelerate electrons to be collected by the anode. The path for current inside a tetrode is from the cathode through the control grid and through the spaces in the screen grid, to the anode.

TETRODE

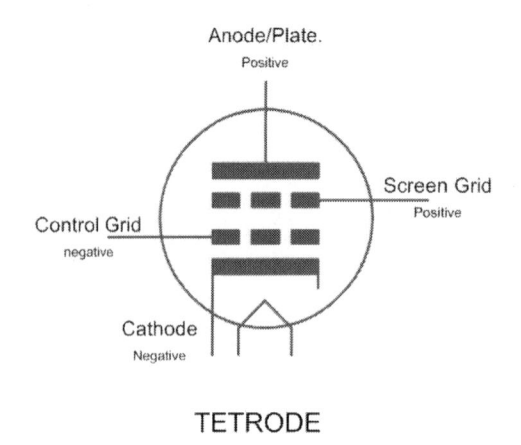

Figure 6-21.

Since the screen grid is positive, it will collect some electrons. Most of the electrons arriving at the screen grid have enough momentum to continue to the anode. Electrons arriving at the screen grid pass through it, attracted by the much higher positive potential on the anode. The screen grid current is wasted current since it is not used in the output circuit. The benefits of the screen grid (reduced control-grid to anode capacitance) outweigh this disadvantage.

More about the triode and feedback

Even though we are talking about electron tubes in this chapter, all that you learn here can be carried over to other devices that amplify, such as transistors.
The diagram in figure 7-21 shows a generic amplifier. The signal to be amplified is applied to the input and a larger signal appears at the output.

All amplifiers work this way regardless of what the amplifying device is. The power for the amplified output is provided by the power supply.

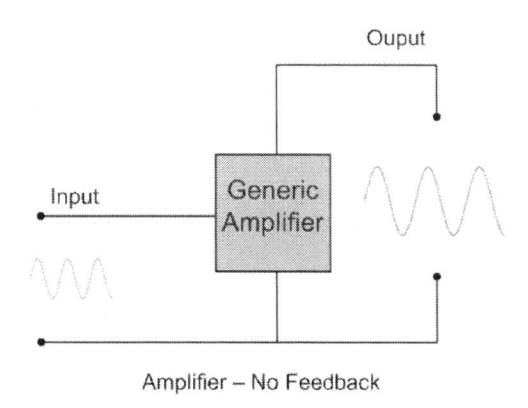

[Figure: Generic Amplifier with Input and Output, labeled "Amplifier – No Feedback"]

Figure 7-21.

Figure 8-21 shows a little more detail of a generic amplifier. Most amplifiers have, as a natural consequence of their construction, a small amount of unwanted capacitance between their output and their input. I have shown this unwanted capacitance on the amplifier below. Remember, this capacitance is not an external physical capacitor, rather an unwanted capacitance inside the amplifying device.

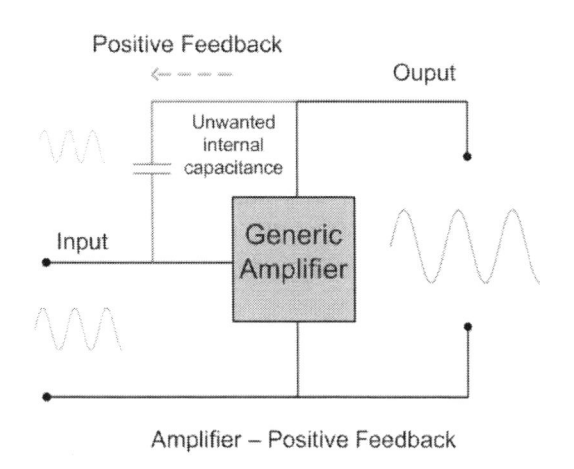

[Figure: Generic Amplifier showing unwanted internal capacitance between output and input, with Positive Feedback arrow, labeled "Amplifier – Positive Feedback"]

Figure 8-21.

The consequence of this unwanted capacitance in any amplifier, between output and input, is that some of the output signal can return to the input. If you like, some of the output signal 'leaks' back to the input. This is called feedback. In the case of unwanted capacitance between output and input, as shown, the feedback is called positive feedback.

Positive feedback means the output signal is fed back to the input in such a way that it adds to, or is in phase with, the input signal. I have shown the feedback signal as a small sine wave. Can you see how this sinewave is in phase with the input signal? Positive feedback adds to the input signal – so it's like taking some of the output and amplifying it all over again. This is called regeneration. The terms regeneration and positive feedback can be used interchangeably.

If there is enough positive feedback the amplifier will stop being an amplifier and it will oscillate. Like a public address system where the microphone is placed too close to the speaker and it howls. An amplifier with too much positive feedback will oscillate (like the howl of a public address) but you may not hear it at all – it could oscillate on a radio frequency above human hearing. It certainly won't be amplifying as it should. The amplifier is said to be self-oscillating and amplifiers are supposed to amplify not self-oscillate. We will cover more of this in detail later. The point I am trying to make here is that unwanted positive feedback is a bad thing and we must prevent it from happening.

THE TRIODE AMPLIFIER

Just to recap for a moment look at a basic triode circuit in figure 9-21.
The input signal is applied to the control grid and cathode. The output signal is taken from the anode and the cathode. (The cathode is common to both input and output). If any of the output signal in the anode were able to get back to the control grid, then this would be positive feedback and is undesirable.

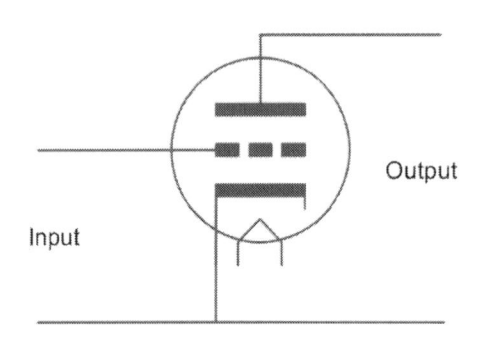

Figure 9-21.

UNWANTED CAPACITANCE IN THE TRIODE

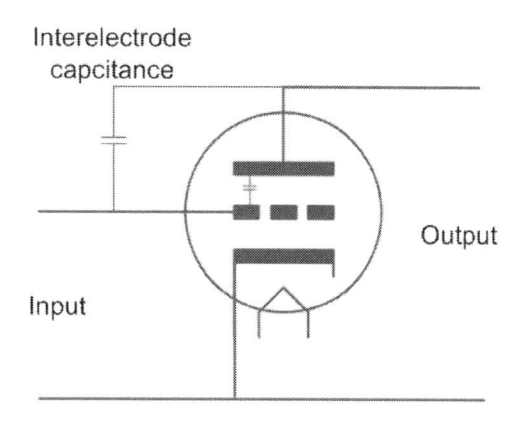

Figure 10-21.

We have an unwanted capacitance formed in a triode between the control grid and the anode. It is there because everything is in place to form a capacitor. The control grid and the anode are the two plates and the vacuum between them is the dielectric. There are other unwanted capacitances in the triode. Between the cathode and the control-grid for example and many others. However, it is the capacitance between the anode and the control grid that will allow positive feedback to occur.

The reactance of a capacitor decreases with increasing frequency. The higher the frequency a triode (or any amplifying device) are operated at, the more positive feedback there is. Remember, enough positive feedback will cause the amplifier to oscillate and prevent it from its proper job of amplifying. This is where the tetrode comes in.

How the tetrode helps

The tetrode breaks the capacitance between the control-grid and the anode into two capacitors in series. This is shown in the diagram of figure 10-21. What happens when you connect capacitors in series? Does the total capacitance go increase or decrease?

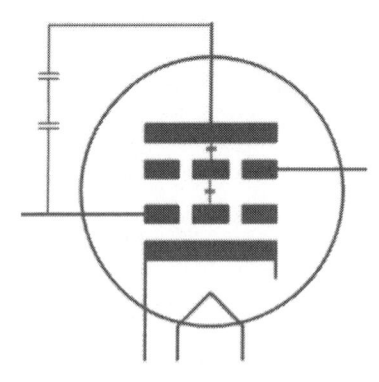

Figure 10-21.

I hope you said the total capacitance decreases. This is the primary function of the screen grid. The insertion of the screen grid reduces the unwanted interelectrode capacitance between the control-grid and the anode and in doing so reduces the likelihood of unwanted positive feedback, enabling the electron tube (or any other amplifying device) to be used at higher frequencies.

The unwanted capacitance is not a real physical capacitor that anyone has deliberately added to the electron tube – it is the nature of the electron tubes construction to have this unwanted capacitance.

When we have done all we can with the amplifying device and we still have positive feedback the only thing we can now do is introduce an equal amount of negative feedback. Negative feedback is out-of-phase feedback. Negative feedback is also called degeneration. The process of deliberately applying negative feedback to cancel out unwanted positive feedback is called *neutralisation*. We will talk more about neutralisation as it is an important concept of amplifiers used at radio frequencies.

PENTODES

Fast-moving electrons (relatively speaking), in an electron tube, may strike the anode with considerable force. Some of the electrons striking the anode will collide with electrons in the anode material and bounce back, or electrons arriving at the anode may dislodge electrons in the anode material and cause them to be emitted from the anode. This phenomenon is called *secondary emission*. Secondary emission is undesirable. Electrons participating in secondary emission are not performing any useful function and can build up a small cloud of electrons somewhat similar to the space charge around the cathode, though nowhere near the same size.

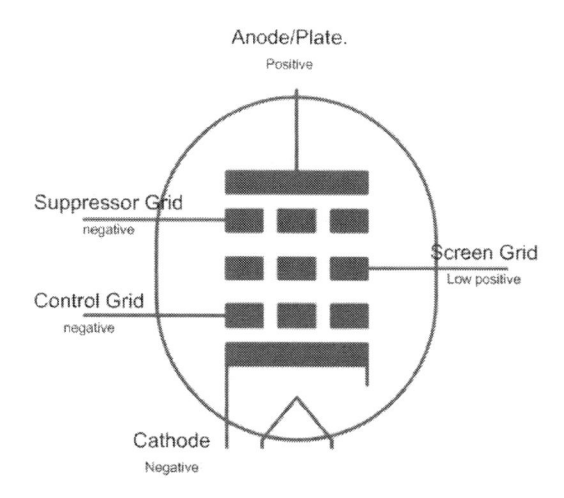

Figure 11-21.

This secondary space charge around the anode will inhibit or restrict the flow of electrons from cathode to anode.

Pentodes (pent = 5) have the same construction as tetrodes but with the addition of another grid, called the suppressor grid, in the space between the screen grid and anode. The suppressor grid has a negative voltage placed on it with respect to the cathode. Now you might think that a negative voltage on the suppressor grid would also inhibit cathode-to-anode current. In fact, it does, but not to the extent of secondary emission. The negative voltage on the suppressor grid does not appreciably reduce the anode-cathode current. Also, by the time electrons come close to the anode they have built up a significant amount of momentum that will carry them through the negative electric field of the suppressor grid.

The secondary electrons, the ones created by secondary emission, do not have a large amount of energy and are repelled back to the anode by the negative potential on the suppressor grid. The pentode has the additional advantage of having still less capacitance between the control grid and the anode. The suppressor grid, like the screen grid, acts like another series capacitor between the control grid and anode reducing the unwanted capacitance even further.

X-RAYS

As a matter of interest, if the cathode/anode current is made very high by an extremely large voltage on the anode relative to the cathode and if secondary emission is encouraged, a by-product of this is that X-ray radiation will be emitted from the anode.

Secondary electrons that are ejected from the anode are at a higher energy level. When they get push back to the anode or just fallback, they discard their excess energy as a photon in the X-ray band. This was how the X-ray machine was developed. In an X-ray machine, the anode is specially shaped at an angle of 45 degrees to the approaching electron stream to deflect the X-ray emission sideways out of the electron tube.

I am amused that X-rays are still called "X" for unknown. Rontgen discovered them. Perhaps 'Rontgen waves' was too hard to pronounce but R-Rays would be cool. X-rays are electromagnetic waves.

We have described all the basic types of electron tubes. There are others, but these are only modifications of the types we have just discussed. An electron tube may have two diodes within one envelope. The two diodes may share one cathode, or they may have separate cathodes - this is called a duo-diode. Other types are duo-triodes and triode-pentodes.

SOFT TUBE

An electron tube can become faulty and have unwanted "gas" or "air" in its vacuum. This means the vacuum is not as "hard" as it should be. Such a fault with electrons tubes is called a gassy or soft tube. The terms 'hard' and 'soft' refer to the integrity of a vacuum.

MICROPHONICS

Another problem which can occur with electron tubes is microphonics. An electron tube is a mechanical device as well as an electronic device and if it is tapped with a screwdriver, or flicked with a finger, it will vibrate. The vibration of the grid will modify the cathode-to-anode current and this vibration will 'modulate' the output.

Modulate or modulation has not been discussed yet. Modulation means imposing the information of something onto something else. In the discussion above the vibrating grids would transfer this vibration 'message' to the output current.

If an audio (sound) system was constructed from electron tubes and some of the audio output sound wave energy from that system was able to reach the electron tubes and caused them to vibrate, then the system would produce microphonics, or perhaps even feedback. In some old 'wireless' sets, if you hit them on the cabinet, you would hear a 'thud' sound from the speaker.

More on Feedback

Feedback is not a problem particular to electron tubes. Feedback is a sub-topic of its own. However, since we have mentioned it in this chapter, we may as well discuss it a little further.

Feedback is only a problem if you don't want it. If you want an electron tube to oscillate, then you may take advantage of the capacitance between the control grid and the anode. Alternatively, you may use external components, usually a capacitor, to take some of the electron tubes output and feed-it-back to its input.

An oscillator is a device for producing electronically, AC voltage. The frequency of the AC voltage may be from audio to microwave and beyond.

We have discussed the two types of feedback, called positive and negative. Nothing to do with polarity. When some of the output of an amplifier is fed back to its input, in such a way that the voltage is in phase with the input voltage, this is called positive feedback. This is the type of feedback we require if we want oscillation to take place.

If some of the output of an amplifier is fed back to its input so that it is out of phase with the input, this is called negative feedback. Negative feedback reduces the amplification (gain) of an amplifier. However, it also has benefit such as improving the linearity. A perfect amplifier (no such thing) would do nothing to the input signal except increase its amplitude. In practice, all amplifiers produce some distortion. The distortion just means that what we are getting out of the amplifier is something we did not put into it. However, there is always unwanted information in the output created non-linearity in the amplifier. Negative feedback can reduce the distortion of an amplifier. The ability of an amplifier to amplify with minimum distortion is called linearity. Negative feedback does improve the linearity of an amplifier.

Remember our old friend direct proportion? If an amplifier's output is directly proportional to its input, it is *linear*. That is the output signal varies in direct proportion to the input signal.

THINGS TO DO

It is not very hard to get your hands on an old electron tube. If you can, please do so. Place it in a cloth bag or towel and smash it and have a look at the construction. You will see that the heater is in the centre with a cylindrical cathode around it; cylindrical spirals around the cathode for the various grids and finally a cylindrical anode.

Nixie Tubes

Figure 12-21.

Before semiconductor 7-segment displays, electron tubes called Nixie tubes, were used to display numbers. They were even made really small and used in some of the very early calculators. Do you need to know about them? No, you don't. Why did I put them here? I just did as they are so cool! Figure 12-21 shows a Nixie tube clock. Now wouldn't you just love to have one?

22. DECIBELS

For some reason, decibels are disliked and misunderstood by many radio experimenters, engineers, technicians and the like. The decibel is used in equipment handbooks and specification sheets so we *really* do need to understand them. Not just understand, but appreciate the advantages of the decibel and want to use it.

Human senses of sight, touch, hearing and others are all logarithmic. Our human senses are not linear. Decibels are also logarithmic and as such more closely match changes that affect our senses.

Early man tiptoed through the jungle listening for:
Sounds indicating something to eat.
Sounds indicating something was going to eat him!

His ears would pick out quite faint sounds. They had to, or he would starve. He also heard some very loud noises from animals bellowing, trees falling, landslides and the like. For shelter, early man lived in caves; yet he had to be able to distinguish members of his family from the noises of predators. And at the same time, he had to handle the harsh bright sunlight when he was out hunting for food. To avoid *overloading* the information processing centre in his brain, his various sensing mechanisms accommodated the signals from almost all his senses into a *compressed range*. Twice as much energy in a sound does not seem twice as loud. There is a lot less energy involved in a leaf rustling, compared to a tree falling. Having two candles lit in the cave is quite a significant change from just one; but lighting a candle at the mouth of the cave, even on a dull day would hardly register on his visual cortex.

If our ears interpreted noises as a *direct proportion*, then to be able to hear very weak sounds, strong sounds would be more than deafening! They would be painful, perhaps lethal. Apart from heat and pressure or touch, most of our senses respond to increments in the signal rather than to its absolute level, i.e. the amount of the change divided by the level already there.

Very early in the development of amplifiers, it was realised that the units for comparing output powers must match the hearing of the operator or listener. These early amplifiers were used for the long distance telephone lines that were starting to criss-cross all the continents and oceans of the world.

There were also economic factors to consider. If one amplifier could increase the signal level so that communication could be heard 1000 miles away, that would be better than one that only carried 100 miles before it required boosting again. Why? Because one tenth the number of booster amplifiers would be required. However, if the 1000-mile amplifier were used, then some of the signal would be heard in adjoining lines, by capacitive and inductive transfer – the crosstalk effect. Also, if all amplifiers were designed for 1000 mile, there would need to be some means of attenuating the signal for nearby connections – or people would have their hearing damaged by the use of the telephone. But how much signal was required for normal hearing, how much could the signal fall or increase before customers complain?

For a sound to be perceived as twice as loud, it seemed that an increase of about 10 times the power at the source was required.

For sound to seem twice as loud, increase the power about 10 times.
For sound to seem twice as loud again, increase the power about another 10 times.
→ The total increase so far is about 100 times.
To seem twice as loud once more, increase the power by about 10 times again.
→ The total power increase is now about 1000 times.

You can see that this system of describing the power of a sound is very cumbersome. This could be written much more conveniently:

To seem twice as loud - increase the power by approximately 10^1.
To seem twice as loud again - increase the power by approximately 10^2.
To seem twice as loud once more - increase the power by approximately 10^3.

($10^1 = 10$, $10^2 = 100$, $10^3 = 1000$)

We can see that a convenient unit to use is the index of 10. Because the Bell Telephone Company was at the forefront in researching long-distance telephonic communications, this unit is called the *Bel*, after Alexander Graham Bell. The Bel is a unit of *comparison*.

We use it, just like with sound levels, to compare current, voltage and power levels. In mathematical terms the Bel is:

$$Bel = \log[\frac{P_1}{P_2}]$$

Equation 1.22

Where P1 and P2 are the two power levels to be compared.

The logarithm [or 'log'] of a number is the power, or index, of which 10 must be raised to get that number.

LOGARITHMS

The mathematical shorthand to indicate that the log of a number is to be found is log(N), where 'N' is the number whose logarithm is to be found. The base of the logarithm used with the Bel is 10. Other bases can be used, but we are not interested in these in this course. If there is any possibility of confusion in texts about which base is being used, the shorthand is written as $\log_{10}(N)$ and log(N) for common (base 10) and natural logarithms respectively.

Common logarithms which express a number as a power of the base 10, are the ones that interest us. Natural logarithms express a number as a power of the base 'e' where 'e' equals 2.71828. For the Advanced exams, you need only concern yourself with the use of common logarithms (base 10). Throughout the rest of this chapter, you can consider log(N) to mean $\log_{10}(N)$.

The laws of indices [plural of 'index'] are:
- to multiply numbers, *add* their indices
- to divide by a number, *subtract* their indices

From the laws of indices, it follows that we can say:
$10^2 = 100$ so log(100) is 2   (10 x 10 = 100)
$10^3 = 1000$ so log(1000) is 3   (10 x 10 x 10 = 1000)

$10^{-1} = 1/10^{th}$ so log(0.1) is -1
$10^{-2} = 1/100^{th}$ so log(0.01) is -2

What is the logarithm of 10,000? In other words, to what power must 10 be raised to give 10 000? The answer is 4 since 10^4 is 10 x 10 x 10 x 10 = 10,000.

The common log of a number between 1 and 10 lies somewhere between 0 and 1. The common log of a number between 10 and 100 lies between 1 and 2. The common log of a number between 100 and 1000 lies between 2 and 3. You can see that using logs simplifies the expression of very large numbers and ratios.

To find out these 'in between' logarithms it was once necessary to look up large tables (books actually) of logarithms. Fortunately, most calculators have a 'log' key. Try the following numbers on your calculator:

log(10)=1:
Enter 10 into the display and press 'log' and you should get 1.

log(50)=1.698:
Enter 50 into the display and press 'log' and you should get 1.698.

log(250)=2.3979:

Try this one for yourself.

Finding the antilog of a number [i.e., an index], converts it back to the original number. On most calculators, an antilog is found by:

1. entering the number [index] into the keypad;
2. pressing 'INV'; (this can be "shift" or 2nd on some calcualtors)
3. pressing 'log'; and
4. the antilog displays, i.e., the original number.

Try the following yourself:

- antilog of 2 = 10^2 = 100
- antilog of 3 = 10^3 = 1000
- antilog of 1.5 = $10^{1.5}$ = 31.622
- antilog of 0.5 = $10^{0.5}$ = 3.1622

For example, to find the antilog of 2:

1. enter 2 into the display;
2. press 'INV';
3. press 'log'; and
4. your display should read 100.

Logarithms are very useful in radio work. Many test and measuring instruments have their scales calibrated logarithmically to enable a greater dynamic range of measurement to be displayed on one scale. On some meter scales, you will find two sets of numbers – one set going from 0 to 10 and the other set going from 0 to 3.16. Why? [Hint: $(3.1622)^2=10$]

Dynamic range means the range between the lowest (weakest) and highest (strongest) level.

BACK TO THE BEL

If P_1 is 30 Watts and P_2 is 2 Watts (this is a change in power from 2W to 30W which is a numerical increase of 15 times – 30 divided by 2 is 15) then this change represents log(30/2) = log(15) = 1.1761Bel (try it).

The Bel as a unit is too large for our purposes. It is more convenient to use one tenth of a Bel and call it a *decibel* or just dB.

10dB = 1Bel.

The formula you will always use for calculating the difference in *power* levels in decibels is:

$$dB = 10\log[\frac{P_1}{P_2}]$$

Equation 2.22

An increase in power level of 15 times is a change of 11.76dB (as we calculated above but multiplied by ten to convert it to decibels).

So, a change of 11.76dB is an increase of 15 times. For example:

From 2 Watts to 30 Watts – is 11.76dB.
From 2 milliwatt to 30 milliwatt – is 11.76dB.

From 10 kilowatts to 150 kilowatts – is 11.76dB.

The 'S' meter on a receiver has (or should have) a logarithmic response. Many radio manufacturers use 6dB as the step between 'S' units.

A decrease rather than an increase is shown by adding a minus sign. A decrease from 15 Watts to 1 Watt (a decrease of 15 times) is a change of -11.76dB. To put it another way, 1 Watt is 11.76dB below 15 Watts.

Using decibels with current and voltage

We know that power is E^2/R or I^2R, so provided the input and output impedances are the same (and they normally are and will be for exam purposes - just remember for the future, that this explanation assumes the above) then:

$P_1/P_2 = E_1^2/E_2^2 = (E_1/E_2)^2 = I_1^2/I_2^2 = (I_1/I_2)^2$

What this means is that to express a change between two voltages or two currents, the log formula must be modified to have a 20 in it rather than a 10.

$$dB(voltage) = 20log[\frac{E_1}{E_2}]$$

Equation 3.22

E_1 and E_2 are the two voltages that we want to express as a change in decibels.

$$dB(current) = 20log[\frac{I_1}{I_2}]$$

Equation 4.22

I_1 and I_2 are the two currents that we want to express as a change in decibels.

As an example, suppose a signal going into the input of an amplifier has a level of 2 Volts. At the output of the amplifier, the signal has a level of 50 Volts. What is the voltage gain of the amplifier in decibels?

dB = 20log(E₁/E₂)
dB = 20log(50/2)
dB = 20log(25)
dB = 20 x 1.3979
dB = 27.95

The amplifier has a voltage gain of approximately 28dB.

*Remember to use '**20**' in the formula for voltage and Current and '**10**' for power.*

Suppose a radio station has an output power of 25 Watts. The operator connects a power amplifier to the transmitter to amplify the power to the antenna from 25 Watts to 200 Watts. What is the gain of the amplifier in dB?

dB = 10log(P₁/P₂)
dB = 10log(200/25)
dB = 10log(8)
dB = 10 x 0.903089
dB = 9.03

The amplifier connected to the transmitter has a power gain of approximately 9dB. The decibel does *not* measure an absolute value. The decibel is a measure of the ratio of two levels. It is ridiculous to say that power or voltage (or anything else) is so many decibels. Such a statement is meaningless unless there is a reference level given.

Amplifiers increase power levels, so we can say an amplifier has a gain of *n*decibels. Attenuators reduce power so we can say they have a loss of *n*decibels. Antennas can increase transmitter power by focusing the radiation into a narrow beam – much like you can increase the water pressure on a hose by placing a nozzle on it. So antennas have a gain expressed in decibels. You may be interested to know that organ pipes are attenuators. Organ tuners go around the pipes adjusting the air flow to each pipe to regulate the sound levels so that all pipes in a 'register' have the same attenuation.

ABSOLUTE MEASUREMENTS USING DECIBELS

Measurements such as Volts, Amps and Watts are absolute measurements.

So far we have dealt with dB as a method of comparing two values. However, there are some related units which can be used with the dB to refer to *absolute values*. To do this, a reference point must be given, such as 1 Watt.

If the reference level is 1 Watt, the unit is called the dBW. So 3dBW is 3db above 1 Watt and -3dBW is 3dB below 1 Watt.

You do not need to know this for assessment purposes. However you will come across it and it is worthwhile to know this for a complete understanding of decibels.

By far the most common absolute and useful unit used in radio and communications is the dBm, which is the number of decibels above or below one milliwatt.

$$dBm = 10\log[\frac{P}{1mw}]$$

Equation 5.22

For example, what is the power of a 50 Watt radio transmitter expressed in dBm?

To express 50 Watts in dBm we use the equation above:
dBm   = 10log(50/1mW)
       = 10log(50 000)
       = 10log(5) + 10log(10 000)
       = 6.989 + 40
       = 46.989dBm (usually rounded to and written as +47dBm).

This is a good figure to remember as many transmitters are 50 Watts. Now, a doubling of power is an increase of 3dB, which means 50dBm is 100 Watts, or 44dBm is 25 Watts. Remember, if you increase power by 3dB, you double the power. If you decrease power by 3dB, you halve the power.

You don't have to be able to recall the following numbers in the exam, but after some experience, you will find these numbers to be quite useful.

-120dBm - typical VHF receiver sensitivity and as voltage across 50 Ohms, = 0.22µV and as power = 1 *femtowatt (1 x 10^{-15}W)*
-30dBm = 1 microwatt
  0dBm = 1 milliwatt
 30dBm = 1 Watt
 47dBm = 50 Watts
 50dBm = 100 Watts
 60dBm = 1000 Watts
 70dBm = 10 kilowatts

See how much simpler it is to use the decibel than the absolute measures?

Another commonly used unit is decibel relative to a Volt or to a microvolt. By now you should have the idea. To express something in decibels relative to a Volt [dBV], you place 1Volt in the denominator. To express decibels relative to a microvolt [dBµV], you place 1µV in the denominator of the equation.

Below is the formula for expressing something in decibels relative to a Volt.

$$dBV = 20\log\left[\frac{E}{1V}\right]$$

Equation 6.22

The symbol E here is used for the voltage in the numerator of the equation – you can use V if you prefer it.

Remember, if you are using voltage or current multiply the log by **20**, for power multiply the log by **10**.

A decibel is a ratio unless it is relative to something. Radio communications experimenters will often use decibels without a reference, which is fine as it is meant to be used either way. Just remember that you can't say, for example, the transmitter power is 30dB - this is a nonsense statement. You should say 30dB higher or lower than what it was, or 'by comparison with', or 'by reference to.' You could say, I am going to decrease the transmitter's power by 3dB (-3dB). Shortly, I will give you an example of a radio system using deciBels and I am sure you will then see the advantages.

For the assessment, you will need to know the basic equations to express either a voltage ratio or a power ratio in deciBels. You may be asked what an increase or decrease of *n*dB means.

For example, we have said that an increase in power by a factor of 2 is 3dB, so how do we convert 3dB back to a ratio, i.e., 2.

dB = 10log(ratio)
This formula can be transposed in terms of the ratio P_1/P_2
(ratio) = antilog(dB/10)

Let's try it for 3 dB:

(ratio) = antilog (3/10)
(ratio) = antilog (0.3)

1. enter 0.3 into your calculator's keypad;
2. press 'INV';
3. press 'log'; and
4. you get 2. [on some calculators you may get 1.9952623 – but for our purposes, this is close enough to 2]

We have worked backward to show that 3dB is an increase of 2 (two times) while -3dB is a halving.

Do remember for voltage or current use 20 and not 10 – I know I harp on this BUT it is a very common mistake made by many students.

Your examiner may not expect you to do this calculation. Though it is in my view the easiest way. What the examiner will expect is that you know some common power and voltage/current ratios expressed in dB.

Table 1-22 has the most important power ratios expressed in decibels.

Power ratio	decibel	Power ratio	decibel
One thousandth	-30	One eight	-9
One hundredth	-20	On quarter	-6
One tenth	-10	One half	-3
Unity	0	Unity	0
Ten times	+10	Twice	+3
One hundred times	+20	Four times	+6
One thousand times	+30	Eight times	+9

Table 1-22.

*This table is **power** ratios (for voltage or current – See table 2-22.).*
So what is an increase in power of 5 times?
We can look at the table and see that 10 times is +10 and halve it (-3dB); or 5 times is 10dB - 3dB = 7dB.

I would much prefer if you could remember the two basic equations for power and voltage/current. I think it is easier to convert 5 times as a power ratio to dB by doing:

$10 \log(5) = 6.989$ dB

Much simpler I think, than remembering tables.

A voltage is increased from 1Volt to 100Volt, what is the change expressed in decibels?

$20 \log(100) = 40$ dB

Table 2.22 has the most important current and voltage ratios expressed in decibels

voltage/current ratio	decibel	voltage/current ratio	decibel
One thousandth	-60	One eight	-18
One hundredth	-40	On quarter	-12
One tenth	-20	One half	-6
Unity	0	Unity	0
Ten times	+20	Twice	+6
One hundred times	+40	Four times	+12
One thousand times	+60	Eight times	+18

Table 2.22

By now you are probably wondering why bother with all this decibel stuff. Well, it's sort of like metric and imperial - once you get used to it, it is just so much easier. Test equipment is frequently calibrated in decibels. How do you measure antenna gain (or loss)? In decibels of course. A filter connected to your transmitter has a loss expressed in decibels. The list goes on and on. For me, saying my transmitter power is +50dBm is just the same as saying 100 Watts and decibels are more practical.

Very skilled people, with their unaided senses, can just about detect a 1dB change. Most of us require 2 to 3dB change before we notice. A 6dB change is quite clear to even the most untrained. So, you can see that unless a new antenna or the tuning an amplifier is going to increase the signal strength at the other end by at least 2dB, nobody will notice.

You may find this strange, but increasing a transmitter's power from 100 Watts to 130 Watts is insignificant. It only sounds significant because 130 sounds a lot more than 100, after all, it is 30 Watts more! It is not significant at all, but for some it gives them a warm, cosy feeling!

Is it really worth boosting a transmitter from a nominal 100 Watts to 130 Watts? Assuming the signal path is unchanged, the increase in power output is only 1.14dB. This would not register on most S meters. To what level would you need to increase the power to achieve 2dB?* [see the end of chapter the answer]

Here is an example of the use of decibels. Refer to the block diagram in figure 1.22, of a transmitter and receiver system.

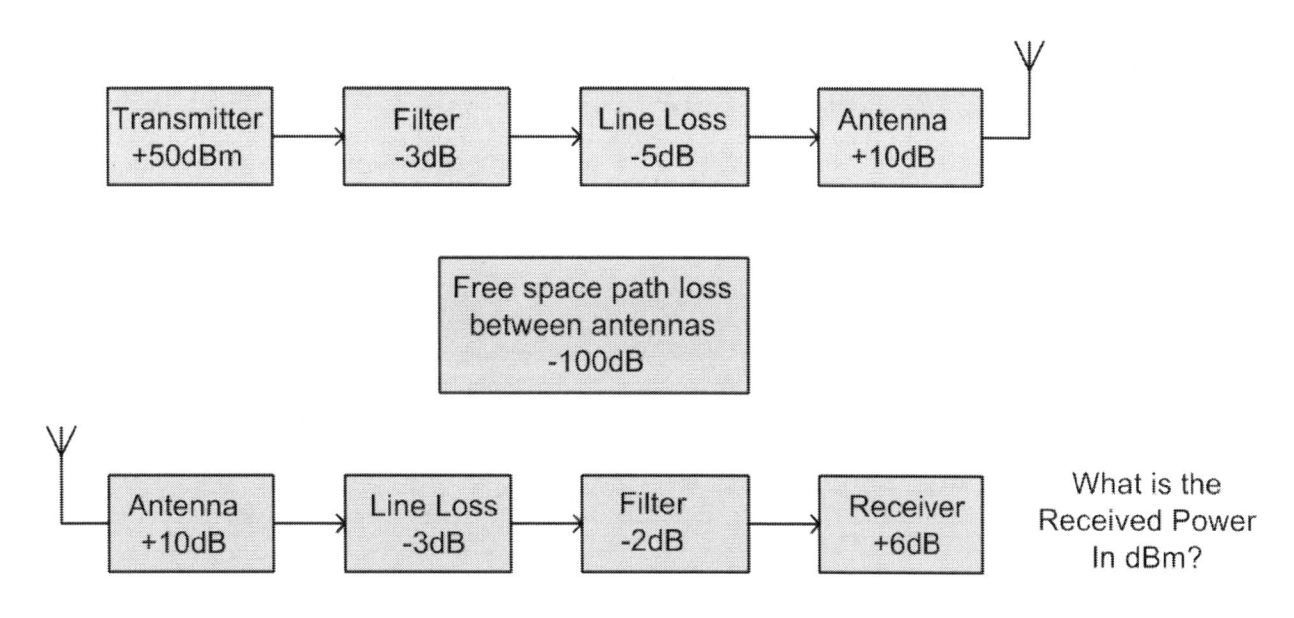

```
[Transmitter +50dBm] → [Filter -3dB] → [Line Loss -5dB] → [Antenna +10dB]

[Free space path loss between antennas -100dB]

[Antenna +10dB] → [Line Loss -3dB] → [Filter -2dB] → [Receiver +6dB]   What is the Received Power In dBm?
```

Figure 1.22

The diagram shows an entire transmitter-receiver system. The path loss block represents the radio propagation path loss. Line loss (coax or whatever) is shown for the receiver and transmitter. Filter losses in dB (it's written on them) are also shown. The receiver has a 6dB gain preamplifier in it. The antenna gains are shown. The transmitter power is shown in dBm (100 Watts). Now, without working in decibels, if I were to ask you what the received power was, this would be a nightmare.

What is the power after the 6dB amplifier in the receiver?

Just add and subtract.

50-2-5+10-100+10-3-2+6 = -36dBm.

Was that easy? -36dBm is a strong signal while -120dBm is verging on a noisy signal.

If you are lost in math and using the calculator, please speak to your Facilitator or Elmer for help. RES does have math video tutorials on DVD.

*Answer:
2dB = 1.58
So, you would need to increase the power output from 100 Watts to 158 Watts. Before embarking on such a step, you might want to consider the rating of the power output device and what safety factor you are willing to sacrifice. Most solid state devices wouldn't like such an increase.

23. ELECTROMAGNETIC RADIATION

THE MECHANISM OF RADIATION

The purpose of this chapter is to describe electromagnetic radiation and how it radiates. We also cover the concept of wavelength, frequency and period. We look at an overview of the electromagnetic spectrum.

Again I am faced with the dilemma of providing some information that you most likely not be examined on. However, such information provides useful insights to the topic, which I believe makes it more interesting.

Though radio waves can have different wave shapes, I would like you to think of them as sine waves for this exercise. Many are just that, sine waves. We have already looked at alternating current and voltage at lower frequencies, for example, the mains, which is 50 Hertz. AC produces radio waves. Yes, the AC you have already learned. The difference is that the frequency of oscillation is much higher than 50 Hz.

We have also learnt that current, either alternating or direct (or any other), has associated with it electric and magnetic fields. We refer to these combined fields as 'electromagnetic fields'.

It was discovered that high-frequency electromagnetic currents in a wire (antenna), which in turn result in a high-frequency electromagnetic field around an antenna, will result in electromagnetic radiation which will move away from the antenna into free space at the velocity of light (approx. 300,000,000 metres per second).

Firstly, for electromagnetic radiation to occur the frequency must be reasonably high - I will explain this. Also, because the drawing of both electric and magnetic fields around a wire in three dimensions is difficult, I will stick to just two dimensions.

All we need to create an electromagnetic wave is to accelerate any charge. The obvious choice of the particle to accelerate is the electron because of its low mass. However, any charge would do.

The simplest way to do this is to get a high-frequency generator (transmitter) and connect it to a dipole antenna. When we do this; electrons will flow in the same direction in each leg of the dipole. When the RF driving force changes polarity then the electrons will flow in each leg of the dipole in the same but opposite to the previous direction. That flow of electrons will not be immediate in all parts of the dipole. The charge movement will begin at the feedpoint (presume the centre) and move in one direction. By this I mean charge will propagate along the length of the antenna and this takes time, it is not instant, even if the electrons had no mass and moved as fast as photons ('c'; 300,000,000m/s) they would still take time to propagate down the antenna.

When electrons are accelerated, they create both magnetic and electric fields. These fields move with the accelerating electrons. The result is an electromagnetic field forms around the antenna. When the RF generator changes polarity, you might expect this field to collapse back into the antenna. The field does not have time to collapse. Another set of fields; electric and magnetic; called the E and H fields respectively is created around the antenna and the direction of the lines of force of this second field is the opposite of the first; and the first field is propelled away at the velocity "c."

If we could see this, it would look like figure 1.23.

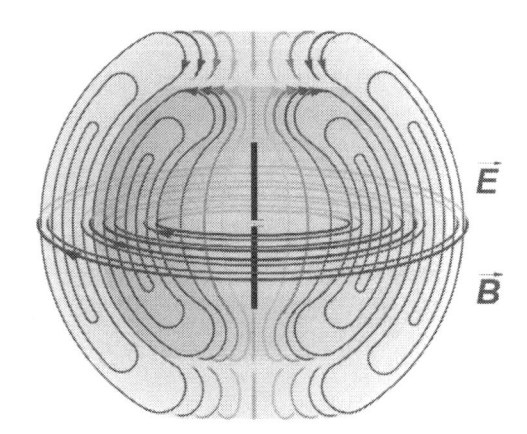

Figure 1-23.

To help visualise this better, we could just look at one field. I chose the magnetic field. Figure 1.23(a) is a frozen moment when the RF generator has stopped and about to change polarity. There have been three reversals of current in the antenna and you see one set of three magnetic loops in cross section. Notice how the direction of the field lines change direction on each alternate loop. This means the loops will repel and cannot merge.

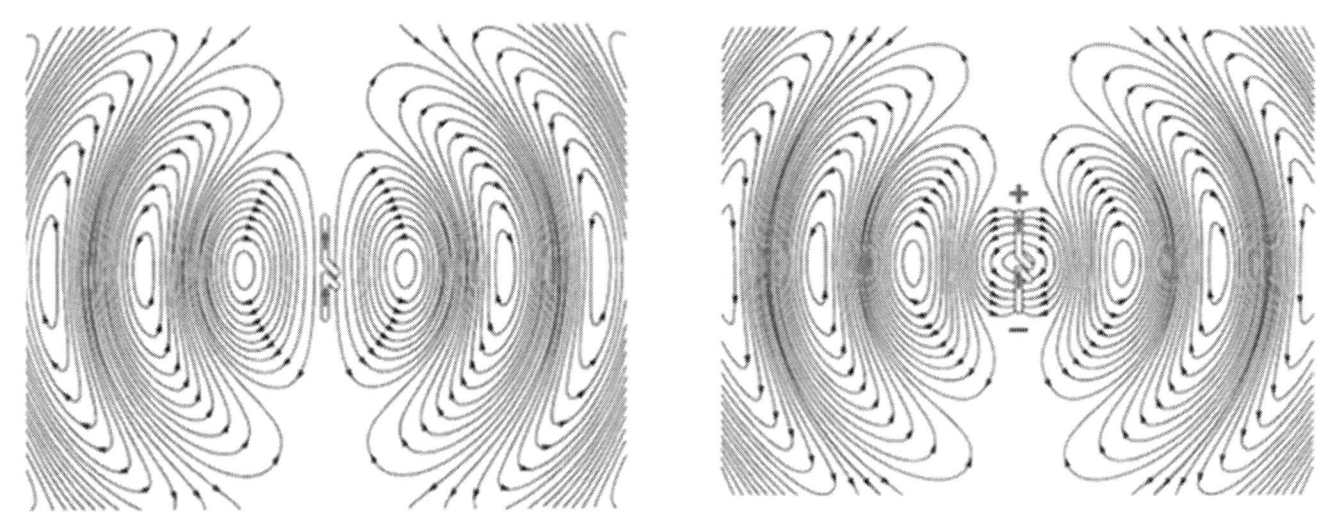

Figure 2-23(a)

Figure 2-23(b)

In figure 1.24(b) a new field is being created with lines of force opposing the last field created. The entire electromagnetic field (both electric and magnetic fields) propagate away from the antenna at the velocity 'c'. The process repeats on the next half cycle but with the polarities on the antenna reversed and the direction of the lines of force also reversed.

EM waves do not need a medium

EM waves; unlike sound waves or water waves do not require a medium to travel in or on. Recall when we discussed tuned circuits that oscillation was sustained in a tuned circuit by the energy of the collapsing electric field in the capacitor created a magnetic field in the inductor. When the magnetic field of the inductor collapses the energy shifts to the electric field. It is the energy of the electric and magnetic fields that do the moving in electrical resonance.

It is the same with electromagnetic waves. The electric and magnetic field propagate at right angles to each other and as one field collapses it generates the other. A collapsing electric field recreates the magnetic field, then when done, the collapsing magnetic field recreates the electric field. An electromagnetic wave is self-supporting. Of course, it gets weaker. There is attenuation in any medium, even a vacuum. Also, as the wave propagates it spreads out in all directions – the same energy in a greater volume means less field strength as we move away from the antenna The wave attenuates at the wavefront in accordance with the inverse square law. If you double the distance from the source the wave intensity decreases by $1/2^2$ or ¼ which is a fourth. At three times the distance the wave intensity is now $1/3^2$ or 1/9 a ninth. Figure 2-.23.

sphere area $4\pi r^2$

source strength S

intensity at surface of sphere

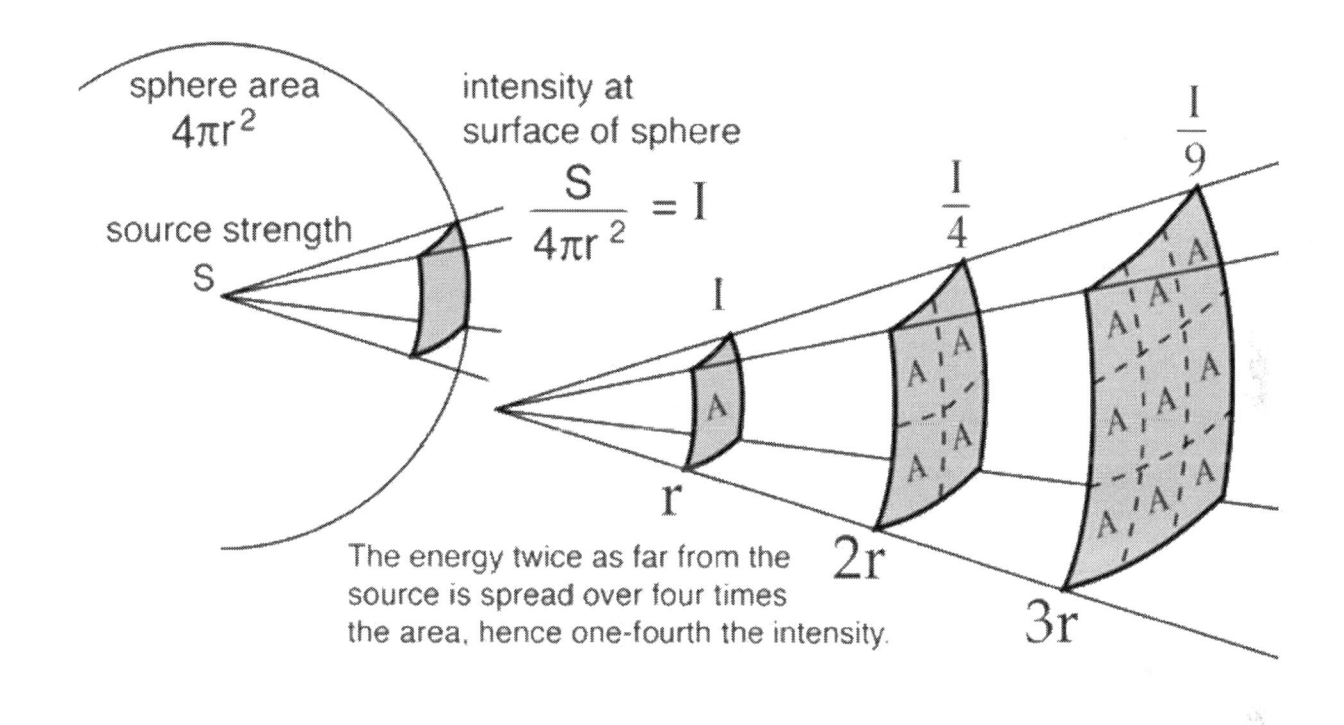

$$\frac{S}{4\pi r^2} = I$$

The energy twice as far from the source is spread over four times the area, hence one-fourth the intensity.

Figure 3-23

Why does an electromagnetic wave travel at the speed of light?

Well, I will give you the simple answer. Light is an EM wave. All waves travel at the same maximum velocity in free space that velocity is denoted by little 'c' and is equal to approximately 300,000,000 metres/second.

A clue to why EM waves travel at 'c' is that they slow down when not going through free space. Why do they slow down?

EM waves propagating through a cable either RF or optical slow down. Why?
Well, it must have something to do with the medium through which they are propagating. This is indeed the case. The velocity of an EM wave in a cable is determined by the

Permittivity and the *Permeability* of that medium.

To remind you;
Permittivity is the ability of any material to concentrate electric lines of force. *Permeability* is the ability of any material to concentrate magnetic lines of force.

We are inclined to think of free space or just 'space' as having no properties. However, this is not the case. Space has properties and is certainly never empty. It was James Clerk Maxwell who worked out the theory of electromagnetism or the combining of electric and magnetic fields into electromagnetic waves and describing their behaviour. Maxwell was the first to calculate 'mathematically' the velocity of 'c.' He did this more accurately than anyone before him.

The Permeability μ_o and Permittivity ε_o can be just measured. These properties were known to Maxwell.

Maxwell discovered that these properties of free space determined the maximum velocity of an EM wave in free space just as they determine the velocity of an EM wave through any other medium. The values for μ_o and ε_o for free space are given below, without units as we do not need them. We just need Maxwell's equation for determining the velocity of 'c.'

Permeability and Permittivity of free space.

$\mu_o = 4\pi \times 10^{-7}$
$\varepsilon_o = 8.85 \times 10^{-12}$

$$c = \frac{1}{\sqrt{\mu_o \varepsilon_o}}$$

Equation 1-23

Inserting the values and doing the calculation we get:

$$c = \frac{1}{\sqrt{(4\pi x 10^{-7}) x (8.85 x 10^{-12}}}$$

$$c = 2.998 \; x \; 10^8$$

That answer we round to 300,000,000 m/s.
Now have we responded to the question posed at the beginning of the section?

Our eyes are biological receivers that can detect EM waves in the region of 400-700 nanometres. We generally call that the visible light spectrum. I often think that it is advantageous that our eyes only detect such a narrow bandwidth of EM radiation. If that were not the case, we would see the world awash with the light of colours yet to be determined. Even you cell phone would glow with light.

USES OF ELECTROMAGNETIC RADIATION

Naturally, since the electromagnetic radiation travels so fast (300,000,000 m/s), we can use it to transfer information from one place to another. In most cases because the velocity is so fast, it *appears* to be instant, though it is important to realise that it is *not instant*. A fact, that is very obvious when we communicate through satellites and with space vehicles. As an example, it takes about eight and a half minutes for light to reach earth from the sun, so you never see the sun where it is, you only see it where it was eight and a half minutes ago.

The process of impressing intelligence or information onto an electromagnetic wave is called *modulation*.

In radio transmission, a radiating antenna is used to convert a time-varying electric current into an electromagnetic wave, which freely propagates through a non-conducting medium such as air or space. In a broadcast radio channel, an omnidirectional antenna radiates a transmitted signal over a wide service area. In a point-to-point communication channel, a directional transmitting antenna is used to focus the wave into a narrow beam, which is directed toward a single receiver site. In either case, the transmitted electromagnetic wave is picked up by a remote receiving antenna and reconverted to an electric current.

Radio wave propagation is not constrained by any physical conductor or waveguide. Radio is ideal for mobile communications, satellite and deep-space communications, broadcast communications and other applications in which the laying of physical connections may be impossible or very costly. The medium through which radio waves propagate is highly variable, being subject to diurnal, annual and solar changes in the ionosphere, variations in the density of water droplets in the troposphere, varying moisture gradients and diverse sources of reflection and diffraction.

All of these aspects we will be looking at in more detail.

Of course with the world communicating so much and using electromagnetic radiation, steps must be taken to have different users share different frequencies so that communications do not interfere with each other. The range of useful frequencies is a finite resource which many governments including ours (Australian) has capitalised on by selling parts of it very much like real estate even to the point of Spectrum Auctions.

THE ELECTROMAGNETIC SPECTRUM

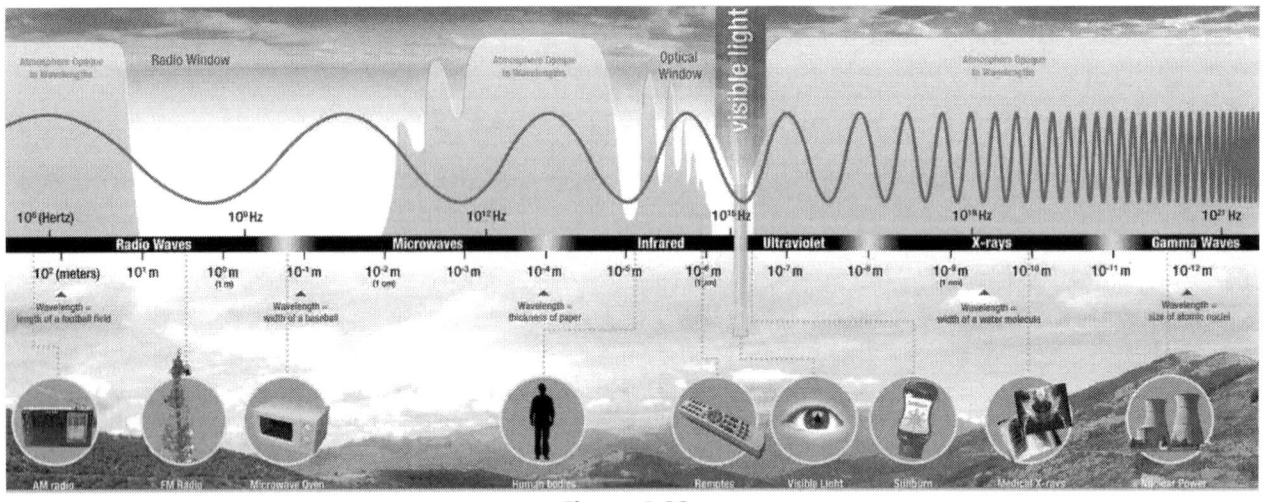

Figure 4-23.

Refer to figure 4-23 showing the electromagnetic spectrum from low frequencies (long wave radio) to exceedingly high values of (gamma rays). Going from the values of radio waves to those of visible light is like comparing the thickness of a piece of paper with the distance of the Earth from the Sun, which represents an increase by a factor of a million billion. Similarly, going from the (frequency) values of visible light to the very much larger ones of gamma rays represents an increase in frequency by a factor of a million billion. This enormous range of values is called the electromagnetic spectrum, together with the common names used for its various parts, or regions.

As frequency increases, it becomes increasingly dangerous to humans. Frequency up to about the top end of ultraviolet are non-ionising. From X-rays and up EM waves are ionising and in this range all exposure should be avoided were possible. Of course, you cannot avoid it all. The Sun drenches the earth in ionising and nonionising radiation. However, to a degree, we have evolved to live in such environments. Additional exposure from X-rays and above should are dangerous. EM waves below X-rays may also be dangerous. They may or may not be but below X-rays, we have safety limits on power, distance from the source and duration (duty cycle) of the source.

Frequency and Wavelength

This conversion must be known for examinations. The relationship between frequency and wavelength is:

Wavelength (λ) = c / F   **Equation 2-23**

Where 'c' is 299,792,458 metres per second. In radiocommunications, it is nearly always rounded to 300,000,000 metres per second.

A more useful form and easier to memorise is:

λ = 300 / F(MHz)

Calculating wavelength

For example, what is the wavelength of 146 MHz, which just happens to be the middle of the 2-metre amateur band?

λ = 300 / 146 = 2.0547 metres.

When describing this 'band', the wavelength is rounded to '2' metres.

More often, accurate measurements of wavelength are required, such as cable lengths and antennas dimensions. Where more accuracy is needed, the wavelength as calculated from the above equation must be expressed using the appropriate number of decimal places.

Frequency bands have the following common names:

3-30 kilohertz	Very Low Frequencies band (VLF)
30-300 kilohertz	The Long Wave band (LW)
300-3000 kilohertz	The Medium Wave band (MW)
3-30 megahertz	The shortwave band (SW) Referred to by Amateurs as the HF Band (High Frequency)
30-300 megahertz	Very High-Frequency Band (VHF)
300-3000 megahertz	Ultra-High-Frequency band (UHF)
3-30 gigahertz	Super High-Frequency band (SHF)
30-3000 gigahertz	Microwave Frequencies
Above	Above 3000 Gigahertz we have infrared, visible light with all its colours, ultraviolet, X-rays and at the very top gamma rays

Table 1.23

POLARISATION

The electric and magnetic fields of an electromagnetic wave are perpendicular (at right angles) to each other.

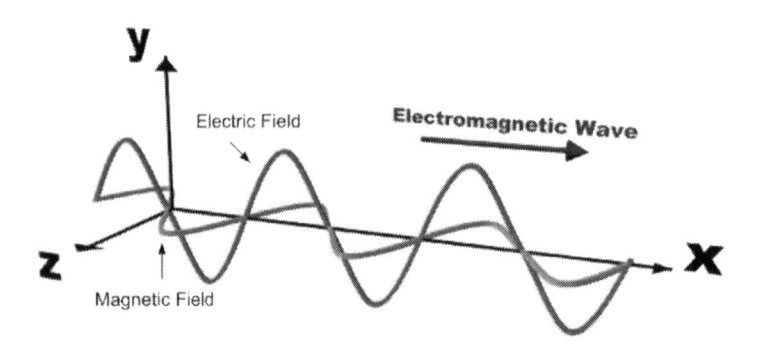

Figure 5.23.

The polarisation of an electromagnetic wave is *always* described with regard to the direction of the electric field relative to the earth. If the electric field is vertical relative to the earth, the wave is said to be vertically polarised. If the electric field is horizontal relative to the surface of the earth, then the wave is said to be horizontally polarised.

If the antenna is horizontal with respect to the earth, the wave is horizontally polarised. If the antenna is vertical with respect to the earth, the wave is said to be vertically polarised. The important factor is that the direction of the electric field is what determines whether a wave is called vertically or horizontally polarised.

When there is propagation via the ionosphere, it is possible for a wave radiated from a vertical antenna to have its polarisation twisted – this is called *Faraday rotation*. So even though a wave may be radiated from a vertical antenna it can have its polarisation rotated by the ionosphere and become horizontal. For this reason, we do not define the polarisation of a wave by the type of antenna which is radiating it. We always define polarisation by the direction of the electric field. Never assume that radiation from a horizontal or vertical antenna will be received at a remote location by ionospheric propagation in the same polarisation that it left the antenna.

Radiation and Reception

How much electromagnetic energy can leave an antenna depends on the relationship of its length to the wavelength of the current.

Just as electric and magnetic fields surround a wire or antenna carrying an RF current, so a wire placed in an electromagnetic field will have a current induced in it. As far as antennas are concerned, transmitting and receiving antennas are basically interchangeable, apart from power handling considerations. In fact, a so-called principle of reciprocity exists, which states that the characteristics of antennas, such as impedance and radiation pattern, are identical regardless of their use for transmission or reception.

To receive a vertically polarised wave (electric field vertical), a vertically polarised antenna should be used.

However as explained earlier Faraday Rotation over a high frequency (HF) circuit can modify the polarisation. Propagation effects such as diffraction, refraction and reflection can alter the polarisation of VHF waves and above.

A different way of looking at impedance

We have talked about impedance in the past as *belonging to something*, i.e. resistors, capacitance and inductance. Impedance is the total opposition to current flow in an AC circuit. When we talk about electromagnetic waves our concept of impedance needs to be broadened. This applies in particular to transmission lines (e.g. coaxial cable) and antennas. Impedance should not be thought of as belonging to a *thing* but as the ratio of voltage over current. From Ohm's law R=E/I and we extend that to impedance with Z=E/I. Frequently we have been talking about the voltage across something and the current through the same thing, to calculate the impedance. This might appear a bit of a leap, but when talking about electromagnetic waves, we need to take the concept of the impedance *belonging* to a component(s) away and think of the impedance as the ratio E/I. This is particularly true of antennas and transmission lines. I am just preparing you for it here. For example, free space has impedance! When an electromagnetic wave travels through free space its electric and magnetic components settle into a particular ratio that just happens to be 120$\Pi\Omega$ or 377Ω - so we say the impedance of free space is 377Ω. Likewise, an electromagnetic wave travelling down a transmission line will cause particular voltage and current ratios to exist at different points. We then have to think beyond the transmission line and reflect on *the ratio of voltage and current* at various points on it to understand or define the impedance at those points.

In other words, electromagnetic waves can be impedance up-setters. A wave travelling down a cable *may* determine the impedance and *not* the cable. This will become clearer as we go (I hope). For now, yes, impedance is the total opposition to current flow in an AC circuit. Introduce electromagnetic waves into transmission lines and antennas and the wave and its behaviour will be what determines the impedance at any given point. Impedance is always the ratio provided by Ohms Law E/I.

We will also be discussing the propagation of waves. It will help if you think about light waves. We are all very familiar with the behaviour of visible light. Whatever we can imagine light doing, we can do it, or it is done with, electromagnetic waves, because they are one and the same thing, just on a different frequency.
Reflection, refraction, absorption and anything else you can think of pertaining to light, pertains equally to electromagnetic waves.

DOPPLER SHIFT

When a car blowing its horn approaches you the pitch (frequency) of the horn sounds higher. As it immediately passes you, you hear the actual pitch of the horn. As it moves away from you, you hear a lower pitch. This effect is called *Doppler Shift*.

The same happens to electromagnetic waves which are being radiated or received from a moving source. Doppler Shift is most noticeable when communicating through satellites. When the satellite is moving at high velocity towards you, the frequency of your receiver has to be tuned a little higher than the actual frequency. When the satellite moves away from the receiver has to be tuned to a lower frequency.

Stars give off visible light. Red light is a lower frequency than blue light. Scientist are able to tell if a stars are moving towards us (or us towards it), by the blue shift (increase in frequency) of the light reaching us. If the star is moving away, then there will be a red shift.

This is also precisely how police Doppler Radar works. An electromagnetic wave pulse is sent to the target (the vehicle). If the reflected wave is echoed on the same frequency, the vehicle is not moving. If the car is moving, there will be a downshift or an upward shift in the frequency of the echo. How much shift there is can be used to compute exactly how fast the vehicle is moving.

Doppler shift only plays a part in radiocommunications when the relative velocity between the receiver and transmitter is high. An example is communication with or through a satellite. When communicating through satellite, the Doppler correction is done manually by the operator or under computer control. We never really have to calculate the actual frequency shift.

The relationship between the observed frequency f and the emitted frequency f_o Is given by:

$$f = \left(\frac{c + v_r}{c + v_s}\right) f_o$$

Equation 3-23

where: c is the velocity of wave in the medium;

v_r is the velocity of the receiver relative to the medium; positive is the receiver is moving towards the source (and negative in the other direction);

v_s is the velocity of the source relative to the medium; positive is the source is moving away from the receiver (and negative in the other direction).

24. SEMICONDUCTORS PART 1

The objective of this chapter is to cover how semiconductors work. Discuss the operation of bipolar junction transistors (BJTs), FETs MOSFETs and SCRs, as well as a few of the characteristic circuits in which they are employed.

Solid state devices

Solid state devices are much smaller physically, more rugged and lighter than vacuum types, but cannot withstand heat as well, changing their characteristics as they warm. However, many newer special solid state devices will operate at much higher frequencies than any vacuum device.

SEMICONDUCTORS

The outer orbit, or *valence* electrons of some atoms, such as metals (conductors), can be detached with relative ease at almost any temperature and may be called free electrons. This is why metals are good conductors. These valence electrons can move outward from a standard outer orbit level into a conduction level or band, from which they can be easily dislodged. Such materials make good electrical conductors. Other substances such as glass, rubber and plastics, have no free valence electrons in their outer conduction bands at room temperatures and are therefore good insulators.

A few materials have a limited number of electrons (4) in the conduction level (outer orbit or valence band) at room temperatures and are called *semiconductors*. Applying energy in the form of photons (small packets of light or heat energy) to the valence electrons moves some of them up into the conduction band and the semiconductors then become better electrical conductors. Energy of some form is required to raise semiconductor electrons to a conduction level. Conversely, if an electron drops to the valence level from a higher conduction level, it will radiate energy in some high-frequency form such as heat, light, infrared, ultraviolet and, if the fall is significant enough, X-rays.

The semiconductors, *germanium and silicon* have four outer valence electrons. Semiconductors can be laboratory grown, this makes mass production of them easy.

COVALENT BONDING

In nature atoms are more stable and 'like' to have eight valence (outer) electrons called an *octet*. Chlorine has seven valence electrons. If two Chlorine atoms are brought near each other, they will snap together and form a Chlorine molecule. Each atom is sharing one of its valence electrons with the other. This covalent bonding makes the Chlorine molecule more stable.

Noble gases have eight valance electrons called an octet, this is what makes noble gases so stable.

In the case of *germanium* and *silicon* which only have four valence electrons. A special arrangement occurs when they rub shoulders. Germanium or silicon in pure form creates what is called a *crystal lattice structure*. The four valence electrons of each atom in a crystal lattice structure share themselves with the adjacent orbits of all other electrons in the crystal lattice structure. In this way by borrowing electrons from neighbouring atoms a type of shared valency of eight is created. Consequently, the valence electron arrangement is very stable and it is difficult to make these electrons participate in current flow. The great breakthrough in physics (electronics) was the discovery of how the characteristics of pure germanium or silicon could be changed dramatically by adding impurities (other atoms) to the crystal. Adding impurities disturbs or upsets the crystal lattice structure.

INTRINSIC SEMICONDUCTOR LATTICE

A perfectly formed intrinsic semiconductor crystal lattice is illustrated in figure 1-24. Such a crystal acts more like an insulator than a conductor at room temperature. I have drawn the crystal in only two dimensions (not being much of an artist). Shown are the valence band (the outer orbit of each atom) and the four valence electrons. The valence electrons are of course not stationary, but orbiting around the atom as if on the surface of a sphere. The sharing of valence electrons is called covalent bonding. This arrangement is very stable electrically.

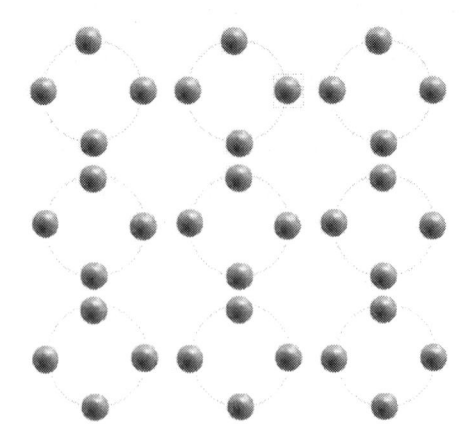

Figure 1-24.

Electrons are locked into the crystal lattice and at normal temperatures the crystal is neither a good insulator or conductor.

DOPING (N-type)

Doping is the process of deliberately adding impurities to the crystal during manufacture. Say we added some arsenic, which has five valence electrons to a pure semiconductor crystal. The ratio of germanium to arsenic is about a million to one.

Arsenic is pentavalent (five outer electrons) and cannot fit into the crystal lattice structure. What happens is four of the arsenic electrons participate in the sharing (covalent bonding) and one is left out!

Extra electron

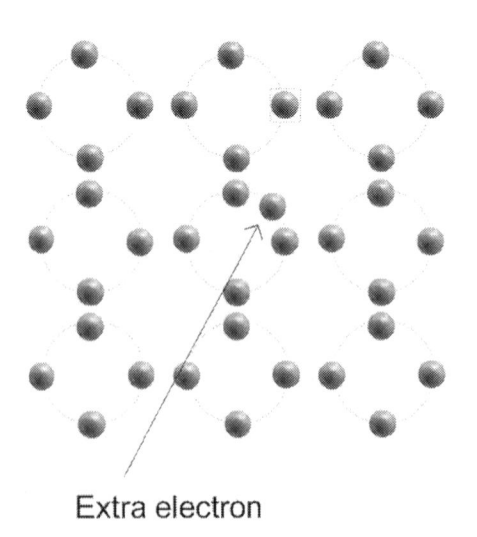

Figure 2-24.

The crystal lattice shown in figure 2-24 has one atom in a million with an excess outer ring electron not being tightly held. This type of doped semiconductor is called N-type.

It should be understood that 'N' type semiconductor does not have extra electrons electrically. N-type semiconductor does not have a negative charge. What is 'extra' in N-type material is electrons which do not fit into the crystal lattice structure. These extra electrons are not locked into the crystal lattice structure, so they are *much easier to move*.

When an electrostatic field (by application of an emf) is developed across such arsenic-doped germanium, a current will flow. The N-type semiconductor is about 1,000 times better as a conductor than the intrinsic semiconductor. Doped germanium with such relatively free electrons is known as N germanium and is a reasonably good conductor. To form N-silicon, phosphorus can be used as the dopant – you do not need to remember the dopant. Merely by adding a small amount of impure pentavalent atoms to a pure crystal, we convert it into a conductor by disrupting the harmony of the crystal lattice structure.

DOPING (P-Type)

When germanium is doped with gallium, which has *three valence electrons* (tri-valent), the crystal lattice is again disrupted. This time, there is an area, or *hole*, in the crystal lattice structure, a region that apparently lacks an electron. While the hole may not actually be positive, at least it is an area in which electrons might be repelled to by a negative charge. This positive *appearing* semiconductor material is called P-germanium or just *P-type*.

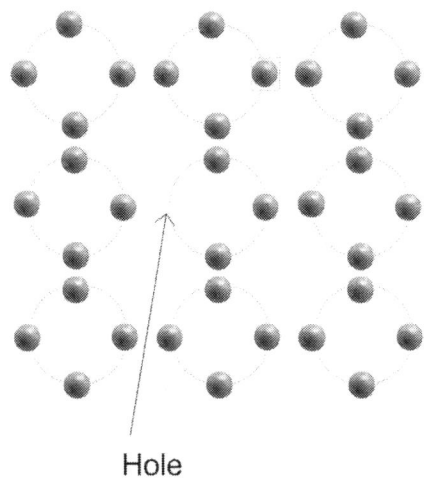

Hole

Figure 3-24

When an electrostatic field is impressed across a P-type semiconductor, the hole areas act as stepping stones for electron travel through the material. It can be said that *hole current* flows in a direction opposite to the electron flow. Note that both N-germanium and P-germanium have zero electric charge because both have an equal number of electrons and protons in all of their atoms. One dopant used to produce P-silicon is boron.

HOLES

I am going to talk about holes for a bit, as it seems to be a stumbling block for many. I have drawn the *hole* as a gap in figure 3-24. The hole is a *missing electron in the crystal lattice structure*, which destroys the crystals insulating properties. It is very easy with N-type material to visualise that electron flow can take place. With a hole, it is a little harder and I find some textbooks a little confusing on this issue. A hole is a hole in the crystal lattice structure. Because the lattice is not complete in the location where a hole is, electrons can move into the hole and in doing so, they create a hole from where they came.

SOME FOOD FOR THOUGHT

A hole is a missing electron in the crystal lattice structure.
A hole is not a positive charge.
A hole like any hole can be filled.
A hole can be filled with an electron.
When an electron does fill a hole – then the filled hole disappears, BUT where the electron came from there is now a hole.

If an electron falls into a hole, then where that electron came from will be a hole.
A hole can be *thought of* as positive for behavioural description purposes.
An electric current is an ordered movement of electrons.
Holes allow electrons to move in the crystal by giving them somewhere to go i.e. filling a hole.
When an electron moves out of the covalent bond to fill a hole it leaves a hole from whence it came.

Figure 4-24.

I think most of us have seen the toy shown in figure 4-24. A flat panel of plastic squares with pictures or numbers on them and the objective is to move the squares around into some order, either to get the numbers in order or to make a picture. Would you be able to slide the squares around if the game was made without a missing square? No, of course not. By leaving a square out, *leaving a hole in the puzzle,* it makes it possible to slide the squares around by moving them into a hole. In moving a square into a hole you create a hole, making it possible to slide other squares into that hole. Think of the squares as electrons and their ability to move is made possible by the presence of the hole (missing square).

Slide 14 (Figure 4-24) could be moved straight up. That would leave a hole where slide 14 was. Now, if you were to put this puzzle on autopilot and sit back and watch it, what would you notice about the way the hole moves? It moves in the opposite direction to the slides. So if the numbered squares are electrons the hole is behaving like a positive charge in that it moves in the opposite direction to electrons. Some references do not explain this very well and go on to talk about "hole current" moving from positive to negative, I believe this to be confusing. Electrons are the only current. Though I will be talking shortly about holes moving, holes do really move when you fill them. However, all of the *real* moving is done by electrons falling into holes in a P-type semiconductor.

The P-type semiconductor material is a conductor because of the presence of holes in the crystal lattice structure. Doped silicon has considerably more resistance than germanium and finds its uses in higher voltage and current applications. Also, it does not change its resistance when heated and it can withstand greater temperatures without its crystalline structure being destroyed.

So what?

Well, we have not seemed to have achieved much have we? We have taken a perfectly good semi-insulator (semiconductor) and turned it into a conductor by adding pentavalent impurities (N-type) or trivalent impurities (P-type). The magic starts when we combine the two, that is, we make one side of the crystal N-type and the other P-type. In reality, they are not made separately and then stuck together; crystals are grown and doped on different sides of the same crystal in the laboratory.

SOLID-STATE DIODES

Before we start, the term *solid state* is only used because the alternative devices before them were the electron tube devices.

Let's take a piece of N-type and P-type semiconductors and join then together. The area in which the N and P substances join is called the *junction*. We don't really take one of each and physically put them together. It is achieved by doping different sections of the same piece of intrinsic semiconductor.

Some of the relatively free electrons in the N-type material at the junction fall into some of the holes in the P-type material at the junction. So, right at the junction, there are no free electrons (in the crystal lattice structure) and no holes, as free electrons from the N material have filled some of the holes in the P material. This creates a region at the junction, which has neither free electrons nor holes. It is a region depleted of free electrons and holes and is called the **depletion zone**.

This develops an area at the junction that is actually slightly negative on the P side (because electrons have filled holes) of the junction and slightly positive on the N side (because electrons have left to go and fill holes). *This produces a barrier to any further electron flow of about 0.2 V with germanium and 0.6-0.7 V with silicon diodes. This small potential is called the **barrier potential**.*

We have created a semiconductor diode (PN junction). In figure 5-24 the depletion zone is shown at the junction drawn as a small band.

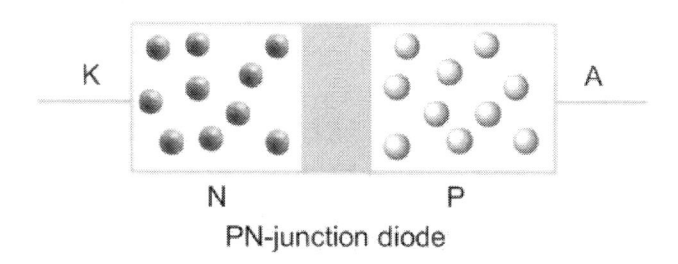

Figure 5-24.

The whole (no pun intended) diagram is exaggerated greatly as the depletion zone is extremely narrow and there are many more holes and electrons in a real PN junction. Don't forget the diode in figure 5-24 right now has a very small potential across it called the barrier potential. We can't do any work with that voltage. It is not a small battery; it is a weak electrical pressure, created by hole and electrons recombining at the junction. However, this barrier voltage is real and affects how the p-n junction diode operates.

Applying a voltage to a PN-Junction diode

The diode of figure 6-24 has the polarity applied to it as shown, The N; end of the diode is called the cathode (K) and the P; the end is called the anode. Remembering electron tubes if we applied negative to an anode and positive to a cathode would we get conduction? No, we would not. This is called reverse bias.

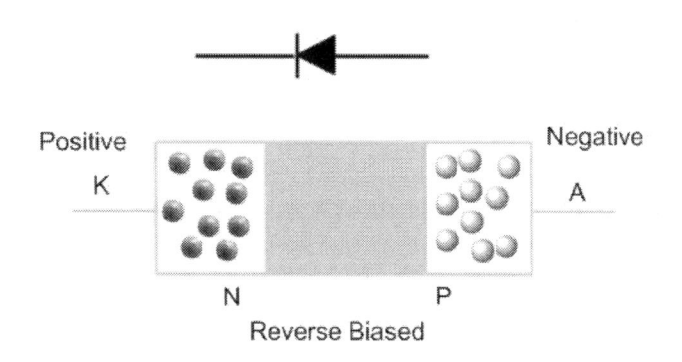

Figure 6-24.

REVERSE BIAS

In figure 6-24, we have applied an external electrical pressure. Electrons are attracted toward the positive terminal whilst holes, thinking of them as if they were positive charges, are attracted to the negative terminal.

No current flows through the PN junction, which from now on, we will call a diode. The depletion zone is widened as electrons and holes are moving away from the junction. The diode is said to be *reverse biased*. I have drawn the schematic symbol of a diode, so you can see that for reverse bias, positive is connected to the cathode and negative to the anode. The left-hand side of the diode shown is the cathode.

Also, recall when we discussed power supplies without really describing how a diode worked. I suggested you look at the diode schematic symbol as an arrow. In figure 6-24, the arrow is pointing to the left. I asked you to remember that conduction was only possible in the opposite direction of the arrow. Electrons can only flow in the direction cathode-to-anode, against the arrow.

We have said that no current flows through the diode when it is reverse biased. There is an extremely low *leakage* current which for most practical purposes can be considered to be zero. Also, if you increase the reverse bias voltage high enough you will blow the crapper out of the diode by exceeding the breakdown voltage (PIV).

The reason why you do get a reverse leakage current in a diode is that some of the N material will have just a few holes in it and some of the P material will have some electrons in it. These are due to extremely small amounts of contaminants. So there is like a minor 'ghost' diode the opposite way around to the 'real' diode. These contaminants are called *minority current carriers*. The real holes and electrons are called the *majority current carriers*. If you like, you have a majority diode one way and a minority diode the other way. When the majority diode is reversed biased the minority diode is forward biased and this accounts for the small leakage current.

The *reverse leakage current* of a PN junction *increases with temperature*. Reverse-biased PN junctions can be used to measure temperature, by amplifying and measuring the reverse leakage current.

FORWARD BIAS

Now let's reverse the polarity. I know you are aware that the diode is going to conduct but let's look at exactly what goes on. We are going to think of the holes as positive charges when we really know they aren't - we covered that issue with the 'puzzle' example earlier. However, do let me know if any part of this is not clear enough.

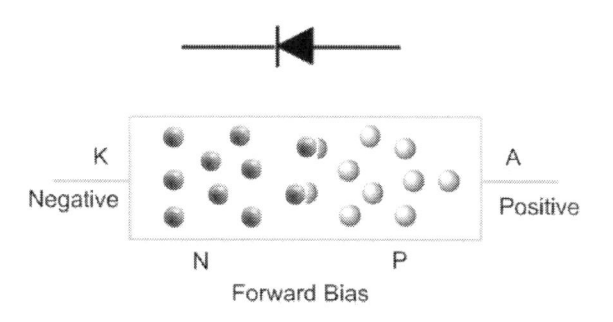

Figure 7-24.

We have now applied a negative potential to the N material (cathode) and positive to the P material (anode). **Provided this potential is greater than the barrier potential** (0.2V germanium, 0.6V silicon) the depletion zone will be flooded with electrons and the diode will conduct as shown in figure 7-24.

You may well ask, "why don't all the electrons move across the junction and fill up all the holes." We all know that current flow is electrons. In figure 7-24 the right-hand side is P-type (holes) and electrons leave the P-type anode and flow to the positive terminal of the battery. Every electron that leaves the anode creates a hole. The holes move toward the junction to be filled by more electrons.

Such a diode could be used in a rectifier circuit. All diodes have a maximum current rating as well as a peak inverse voltage rating. A diode, for example, may be rated at 1Amp 400V PIV. Diodes come in many shapes, sizes and package types. All of them more or less work on the principles we have described here. Even LED's (light emitting diodes) are just special PN junction diodes.

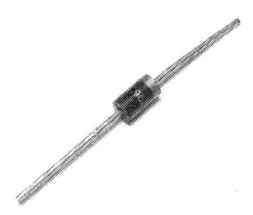

Figure 8-24

Figure 8-24 is a typical rectifier diode that could be used as part of a power supply. The band around one end marks the cathode. A group of four such diodes could be packaged into a bridge rectifier.

CHARACTERISTIC CURVE of a DIODE

The graph in figure 9-24 shows the operating characteristics of a diode. The forward voltage is the voltage that biases the diode 'on' and it conducts. The reverse voltage causes the diode not to conduct or block current. However, if the reverse voltage is made high enough, the diode will breakdown and conduct in the reverse direction. The blue line shows the forward and reverse current versus the forward and reverse voltage.

This (break down) is either the *Zener effect* if the emf value at breakdown is less than about 5V, or A*valanche effect* if it is more than about 5V. This is not a normal operating condition for most semiconductor diodes and may cause lattice damage, ruining the diode.

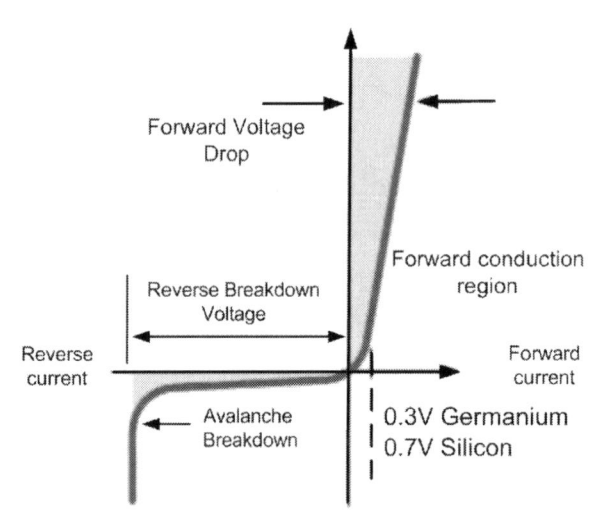

Figure 9-24

ZENER DIODES

The reverse voltage breakdown effect, however, is used in special Zener diodes. These diodes are deliberately operated under enough voltage to cause them to conduct in the reverse direction. A resistor must be connected in series with a Zener diode to prevent the junction from being destroyed. The circuit across which the diode is connected will not increase in voltage over the Zener breakdown voltage. For this reason, Zener diodes are used as shunt (parallel) *voltage-regulating* devices. Zener diodes are used to provide a regulated DC voltage in low power applications. So if some part of a 12-Volt DC circuit required 5V DC at low power, then a Zener diode with an appropriate series resistor could be used to provide a regulated 5-Volt DC output in spite of variations in the 12-Volt DC input voltage. A good way to remember the symbol of a Zener diode, as shown in figure 10-24, is to note the shape of the cathode line on the diode as representing the forward and reverse current characteristics of a diode shown earlier in figure 9-24.

Figure 10-24

THE CIRCUIT OF A ZENER VOLTAGE REGULATOR

The circuit diagram of a Zener used as a voltage regulator is shown in figure 11-24. The unregulated DC to the circuit varies from 8 to 12 Volts. R_S is a current limiting resistance to prevent the Zener from being destroyed.

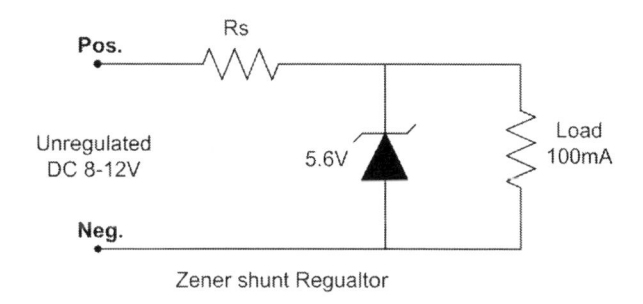

Zener shunt Regualtor

Figure 11-24

Remember, a Zener regulator is operated with reverse bias and to the point where it breaks down and conducts in the reverse direction. For some, the term 'breakdown' is confusing. Breakdown does not mean the Zener is destroyed or busted! Break down means it is forced to conduct in the opposite direction (from the anode to cathode). Normally a diode operated beyond breakdown is destroyed. However, Zener diodes are designed to operate in the breakdown region with a small breakdown current. A Zener would be destroyed just like any other diode except for the current limiting resistance R_S. Under these conditions the voltage across the Zener is constant. In figure 11-24 the voltage across the Zener will be 5.6 Volts (it is a 5.6 Volt Zener – you buy them with a voltage rating). Irrespective of input voltage fluctuations, the voltage across the Zener will be a very constant 5.6 Volts.

R_S and the Zener form a series circuit. The sum of the voltage across the Zener (5.6V) and across R_S is equal to the input voltage. Suppose the unregulated input voltage was at a maximum (12V), then the voltage across R_S would be 12-5.6 = 6.4 Volts.

As an important rule-of-thumb, the reverse Zener current is about $1/10^{th}$ of the maximum current drawn by the load – the load draws 100mA so one could expect the reverse current of the Zener to be 10mA. The Zener and the load form a parallel circuit – so the sum of the branch currents is equal to the current through R_S. The current through R_S must then be 100 + 10 = 110mA. So we know the maximum current through R_S is 110mA. What is the maximum voltage across R_S? The maximum input voltage is 12V, so the maximum voltage across R_S will be 12-5.6 = 6.4 Volts. We can now calculate the resistance of R_S from Ohms Law:

R_S = E(across R_S) / I(through R_S) = 6.4 Volts / 110 mA = 58 Ω.

Just as important is the power rating of R_S:

Power (of R_S) = E(across R_S) x I(through R_S) = 6.4 Volts x 110 mA = 0.704 Watt.

That's all there is to the design of a simple voltage regulator using a Zener. Zener diodes come in a range of voltages up to about 18 Volts.

VARACTOR OR VARICAP

A reverse biased diode (or PN junction) does not conduct. If you refer to when we applied a reverse bias to a PN junction, we saw the width of the depletion zone increased. If we increase the amount of reverse bias further (without reaching breakdown), the width of the depletion zone would widen even further. If we reduce the reverse bias, the depletion zone will narrow. Now, if we continually varied the amount of reverse bias, the width of the depletion zone would also continuously vary. Each time the depletion zone changes in size there must be some movement of electrons in the circuit. However, electrons never move across the junction. The junction, under reverse bias, is an insulator.

Figure 12-24

Does this movement of current on each side of an insulator remind you of capacitance? It should because a reversed biased diode will act just like a small capacitor whose capacitance can be changed by altering the amount of reverse bias. If you like to think of it another way, the depletion zone is the dielectric, which can be made to change in thickness or width by the amount of reverse bias.

REVERSE VOLTAGE

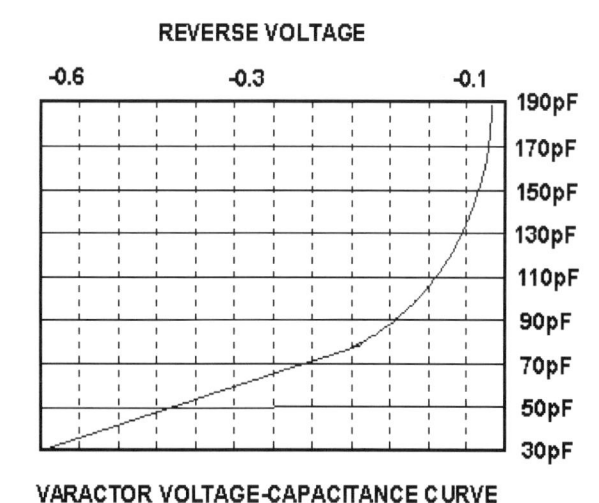

Figure 13-24.

So a reverse biased diode can be used to create a voltage variable capacitor, the symbol of which is shown in figure 12-24. There are purpose made diodes for use as varactors, though in practice almost any diode can be used for this effect. This ability of a diode to behave as a variable capacitance is extremely useful. The frequency of a tuned circuit or a quartz crystal can be made to vary by using a varactor diode. By using a variable resistor to adjust the reverse bias, the capacitance of the diode can be made to vary, which in turn will affect the frequency of the tuned circuit.

Though we have not discussed single sideband (SSB) reception yet, many readers will know that with an SSB receiver you have to tune the radio finely to the received stations with a control most often called a BFO (Beat Frequency Oscillator) or clarifier. The BFO is usually a variable resistor in combination with a varactor to make small changes to the receiver's frequency. On amateur transceivers, the 'clarifier' is called the 'RIT' (Receiver Incremental Tune).

LIGHT EMITTING DIODES

In any forward biased diode, free electrons cross the junction and fall into holes. When electrons recombine with holes, they radiate energy. In the rectifier diode, the energy is given off as heat. In the light emitting diode (LED), this energy radiates as light.

It takes energy from the source to move an electron from the valence level to the conduction level. When an electron drops back to the valence level, it will emit energy in the form of photons (light).

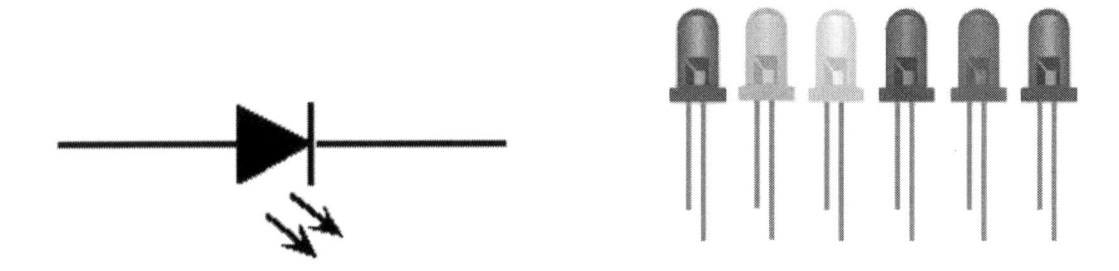

Figure 14-24.   **Figure 14-25.**

An electron moving across the PN junction moves to a hole area. This can allow a nearby conduction electron to fall to its valence level, radiating energy. In common diodes and transistors made from germanium, silicon, or gallium arsenide, this *electromagnetic radiation is usually at a heat frequency, which is lower than visible light frequencies*. With gallium arsenide phosphide the radiation occurs at red light frequencies. Gallium phosphides produce still higher frequency (yellow through green) radiations. Gallium nitride radiates blue light.

Light emitting diodes come in many colours. The short leg on the diode identifies the cathode so if you want the diode to make light it needs negative on the cathode for forward bias. Typical forward current for the conventional LED is 10-20mA

There are several photodiodes and photosensitive devices. The photodiodes convert photons to electric emf. There are also a number of other specialist diodes. For examination purposes, we have more than covered enough material here. Also, the depth of the material is more than adequate. You will *not* be asked to describe the operation of a diode at the electron/hole level, although you should remember the terms used thus far and what they mean.

AN APPLICATION FOR A DIODE

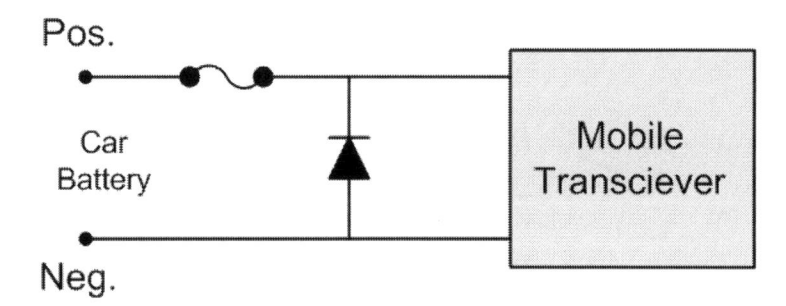

Figure 16-24.

Power diodes are made from silicon. Besides being used as rectifiers, one very useful application for a single power diode is reverse voltage protection in radio.

All mobile (used in vehicles) radio equipment is very much prone to being connected to the source of power the wrong way. I have first-hand experience of doing this myself. Connecting the power supply or battery to a mobile radio (or any radio) the wrong way around would be devastating to the radio if some sort of reverse polarity protection did not exist. All mobile radio equipment has an inline fuse. Where the negative and positive leads of the power supply enter the equipment, there is a reverse biased silicon diode. When the power supply polarity is connected the correct way, this diode is reverse biased and acts as an open circuit. If the user accidentally connects the radio to the wrong polarity, the diode becomes forward biased and conducts heavily, blowing the fuse. Most times when this happens, the diode is destroyed and remains as a short circuit across positive and negative. Backyard technicians will often fix the problem by just cutting one lead of the diode removing it from the circuit and replacing the fuse. Of course, they will charge you too many dollars for this two-minute job! With the diode out of the circuit, the next time the radio is connected to reverse polarity you can say goodbye to the radio for good.

TRANSISTORS

The basic transistor can be thought of as a two diode junctions constructed in series. From the bottom, there is a contact against an N-type emitter element. Next to this is a *thin* P-type base element, with a metal electrode connected to it, forming the first PN junction. A second N-type collector element is added, with a contact on it, forming the second PN junction. This produces an NPN transistor.

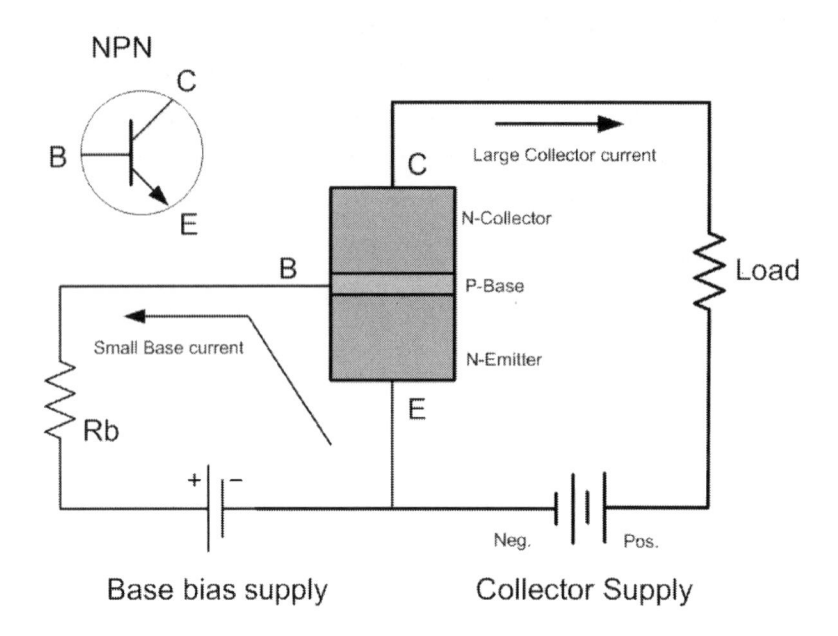

Figure 17-24.

This type of transistor is called a bipolar junction transistor (BJT). The important characteristics of construction are:

The base region is very thin and lightly doped.
The emitter region is heavily doped.
The collector is large and usually connected to the case as a heat sink, so that *heat* can be removed from the transistor.

With just the collector supply connected, no current will flow in the collector circuit as the top PN junction - the one between collector and base - is reverse biased.

Consider when the base-emitter junction is forward biased as shown in figure 17-24. The heavily doped emitter region floods the thin base region with charge carriers (electrons). The base region is lightly doped compared to the emitter. All of the electrons passing across the forward biased base-emitter junction are looking for a hole to fall into. There are more electrons crossing the lower junction than there are holes available on the other side (in the base) to meet them. I like to think of the base region as becoming saturated with charge carriers (which are electrons for an NPN transistor). The excess of electrons come under the influence of the collector voltage (which is higher than the base voltage) and consequently electrons flow in the collector circuit.

The important aspect of the transistor is that *small amounts of base-emitter current can control large amounts of collector current.* If the base current was made to vary, say by the insertion of a carbon microphone (which is a sound dependent resistor), then the collector current will be an amplified representation of the base current. The transistor is an amplifier. Since the BJT transistor uses a small base input current to control a larger output collector current, it is called a current amplifier. Where there is current, there is voltage. Z=E/I and because the current is significant the BJT transistor is a low input impedance device.

The symbol for a PNP transistor (figure 18-24) is the same as the NPN except the arrow pointing IN rather than OUT and all the polarities would be reversed.

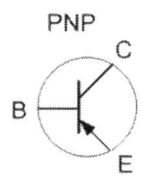

Figure 18-24.

It is interesting to note that the name transistor comes from "transfer resistor." Another way of looking at the operation is that without the base-emitter junction being forward biased there is no collector current.

When the base-emitter junction is forward biased the resistance between collector and emitter decreases from infinity (or some very high value) and current flows in the emitter circuit. Small variations in the amount of base-emitter current cause the resistance between collector and emitter to vary greatly, but in proportion. Therefore, the collector current faithfully follows the base current but is much larger and supplied with a higher voltage. The collector current is greater than the base current, so the transistor has amplified.

The amount by which an amplifier amplifies is called the gain. The amount by which a transistor amplifies is called Beta and has the Greek symbol β. The beta of a transistor is calculated from:

$$\beta = \frac{\Delta I_c}{\Delta_b}$$

Equation 1-24

The triangle symbol Δ is the Greek letter delta and is the mathematical shorthand for 'change in'. Thus, ΔI_C and ΔI_B are the change in collector and base currents respectively.

So the Beta (β) or gain of a transistor, is the change in collector current divided by the change in base current.

A small variation of base current can control 50 to 150 times as much collector current. Thus, the transistor is an ideal control and amplifying device. A junction transistor of this type can be called a bipolar junction transistor (BJT) to differentiate it from a field-effect transistor (FET) discussed later.

If a Junction transistor has its base-emitter current changed from 10 to 30 milliamps and this causes the collector current to change from 50 to 250 milliamps, what is the current gain or Beta of the transistor? The change in base-emitter current is from 10mA to 30mA = 20 mA. The change in collector current is from 50mA to 250mA = 200 mA. The Beta is therefore 200/20 = 10. The transistor has a current gain of 10.

Comparing a triode and BJT

A triode electron tube is also an amplifier as we have learnt. There is one significant difference between a triode and the BJT, which I would like to mention. Firstly, just to refresh your memory. A triode amplifies by adjusting the negative voltage on the control grid, which in turn can control the large cathode to plate current, resulting in amplification.

A triode is often called a *voltage amplifier* because it is voltage on the control grid which controls the anode current. Also, because there is no input current to a triodes control grid (Z=E/I) the input impedance of any electron tube is high. A BJT is called a *current amplifier* because it is base current which controls the much larger collector current. A BJT is a low input impedance amplifier.

The control-grid cathode circuit of a triode does not have any current flowing in it (there are exceptions). This is important. Since a triode can amplify with voltage alone, a triode consumes no power from the input source. From P=EI, if you have no 'I' you have no 'P'. A triode is a high input impedance amplifier, whereas a BJT is a low input impedance amplifier.

Whether a device is a voltage amplifier or current amplifier is irrelevant to the final amplification, though sometimes the input impedance is important. The triode has a significant advantage in being able to amplify a weak signal from a low power source without taking any power from it.

The advantages of the BJT though, are enormous: size, low heat, lower voltages, easier construction and much more.

Amplifier Input Impedance

What is the big deal about input impedance of an amplifier? Take a high input impedance like the electron tube. Only voltage is used on the control grid to control the output current. There is no input current. P=EI; without I there is no P. Suppose an antenna was connect to the control grid. There is only a very small amount of power arriving at the input of a receiver. You cannot cook potatoes with the power coming off an antenna. The power from an antenna is measured in femto-watts; that's 10^{-15} Watts. So an amplifying device that does not need power (high input impedance) is the best to use. If you do take power from an antenna, remembering that it is a resonant circuit, you would drastically lower the Q of the antenna. BJT's are low input impedance; so we have to do tricks with transformers to make them look like a higher input impedance.

Memory Jogger:

Always remember, if you are trying to work out, or asked to work out if a transistor has the correct polarity voltages to operate:

> The base must have forward bias.
> If the bias voltage is correct, current will flow *against the arrow* in the symbol.
> The collector voltage must also permit current flow against the arrow.

A LITTLE ABOUT NOISE

We tend to think of electricity (electron flow) as being fluidic or smooth. This is not correct; electricity is made up of lumps – very small lumps called electrons. Noise is produced whenever an electron does not do what it is supposed to do when it is supposed to do it. In the Electron tube, electrons might collide with secondary electrons emitted from the anode. Imagine a situation in a PN junction where an electron is ready to fall into a hole, but no hole is to be found! For that very small instant that electron represents noise. Any random or unwanted electron motion, or lack of motion, in semiconductor devices, is noise.

Don't get the wrong idea – PN junctions in transistors are wonderful low noise amplifiers, however when it comes to super sensitive receivers like those used for radio astronomy, PN junctions and hole-electron recombination is just too noisy. We shall see later that there are semiconductor devices that amplify and don't have a hole-electron recombination, making them lower noise devices.

POINT CONTACT DIODE

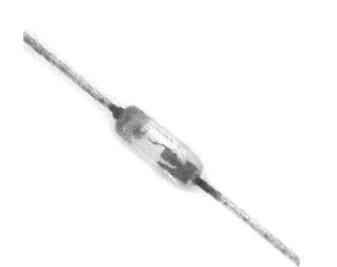

Figure 19-24.

A point-contact germanium or silicon diode is a semiconductor pellet (or germanium of silicon) with fine gold-plated tungsten wire with a diameter of about 80 to 400 microns (millionths of a metre) and a sharp point, makes contact with the polished top of the semiconductor pellet and is pressed down on it slightly from a spring contact. This cat's whisker, as it is known, is connected on the right-hand side to a brass plate which is the cathode. The semiconductor injects electrons into the metal. The energy level between the valence electrons in the semiconductor pallet and the tip of the wire produces a diode action. The contact area exhibits extremely low capacitance.

Because of the low capacitance, point contact diodes can be used for applications at extremely high frequencies. Ordinary PN-junction diodes have too much junction capacitance for use at these high frequencies.

SCHOTTKY DIODE

The Schottky diode (named after the inventor) uses a metal such as gold, silver, or platinum on one side of the junction and doped silicon (usually N-type) on the other side. When the Schottky diode is unbiased, free electrons on the N side are in smaller orbits than the free electrons in the metal. This difference in orbit size is called the Schottky barrier.

When the diode is forward biased, free electrons on the N side gain enough energy to travel to larger orbits (it takes energy to make an electron move to a larger orbit). Because of this, free electrons can cross the junction and enter the metal, producing a large current. Because the metal has no holes, there is no depletion zone. In an ordinary diode, the depletion zone must be overcome before a diode can conduct – this takes time - a very short time, but time nonetheless. The Schottky diode can switch on and off faster than an ordinary PN junction. In fact, a Schottky diode easily rectifies frequencies above 300 MHz. The Schottky is also called the hot-carrier diode. The current carriers are 'hot to trot' as it were and conduction (unlike a conventional diode) take very little electric pressure (voltage) in the forward direction to conduct. This is an advantage with very weak signals.

The schematic symbol of a Schottky diode:

Figure 20-24.

25. SEMICONDUCTORS PART 2

TRANSISTOR CONFIGURATIONS

There are different ways of connecting a transistor so that it will amplify. The principle of small changes in base current controlling larger changes in collector current remains the same. The configurations we are going to discuss cause the amplifier to have different characteristics.

I am not going to go very deep into this subject, as there is no need to for examination purposes. For the exam, all you really need to able to do is identify which configuration is illustrated and remember some basic characteristics of each configuration.

The three configurations are:

Common-emitter
Common-base and
Common-collector (also called emitter-follower)

I will give you the basic circuits of each. For the exam, you need to identify which circuit is which and that is all. Please do remember that with all transistor amplifier configurations, it is still the small base current controlling the larger collector current which results in amplification. At the end of this section, I will also provide a table of the different characteristics of each configuration. A detailed description is not necessary as this would be rather lengthy and contain much information which is not required.

COMMON-EMITTER CIRCUIT

The basic transistor amplifier circuit is the common-emitter. It is also the most common configuration you will see and is easy to remember. The circuit shown in figure 1-25 is a basic common-emitter amplifier using an NPN transistor.

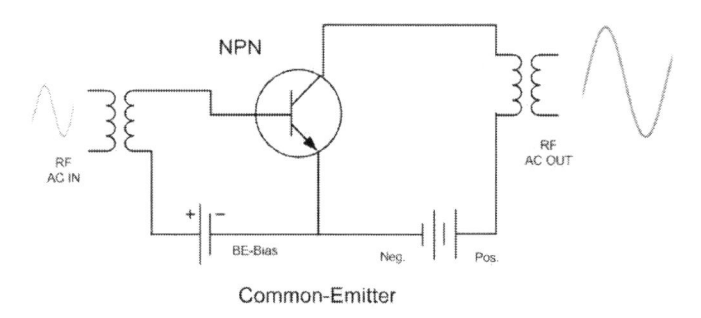

Figure 1-25

If a PNP transistor were to be used the polarity of the batteries would be reversed.
This amplifier circuit uses transformer coupling at both the input and output. The transformers are air cores, so this indicates that the circuit is used at radio frequencies, as iron core transformers have too much loss in the laminated iron core at radio frequencies. The type of coupling *does not* determine the configuration.

The reason the configuration of figure 1-25 is called common-emitter is because *the emitter is the common leg (or lead) between input and output*. This is how you will identify the common-emitter circuit.

The identifying feature of this circuit as common-emitter is the emitter leg being common to input and output. One thing worth noting is that the signal being amplified undergoes an 180° phase change. Battery bias is used here more complex and practical circuits would not use battery bias.

COMMON-BASE

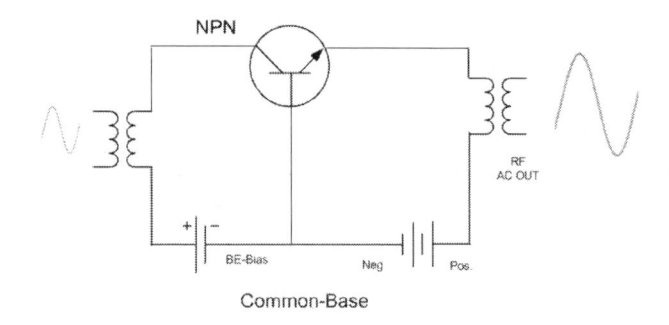

Figure 2-25.

The common base circuit is identified by the *common base leg between input and output*. In this circuit transformer coupling is once again used.

Transformer coupling in these two circuits is for impedance matching. We have learnt that one of the functions of a transformer if required, is to match two unequal impedances. For example, if you had an audio amplifier with an output impedance of 1000Ω, it would not work well if connected to an 8Ω-speaker. An audio output transformer could be used to perform an impedance conversion from 1000 to 8Ω.

Just a note, in all the transistor circuits we have seen so far the bias voltage (between base and emitter) and the collector voltage (between collector and emitter) has used separate batteries. In practical circuits, one source of supply is used for base bias and collector voltage. The input signal for both circuits so far is AC. BJT's do not work on AC. However, the base-emitter battery bias not only turns the base-emitter junction on; it converts the incoming AC to VDC (varying DC).

EMITTER-FOLLOWER

Figure 3-25 shows an emitter-follower, also called a common collector configuration. I have shown how one source of supply can be used for both base-emitter forward bias and collector voltage. Unlike the previous two circuits, this one is a little different. The key to the identification of the *common-collector (emitter-follower) is that the output is taken from across a load or resistor in the emitter circuit.* The output of this amplifier and all amplifiers of this configuration is taken from across the emitter load resistor; R_L.

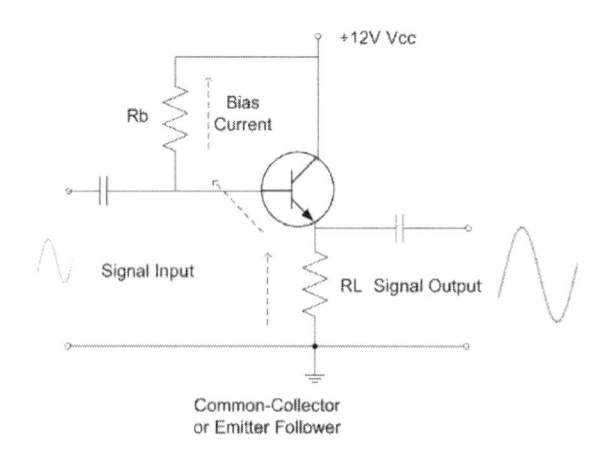

Figure 3-25.

The only reason the different configurations are used is that the amplifiers *provide different characteristics*, a summary of which I will provide in table form shortly. They are all just amplifiers, relying on small changes in base current to control larger changes in collector current.

The coupling between the input and output of the circuit in figure 3.25 is called capacitive coupling. Coupling is the method by which signal is fed into and the output is taken away from a transistor amplifier. There will be more on coupling later.

Notice also how a single resistor R_B is used to forward bias the base-emitter junction of the transistor, doing away with having to use two power sources. The resistance of R_B determines the amount of bias current that flows from the negative ground up into the emitter and out of the base, through R_B and back to the positive supply. This bias arrangement is called fixed-bias. I would also like to point out that a number of transistors could be connected via coupling to produce a multistage amplifier.

There you have it. The three basic transistor configurations and an illustration on biasing with a single voltage in the last circuit. We have been using BJT's here. We will be talking about other types of transistors soon and we have already spoken about the electron tube. All of these active devices can be connected in the above configurations.

Summary of the characteristics of each configuration:

Parameter	CE	CB	CC
Input impedance	1000 Ohms	60 Ohms	40,000 Ohms
Output impedance	40,000 Ohms	200,000 Ohms	1000 Ohms
voltage gain	500	800	0.96
Current gain	20	0.95	50
Power gain	10,000	760	48
Phase in/out	180°	0°	0°

A *common-emitter* has a low input impedance and a high output impedance and inverts the phase of the incoming signal.

A *common-base* also has a low input impedance and has a high output impedance and does *not* create a phase inversion between input and output.

A *common-collector* has high input impedance and a low output impedance and provides no phase change. A common collector could drive a low impedance load such as a 50Ω antenna or an 8Ω speaker. The pass-transistor in a low voltage high current power supply are common-collector as the impedance of most things (like a transceiver) connected to a power supply are low impedance. Think about it 13.8V at 20A for a typical transceiver means a low input impedance R=E/I.

BIASING TRANSISTORS

Biasing transistors is another broad topic that much could be written about. However I am again going to confine this chapter to what I think is the 'need to know' and even that may extend further than necessary.

Biasing a transistor means applying a fixed DC voltage between the base and emitter. By varying the amount of bias, we can make the transistor amplify in the required class of operation i.e. A, B, AB or C.

If you have trouble understanding the need for bias (and many students do) imagine using a transistor without any bias and feeding an AC signal to it to be amplified. Half of the AC signal would not forward bias the base-emitter junction and there would be no output from the transistor during this half cycle. Transistors (including FET's) and electron tubes are not AC devices – they work on varying DC. When we bias the input of an active device at a DC potential, we can then feed an AC signal to that input. The combination of the AC superimposed on top of the DC bias produces varying DC at the input of the transistor or other active device.

We have already seen one method of biasing using a single resistor (R_B) in the emitter-follower circuit shown earlier. This type of biasing is called fixed bias. I suggest you go back and have a look at it now.

Fixed bias is simple. However, it is quite unstable thermally. If the transistor warms for any reason, due to a rise in ambient (surrounding) temperature or due to current flow through it, the collector current increases. The higher the current gain of the transistor the greater the instability of the circuit.

A far superior and the most common biasing arrangement for BJT's is shown in the schematic circuit of figure 4-25 and is called *voltage-divider bias*. I have shown typical values for a practical audio amplifier. You will not have to perform any calculations on the circuit, but you may need to identify the circuit, not as a whole but in separate parts so we will have a look at it for the necessary detail.

The common-emitter circuit with voltage divider bias.

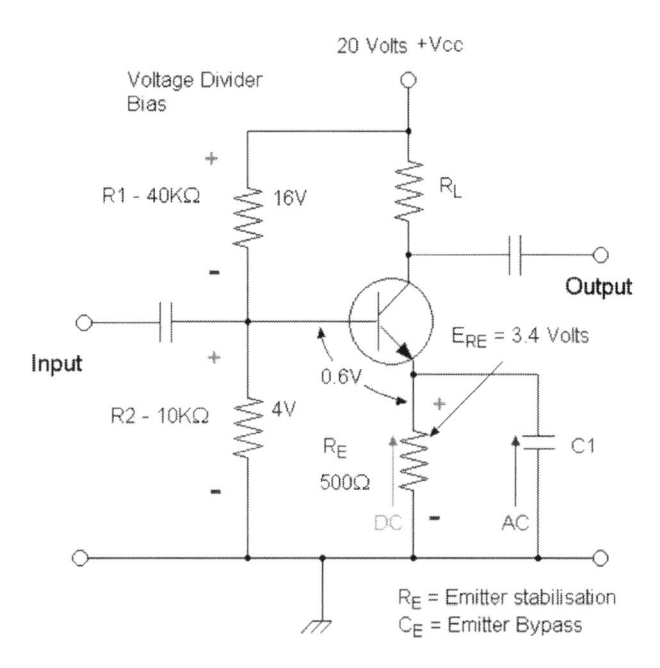

Figure 4-25

The purpose of bias in any BJT (transistor) circuit is to forward bias the base-emitter junction of the transistor. In the circuit of figure 4-25, this is achieved primarily by the series circuit consisting of R_1 and R_2 in series with the supply voltage (in this case 20 Volts) The transistor is an NPN type which requires the base to be positive with respect to the emitter for forward bias. R_1 and R_2 in series are across the 20 Volt supply which is positive at the top (collector) and negative at the earth symbol. Now you could use Ohms law to approximate the voltage across R_1 and R_2 knowing they are across a 20 Volt supply and have the values shown. Their total resistance is 50K. 40/50 of 20 Volts will be dropped across R_1 and 10/50 (1/5) of the voltage will be dropped across R_2. One fifth of 20 Volts is 4 Volts. So the approximate voltage across R_2 is 4 Volts. The top of R_2 is positive and the bottom negative. So neglecting R_E, there is about 4 Volts applied between the base and the emitter of the transistor, positive at the base and negative at the emitter. The polarity is correct, so we have a forward biased base-emitter junction. This is how voltage divider bias works. R_1 and R_2 form the voltage divider.

Let's just have a look at some of the other components in the circuit and what they do. Resistor R_E you will notice a relatively low value (500 Ohms) and is connected between the emitter and ground. R_E provides thermal stability to the circuit. An increase in DC voltage through R_E caused by a reduction in the transistor collector-emitter resistance will cause an increase in voltage across R_E. *The voltage across R_E is in opposition to the bias voltage.* If the voltage across R_E was 1 Volt, then this opposes the voltage across R_2 and so the effective bias voltage is only 3 Volts. The voltage across R_E will only increase if the transistor heats up and as a consequence its collector-emitter resistance decreases. Without R_E the transistor could go into *thermal runaway*.

Let's look at the situation *without* R_E. The transistor is operating. It heats up due to the current through it or an increase in the ambient (room or surrounding) temperature. If the transistor heats up, its emitter-collector current increases which causes it to heat up more, which causes its resistance to decrease more, which increase the current more, which causes it to heat up more, etc. - this is a thermal runaway and results in the transistor being destroyed.

R_E provides what is called *degeneration*. Degeneration is a single word, which explains the operation in the second last paragraph. The degeneration used here is also known as *inverse current feedback*. If the current increases through R_E (due to the transistor heating up for whatever reason) the voltage across R_E reduces the bias voltage, that decreases the emitter-collector current, allowing the transistor to cool and not go into thermal runaway.

Now, it is the DC current flowing through R_E that we are using to provide thermal stabilisation and prevent thermal runaway. There is an AC component in the circuit. This AC component is the signal we are trying to amplify. An AC voltage is applied to the input of the transistor via capacitive coupling and that AC signal will appear in the emitter-collector circuit, amplified. We do not want the AC signal to flow through R_E. If the AC component (the signal to be amplified) did flow through R_E, it would affect the bias of the transistor. So the simple approach is to provide an AC path around R_E for the AC component. This is accomplished by adding the parallel capacitor across R_E (C_1). So, the signal which is being amplified flows through C_1 and does not affect the biasing of the transistor, while the DC current, which is what affects the operating characteristic of the transistor and can cause thermal runaway, is allowed to flow through R_E.

If this is difficult to see, let me add the following points to refresh your memory. The circuit of the transistor has a current, this current without any input signal is just DC. When an AC (or varying DC) signal is applied to the transistor, the current through the transistor now has two components (or parts): a DC component, which is responsible for bias; and an AC current which is the signal being amplified. The DC component is used to bias the transistor and prevent thermal runaway - the degenerative voltage across R_E must be treated differently by the transistor than the AC component, which is the signal we are trying to amplify.

If we were to allow the AC component of the current to flow through R_E, then any increase in the signal to be amplified would reduce the gain (amplification) of the transistor, defeating the whole purpose of amplification.

A capacitance will pass an AC current, but block a DC current. So the bias current flows through R_E (DC), while the signal to be amplified passes through C_1. For this to work effectively the capacitive reactance of C_1 must, as a rule of thumb, be about $1/10^{th}$ the value of R_E. So in the circuit of figure 4-25, the X_C of C_1 would be about 50 Ohms.

Since C_1 bypasses the AC component of the emitter-collector current around R_E, C_1 is called an *emitter bypass capacitor*, or just a *bypass capacitor*. R_E is referred to as the *emitter stabilisation resistor*.

You do not have to explain how this works as I have done. But you do need to know the purpose of R_1, R_2, R_E and C_1 (which may be labelled differently).

R_1 and R_2 provide voltage divider bias.
R_E provides thermal stability.
C_1 bypasses the AC component (i.e. the signal to be amplified around R_E).

R_L is just a load resistor. If it were not there, the output would be connected to the +20 DC rail and there would be no output other than +20DC - hardly a useful amplifier.

JUNCTION FIELD EFFECT TRANSISTOR (JFET)

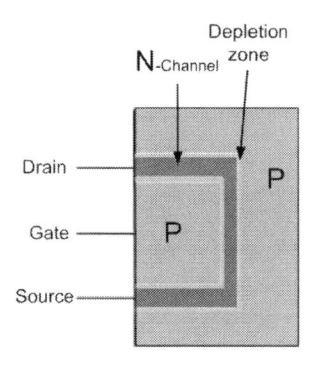

Figure 5-25

Field-effect and bipolar junction transistors are entirely different devices. FET's are solid-state amplifying devices that have operating characteristics very similar to those of triode electron tubes. There are three types of FET's, the junction (JFET) and two metal oxide semiconductor types known as MOSFET's. One is the depletion MOSFET; the other is an enhancement MOSFET.

The essential construction of a JFET is shown in figure 5-25. A block of P-material called the *substrate* has an N-type channel running through it. The figure shows a cross section. Think of the N-channel as a wormhole in the P substrate. As with any PN junction, electrons fall into holes at the barrier and a depletion zone is created. You can see the N-channel running through the P material (substrate, shown in orange) - the depletion zone is drawn as light blue. The connections are made as shown. The drain can be thought of as similar to the anode, the source similar to the cathode and the gate similar to the control grid of a triode valve. Though this device is fully manufactured from semiconductors, the purpose of the JFET is to amplify, as is the purpose of the BJT and the triode. If a supply voltage is connected between drain and source, there will be a current flow as there is no PN-junction either forward or reverse biased.

So we have connected a supply voltage between drain and source and you can imagine a current flowing from source to drain through the N-channel.

Notice also that the depletion zone occupies part of the channel.

To refresh your memory. A depletion zone is called a depletion zone because it is depleted of *charge carriers*. A charge carrier is either an electron or a hole. A PN junction can be reverse biased increasing the size of the depletion zone and how much voltage is applied to a PN junction in the reverse bias condition will determine the *size* of the depletion zone.

If we wanted to reverse bias the JFET in the previous diagram, we would apply a negative potential to the substrate that would attract the holes and a positive potential to the channel that would attract electrons.

If we reverse bias the PN junctions on each side of the channel, we will increase the width of the depletion zone. Since the depletion zone would widen, it would occupy more of the channel. In fact, if we applied enough reverse bias the depletion zones could be made to occupy the entire channel and no current would flow from source to drain. This is called *pinch-off*.

We have a way of controlling the amount of current that flows in the channel by changing the amount of reverse bias between the gate connection and the channel (the substrate and the gate are connected together).

The gate is always biased negative in the JFET (N-channel), then the signal to be amplified can be applied to the gate. The signal to be amplified would change the reverse bias and hence the size of the depletion zone. Since the depletion zone occupies part of the channel, the resistance of the channel will alter in sympathy with the input signal and in doing so control the source-drain current.

So the JFET is an amplifier. Since the gate connection is always reverse biased (no current), the input impedance is very high. It is the changes in the negative potential on the gate that controls the width and conductivity of the depletion zone that controls the much larger drain current.

This is, in essence, exactly how a triode works - by changing the negative potential of the control-grid of a triode we can control the much larger anode current. If the control-grid of a triode is made negative enough, no electrons will flow from the cathode to the anode. This is called *cut-off*.

Similarly, if the negative potential of the gate of an N-channel JFET is large enough, the depletion zone will occupy the entire channel and there will be no source-drain current. This is called *pinch-off*.

To run through the operation of an N-channel JEFT one more time, the gate is biased negative with respect to the source, causing the depletion zone to occupy part of the channel. The width of the depletion zone in the channel controls the channels resistance. If a signal is applied between the gate and the source, it will control proportionately a much larger drain current. The drain current will be an amplified version of the signal applied to the gate.

The JFET is a *voltage amplifier* like a triode. By this, we mean that changes of voltage only on the gate are required to control the drain current. Since voltage alone is needed to control the JFET, the input signal does not have to deliver any power. You can have all the voltage in the world, but without any current you have no power (P=EI). This is very different to the BJT, which is a current amplifier. You can have voltage without current, but not current without voltage - think about that. The JFET and all FET's are then high input impedance devices.

You could also have a P-channel JFET that would have an N-substrate. The operation would be the same except you would have to bias the gate positive.

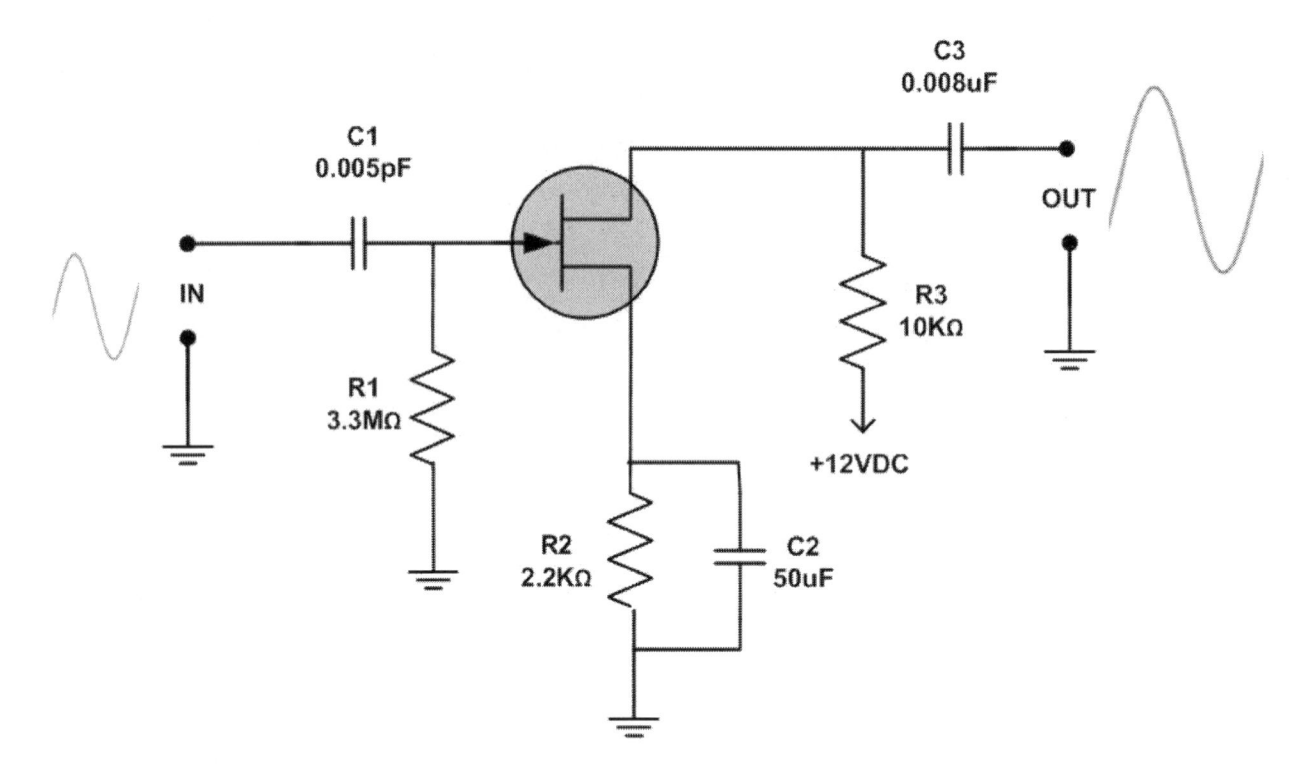

N-Channel JFET Amplifier

Figure 6-25.

Important points to remember: Symbol; arrow in for N-channel; arrow out for p-channel High input impedance; voltage Amplifier; the semiconductor version of a triode valve; why high input impedance can be an advantage. There are no forward biased PN junctions in a FET; therefore, no electron-hole recombination and therefore less noise.

MOSFET'S

Metal Oxide Semiconductor Field Effect Transistors.

When a FET is constructed as shown in figure 7-25 with the gate insulated (silicon dioxide) from the very narrow N-channel, it is called an insulated gate FET, IGFET, or MOSFET, the latter being the most common term. Combinations of these transistors as a single unit are referred to as CMOS, MOS, VMOS.

When a voltage is applied, negative to the gate and positive to the source, its electrostatic field extends into the N-channel. This reduces its ability to carry current by repelling channel electrons, depleting the channel of its carriers.

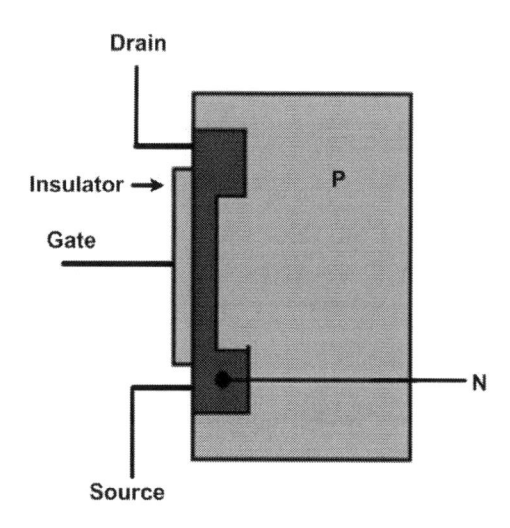

Figure 7-25.

This is a depletion mode MOSFET. Sufficient reverse bias can pinch off the drain current completely. Conversely, a forward bias can increase drain current up to a certain point. As a result, this device can operate with no bias at all, which simplifies circuitry. Unlike in vacuum tubes and JFET's, a forward bias does not produce any input circuit current due to the insulation between gate and channel. MOSFET's have an extremely high input impedance in the order of 10 Megohm. This extremely high input impedance is the MOSFET's great advantage apart from being less noisy than BJT this extremely high input impedance means virtually zero drive power. So the source (such as an antenna) does not have to deliver power to a MOSFET to have that signal amplified.

The input impedance of MOSFET's (as much as 40 Megohm) and similar devices are so high that they require special handling, as normal handling can result in a static charge being created which can destroy the device. You may find some wrapped in metal foil to protect them from static voltages.

The MOSFET was a significant breakthrough for semiconductor technology and led to combinations of them being fabricated within a single package called an integrated circuit. Of course, even BJT'S and JFET'S can be fabricated in conjunction with MOSFETS into a single package. It is a simple matter to create resistors in an integrated circuit by controlling the width and doping level of semiconductor strips.

JFET

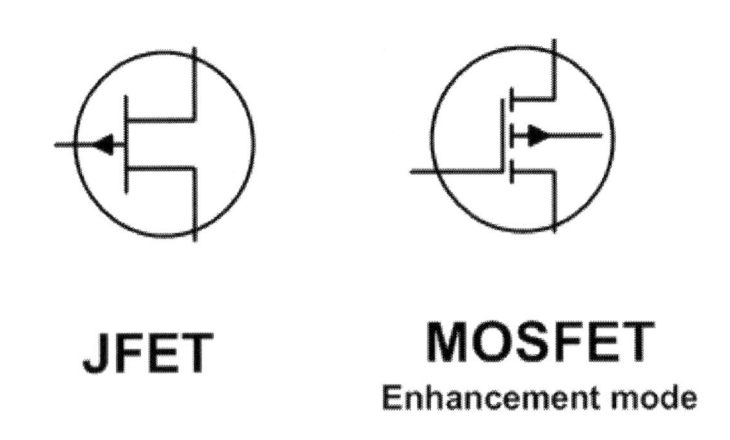

MOSFET
Enhancement mode

Figure 8-25 – both P-channel

Capacitance in an integrated circuit can make use of the fact that a reverse biased PN junction acts as a capacitor. For your radio assessment you should remember the symbols; identify P-channel (arrow out); and N-channel (arrow in); and that these devices are very high input impedance (2MΩ for the JFET and 10MΩ for the MOSFET In figure 8-25 both of these FET's are P-channel. MOSFET's can have more than one gate. This makes them ideal for use as mixers where two signals are mixed (with the MOSFET being non-linear) to produce new frequencies – covered later when we look at mixers.

PROTECTED GATE MOSFETS

MOSFETS have very high input impedance 10 Megohms or higher. Because of this high input impedance, even static electricity from handling them can build up a high voltage on the gate (or gates) and rupture the Metal Oxide insulator. Some MOSFETS are made with internal Zener diodes connected in parallel and in opposite directions between the gate and the substrate. Under normal operation, neither of these Zener's will conduct. However, if a high static voltage is created on the gate, one of the Zener's will conduct and discharge the voltage. These are called *protected gate* MOSFETS. The symbol is that of an ordinary MOSFET, but within the symbol, you will see one or more Zener diodes.

SILICON CONTROLLED RECTIFIERS (SCR'S)

Identify the symbol and external operation and an application of an SCR.

For practicality reasons, I will cover the internal operation of the SCR briefly.

Silicon controlled
Rectifiers

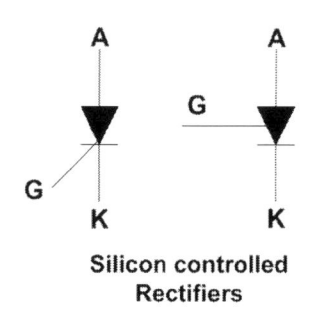

Figure 9-25

The SCR construction shown in figure 10-25 illustrates one of the many NPNP or multi-layer semiconductor devices. The layer structure could also be PNPN. This one is called an SCR. "A" is the anode, "K" is the cathode and "G" the gate. If a voltage, even a large one, is placed across anode and cathode (of any polarity), this device will not conduct because at least one junction will be reverse biased. The main current path of an SCR is between cathode and anode.

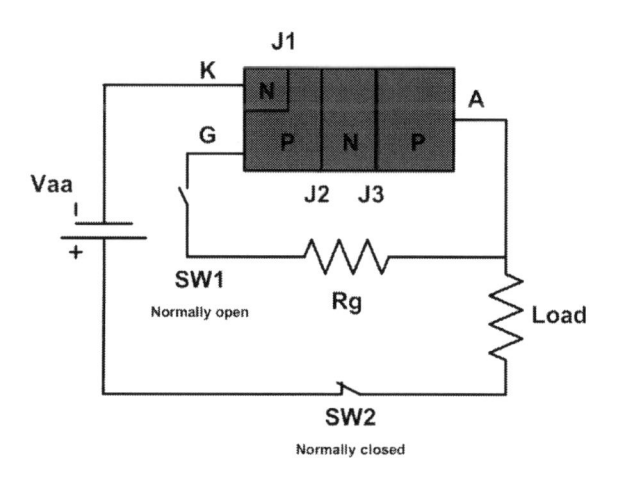

Figure 10-25.

Switch SW1 is normally open (NO). When SW1 is closed, the first P-Type region, acting as the base of an NPN BJT, is forward biased through R_G and "Load" which allows current to flow through J2. Since J3 is forward biased already, current flows through the whole device from K to A and the load.

The gate switch SW1 can now be opened and current will continue to flow because J2 loses control of its current carriers. Even reverse biasing the gate will not stop the cathode-anode current. It is necessary to reduce the applied voltage (Vaa) to almost zero or open SW2 to stop current flow. No current will flow if the source potential (Vaa) is reversed, even with SW1 closed.

Therefore, with an AC source rather than Vaa, which is DC, the SCR acts like a rectifier in that it allows current to flow in one direction only. The SCR is a unidirectional switch.

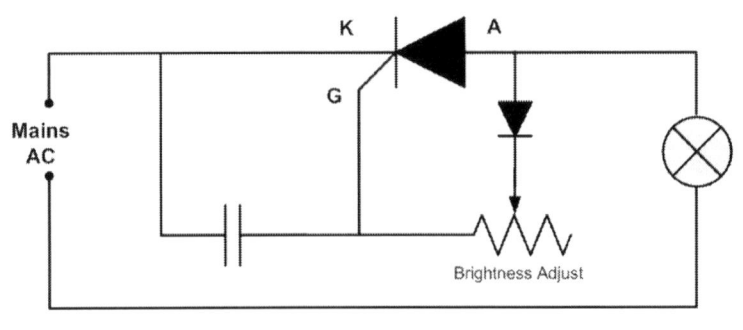

Simple light dimmer

Figure 11-25.

An SCR acts like an ordinary rectifier except that it can be switched ON and OFF. When OFF it is an open circuit and does nothing. When ON it acts like any rectifier would. The switching ON can only be done by the gate; the switching OFF can only be done by removing the supply voltage. If the supply voltage is the mains, then on the change of the half cycle on the sinewave the SCR will automatically switch OFF. SCR's can be made to handle medium to very high currents. A typical application could be a light dimmer or motor speed control circuit (though a TRIAC is better for the latter). The gate of the SCR is not switched (usually) with a mechanical switch, some of the mains AC that the SCR is going to control is used to switch or 'fire' the SCR's gate. An SCR is still a rectifier like any other rectifier, though when it starts to behave as a rectifier depends on the voltage applied to its gate. If you want to control both halves of an AC waveform you could use two SCR's. A device in one package which does this and which has mostly replaced the SCR is called a Triac. Triacs are not in the current (Australian) syllabus.

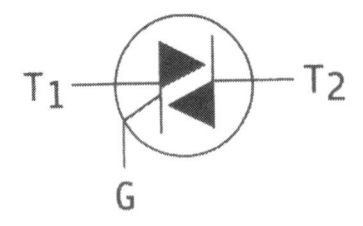

Figure 12-25

Triacs can be considered as back-to-back SCRs and perform as a full wave SCR for alternating current. An SCR is a unidirectional switch and TRIAC is a bidirectional switch. The gate control voltage of a TRIAC, unlike the SCR, can either be positive or negative in polarity. Figure 12-25 is the schematic symbol of a triac.

26. AMPLIFIERS

DEGREES

The basic purposes of an amplifier is to do just that, to amplify (a signal). However, the way different types of amplifiers do this can vary greatly and an amplifier can do other tasks besides amplifying. When we speak of amplifying, we mean, to increase the amplitude of a signal. We are starting to get into the building blocks that make up part of a radio system.

Signals that we actually amplify in radio and communications are very often not sinewaves, though many times they are. For our discussion, we will consider that we are amplifying a sinewave(s) of alternating voltage or current.

When discussing alternating signals, it is usual to speak of a *full cycle* as being equivalent to *360 degrees*. Fractions of a full cycle can then be more easily described. As an illustration, if a full 360 degrees sine wave of alternating current is sent to a rectifier, only 180 degrees of current will appear at the output. A halfwave rectifier or diode could be said to have an *operating angle* of 180 degrees. Similarly, different types of amplifier will provide at their output *all or only a portion of the amplified input signal*. An amplifier with an operating angle of 180 degrees will behave similarly to a halfwave rectifier except that the output is amplified.

This is usually the hardest concept about amplifiers to grasp. It is incorrect to make the assumption: "what goes into an amplifier is the same as what comes out, only bigger". True in many cases, audio amplifiers, for example, we want exactly what goes in to come out. We just want it to be louder or amplified. However, in radio circuits, there are all sorts of other considerations. One of the main ones is the efficiency of the amplifier. A number of techniques are used to improve the efficiency. Typically, audio (sound) amplifiers have a very low efficiency.

We have covered the actual principle of amplification in the last two chapters. A smaller voltage or current is used to control a much larger voltage or current. Most of the time this is done using one of the *active devices*, BJT, JFET, Triode, etc. Now we cannot get power for nothing. The power supplied to the amplifier comes from the power supply, so if an amplifier has an efficiency of only 25%, that means for 25 Watts of useable signal output we have to provide it with 100 Watts of power from the power supply. So efficiency, particularly at radio frequencies where the power tends to be high. perhaps 100,000 Watts for a TV station, becomes important. Efficiency is not just important in terms of say running down batteries in portable equipment, but also *heat*. If a power supply is only 25% efficient, then 75% of the power is being converted into heat and this heat brings with it all sorts of other problems, particularly with semiconductors and keeping them cool.

A little more on efficiency. An active device (in practice several active devices in a multistage amplifier) say a BJT, is forward biased which is normal. Even with no input signal there is still emitter-collector current. So even though we are not doing any amplifying, the active device is consuming power from the power supply.

So accept just for now, that not all amplifiers actually amplify the entire input signal fed to them. Like the example given above, if an amplifier only amplified half of one cycle it would have an operating angle of 360/2 = 180 degrees. It may seem that if an amplifier does not have an operating angle of 360 degrees then what we put in (the input signal) will not be what we get out and this is true. An amplifier that does not have an operating angle of 360 degrees distorts, or is *non-linear*. Sometimes this does not matter, other times it does.

Amplifiers fall into four basic categories called Classes and you need to remember them. It is important you remember the classes and their approximate efficiency and the basic principles and in a couple of cases, you will need to identify a circuit. This does not mean you need to know the complete circuit operation.

CLASSES OF AMPLIFIER AND THEIR OPERATING ANGLES

Class A 360 degrees
Class B 180 degrees
Class AB 270 degrees
Class C 90-120 degrees

At the outset you should see that a Class A amplifier with an operating angle of 360 degrees amplifies *all* of the input signal - this is the basic definition of a *linear amplifier*, though more formally "without distortion" should be added. However, Class A operation is the least efficient. We will discuss efficiency later. Class A is used for small signal applications because of its poor efficiency, but nevertheless, it is a quality amplifier. However, we can do tricks with the others as well, to make linear amplification possible. In some cases, we don't care if we have non-linear amplification.

A LOOK AT HOW THE CLASSES DIFFER

Take an active device such as an electron tube. If the control grid is made negative enough, the anode current will fall to zero. This point is called **cutoff**. Decreasing the negative bias applied to the control grid increases the anode current linearly, until at zero bias the anode current is almost at maximum. Making the control grid positive will increase anode current slightly, but the electron tube quickly *saturates*. Saturation is when an active device is conducting its maximum current.

A fixed DC voltage applied to the control grid (bias) determines the class of operation. That's so important I will say it again - *bias* determines the class of operation. For Class A operation, the electron tube (or any active device) is biased at the centre of the linear portion (straight-line section) of the operating curve. Class B is biased at, or near, cutoff. In Class C operation the bias is about twice cutoff, though it can be higher. For Class AB, bias is about halfway between cutoff and the centre of the linear portion of the operating curve.

Radio experimenters and professionals frequently use the term "linear amplifier" as a synonym for "power amplifier". Such usage is incorrect as power amplifiers can be linear or non-linear, depending on their application. For audio, Class A operation is the only single-ended amplifier that provides linear amplification. Single ended means only one active device employed. Linear amplifiers do not distort the amplitude of the signal being amplified within specified practical limits. It is possible to use Class AB and B as linear radio frequency (RF) amplifiers for amplitude modulated signals since the output power is proportional to the input power. The missing cycle can be restored in the tank circuit (flywheel effect). Amplitude modulated signals are those which contain intelligence (or information) in their amplitude.

Since a Class B amplifier has an operating angle of 180 degrees, it is possible by using a pair of active devices and to arrange for each device to amplify half of the sine wave, to combine the two amplified halves in the output to achieve linear amplification. Such an arrangement is called a class B push-pull and is most useful for audio and RF amplifiers.

Class B is non-linear and has an operating angle of 180 degrees. However, we can arrange two such amplifiers to look after each half of the input and combine them after amplification in the output, to obtain the full signal with *linear* amplification

Why bother? Well we discussed that Class B was biased at or near cutoff, so with no input signal (the signal to be amplified), little or no current flows in the anode, collector or drain circuit - no power taken from the supply when no amplification is taking place means higher efficiency

AMPLIFIER EFFICIENCY

CLASS **A**   25-30%
Use:
Audio - low distortion
RF: Very inefficient
CLASS **B**   30-35%
Use:
Audio Power (push-pull)
RF power - all modes - can be linear with two active devices
CLASS **AB**   50-60%
Use:
RF Power Amplifiers all AM modes - can be linear with two active devices
CLASS **C**   60-80%
Use:
RF power - FM

We have not covered transmission modes. Most will be familiar with AM and FM radio stations. AM is amplitude modulation, the information, music, intelligence, is in the amplitude of the signal. FM is frequency modulation, the information, music; intelligence is in the frequency of the wave. Remember, a sine wave has two characteristics - think back to your basic sine wave. These characteristics are amplitude and frequency. Either can be modified to carry information (modulated), the amplitude or the frequency and we can disregard the other without losing any information.

INPUT-OUTPUT CURVES

CLASS A – OPERATING CURVE

The characteristic curve diagram of figure 1-26 is that of an electron tube Class A amplifier. It could just as easily be for a FET or BJT. In an electron tube, a negative bias is applied to the grid. In the diagram, I have labelled the grid bias (above zero) as shown. The actual value in Volts of the grid bias is of no significance, as it will vary from device to device.

The graph is a plot of grid bias versus plate (anode) current. It is important to understand this graph. Forget the waveforms for a moment and just concentrate on the lines on the graph, which is a plot of grid bias verses plate current. The plate (anode) current is on the vertical axis while the negative grid Volts are on the horizontal axis. You can see from the graph that there is a point where the grid bias is so negative that no plate current will flow and this point is labelled **cutoff**. Further along the curve is a spot marked **operating point**. This operating point is the amount of bias required for Class A operation.

Note that the characteristic curve is really a curve, for if I had extended the curve a bit further upward it would have quickly flattened out, this is called **saturation**. Importantly, notice that a portion of the characteristic curve is a straight line. If the active device (here an electron tube) is operated on the straight line portion of its characteristic curve, then variations to the bias by the input signal superimposed on it will cause proportional variations in the larger anode current.

Operating an amplifier on the linear portion (straight-line portion) means the amplifier will be linear and faithfully reproduce the output signal the same as the input signal.

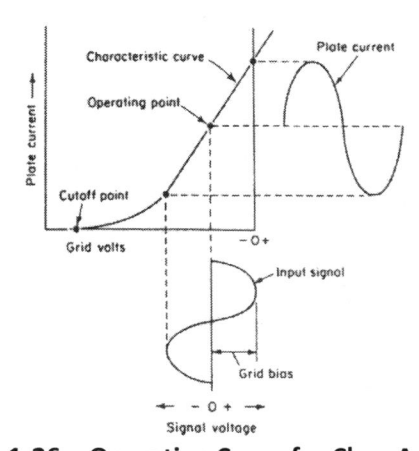

Figure 1-26 – Operating Curve for Class A Amplifier

If the input signal drives the bias point into cutoff or saturation, then the amplifier will produce distortion and the output anode current will not be an exact representation of the input signal.

As an example, audio amplifiers often operate in Class A. Think of your TV or stereo system. If you turn the volume control up too high, the music will be louder, but it will begin to sound distorted.

The volume control adjusts the level of the input signal into the audio amplifier. When you turn it up too high, you overdrive the amplifier causing it to go into the cutoff and saturation regions of the curve and you hear this as distortion.

So figure 1-26 demonstrates Class A operation. The active devices are operated on the linear portion of the operating curve. Since all of the input signal is amplified in a Class A amplifier, it is said to have an operating angle of 360 degrees.

CLASS B - OPERATING CURVE

Have a look at the operating curve of a Class B amplifier in figure 2-26. Notice the operating point is at, or near, cutoff

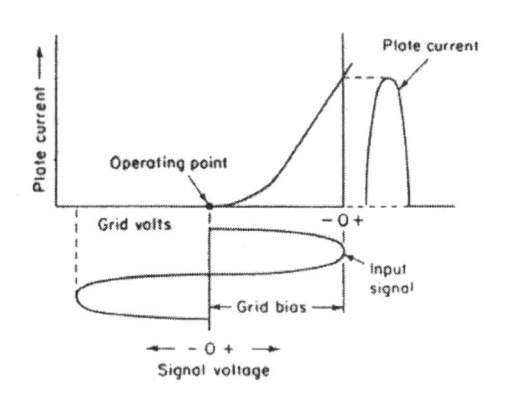

Figure 2-26 – Operating Curve for Class B Amplifier.

Cutoff is when the grid voltage is so negative that it prevents any electrons from flowing from cathode to anode. If the input signal is negative going, it takes the electron tube beyond cutoff and no anode (plate) current flows. However, if the input signal goes positive it will take the electron tube out of cutoff (to the right on the horizontal axis) and current will flow in the output and be amplified. As you can see by the diagram, only one-half cycle of the input signal is amplified. A Class B amplifier is therefore *not* a linear amplifier. However, unlike a Class A, with no input signal, there is no anode current in a Class B, so it is a more efficient amplifier. A Class B amplifier has an operating angle of 180 degrees. A combination of two Class B amplifiers configured correctly, one to amplify one-half cycle and the other the other half cycle, can be used to produce linear amplification.

CLASS C - OPERATING CURVE

Have a look at the operating curve of a Class C amplifier in figure 3-26. Notice the operating point is well beyond cutoff.

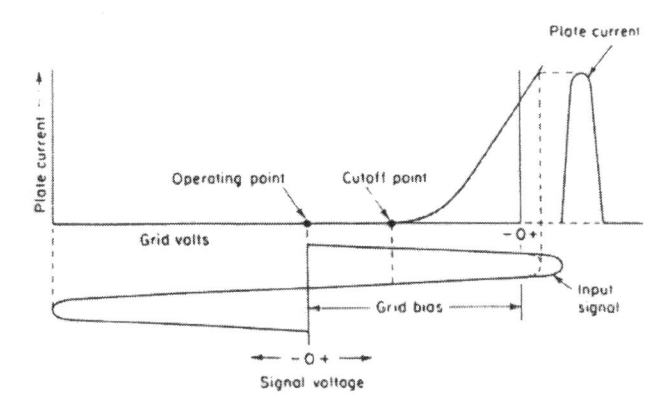

Figure 3-26 – Operating Curve for Class C Amplifier

Not even one-half cycle of the input signal is amplified (180°). About 90° to 120° of the input signal is amplified. A Class C amplifier causes massive distortion to the amplitude of the input signal. Don't be confused by the size of the input signal and the output signal (plate current) shown on any of these curves, their physical size on the curves does not represent their power levels. On all the curves, plate current is much larger and at a much higher voltage than the input signal.

If there is information in the amplitude of the input signal it will be severely distorted. However, if there is information in the frequency of the input signal, there is no distortion to frequency, as the frequency of the signal that goes in, is the same as that which comes out. So if we have coded some information into the frequency (frequency modulation), the Class C amplifier is the only way to go, as it is the most efficient amplifier of all. A Class C amplifier can only be used at radio frequencies (RF).

We can restore the full cycle in the output of a Class C amplifier by using a tuned circuit in the plate of the electron tube. Such a circuit is called a tank circuit and it will restore the other half cycle by utilising the flywheel effect.

CLASS AB

I will just mention that there is another Class of amplifier called Class AB - it is not in the syllabus. It is an improvement of the Class B. For linear amplification, two active devices like a Class B amplifier, are required. One problem with the Class B is that the operating curve is not quite straight for a little bit near cutoff. Even if you use two Class B amplifiers, one for each half cycle, you get some distortion when one amplifier turns off (after one-half cycle) and the other Class B amplifier turns on. This is due to the non-linear part of the operating curve near cutoff. The distortion that results is called *crossover distortion*. The Class AB is a minor modification of the Class B to overcome this problem. The Class AB in push pull is an excellent practical RF Linear amplifier.

ACTIVE DEVICES

I have used graphs of an electron tube to illustrate the operation of Class A, B and C, as I think it is easier to visualise. Any active device can be used for all of the classes of amplifier. For a FET we would have to speak of gate bias instead of grid bias and drain current instead of plate current. The polarity of the bias voltage will be different depending on the type of active device used. Polarities are not important for this chapter; I just used them because I was explaining the operation of the curves using an electron tube.

You may be presented with a circuit diagram of different amplifiers in the exam and basically have to identify them. Most exams are multiple-choice, so you are presented with a circuit (schematic) diagram and asked to pick one of four choices of what it is. I have seen dozens of actual exam papers, so I will show you a couple of schematics of amplifiers and give you the clues to identification. I may cover some of the operation only for the purpose of reinforcing some earlier material - a type of revision.

CLASS A, B and C on one graph

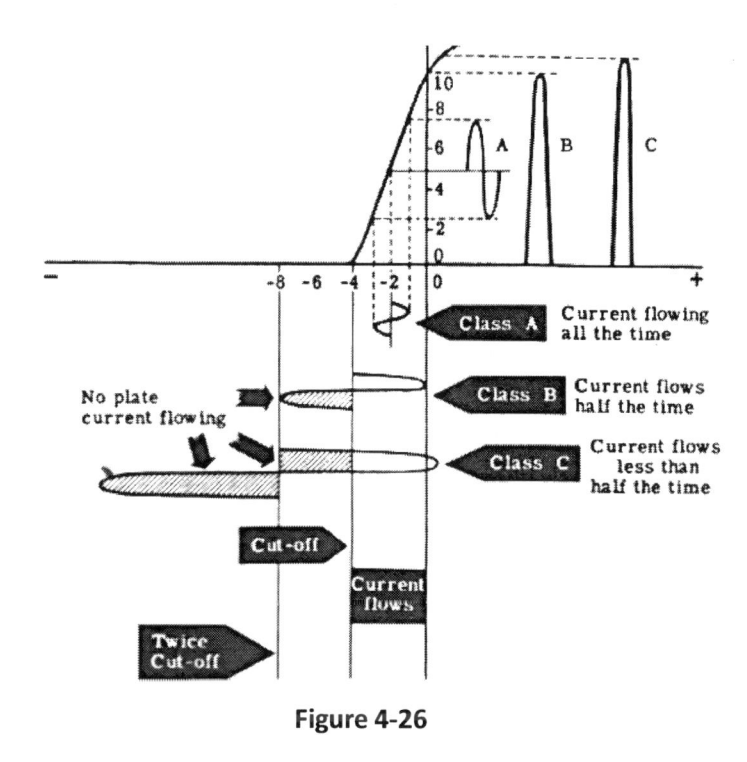

Figure 4-26

Figure 4-26 shows the three main classes of amplifier and their operating curves on the one graph. This is a U.S. Navy diagram. I have left it untouched. It is a very useful diagram that shows in an easy way how the main three classes differ.

PUSH-PULL AMPLIFIERS

We spoke of a Class B amplifier as having an operating angle of 180°. To obtain linear amplification we could use two Class B amplifiers, one to look after and amplify each half cycle. This type of amplifier is called a *push-pull*. Class A amplifiers can also be used in push-pull. The diagrams below are labelled Class A push-pull, but they are virtually the same. If you do see a diagram like this it won't be ambiguous, the examiner will be testing you to see if you know what a push-pull configuration looks like. The bias would be set to about 0.66 of cutoff for Class A operation. For Class B operation (the most likely use) the bias is about 0.95 of cutoff.

Following is a brief description of the operation that will give you a bit more of an idea of what push-pull is, rather than just circuit identification.

Refer to the vacuum (electron) tube circuit of figure 5-26 and assuming Class A operation (established by the Cathode resistor). With no signal applied to the grids, equal anode current will flow through V1 and V2 since the bias is the same to both tubes.

Both tubes are biased to the centre of the linear portion of their operating curves (Class A). Assume that I1 and I2 each register an anode current of 50 mA; then I3 will be I1 + I2 = 100 mA. Now suppose an input signal is applied via T1, which makes the top of the secondary 5 Volts positive with respect to the centre tap and the bottom five Volts negative with respect to the centre tap.

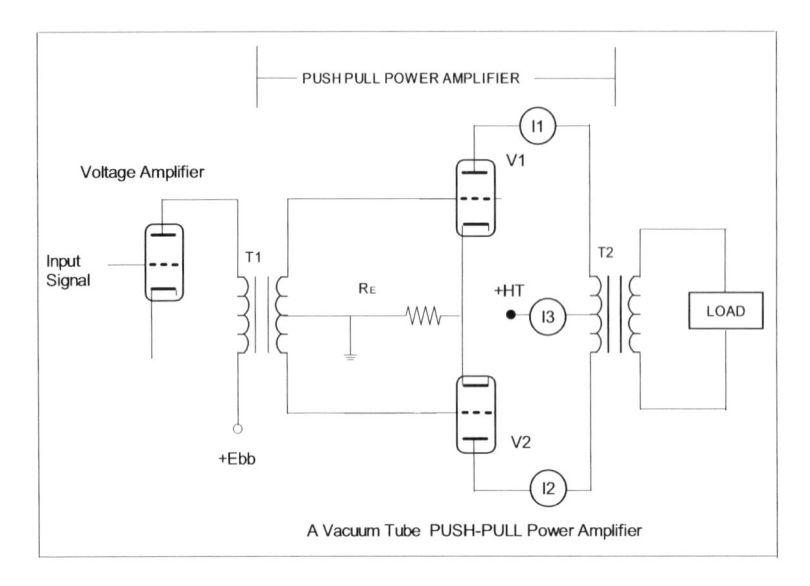

Figure 5-26

In other words, the grid voltage on V1 will be 5V *less* negative while on V2 the grid voltage becomes 5V *more* negative. Assume that the reduced negative bias on V1's grid causes V1's anode current to *increase* by 10 mA, then the increased negative bias on V2's grid will cause the anode current to *fall* by 10 mA. I1 will now be 60 mA and I2 will be 40 mA. Notice that as one tube draws more current, the other tube draws less and the current I3 will remain constant.

On the next half cycle, V2's anode current will increase and V1's anode current will decrease by a corresponding amount. T2's secondary voltage will be the resultant of each of the top and bottom primary currents. T2 also serves to match the high output impedance of the vacuum tubes to the low impedance of the speaker.

The operation of the transistor circuit in figure 6-26 is the same, except that voltage divider bias is shown. With Class A operation the signal currents that flow through R_E are equal and opposite and as such, net degenerative signal voltage will appear across the resistor, making the usual emitter bypassing unnecessary.

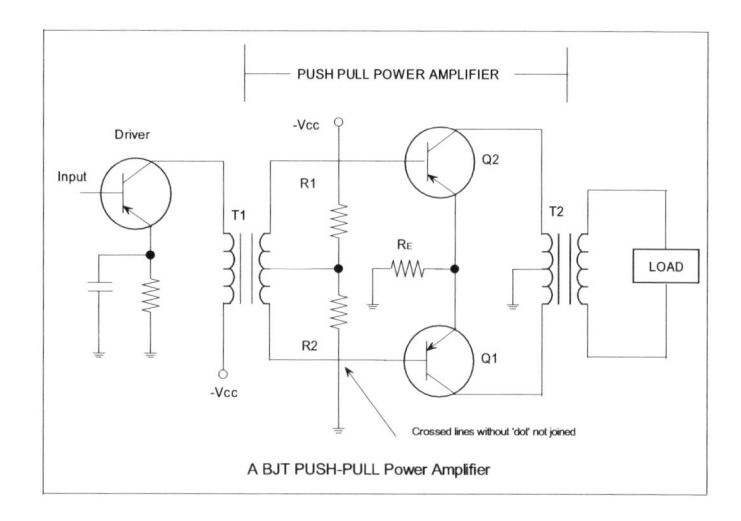

Figure 6-26

An important advantage of push-pull operation over the single-ended circuit is the tendency to cancel *even order harmonics* that may develop in the stage. Such even order (i.e. 2nd, 4th, 6th) harmonics are 180° out of phase in the output and completely cancel.

HARMONICS

Harmonics are distortion products (signals) produced on multiples of the operating frequency. So a radio frequency amplifier operating say on 10 MHz will produce some output on 20, 30, 40MHz etc. It is a fact that harmonics always exist, are caused by non-linearity in an amplifier and that *every amplifier has some non-linearity*. The big advantage of the push-pull circuit is that even order, 2nd, 4th, 6th, etc. harmonics are cancelled in the output.

COMPLEMENTARY SYMMETRY AMPLIFIER

For identification purposes only, a complementary symmetry amplifier can look at first glance like a push-pull amplifier.

The schematic of a transistor complementary symmetry amplifier is shown in figure 6-26.

[Figure 6-26: Complementary Symmetry amplifier circuit diagram showing PNP transistor Q1 and NPN transistor Q2 with resistors R1, R2, R3, R4, supplies V_{cc1} and V_{cc2}, and LOAD.]

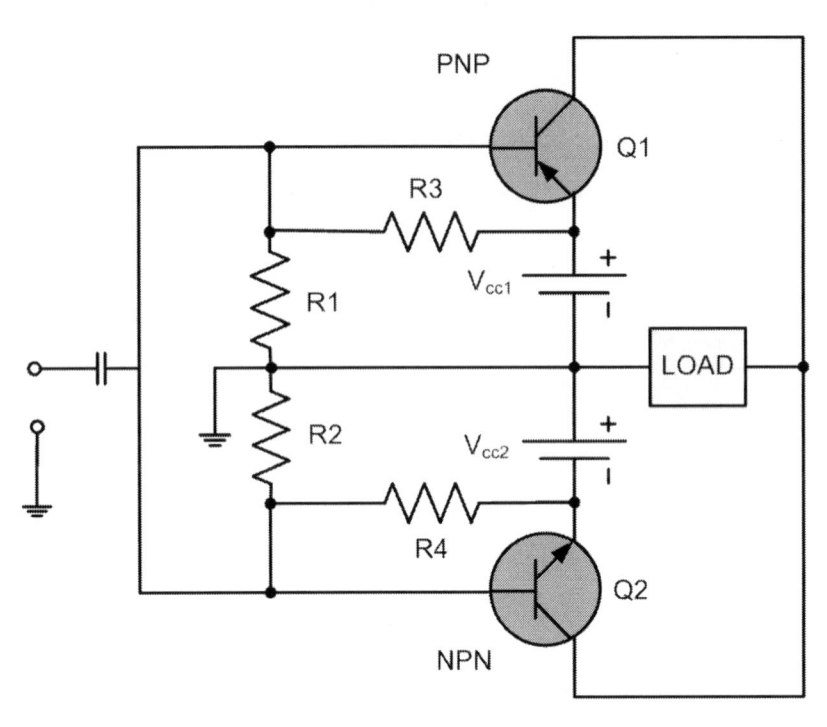

Figure 6-26.

This amplifier takes advantage of semiconductor devices that are fabricated as the complement of each other and a matched pair in terms of characteristics. An NPN transistor is the complement of a PNP transistor. A push-pull amplifier normally has an input transformer to provide out-of-phase signals to the input elements of two active devices. With complementary symmetry, phase inversion is unnecessary.

In the figure 6-26 notice that the base connections of the two transistors are connected directly together. If the input goes negative, the PNP transistor (Q_1) will be biased harder and its collector current will increase. However, a negative-going input signal will reduce the bias current of the NPN transistor (Q_2) causing its collector current to decrease by the same amount.

When the input signal swings positive, Q_1's base and collector current is decreased, while Q_2's base and collector currents undergo a corresponding increase.

The resistor networks on each transistor look complicated, but close inspection shows that R_1 and R_3 provide voltage divider bias for Q_1, while R_2 and R_4 provide the same function for Q_2.

AN INTERMEDIATE FREQUENCY AMPLIFIER

When we come to cover receivers and transmitters, we will come across a stage called the intermediate frequency amplifier. The intermediate frequency or IF amplifiers are a little different from other amplifiers, in that they are designed to *operate on one narrow band of frequencies*. Typical intermediate frequency amplifiers operate on 10.7 MHz and 455 kHz.

Because this amplifier operates only on one frequency, it can be designed with tuned circuits only to pass that frequency, actually a *narrow band of frequencies*. An intermediate frequency amplifier is just like any other amplifier. However, you may need to identify a circuit similar to this in and Advanced assessment. See figure 7.26.

The clue to the identification of the circuit is the use of the transformers (radio frequency transformers) T_1 and T_2. They are ferrite slug tuned (indicating radio frequencies) and shielded (the dashed box). This is a circuit of a fixed tuned radio frequency amplifier, or an intermediate frequency amplifier. The tuned transformer coupling provides greater selectivity. The transformer primaries form part of a parallel tuned circuit and their bandwidth determined by their Q.

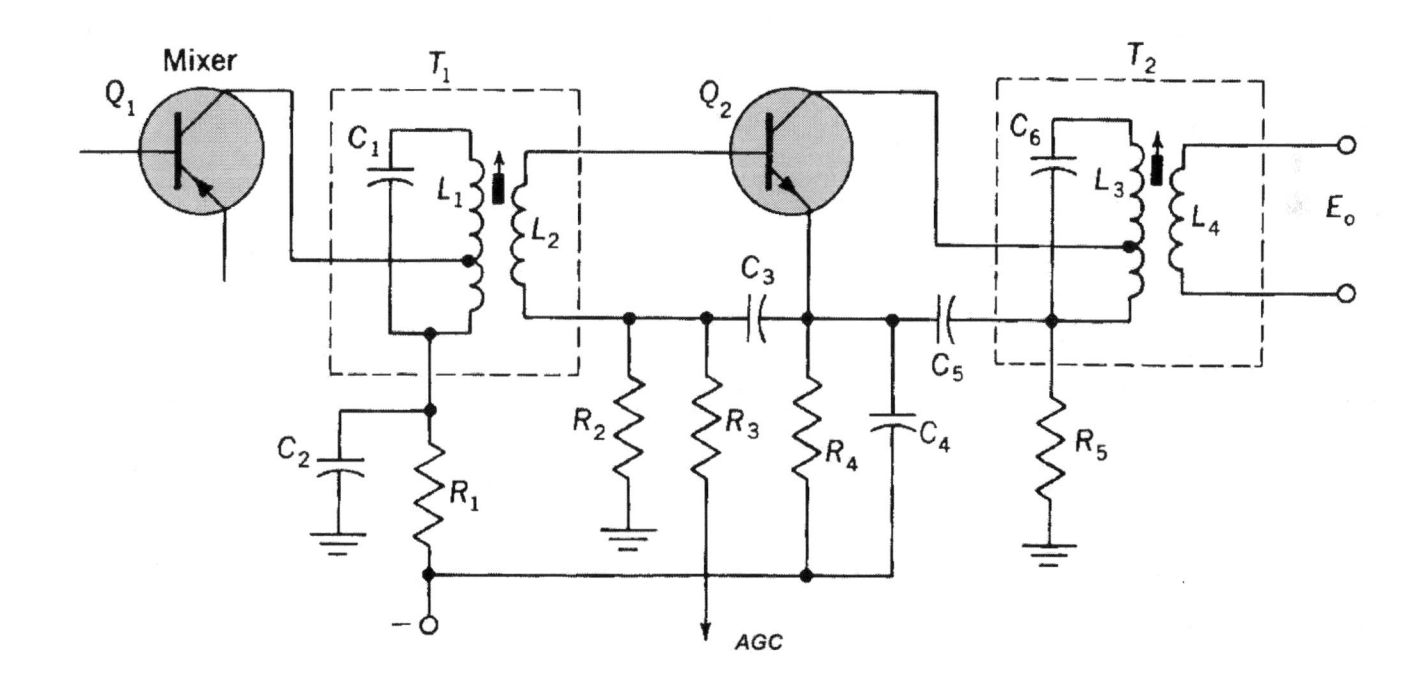

Figure 7-26 IF Amplifier

The IF amplifier in figure 7-26 would be selective enough for say a broadcast band receiver. With narrow band radio transceivers, we need bandwidths as narrow as 2-300Hz for CW, for SSB 2.8KHz and DSB 6kHz. Such high selectivity can only be achieved by the use of ***crystal*** or ***ceramic filters*** that are switched in when the mode switch is operated.

Out of interest and some reinforcement, notice how the collectors are connected to a tapping on the transformer primaries. This is to prevent the low impedance of the transistors loading the tuned circuits and reducing Q and hence the selectivity. The turns ratio of T_1 and T_2 also serve to match the output and input impedances of each stage. R_2 and R_3 provide voltage divider bias for Q_2. R_4 is the emitter stabilisation resistor and C_4 is the emitter bypass capacitor.

This circuit really has much more than you need to know and as long as you can identify it as an intermediate frequency amplifier by the clues given, you will do fine.

FREQUENCY MULTIPLIER

While frequency multipliers are not amplifiers, it seems like an appropriate place to discuss them. The purpose of a frequency multiplier is to do just that, multiply frequency. A 1MHz signal can be converted to a 2MHz signal using a frequency multiplier, or to any multiple of 1MHz.

Frequency multipliers should not be confused with mixers, which is another method of converting a signal from one frequency to another.

Frequency multipliers can only convert to multiples of the input signal.

If you remember (if not go back and have a look at the chapter on tuned circuits) that a parallel tuned circuit has a property called the *flywheel effect*. A parallel tuned circuit in the output of an amplifier is commonly referred to as a tank circuit. Frequency multipliers are designed with output tuned circuits (tanks) that are resonant on either the second, third and less frequently the fourth, harmonic of the input frequency.

The output waveform of a multiplier is a sine wave produced by the " flywheel effect" of the output tank. Multipliers are also referred to as doublers, triplers and quadruplers. Class C operation is essential, as current pulses into the output tank are required to keep the tank oscillating. In a doubler, the output tank is tuned to twice the input frequency and will receive one input current pulse and then go through two complete oscillations before the next current pulse arrives just in time to overcome the lost energy (remember damping). Output tanks in quadruplers require a much higher Q since they must sustain four full cycles of oscillation without appreciable attenuation or damping, before receiving a new current pulse.

Multipliers can only be used to multiply signals that have no intelligence in their amplitude, as the stage is nonlinear. I emphasise again that multipliers operate in Class C, which is nonlinear.

BUFFER AMPLIFIER

An oscillator is a circuit (to be discussed) which provides weak low power signals on a certain frequency. The power output of an oscillator is very low. If we were to connect something like a BJT amplifier after an oscillator, the oscillator would probably cease to function as a BJT requires current from the source to be amplified.

A buffer amplifier is any amplifier (typically after an oscillator) designed to amplify the input signal, without placing any great demands, power-wise, on the stage supplying the signal. A buffer would normally be a high input impedance device like a FET.

In other words, a buffer amplifier requires very little or no power to drive it and it prevents excessive loading of the previous stage.

Some microphones have a small pre-amplifier actually built into them, as the microphone insert (crystal, dynamic, etc.) cannot provide much power out. A pre-amplifier after a microphone is never called a buffer, but in effect, it is doing the same job as a buffer amplifier.

OTHER CLASSES OF AMPLIFIER

There are many other classes of amplifier. Probably the one you might come across is the Class D, though at the time of writing it is not part of the syllabus. If and when it is; it will be added. These classes are not a knowledge requirement now.

- Class D Amplifier – A Class D audio amplifier is basically a non-linear switching amplifier or PWM amplifier. Class-D amplifiers theoretically can reach 100% efficiency as there is no period during a cycle where the voltage and current waveforms overlap as current is drawn only through the transistor that is on.

- Class F Amplifier – Class-F amplifiers boost both efficiency and output by using harmonic resonators in the output network to shape the output waveform into a square wave. Class-F amplifiers are capable of high efficiencies of more than 90% if infinite harmonic tuning is used.

- Class G Amplifier – Class G offers enhancements to the basic class AB amplifier design. Class G uses multiple power supply rails of various voltages and automatically switches between these supply rails as the input signal changes. This constant switching reduces the average power consumption and therefore power loss caused by wasted heat.

- Class I Amplifier – The class I amplifier has two sets of complementary output switching devices arranged in a parallel push-pull configuration with both sets of switching devices sampling the same input waveform. One device switches the positive half of the waveform, while the other switches the negative half similar to a class B amplifier. With no input signal applied, or when a signal reaches the zero crossing point, the switching devices are both turned ON and OFF simultaneously with a 50% (Pulse Width Modulation) PWM duty cycle cancelling out any high-frequency signals.

   To produce the positive half of the output signal, the output of the positive switching device is increased in duty cycle while the negative switching device is decreased by the same and vice versa. The two switching signal currents are said to be interleaved at the output, giving the class I amplifier the named of: "interleaved PWM amplifier" operating at switching frequencies in excess of 250kHz.

- Class S Amplifier – A class S power amplifier is a non-linear switching mode amplifier similar in operation to the class D amplifier. The class S amplifier converts analogue input signals into digital square wave pulses by a delta-sigma modulator and amplifies them to increase the output power before finally being demodulated by a band pass filter. As the digital signal of this switching, amplifier is always either fully "ON" or "OFF" (theoretically zero power dissipation), efficiencies reaching 100% are possible.

   Class T Amplifier – The class T amplifier is another type of digital switching amplifier design. Class T amplifiers are starting to become more popular these days as an audio amplifier design due to the existence of digital signal processing (DSP) chips and multi-channel surround sound amplifiers as it converts analogue signals into a digital pulse width modulated (PWM) signals for amplification increasing the amplifiers efficiency. Class T amplifier designs combine both the low distortion signal levels of class AB amplifier and the power efficiency of a class D amplifier

27. OSCILLATORS

WHAT IS AN OSCILLATOR?

Oscillators are used in radio circuits to produce radio and audio frequency energy, generally with a sinewave output, though the waveform can be many shapes such as a square wave or saw tooth. The sinusoidal waveform may be AC or VDC. Oscillators used in radio frequency circuits are always very low power devices, in contrast to AC generators in a power station. Nevertheless, the AC power generator and the electronic oscillator are related, in that they both produce sinusoidal electrical energy. Unlike the AC generator, though, the electronic oscillator can produce output on frequencies measured in tens of megahertz. Special oscillators can produce output at microwave frequencies.

An oscillator producing a radio frequency output is actually a low power transmitter in its most basic form. In a radio transmitter and receiver, up to several or more oscillators may be employed.

We are going to look at a number of different types of oscillators and their circuits in this chapter. Don't be put off by the number of circuits as you don't have to remember them in detail. You *do* need to learn the identifying features of each oscillator circuit. In the assessment, you will be asked to name the type of oscillator shown in the circuit. The fundamental principles of oscillator operation will be explained for each of the types. You will find a repeating theme across all of the oscillator types.

REQUIREMENTS FOR OSCILLATION

If any circuit has the following properties in the required amount, then the circuit will oscillate whether it is supposed to or not: -

a) Amplification.
b) A frequency determining device.
c) Positive feedback (regeneration).

In oscillators, the factors above are designed into the circuit intentionally. The requirements (a) and (c) also occur in many amplifiers. For this reason, care must be taken with amplifiers to prevent or control, in particular, the third requirement for oscillation, positive feedback. Any amplifier provided with sufficient positive feedback will begin to *self-oscillate*. Amplifiers are *not* supposed to oscillate, they are meant to amplify, though many amplifiers can easily become undesirable oscillators.

An amplifier which gets unwanted positive feedback will become an oscillator and potentially cause interference. Amplifiers that oscillate generate a signal, rather than amplify one. Such unwanted signal generation can cause interference.

A generic-oscillator (figure 1-27) is any amplifier with positive feedback. When an amplifier has enough positive feedback it 'takes off' and oscillates – not a good thing if this is not intended.

Example: When sound from the speaker of a public address system gets back into the microphone(s) of that system, oscillation will occur. In audio circles, it is called feedback. The amplifier "squeals". When this happens at a radio frequency, you can't hear it. The oscillation is well beyond human hearing, but the effect is the same.

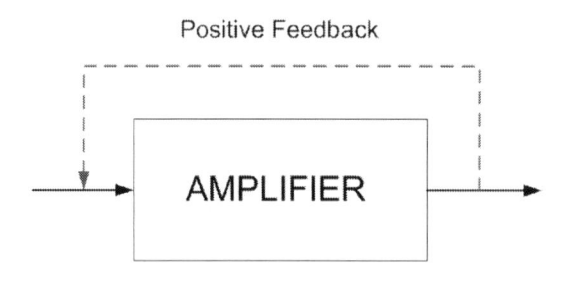

Figure 1-27.

FREQUENCY DETERMINING DEVICE

The frequency-determining device is usually a resonant LC circuit or a quartz crystal. Slices taken from quartz crystals make the most stable oscillators.

STABILITY

To ensure good stability, an LC oscillator should:

a) Have a high C-to-L ratio.
b) Have a well-regulated power supply.
c) Have good isolation between the oscillator and its load.
b) Employ components which have low-temperature coefficients.
e) Not be exposed to large changes in temperature.
f) Have all components mechanically rigid.

DRIFT

Drift is an unwanted *slow* change in the frequency output of an oscillator.

One of the main causes of drift in LC oscillators is unwanted capacitance changes in the circuit. These capacitance changes are mostly due to temperature effects. If the tuning capacitance is made *high* compared to the inductance in the frequency-determining-circuit, then such unwanted capacitance changes will cause a smaller percentage change than if the tuning capacitance were *smaller*. Simply having a large capacitance compared to inductance produces a more stable oscillator. We say the stability is better with a higher C-to-L ratio; this means a high value of capacitance compared to inductance in the resonance equation.

BUFFER AMPLIFIER

A buffer amplifier improves the frequency stability of the oscillator by isolating it from the load. An oscillator is not able to deliver much, if any, power to other circuits. If too much power is taken from an oscillator, then it may be 'pulled' off frequency, or even damped so badly that it fails to oscillate. The buffer amplifier is placed immediately after the oscillator. The buffer amplifier has high input impedance and as such draws little or no power from the oscillator. The buffer has just enough gain to supply the following stage with usable power without loading the oscillator.

CHIRPING

In a telegraphy (Morse code) transmitter, the stage which is being keyed (by the Morse key) should never be too close to the oscillator as this can result in oscillator chirping.

Chirping sounds like rapid short changes in frequency, very much like a canary chirping. What is happening is that sudden changes of load on an oscillator are occurring when the telegraphy key is closed, pulling the oscillator and hence the output of the transmitter off frequency. The chirp is the oscillator stabilising to the new frequency.

As we go through some different types of oscillator circuits, you will notice a common theme. I would like you to take notice of the:

(a) Type of active device employed (however the type of oscillator is *not* determined by the active device used)
(b) The polarity of the power supply and whether it is correct!
(c) How feedback (regeneration) is achieved
(d) What is the frequency-determining device?
(e) The bias arrangement for the active device.

THE ARMSTRONG OSCILLATOR

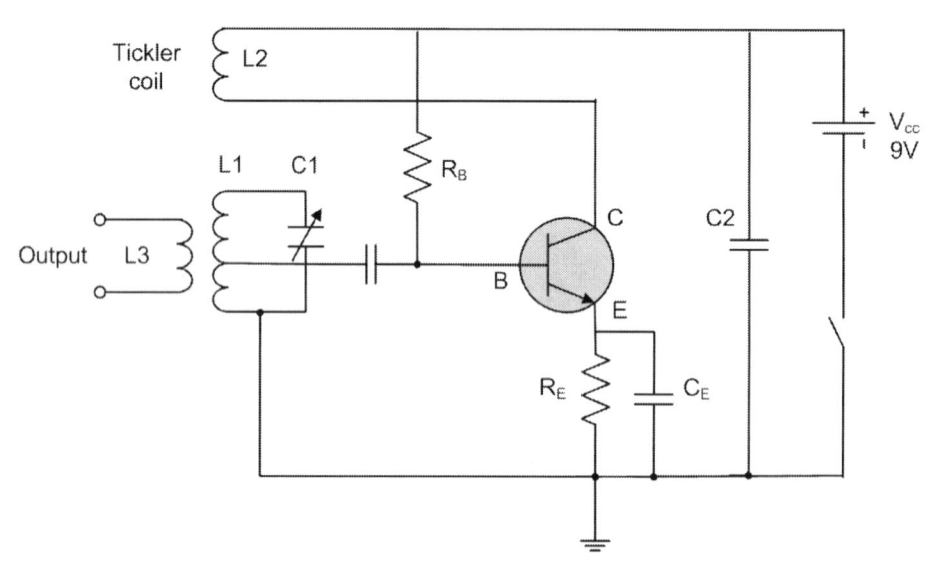

An Armstrong Oscillator using an NPN transistor

Figure 2-27.

In figure 2-27 an NPN BJT is used as the active device. L2 is called the ***tickler coil*** and is the distinguishing feature of an Armstrong oscillator. L2 provides regeneration to the input of the BJT. L2 does this by being inductively coupled to L1. Some of the Signal in the output circuit is inductively coupled back to the input circuit by L2. The base circuit of the transistor contains a parallel tuned circuit consisting of L1 and C1. This circuit determines the frequency of operation. C1 is variable to change the frequency of oscillation. Provided the connections to the tickler are the right way around, then feedback is positive (regenerative) and oscillation will be sustained. Connecting the tickler coil the wrong way would produce negative feedback (degeneration) and the circuit would not operate. R_B provides for the correct amount of bias current. DC bias flows from earth (or negative) through R_E, into the emitter, out of the base, through R_B and then back to positive. The value of R_B and to a lesser extent R_E determines the amount of DC bias current.

R_E provides emitter stabilisation to prevent thermal runaway and C_E is the emitter bypass capacitor. We do not want the oscillating signal to flow through R_E, as any signal current that flows through R_E will produce negative feedback. The operation of R_E and C_E was discussed in an earlier chapter.

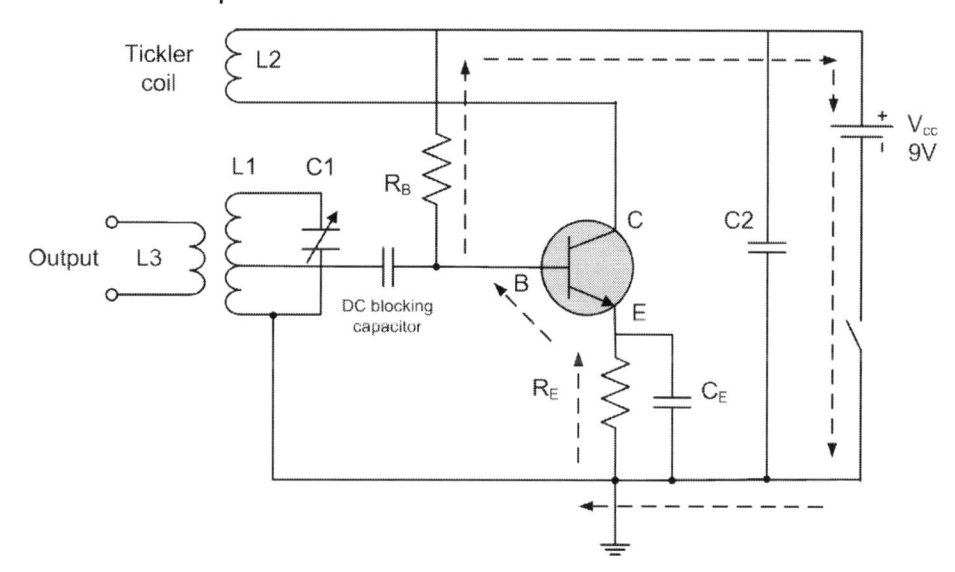

DC Bias – primarily determined by R_B

Figure 3-27.

The figure 3-27 schematic shows where the DC bias current flows in our Armstrong oscillator. This is fairly straightforward. The amount of DC bias current is primarily determined by the value of the resistor R_B.

The capacitor in series with the base is a DC blocking capacitor. This capacitor is also known as a coupling capacitor as it couples signals into the active device and prevents the DC bias from being affected. For example, L1 would short circuit the base to ground if this capacitor were not present. This capacitor will block the DC bias current from flowing into L1 but allow the signal coming from L1-C1 to pass to the base.

The figure 4-27 circuit shows the DC output circuit of the transistor (dashed lines). Because the transistor is forward biased in its base-emitter circuit, then, emitter-collector current will flow. This circuit shows the DC emitter-collector current.

Figure 4-27

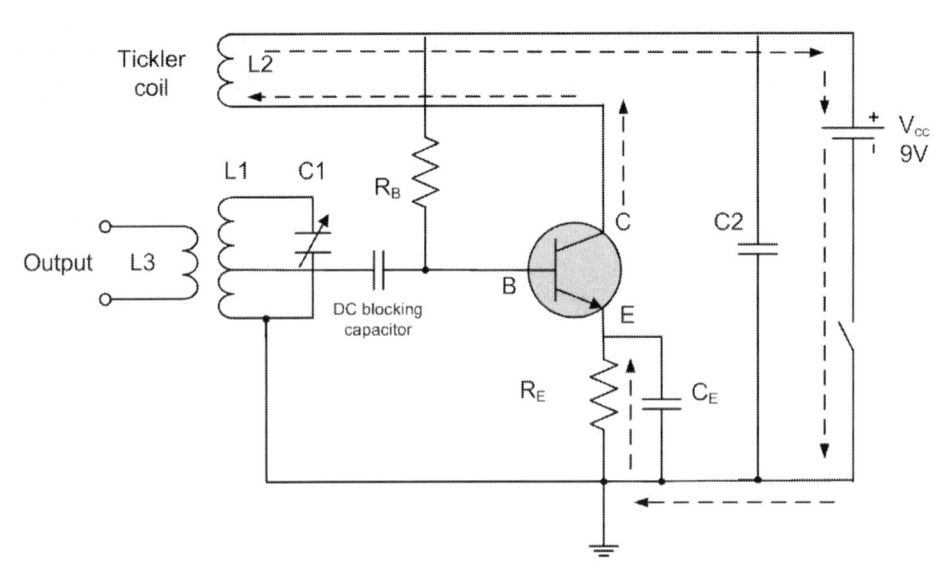

The DC Collector-emitter circuit

A much larger current than the base current flows from the negative terminal of the battery – up through R_E, into the emitter out of the collector and back to the positive terminal of the battery.

The figure 5-27 circuit shows where the *signals* would flow in this oscillator. Assume that the oscillator is meant to produce a sinewave on 1 MHz. This will be a sine wave of varying DC, not AC. Most active devices do not work on AC.

When the oscillator is turned on, L1 and C1 start producing an oscillation on 1 MHz. This oscillation would normally die down due to losses in the components of the resonant circuit.

The oscillating voltage across L1 and C1 is superimposed on top of the DC bias current in the base circuit. So a 1 MHz signal current flows in the base circuit as shown as a dashed line. Notice how the signal flows through C_E and not R_E. (A little bit of signal current does really flow through R_E but not enough to be significant). The capacitive reactance of C_E at 1MHz would be 1/10th the value of R_E.

Now this 1 MHz signal in the base circuit causes a 1 MHz signal in the collector circuit. The signal in the collector circuit is much stronger and flows as shown with a different dashed line. The capacitor across the battery bypasses the signal around the supply. We never want signal currents to flow through a battery or power supply. For one reason, the power supply is common to all stages.

If we allow signals from any stage into a power supply, they (the signals) can affect the operation of other stages via the power supply. You will nearly always see a power supply bypass capacitor and often an RFC (radio frequency choke) in series with the power supply just to make it all that much harder for signal currents to get into the power supply.

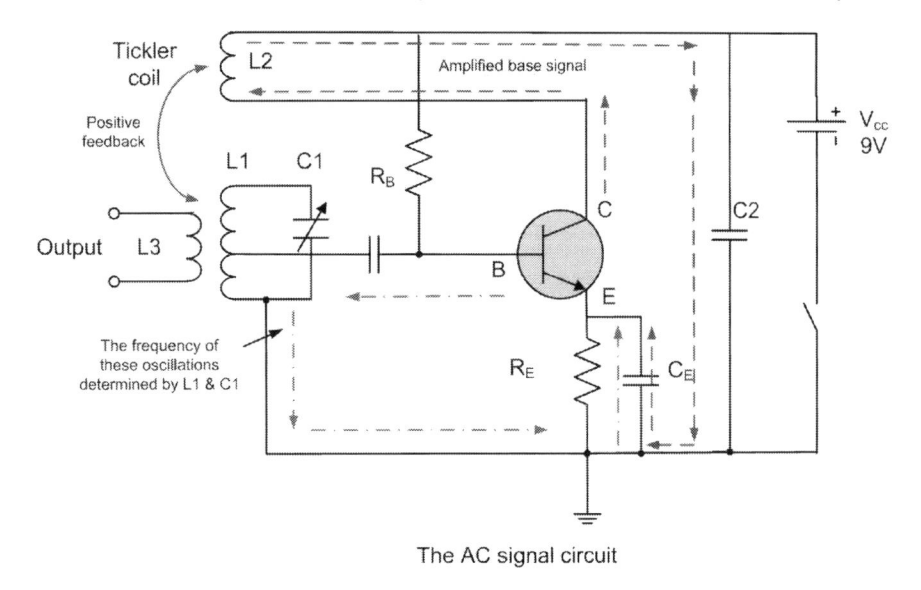

Figure 5-27

Notice that the amplified signal flows in the tickler coil. The tickler coil (L2) is inductively coupled to L1. If you like, think of the tickler coil as the primary of a transformer and L1 as the secondary. We have positive feedback from the tickler coil into L1 – so the oscillations are sustained.

L1 is also inductively coupled to L3, so we can take some of the signal current away from the circuit for use elsewhere. An oscillator would not be of much use if it did not provide us with output – L3 is the output. We will discuss what we can do with oscillators in a future chapter.

AN ELECTRON TUBE ARMSTRONG OSCILLATOR

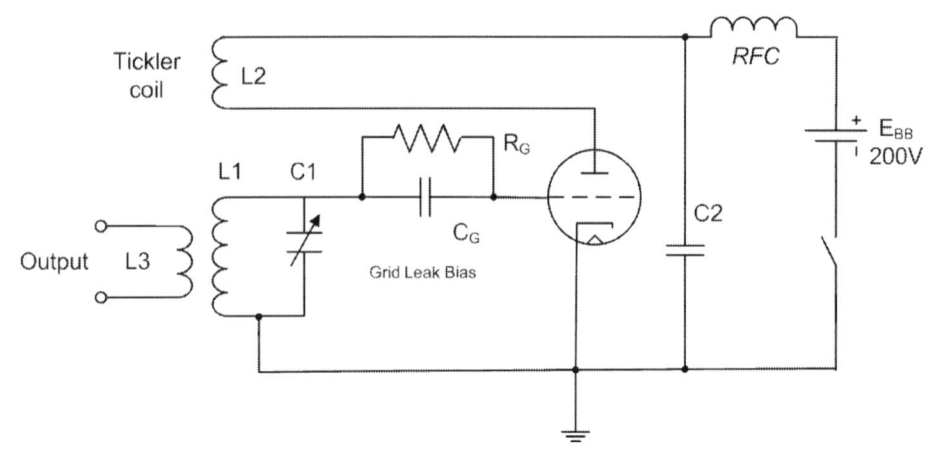

Armstrong Oscillator using a Triode

Figure 6-27.

As with the BJT circuit, the *tickler coil* provides feedback. The *tickler coil* is the most identifying feature of the Armstrong oscillator.

L1 and C1 determine the frequency of oscillation. The output is taken across L3, which is inductively coupled to L1.

The oscillator is operating at a high frequency (radio frequencies). The one stage that is common to all other stages in a radio system is the power supply. Therefore, *the power supply is a potential path for each stage to interfere with the other stages*. C2 and the RFC (radio frequency choke), either alone or together, will be seen in many RF signals and be choked by the RFC and bypassed by C2. Because of the high frequency of the oscillator, C2 has a low reactance to RF energy created by the oscillator and bypasses this energy around the power supply. The power supply is only shown as a battery for simplification. On the other hand, the RFC has a high reactance to high frequencies and blocks radio frequencies from entering the power supply. C2 and the RFC have no affect on the DC from the power supply getting to the oscillator. Looking at it another way, C2 and the RFC form a low-pass filter, allowing DC to pass from the supply to power the oscillator, but blocking RF from getting from the oscillator into the supply.

GRID LEAK BIAS

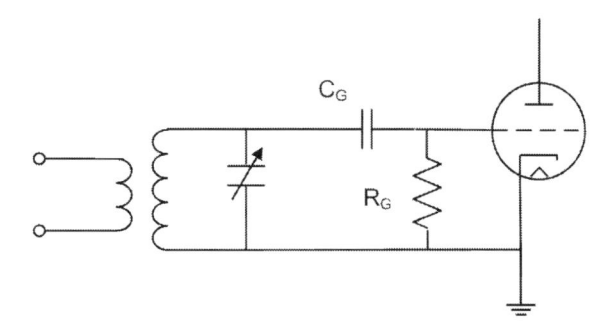

Figure 7-27. Grid Leak Bias – drawn differently

Bias for the electron tube is obtained by C_g and R_g. I have not discussed this bias arrangement before. In the case of an electron tube, it is called **grid leak bias**. This type of bias will also be found on a FET, in which case it is called **gate leak bias**.

By now you should well and truly know that the purpose of bias on any active device is to place the correct DC operating potential onto the input element of the active device, to set the correct operating point and class of operation. However, in the case of BJT's it is bias current rather than voltage.

With grid or gate leak bias, you will always see the same configuration of R_g and C_g at the input of the device (the control-grid or gate). When the oscillator is first turned on, it will have no bias.

Oscillations in the parallel LC circuit (L1 and C1) will place an AC voltage on the grid. When the grid is positive, it will draw current from the cathode (space charge) and C_g will be charged negative on the right and positive on the left. After a few cycles, the negative voltage on the right will remain constant and provide the required negative bias.

The circuit of figure 8-27 shows how grid leak bias works. L1 and C1 are replaced with a source of AC. A diode rectifier is formed by the control-grid and the cathode. I have just shown a semiconductor diode here.

When the oscillator first starts up (when it is turned on), for a while positive half cycles will cause current to flow in the circuit as shown and C_g will charge negative on the right-hand side. It is this negative voltage that biases the control grid.

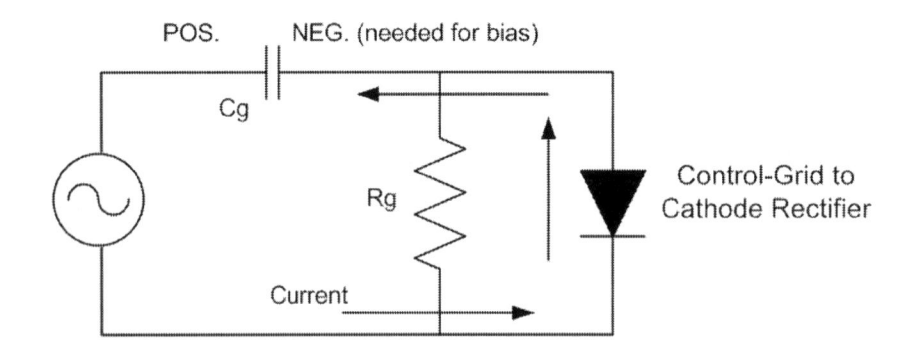

Figure 8-27.

THE HARTLEY OSCILLATOR

The Hartley is a simple extension of the Armstrong. I suspect it came about because Hartley was lazy (not really) and worked out a way to make the Armstrong work without having to wind a separate tickler coil.

The tickler coil is now incorporated into part of the resonant tank inductor L1. The output in either circuit flows through a portion of L1 providing the necessary regeneration.

The **tapped inductance** most easily identifies the Hartley oscillator. The tap position is adjusted to control the amount of feedback - it is not a centre tap as some diagrams suggest.

There is no difference between an Armstrong and a Hartley except the tickler coil is made part of L1. Instead of being mutually coupled pair of coils, L1 is an autotransformer, the primary of which is the tickler coil. All other operation of this circuit is the same.

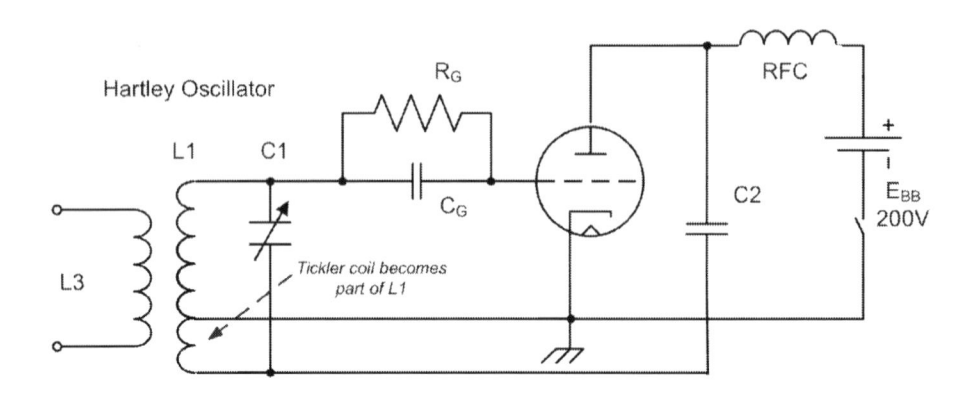

Figure 9-27 – Hartley Oscillator using Triode

A BJT HARTLEY OSCILLATOR

Another Hartley oscillator (tapped inductance doing away with the tickler). L1 and C1 determine the frequency of oscillation. R_e and C_e are for emitter stabilisation and bypassing. See figure 10-27.

The input signal from the oscillatory circuit is taken from between the tapping and the top of L1. Feedback is injected back into the circuit because the output flows between the bottom of L1 and the tapping.

About the only other difference with this circuit is that bias is now determined by R1 and R2. This is called *voltage divider bias*.

C2 is a DC blocking capacitor and C3 is a power supply bypass.

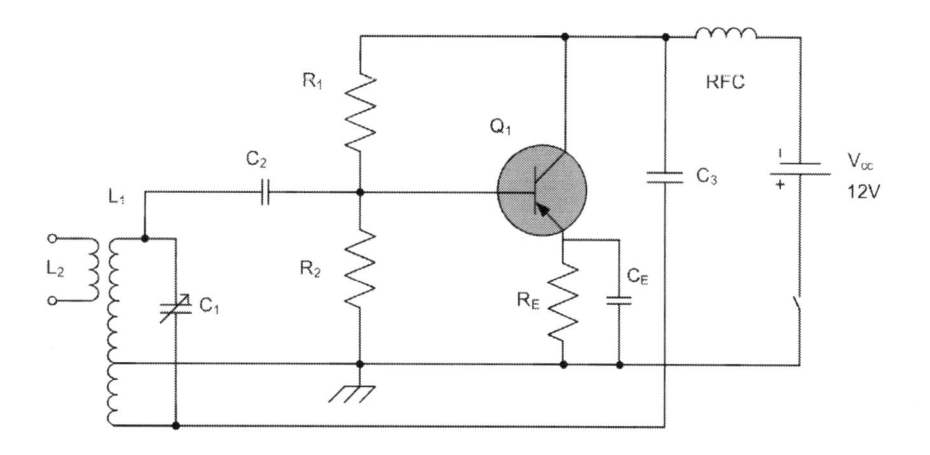

Figure 10-27

Disconnecting the power supply from the 'signal' using RFC's and a bypass capacitor like C3 is sometimes referred to as *power supply decoupling*. So, you may see C3 and the RFC described as 'power supply decoupling components.' This is just a fancy way of saying that the RFC's and bypass capacitors 'decouple' or disconnect the power supply from the point of view of the signal.

THE COLPITTS OSCILLATOR

The configuration of the Colpitts oscillator resembles that of a Hartley in operation and appearance. The difference is that the tapping to the resonant circuit is now made with a ***capacitive voltage-divider*** *rather than with a tapped inductance.* The output voltage is applied to the input via the voltage divider. The ratio of C1 and C2 controls the amount of feedback, a lot easier than fiddling with an inductance. Sometimes C1 and C2 are ganged to provide a fairly constant amount of feedback over a wide range of operating frequencies.

Ganged capacitors are two or more variable capacitors on the one shaft, with their movable plates connected to that shaft.

A good feature of the Colpitts oscillator is its comparatively good wave purity. This is because C1 and C2 provide a low impedance path for harmonics, effectively shorting them to the emitter. The Colpitts is an exceptionally fine high-frequency oscillator and has been used as the VFO (variable frequency oscillator) in many radio experimenters transceivers.

The Colpitts and the Hartley are the same – except the Colpitts uses a *tapped capacitance* rather than a tapped inductance to provide feedback.

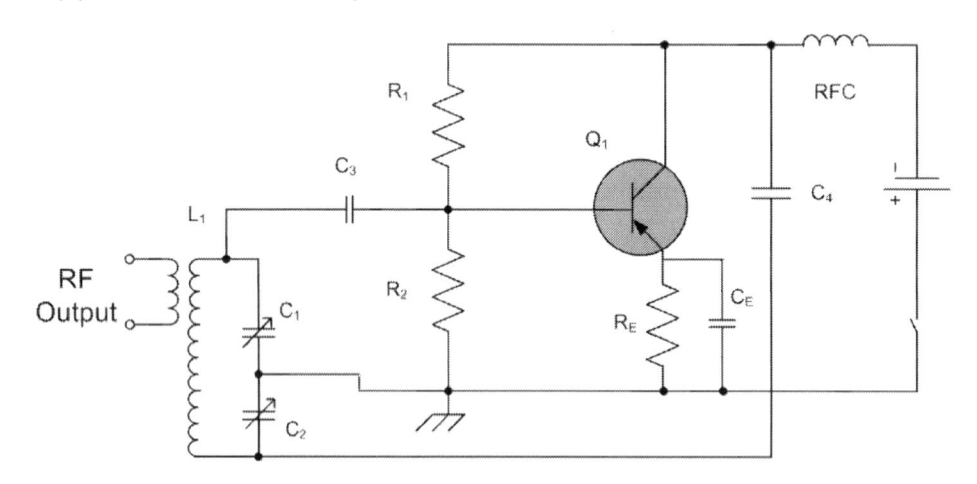

Figure 11-27 – Colpitts Oscillator

The inductor L1 in 11-27 can be replaced with a crystal – in which case the circuit become a 'Crystal Colpitts Oscillator.'

THE CLAPP OSCILLATOR

All of the oscillators we have discussed up to now, have contained a *parallel* LC circuit at the input of the active device to determine the frequency of operation. The Clapp oscillator uses a **series LC circuit**.

In figure 12-27, L1 and C1 form a series resonant circuit to determine the frequency of operation.

The voltage-divider capacitors C2 and C3 perform the same function as in the Colpitts oscillator. The frequency of oscillation is slightly higher than the series-resonant frequency. Because a series circuit has *low impedance*, the Clapp oscillator is less affected by variations in load conditions. The Clapp oscillator has excellent frequency stability and has frequently found applications in electronics and radio.

The Clapp Oscillator

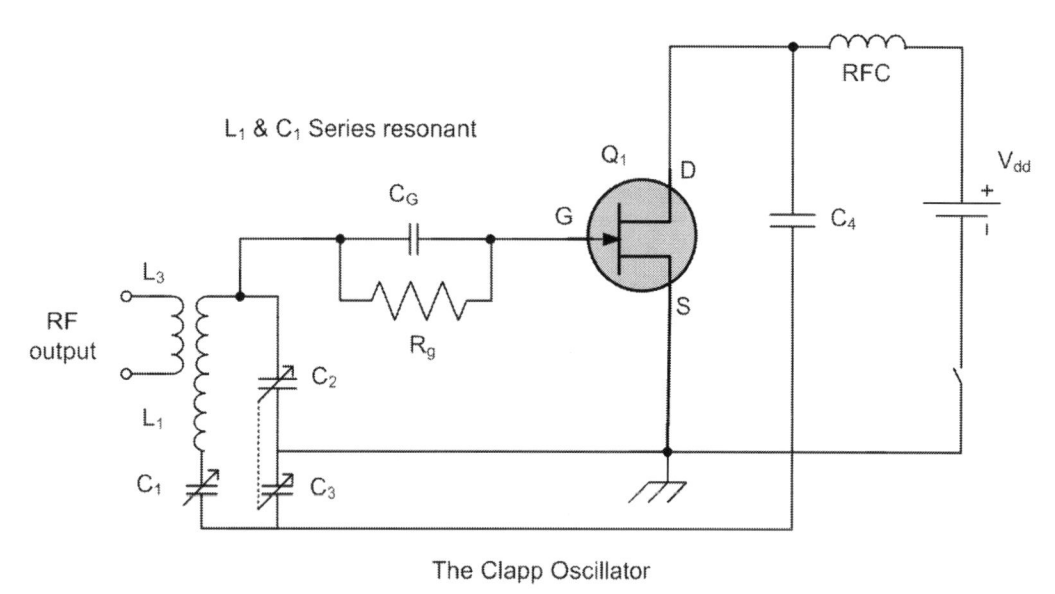

Figure 12-27

Gate leak bias is provided by Cg and Rg. The purpose if Rg is to discharge Cg.

You have to be a little careful not to confuse a Clapp and a Colpitts. The Clapp has a *series* LC circuit as shown by L1 and C1 in figure 12-27. If C1 was removed in this circuit, then it would be a Colpitts.

QUARTZ CRYSTALS

Quartz crystals or better put, thin slices or quartz cut from a larger crystal, exhibit the piezoelectric effect. They will oscillate just like a tuned circuit. The accuracy of the frequency of oscillation is extremely stable. Hence, quartz is used in watches and many other timing devices.

Figure 13-27

FREQUENCY OF A QUARTZ CRYSTAL

The physical dimensions primarily determine the resonant frequency of a quartz crystal. However, cuts (the plane or angle of the slice through the main crystal) from the natural crystal will provide different frequency ranges and characteristics.

By proper selection of the type of cut, dimensions of the plate (the plates are the electrical contacts to the crystal) and mode of vibration, it is possible to obtain crystals with resonant frequencies from as low as 6 kHz and as high as 75 MHz. For higher frequencies, the slice of crystal becomes too thin and fragile and is more susceptible to frequency changes with temperature.

MORE THAN ONE RESONANT FREQUENCY

A quartz crystal actually has two resonant frequencies, a *series-resonant* frequency and a *parallel-resonant* frequency.

A quartz crystal is capable of acting as a parallel resonant circuit or a series resonant circuit. The electrical equivalent circuit of a quartz crystal along with its two resonant frequencies is shown in figure 14-27.

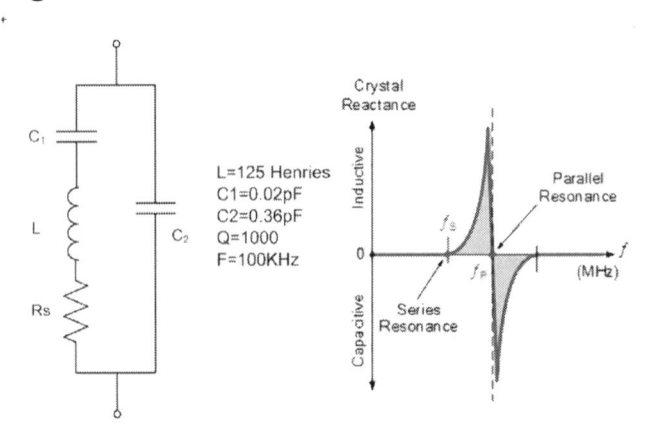

Figure 14-27

The ***equivalent circuit*** is a combination of a series and parallel-tuned network. It is *not possible* to construct the electrical equivalent circuit of a crystal as any man-made inductor of the magnitude shown would have very large losses indeed. If a quartz crystal is placed in *series* with a signal path, then signals on the series resonant frequency will be passed easily through the low-impedance offered by the crystal. That is, it behaves like an LC series circuit.

TEMPERATURE COEFFICIENT

The term " temperature coefficient " defines the way in which the crystal frequency will vary with temperature change. Crystals are usually rated in Hertz-per-megahertz-per-temperature change in degrees Celsius.

A crystal might have a positive, negative, or zero temperature coefficient.

If a crystal has a negative temperature coefficient (NTC) a temperature increase results in a frequency decrease. A positive temperature coefficient indicates that a temperature increase results in a frequency increase. A crystal with zero temperature coefficient will maintain a relatively constant frequency within the manufacturers stated frequency limits.

Temperature coefficients are usually expressed in Hertz per Megahertz per Degree Celsius (or centigrade), or more simply, in parts per million, with the degrees Celsius understood.

When crystal stability is of paramount importance, the crystal is enclosed in a temperature controlled 'oven'. The crystal is then kept at a constant temperature via heating and a feedback mechanism to maintain the constant temperature. Crystal ovens were once used even in standard communications equipment. In modern times, crystal stability has been much improved and ovens are not as common. However, they are still used where equipment is likely to be exposed to great temperature extremes (space, polar environments and the like).

OVERTONES

An overtone crystal is specially ground to obtain enhanced oscillation on *odd harmonics* of its fundamental frequency. A crystal cut with a fundamental frequency of 10 MHz can be cut so as to enable oscillation on 30 and 50 MHz, which are the third and fifth overtones respectively. The use of a crystal on overtone frequencies makes stable oscillator operation possible up into the VHF range. In many VHF/UHF transceivers, the conversion of the incoming signal to the first intermediate frequency (IF) is accomplished by a local overtone oscillator.

Crystals that are to be used on overtone frequencies are always connected in series with a signal path because, for overtone operation, the crystal must operate on its series resonant frequency. For this reason, an overtone is more accurately defined as an odd multiple of the series resonant frequency. Most ordinary crystals can be used on the third or fifth overtone. When a crystal is used on an overtone, the crystal vibrates at the overtone frequency. Many things can vibrate (oscillate) in different planes. Take a wobble board. It can vibrate long ways, cross ways and from corner to corner all at different frequencies. Similarly, overtone crystals are cut to enhance vibration in various planes.

CIRCUIT OF AN OVERTONE OSCILLATOR

In the circuit of an overtone oscillator shown in figure 15-27 – the crystal frequency is shown as 43 MHz – this is very high for a crystal frequency. The fundamental frequency of the crystal would really be much lower for stability. The crystal is operating on an overtone frequency. The actual fundamental frequency of the crystal could be 14.333MHz. However, the crystal has been cut in such a way that it will physically vibrate on an overtone. In this case 43 MHz, which is an odd multiple of the series resonant frequency of the crystal.

You can tell the circuit is not that of a harmonic oscillator, because the crystal is labelled with the same frequency as the output.

The variable capacitor at the output of the JFET would be used to tune the primary of the RF transformer to 43 MHz. The 100-Ohm resistor and the 0.01µF capacitor provide power supply decoupling. That is, they form a simple low-pass filter to prevent RF from getting into the power supply.

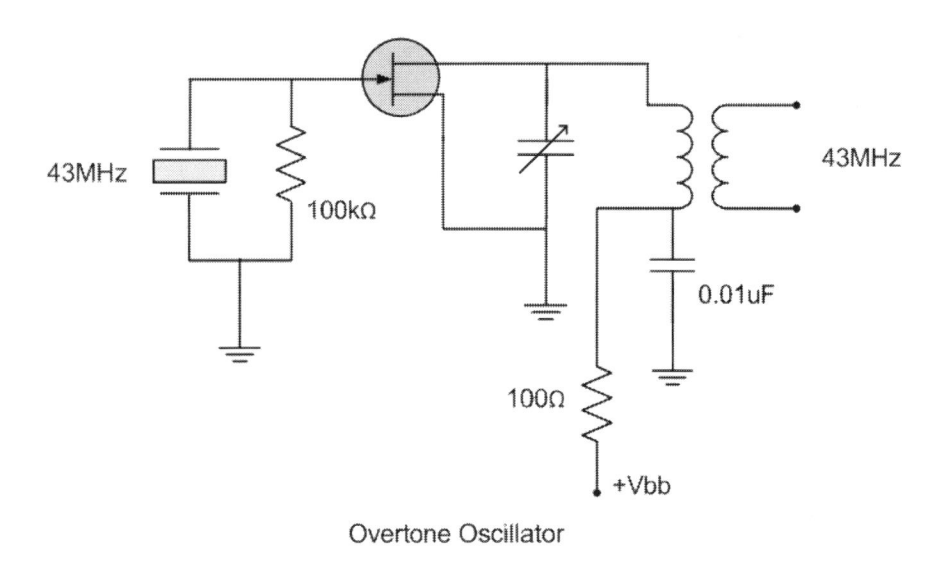

Figure 15-27 – Overtone Oscillator.

THE HARMONIC CRYSTAL OSCILLATOR

A crystal oscillator with its output circuit tuned to any harmonic of the fundamental frequency of the crystal is called a *harmonic oscillator*. The harmonic oscillator takes advantage of the flywheel effect of the output stage to maintain oscillation in the same way as a frequency multiplier.

The harmonic oscillator operates on an entirely different principle to that of the overtone oscillator. In the harmonic oscillator, the crystal is physically vibrating on its fundamental frequency. In the overtone oscillator, no fundamental vibration or frequency is present anywhere in the circuit.

THE PIERCE CRYSTAL OSCILLATOR

The Pierce crystal oscillator *has no resonant tank circuit*. The crystal is connected directly between the output and input of the active device used. The crystal operates on its series-resonant frequency.

The Pierce oscillator is a crystal oscillator. It must have a crystal. An interesting thing about the Pierce is that the crystal provides the feedback path. The crystal shown is connected between drain and gate. The crystal is series resonant meaning the crystal is operating on its series resonant frequency and has a low impedance. Any signal that can travel from the drain to the gate is positive feedback.

The resistor between gate and source provides the small bias voltage for the JFET.

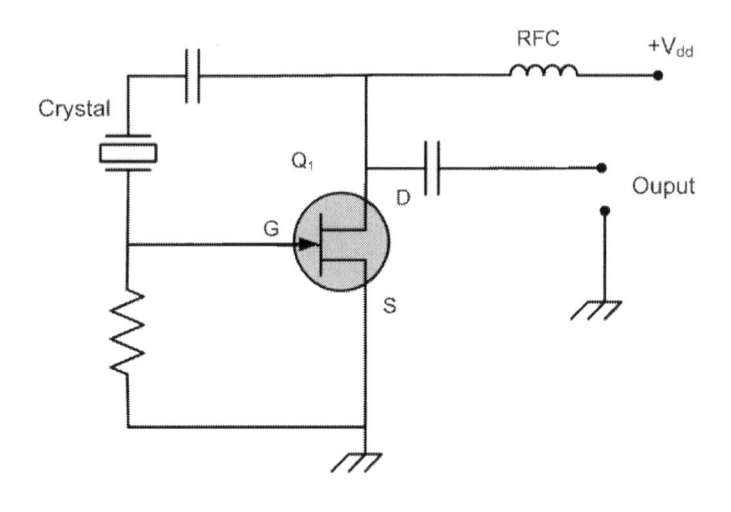

The Pierce Oscillator

Figure 16-27

THE COLPITTS QUARTZ CRYSTAL OSCILLATOR

The Colpitts crystal oscillator is like any other Colpitts except a crystal is installed in place of the LC tank. As with a normal Colpitts, feedback is provided by the capacitive voltage divider. The operation of the circuit is the same except that the crystal now determines the frequency of oscillation. The crystal operates on its parallel-resonant frequency. Since the Q of the circuit is high, the feedback required is considerably less than with standard LC Colpitts.

A small 'trimmer' capacitor may be placed in series with the crystal to enable *small adjustments to the crystal frequency* (doing this is not a unique feature of the Colpitts). This capacitor would typically be 30 to 100 picofarads and one should not expect to vary the frequency by more than +/- a few kilohertz if predictable operation is to be maintained. Adjusting a crystal frequency in this way is called *pulling* the crystal. Try to change the frequency too much using this method and you may find that the crystal will jump to some unpredictable frequency and operation becomes unreliable and unstable.

This excessive pulling is sometimes done illegally on 'pirate' radio. Often referred by the users as a slider control since you can slide-off-channel between channels. It is a bad engineering practice and illegal as it can cause the transmitter to operate on unauthorised channels (which is the intention). However, more important than the legalities is the interference to other radio services, which has the potential to endanger life.

There is nothing new in the circuit of figure 17-27. The LC circuit has been replaced by the crystal. There is a variable capacitor in series with the crystal to enable small frequency adjustment.

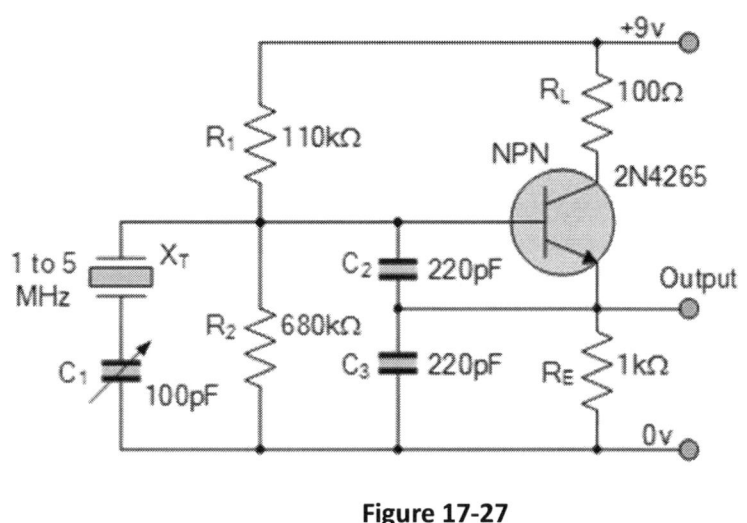

Figure 17-27

A VOLTAGE CONTROLLED OSCILLATOR

Figure 18-27 shows one method of adjusting a crystal oscillator's frequency with a varactor diode.

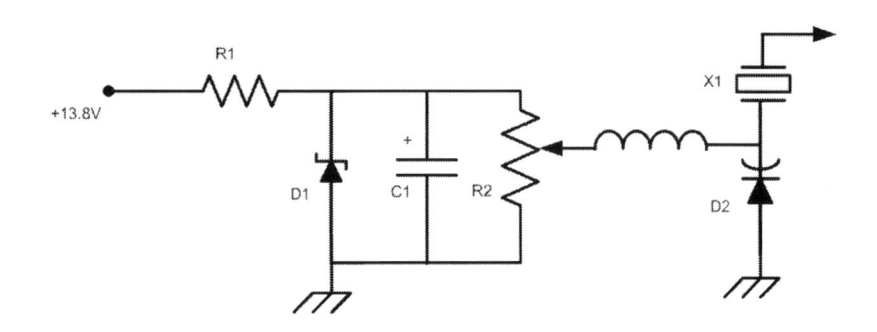

Voltage Variable Oscillator

Figure 18-27

This technique is frequently employed in the BFO/clarifier (single sideband is to be discussed) circuits of transceivers. The crystal would form part of a standard crystal oscillator. The reverse-biased varactor diode is connected in series with the crystal and the reverse bias voltage (and hence the junction capacitance) is adjusted by the potentiometer (variable resistor) R2. R1, D1, C1 and R2 provide an adjustable and regulated reverse voltage for D2.

You might ask why bother with such an arrangement? The alternative would be to have a variable capacitor in place of D2. In addition, this capacitance will, by necessity, have to be mounted on the front panel of the radio. This creates all sorts of engineering difficulties. The wiring between the variable capacitor and the crystal all form part of the total capacitance in the oscillator circuit. However, with the varactor method just described, the variable capacitance is right at the crystal and R2 can be mounted in any convenient position on the front panel, as the wiring between R2 and D2 has no appreciable effect on the operation of the oscillator. This is an excellent method to overcome the unwanted stray capacitance effects of wiring.

The Piezoelectric Effect

Piezoelectricity is a complex subject, involving the advanced concepts of both electricity and mechanics. The word piezo-electricity takes its name from the Greek piezein "to press", which means pressure-electricity. Certain classes of piezo-electric materials will, in general, react to any mechanical stresses by producing an electrical charge. In a piezoelectric medium, the strain or the displacement depends linearly on both the stress and the field. The converse effect also exists, whereby a mechanical strain is produced in the crystal by a polarising electric field. This is the basic effect that produces the vibration of a quartz crystal.

Quartz resonators consist of a piece of piezoelectric material precisely dimensioned and orientated with respect to the crystallographic axes. This wafer has one or more pairs of conductive electrodes, formed by vacuum evaporation. When an electric field is applied between the electrodes, the piezoelectric effect excites the wafer into mechanical vibration. Many different substances have been investigated as possible resonators, but for many years quartz has been the preferred medium for satisfying the needs for precise frequency generation. Compared to other resonators e.g. LC circuits, mechanical resonators, ceramic resonators and single crystal materials, the quartz resonator has proved to be superior by having a unique combination of properties. The material properties of quartz crystal are both extremely stable and highly repeatable. The acoustic loss or internal fraction of quartz is particularly low, which results in a quartz resonator having an extremely high Q-factor. The intrinsic Q of quartz is 107 at 1 MHz. Mounted resonators typically have Q factors ranging from tens of thousands to several hundred orders of magnitude better than the best LC circuits. The second key property is its frequency stability with temperature variations.

Phased Locked Loop Frequency Synthesis

The Phased Locked Loop (PLL) method of frequency synthesis is now the most commonly used method of producing high-frequency oscillations in modern communications equipment. There would not be an amateur or commercial transceiver of any worth today that does not employ at least one if not several, phase locked loop systems, to generate stable high-frequency oscillations.

PLL circuits are now frequently being used to demodulate FM signals, making obsolete the Foster-Seerley and ratio detectors of early years. Other applications for PLL circuits include AM demodulators, FSK decoders, two-tone decoders and motor speed controls.

The PLL technique has, surprisingly, been around for a long time. In the 1930s, the superheterodyne receiver was in its heyday (and it's still going strong today). However, attempts were made to simplify the number of tuned stages in the superheterodyne.

The Homodyne Receiver

Around 1932, British radio engineers developed a new type of receiver to challenge the superheterodyne - it was called the Homodyne or Synchrodyne.
The idea was simple: the receiver consisted of a mixer and a local oscillator, followed immediately by an audio amplifier!

When the input signal and local oscillator are mixed at the same phase and frequency, the output is an exact audio representation of the modulated carrier.

Problems, however, occurred in trying to keep the local oscillator on the same phase and frequency as the input signal. To counteract local oscillator drift, the output of the local oscillator was fed together with a sample of the input signal to a phase detector. The output of the phase detector was a correction voltage which was then applied to the local oscillator to keep it on frequency.
It was this type of feedback circuit, which led to the evolution of the phase locked loop. The homodyne receiver was superior to the superheterodyne, but the cost of the PLL circuit outweighed its advantages, so the idea did not take off.
In the late 80's some single-chip receivers were developed using the homodyne principle. Such receivers are referred to by some manufacturers as zero intermediate frequency receivers since there is a direct conversion from RF to audio.

Developing a PLL

Let's have a look at how we can develop our own PLL system, step by step from scratch and also look at a practical application.

About 100 years ago, Pierre Curie along with his brother Jacques, discovered the effect that an electric field has on certain crystals. The Curie's discovered that when physical pressure was applied to certain crystals, electric charges were produced on the opposite faces. Conversely, if an electric field was applied, the crystal structure itself would distort. We know this phenomenon today as the Piezoelectric Effect.

With the invention of 'wireless', there came a need for an accurate frequency-determining device. So evolved the crystal oscillator with its excellent frequency stability.

The crystal oscillator had one major limitation; it could only be used for a narrow range of frequencies. The crystal frequency can only be varied slightly by use of a variable capacitor or the use of circuits designed to respond to harmonics of the crystal's fundamental frequency, as is done with frequency multipliers or overtone oscillators.

For a larger range of frequencies, these methods are not very practical. To obtain a larger frequency range and maintain essential stability, two crystal oscillators could have their outputs heterodyned (mixed) to produce two new frequencies - i.e. the sum and the difference of the two original frequencies.

Master/Slave

As a practical example: a 24-channel system could use four master crystals, heterodyned one at a time with each of six other slave crystals, to give us four times six or 24 output frequencies.

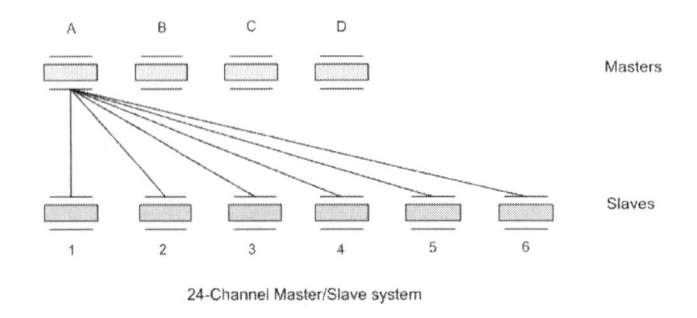

Figure 19-27

So, with the use of only 10 crystals, we can obtain 24 frequencies. This appears okay, but the inhibiting factor in such a system is the cost of the crystals. Imagine the cost of a 400-channel system using crystals!

For those of you who were around at the time of the "illegal" 23 channel CB radio. This master/slave technique was used to generate the 23 channels, though some enterprising operators worked out they could get an additional channel by switching in the last master/slave combination.

Doing more with a VCO

Voltage controlled oscillators allow us to tune easily to a large range of frequencies. The big problem is the VCO's are nowhere as stable as a crystal oscillator. Any slight voltage variation in the circuit will cause the frequency to shift. If there was some way, we could combine the flexibility of the VCO with the stability of the crystal oscillator; we would have the ideal frequency synthesis system.

Let's try something. Suppose we feed the output of a VCO and a Crystal Oscillator into a phase detector. What's a phase detector? Well, it is similar to a discriminator or ratio detector used in frequency demodulation, or it could be a digital device (an exclusive-or gate).

If two signals are fed into a phase detector and these signals are equal in phase and frequency, there will be no output from the detector. On the other hand, if the signals are not in phase, the difference is converted to a DC output voltage. The greater the frequency/phase difference in the two signals, the larger the output voltage.

Now, where were we? Right, take a look at Figure 19-27. The outputs of the VCO and the crystal oscillator are combined with a phase detector and any difference will result in a DC voltage output. Now, suppose this DC voltage is coupled back to the VCO in such a manner that it drives the output of the VCO towards the crystal oscillator frequency - eventually, the VCO will LOCK onto the crystal oscillator frequency. This is a PHASE LOCKED LOOP in its most basic form. Only part of the VCO output need be sent to the phase detector. The rest can be usable output.

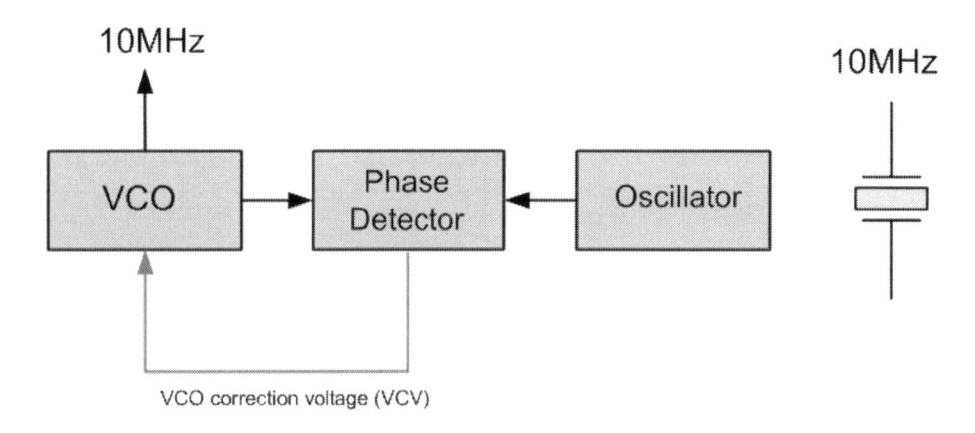

```
                    10MHz                              10MHz
                      ↑                                  │
                      │                                 ═══
  ┌─────┐    ┌──────────┐    ┌──────────┐              ┌─┐
  │ VCO │───▶│  Phase   │◀───│Oscillator│              └─┘
  │     │    │ Detector │    │          │              ═══
  └─────┘    └──────────┘    └──────────┘               │
     ▲            │
     │            │
     └────────────┘
    VCO correction voltage (VCV)
```

Using a phase detector to obtain phase lock

Figure 19-27

Further Development

Wait a minute! The VCO is locked onto the crystal oscillator and is, therefore, behaving as if it were a fixed frequency oscillator. This gives us the stability of a crystal oscillator, but we have lost the flexibility we were aiming for. We may just as well use the crystal oscillator alone for all the good this arrangement has done us. It certainly doesn't appear as if we have accomplished anything at all.

Okay, let's investigate how this problem may be overcome. Suppose our crystal frequency was 10MHz, but we wanted the VCO to operate on 20MHz. The phase detector will, of course, detect a frequency difference and pull the VCO down to 10MHz, but what if we could fool the phase detector into 'thinking' the VCO was really only operating on 10MHz when in reality it is operating on 20 MHz. Take a look at Figure 20-27.

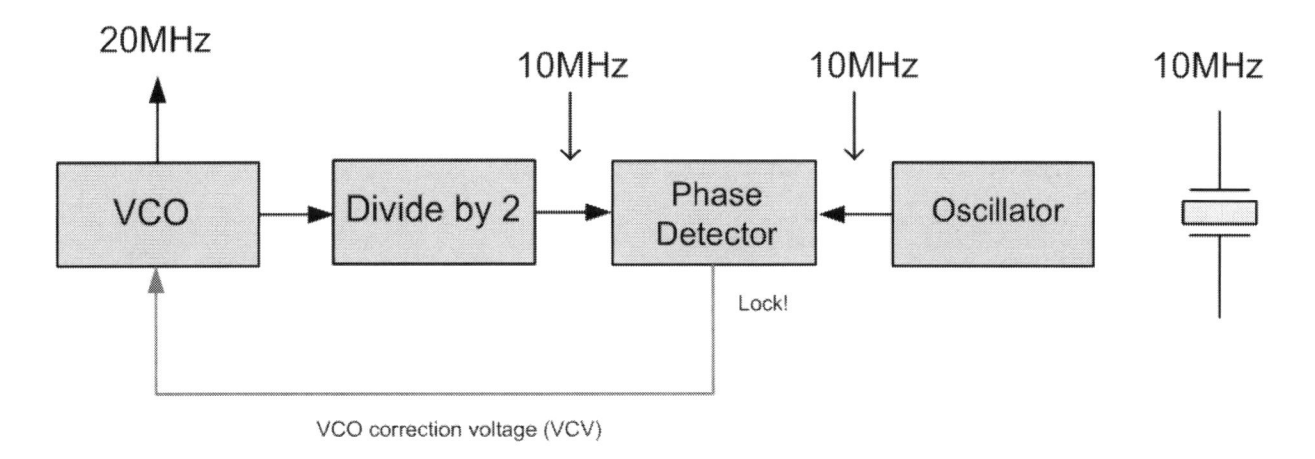

Using a divider to fool the Phase Detector

Figure 20-27

For example, suppose in Figure 20-27 we used a divide-by-four instead of the divide-by-two. Then, at lock, the VCO would be oscillating at 40MHz yet will still be as stable as the crystal reference frequency.

Counters/Dividers

If we could design a divider that would enable us to change the 'divide by' figure at will, we would certainly have a very versatile frequency generating system.

Such frequency dividers are available and are known as Programmable Frequency Dividers. The internal operation of such circuits is not within the scope of this book, but let's have a look at the basic principle involved.

Programmable dividers are digital circuits in that they operate with two voltage levels, a low-level (0) and a high level (1). Ordinary circuits which operate on all voltages are called analogue. As the VCO is an analogue device, the sine waves it outputs to the frequency divider needs to be converted into square waves - changing the wave shape, but not the frequency.

The divider is really a pulse counter. It can be programmed to count to two and then reset, count to two again and reset and so on. At each reset, one pulse is output from the divider - this is a divide-by-two action. This divider would be suitable for our circuit of Figure 20-27.

More elaborate counter/dividers can be designed to count to different amounts before resetting - e.g. a count-to-four divider would be a divide-by-four; a count-to-35 would be a divide-by-35; and so on. Remember that no matter what the reset level is, we only get one output pulse from each reset. The counter/divider may have several inputs which can be programmed in binary or some other digital code to determine its divide-by number. For example, if the divider had four inputs to be coded in binary and we applied a high level on inputs 1 and 3, a binary code of 1010 (which is the equivalent of decimal 10), we would have a divide-by-10 action. More specialised programmable dividers provide for a large range of frequency divisions.

A PRACTICAL PLL

Now we have assembled all the necessary components for a workable PLL system: A voltage Controlled Oscillator, a Crystal Oscillator, a Programmable Divider and a Phase Detector.

One of the most versatile PLL systems, seen particularly in Citizens Band transceivers, is the UNIDEN 858. Figure 21-27 shows a block diagram of the system. Let's work through it step by step and see what we can discover.

Uniden 858 PLL System

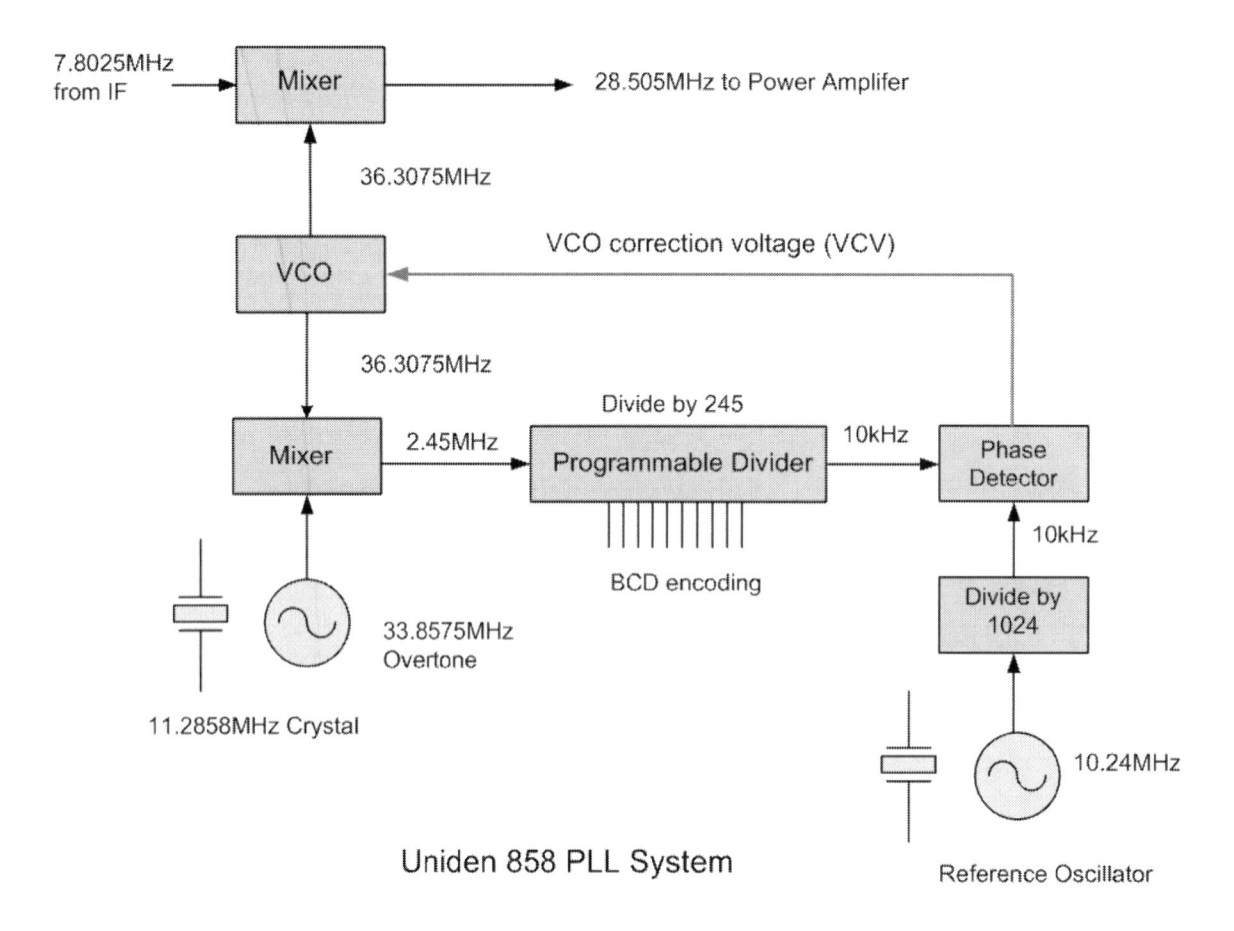

Figure 21-27

The reference crystal, in this case, is 10.24MHz but note that in this instance the reference crystal is not oscillating at the reference frequency - its signal is passed through a 1024 divider to give us a reference frequency of 10kHz. This 10kHz reference signal is passed to the phase detector.

Now we know that the signal coming from the VCO must be divided to 10kHz before being applied to the phase detector - but notice one thing: in this system, our VCO must oscillate at around 36 MHz to give us the correct output frequency. This frequency is going to take a lot of dividing to get it down to the 10 kHz reference frequency.

So here a cunning method has been used to convert the VCO frequency to a workable value before division. This is where the 11.2858 MHz crystal oscillator comes into play. It's an overtone oscillator producing an output on the third overtone of the crystal's fundamental frequency - i.e., 33.8575 MHz. This signal is then mixed with the VCO output, the difference frequency being around 2 to 3 MHz. This signal can then be divided to 10 kHz quite simply and applied to the phase detector.

Before we get too involved in some actual circuit frequencies, let's look more closely at the frequency divider.

We see that it is a programmable divider; it can be set to divide by any amount from 1 to 399. The input data to the divider is binary coded decimal (BCD) and is applied to pins 13 through to 22 on the actual integrated circuit.

What's BCD? It's a method of expressing a decimal value as a four-bit binary number. The units in the decimal number are expressed as a four-bit binary number, as are the tens, then the hundreds, etc. For example, decimal 251 would require three four-bit conversions: 1 is expressed as 0001; 5 becomes 0101 and 2 equals 0010. The final figure is the chain of the three four-bit numbers: 251 equals 0010 0101 0001 in BCD.

The divider has 10 inputs. These inputs can be set with a BCD code in the range 0 to 399 which equates to 400 channels. For channel 399 the four bit BCD code would be 0011-1001-1001 (3-9-9).

So there are a total of 400 possible input combinations to the programmable divider, which theoretically means 400 operating frequencies, provided the radio frequency stages of the transceiver remain in tune or are tuned accordingly.

A logic (1) is represented by a +5V input level and a logic (0) by zero Volts. If the input to the divider is supplied with the BCD code for 146 (0001 0100 0110), then the input signal will appear at the output divided by 146.

Another point worth mentioning is that the VCO signal is not the final transmit frequency. All necessary processing of the signal to be transmitted is performed before the PLL stage is reached. The output of the VCO is then mixed with the intermediate frequency (IF) and the difference frequency produced is the actual transmit frequency.

Conversely, during receive operation, the VCO and the incoming signal are heterodyned to produce the IF frequency.

A WORKED EXAMPLE

Let's take a desired output of 28.505MHz and develop the conditions which would need to exist around the circuit. Since the output is 28.505MHz, the VCO will be 7.8025MHz higher (the IF frequency). The VCO is therefore 36.3075MHz. This is fed to the mixer along with a 33.8575MHz signal derived from the third overtone of the 11.2858MHz crystal oscillator.

The 36.3075 MHz and 33.8575MHz signals mix to produce a difference frequency of 2.45 MHz. Now the divider must be set to divide the 2.45MHz frequency down to 10kHz. This requires a BCD code on the data inputs equivalent to 245 decimal (divide-by 245). The logic levels applied to the 10 BCD inputs would be 10 0100 0101

The 10kHz signal produced by the divider is phase detected with the stable 10kHz reference frequency from the crystal oscillator and a control voltage applied back to the VCO. After a very brief period, the circuit will establish a lock.
Changing the divide-by number will change the output frequency in 10kHz steps over a maximum frequency range of 400 x 10 kHz = 4 MHz. Incrementing or decrementing the divider will alter the VCO frequency by 10kHz. So the channel spacing, in this case, is 10kHz.

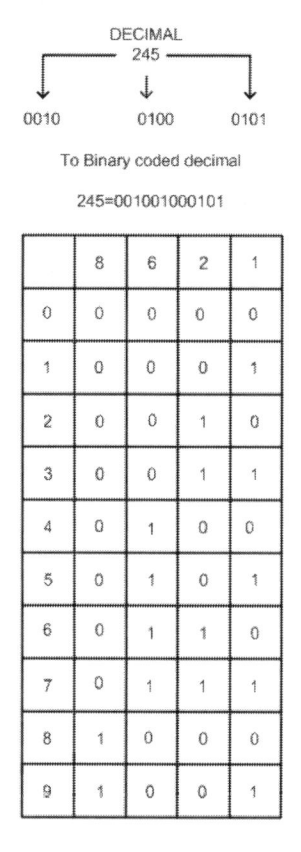

```
            DECIMAL
          ┌─── 245 ───┐
          ↓     ↓     ↓
         0010  0100  0101

       To Binary coded decimal

        245 = 001001000101
```

	8	6	2	1
0	0	0	0	0
1	0	0	0	1
2	0	0	1	0
3	0	0	1	1
4	0	1	0	0
5	0	1	0	1
6	0	1	1	0
7	0	1	1	1
8	1	0	0	0
9	1	0	0	1

Table 1-27

If you think there is some relationship between the 10kHz reference frequency and the 10kHz channel steps, then you're right! If 5kHz channel spacing was required, all that would be necessary is to change the 1024 divider to a 2048 divider. The 858 chip has a provision on board for this - pin 7 can be toggled between high and low to change from 1024 to 2048.

A microprocessor can be added to the PLL circuit to control the logic fed to the programmable divider. Such systems are common in modern communications receivers/transmitters. See figure 22-27. Now a whole new scope of functionality is available. The microprocessor can be told what to do from a keypad - and its internal program can do things like change frequency, memorise frequencies, scan a range of frequencies, or operate between different programmable transmit and receive frequencies and much more. The microprocessor can even re-tune the VCO and other parts of the transceiver to give extremely broadband coverage. An alternative to re-tuning the VCO is to have several VCO's and have the microprocessor switch the appropriate one into the circuit for the desired frequency range.

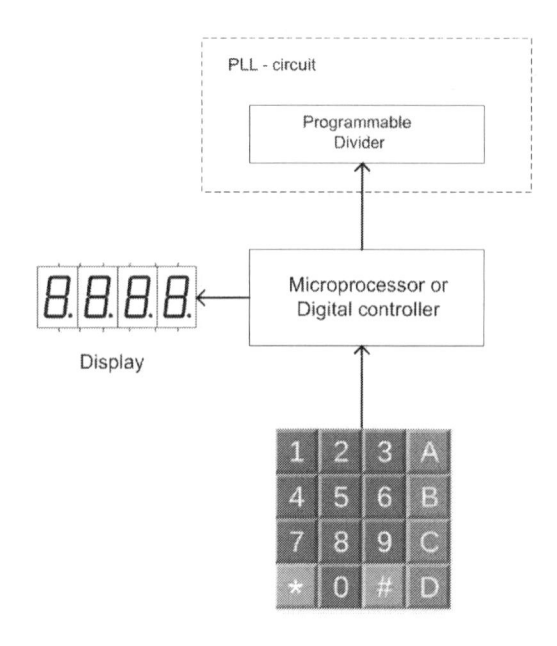

Figure 22-27

You now should have a very good idea of the way PLL frequency synthesis systems work. More complex systems do exist, but they are merely extensions of what has been described. Some modern communications equipment may use not one, but several PLL's and even PLL's within PLL's.

Direct Digital Syntheses (DDS)

There is currently no assessment requirement for DDS though this may change.

We are used to digital signals. Many things today are digital such as DTV, mobile phones, broadcast radio and many others. We are not digital so there has to be a digital analogue conversion (DAC) so we can hear it. The types of information encoded into digital is usually complex. A TV signal contains an enormous amount of data, but vision and sound can be encoded to digital and transmitted and decode into sound a picture at the receiver, though the picture can remain digital. It is the TV screen that produces light output that we perceive as analogue.

So what if we were to store just one sinewave only – one cycle in digital format. We could do this easily with just a small amount of memory. Then what if we were to read that memory and perform a digital to analogue conversion we would get one cycle of the sine wave.

Not that interesting really. However, what if used a clock to access the stored digital data for a sine wave repeatedly? Read that data do a DAC and repeat endlessly. If we did this, we would have a continuous sinewave – we would have a signal that could be an audio or RF signal of almost any frequency. We would have created a continuous sinewave signal using only one digitally stored single cycle. We would need a DAC and some analogue filtering after it. The point is though we have created an RF signal without tuned circuits. We would need a stable frequency source for the clock reading the data. This would be a crystal. However, using some tricks we can not only produce one frequency from the digitally stored single cycle we can produce a very wide range of frequencies. This is the essence of direct digital syntheses.

From the start – unlike PLL – DDS has been integrated into specialised integrated circuits. The same as PLL has today – so there is no need to build discrete component circuits for DDS or PLL.

What is Direct Digital Synthesis?

Direct digital synthesis (DDS) is a method of producing an analogue waveform—usually a sinewave—by generating a time-varying signal in digital form and then performing a digital-to-analogue conversion. Because operations within a DDS device are primarily digital, it can offer fast switching between output frequencies, fine frequency resolution and operation over a broad spectrum of frequencies. With advances in design and process technology, today's DDS devices are very compact and draw little power.

The ability to accurately produce and control waveforms of various frequencies and profiles has become a key requirement common to a number of industries. Whether providing agile sources of low-phase-noise variable-frequencies with a good spurious performance for communications, or simply generating a frequency stimulus in industrial or biomedical test equipment applications, convenience, compactness and low cost are important design considerations.

Many possibilities for frequency generation are open to a designer, ranging from phase-locked-loop (PLL)-based techniques for very high-frequency synthesis to dynamic programming of the digital-to-analogue converter (DAC) outputs to generate arbitrary waveforms at lower frequencies. But the DDS technique is rapidly gaining popularity for solving frequency-(or waveform) generation requirements in both communications and industrial applications because single-chip IC devices can generate programmable analogue output waveforms simply and with high resolution and accuracy.

Furthermore, the continual improvements in both process technology and design have resulted in cost and power consumption levels that were previously unthinkably low. For example, the AD9833, a DDS-based programmable waveform generator (Figure 23-27), operating at 5.5 V with a 25-MHz clock, consumes a maximum power of 30 milliwatts.

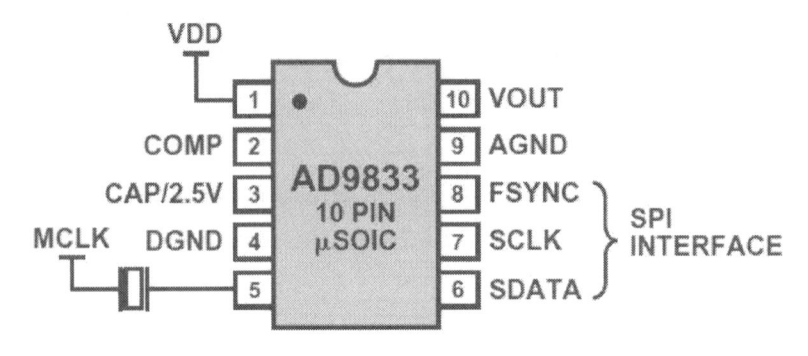

Figure 23-27

How does a DDS device create a sinewave?

Here's a breakdown of the internal circuitry of a DDS device: its main components are a phase accumulator, a means of phase-to-amplitude conversion (often a sine look-up table) and a DAC. These blocks are represented in Figure 24-27.

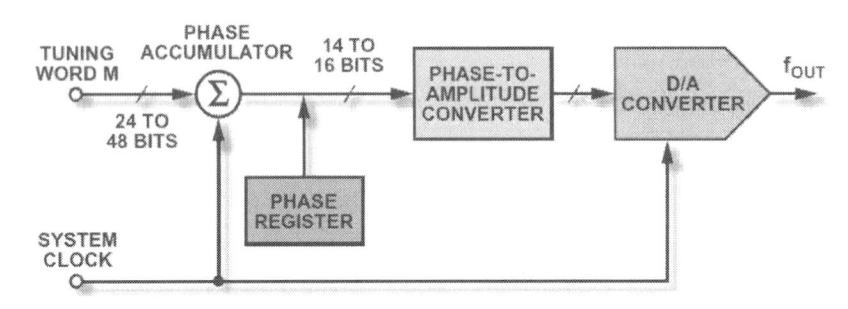

Components of a direct digital synthesizer.

Figure 24-27

DDS produces a sinewave (or any other wave shape) at a given frequency. The frequency depends on two variables, the reference-clock frequency and the binary number programmed into the frequency register (tuning word).

The binary number in the frequency register provides the main input to the phase accumulator. If a sine look-up table is used, the phase accumulator computes a phase (angle) address for the look-up table, which outputs the digital value of amplitude, corresponding to the sine of that phase angle, to the DAC. The DAC, in turn, converts that number to a corresponding value of analogue voltage or current.

To generate a fixed-frequency sine wave, a constant value (the phase increment—which is determined by the binary number) is added to the phase accumulator with each clock cycle. If the phase increment is large, the phase accumulator will step quickly through the sine look-up table and thus generate a high-frequency sine wave. If the phase increment is small, the phase accumulator will take many more steps, accordingly generating a slower waveform.

Continuous-time sinusoidal signals have a repetitive angular phase range of 0 to 2π. The digital implementation is no different. The counter's carry function allows the phase accumulator to act as a phase wheel in the DDS implementation.

To understand this basic function, visualize the sine-wave oscillation as a vector rotating around a phase circle (Figure 25-27). Each designated point on the phase wheel corresponds to the equivalent point on a cycle of a sine wave. As the vector rotates around the wheel, visualize that the sine of the angle generates a corresponding output sine wave. One revolution of the vector around the phase wheel, at a constant speed, results in one complete cycle of the output sine wave. The phase accumulator provides the equally spaced angular values accompanying the vector's linear rotation around the phase wheel. The contents of the phase accumulator correspond to the points on the cycle of the output sine wave.

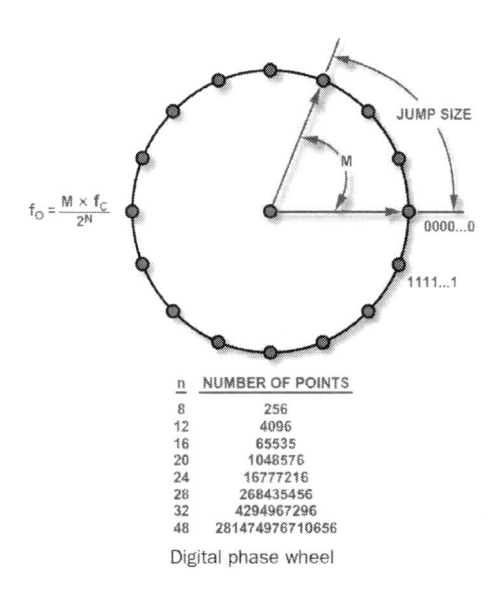

$$f_O = \frac{M \times f_C}{2^N}$$

n	NUMBER OF POINTS
8	256
12	4096
16	65535
20	1048576
24	16777216
28	268435456
32	4294967296
48	281474976710656

Digital phase wheel

Figure 25-27

The phase accumulator is a modulo-M counter that increments its stored number each time it receives a clock pulse. The magnitude of the increment is determined by the binary-coded input word (M). This word forms the phase step size between reference-clock updates; it effectively sets how many points to skip around the phase wheel. The larger the jump size, the faster the phase accumulator overflows and completes its equivalent of a sine-wave cycle. The number of discrete phase points contained in the wheel is determined by the resolution of the phase accumulator (n), which determines the tuning resolution of the DDS.

Creating the analogue sinewave

A phase-to-amplitude lookup table is used to convert the phase accumulator's instantaneous output value (28 bits for AD9833)—with unneeded less-significant bits eliminated by truncation—into the sine-wave amplitude information that is presented to the (10-bit) D/A converter. The DDS architecture exploits the symmetrical nature of a sine wave and utilizes mapping logic to synthesize a complete sine wave from one-quarter-cycle of data from the phase accumulator. The phase-to-amplitude lookup table generates the remaining data by reading forward then back through the lookup table. This is shown pictorially in Figure 26-27.

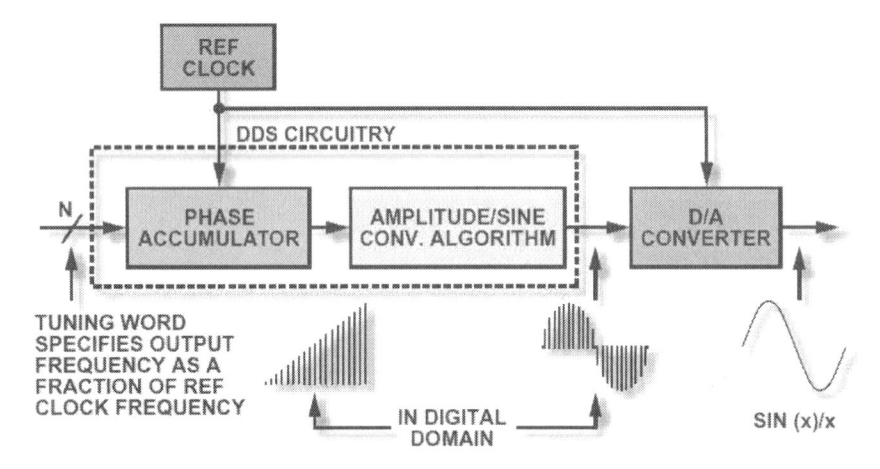

Signal flow through the DDS architecture

Figure 26-27

A modern radiocommunications system today will mostly likely use DDS and PLL in all manner of combinations and configurations.

28. PROPAGATION

THE IONOSPHERE

The ionosphere is a region of the upper atmosphere extending from a height of about 50 km to greater than 500 km. Within this region, some molecules of air become ionised by *ultraviolet radiation* from the sun. The ionised gas within the ionosphere is called plasma.

Ionisation is a process by which electrons, having a negative charge, are stripped from neutral atoms or molecules to form positively charged ions. It is these ions that give their name to the ionosphere. The electrons that have been stripped off are light and free to move. These electrons under the influence of an electromagnetic wave will absorb and re-radiate energy and modify the direction of an electromagnetic wave front. If the electron density is sufficient and the frequency of the electromagnetic wave is neither too high nor low, then total wave refraction will occur.

There are four regions of the ionosphere called layers.

D   50 - 90 km
E   90 - 140 km
F1 140 - 210 km
F2 > 210 km

The regions of the ionosphere are not sharply defined and merge smoothly from one into the other, each region containing similar chemical and physical properties. Ionisation of the D region is weak because it is low down in the atmosphere. The D region can refract signals at low frequencies (less than 5MHz). High frequencies pass right through it with only partial attenuation. After sunset the D region disappears rapidly due to recombination of its ions.

The E region is also known as the Kennelly-Heaviside layer. The rate of ionic recombination is rapid after sunset and the region has usually disappeared by midnight. The E region can refract higher frequency signals than the D region. In fact, the E region can refract signals as high as 20 MHz. The ionospheric layers during the day and the night are shown in Figure 1.28.

During *daylight hours,* the F, region separates into two distinct regions called the F1 and F2. The ionisation levels in these regions can be quite high and varies greatly depending upon the degree of ultraviolet radiation from the sun, which in turn is primarily determined by the altitude of the sun (i.e. time of day). At the height of the F region the density of the atmosphere is low and because of this, recombination of ions occurs slowly after sunset. The F2 region will remain throughout the night at a fairly constant ionisation level. The F regions are responsible for long-distance communications due to their ability to refract signals up to 30 MHz **and** also due to the long skip distance provided by refraction from such a high elevation.

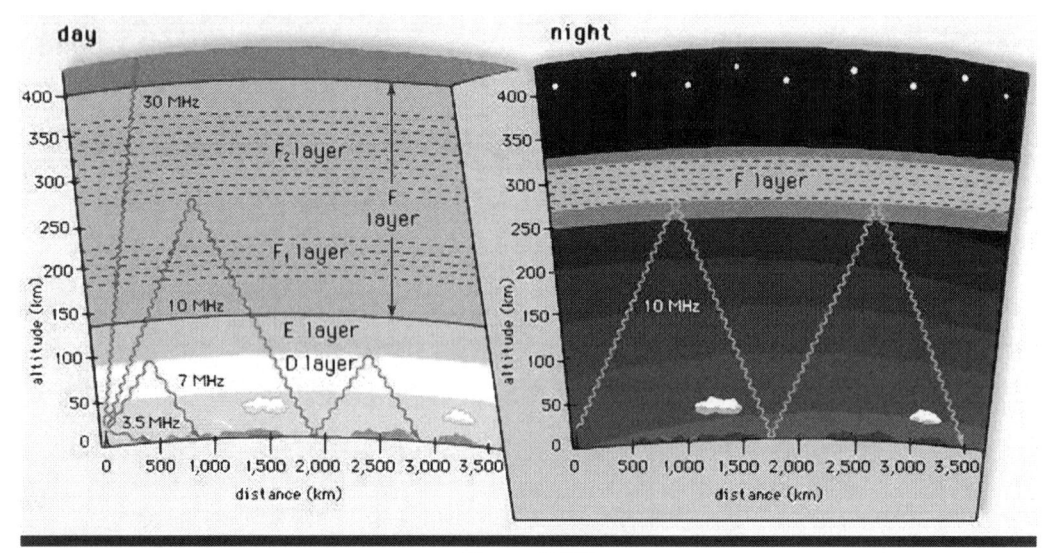

Figure 1-28

CAUSES OF IONISATION

Two kinds of *solar radiation* cause ionisation in the ionosphere, namely X-rays and extreme ultraviolet (EUV). The X-ray output from the sun is irregular, increasing greatly during large solar flares. However, by far, the most important ionisation radiation is EUV. It is generated by the chromosphere in hot plage regions that overlie sunspot groups.

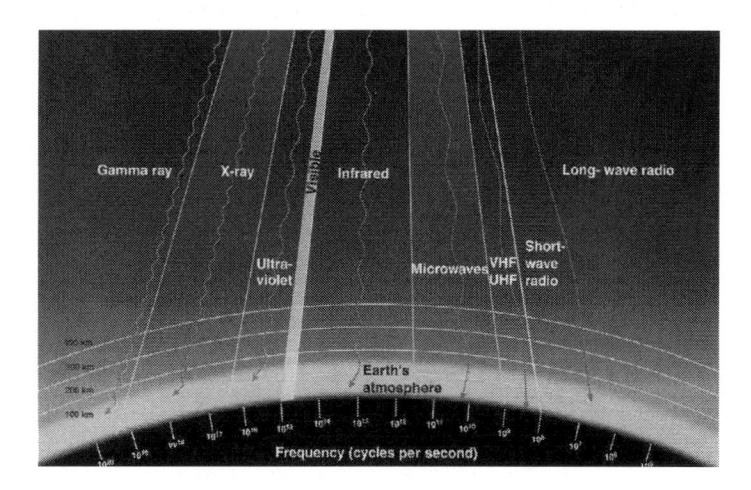

Figure 2-28 – EUV and X-rays primary ionisation

Meteor storms occur often and are predictable. They can cause considerable ionisation and are used for long distance HF communication.

A *plage* is a bright region in the chromosphere of the Sun, typically found in regions of the chromosphere near sunspots. The term itself is poetically taken from the French word for "beach". The plage regions map closely to the faculae in the photosphere below, but the latter have much smaller spatial scales. Accordingly, plage occurs most visibly near a sunspot region.

At any time, the EUV output from the sun is approximately constant, but varies from month to month and year to year, as the number of sunspots varies.

11 YEAR SUNSPOT CYCLE

The most important slow changing factor in high frequency (HF) propagation prediction is the *11-year sunspot cycle*. Solar activity follows a cycle, which has a period of 11 years. The best indication of solar activity is the sunspot number. Sunspots are always accompanied by plage areas, which are the source of extreme ultraviolet radiation. It has been found that the solar radio flux at a wavelength of 10.7cm (10cm flux) is closely related to the sunspot number and is more easily measured. The sunspot number and the 10cm flux may vary greatly from day to day so that monthly or yearly averages are usually employed to show the progress of the solar cycle.

The graph in Figure 3-28 was produced by the Ionospheric Prediction Service in 2016. The graph shows all the sunspot cycles since accurate recording began. Sunspot cycles have been traced further back by means other than direct solar radiation. There is some evidence that the 11-year sunspot cycle is not set in stone. Earth's cycles take centuries to be completely mapped and established. For all we know there may be cycles within cycles. It would be naive of us assume that after just 100 years of solar observation that solar cycles are firmly established.

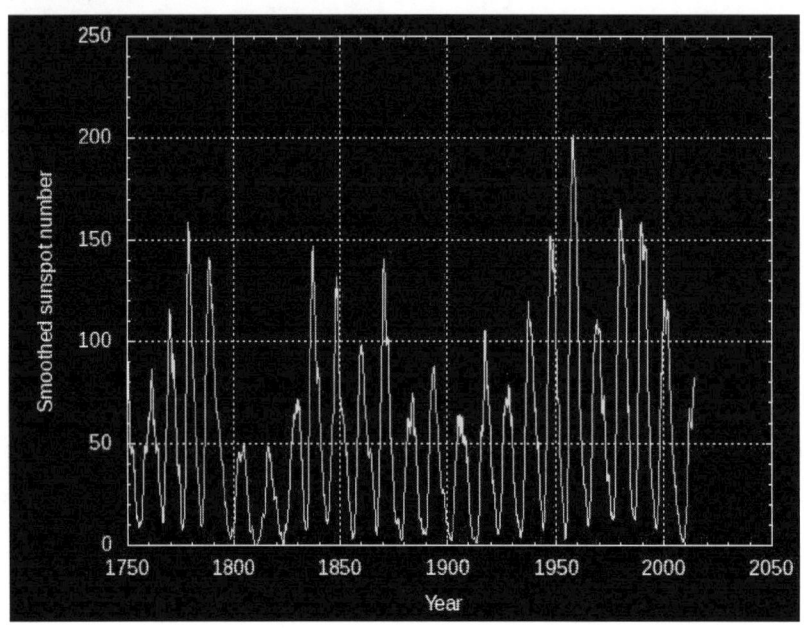

Figure 3-28

SPORADIC-E

Sporadic-E layers are only a few kilometres thick and can have peak electron density comparable to the F regions maximum density. Sporadic-E layers, like the higher F-Layer, are very efficient reflectors of HF signals. Sporadic-E is formed mainly by the concentration of winds of ions and electrons into a thin layer. At mid-latitudes, Sporadic-E is most prevalent during summer and at midday. In the equatorial zone, it is nearly always present at midday.

CRITICAL FREQUENCY - MAXIMUM USABLE FREQUENCY - ABSORPTION LIMITING FREQUENCY

At vertical incidence (straight-up), the highest frequency which an ionospheric layer will reflect is called the *critical frequency* (f_c).

The highest frequency which is returned to earth at a given angle of radiation is referred to as the maximum usable frequency (MUF).

The *absorption limiting frequency* is the lowest frequency for reliable radio communication via the ionosphere.

Operating a radio transmitter right on the MUF will not produce the most reliable communications, as the precise MUF does fluctuate somewhat. The *optimum working frequency* (OWF) is the one that provides the most consistent communications. For transmission using the F2 region, the optimum working frequency is about *85% of the MUF*. However, when the E layer is used, propagation is quite consistent if a frequency very near but less than the MUF is used. Since ionospheric attenuation of radio waves is inversely proportional to frequency, using a frequency as close to the MUF as possible results in the maximum signal strength.

VIRTUAL HEIGHT

The virtual height is the height of the ionosphere when regarded as a plane surface reflector such as a mirror.

Until it reaches the ionosphere, a radio wave propagates in a straight line. Once in the ionosphere, it is *refracted* back to earth. Refraction is a process whereby the wave front is slowly bent downward toward the earth. Although a wave is actually refracted in the ionosphere, it is permissible to substitute a simple triangular path for the real ray path, as if the ray were mirror-reflected.

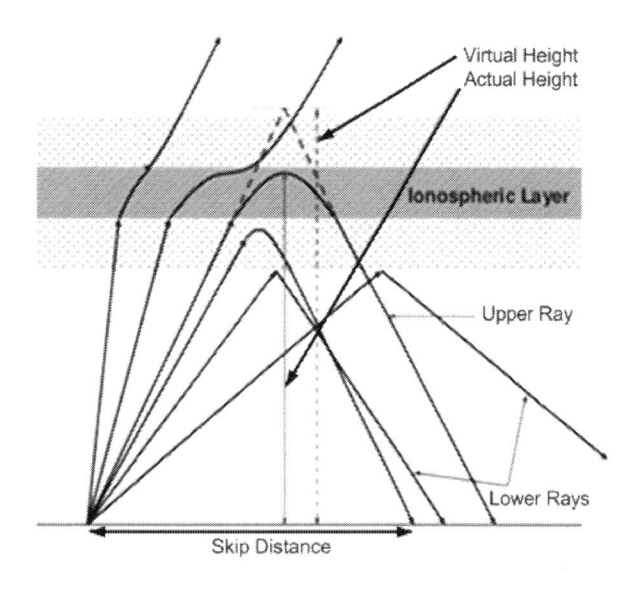

Figure 4-28

Radio waves are often spoken of as being reflected from the ionosphere when in fact they are refracted by the ionosphere and reflected by the Earth.

CRITICAL ANGLE

The critical angle is the angle measured between the wavefront incident on the ionosphere and a line extended to the centre of the earth. For each frequency, there is a critical angle and if the wavefront arriving at the ionosphere is less than the critical angle then wave refraction will not occur.

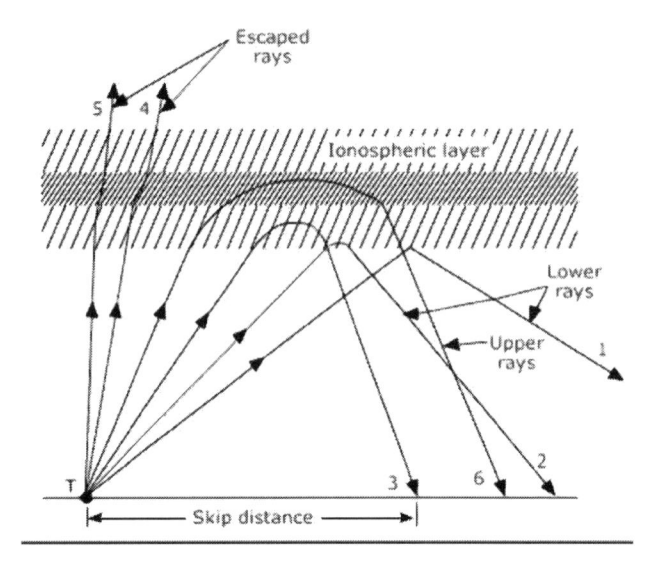

Figure 5-28

The shortest distance through an ionospheric region is when the wavefront strikes the region at vertical incidence. The *critical frequency is the highest frequency returned to earth when the wave is transmitted at vertical incidence*. However, if the wavefront strikes the ionosphere obliquely then the apparent depth of the region is greater and frequencies higher than the critical frequency will be returned to the earth. For any particular frequency higher than the critical frequency there will be a minimum critical angle if refraction is to take place. A critical angle of zero corresponds to vertical incidence.

GROUND WAVE

Ground wave is the wavefront, which upon leaving the antenna, propagates close to the earth's surface. Severe signal losses due to ground conductivity limits the range of ground waves to about 100 km overland and 300 km over water for the lowest HF frequencies.

The higher the frequency, the shorter the ground wave coverage of the transmitter. Ground wave distance drops off very quickly with increasing frequency. Long ground wave is best under 5MHz. Even on the 80M band (3.5MHz) when communication over a short distance, say across town, you may very well be communicating using skywave.

SKIP DISTANCE

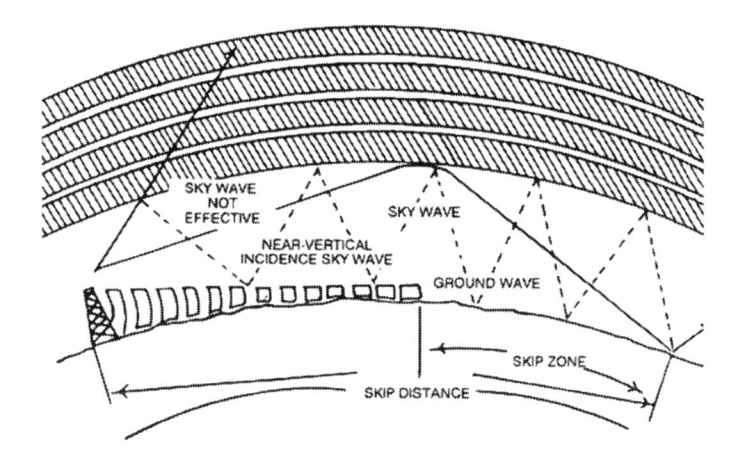

Figure 6-28

The skip distance is the distance from the transmitter to where the sky wave first returns to the earth. Within the skip distance, only ground wave communication is possible. The skip distance is shorter for lower operating frequencies. Between the point where the ground wave ends and the point where the sky wave first returns to earth, no signals will be heard. This area is called the *skip zone*. The skip zone is a no signal area.

FACTORS DETERMINING SKIP DISTANCE

The main factors affecting range are frequency, radiation angle and the height of the reflecting region.

If the height of the reflecting region is increased, then the skip distance will increase. As the frequency of the transmission is increased the skip distance will increase until the frequency exceeds the MUF, above which the signal is not refracted back to earth.

If the angle of radiation above the earth's surface is increased, then the skip distance is decreased. If the transmission frequency is greater than the critical frequency, then making the angle of radiation too high will cause the wave front to penetrate the refracting region.

LONG-DISTANCE COMMUNICATIONS ABOVE THE HF

Temperature inversions in the earth's troposphere make it possible for radio waves to be ducted or guided over large distances. This mode of propagation is called *tropospheric ducting.* This propagation mode occurs only on VHF frequencies and above.

The troposphere is the layer of the atmosphere that lies directly above the surface of the earth; it extends to a height of about 10.5km. Under normal conditions, the density of the troposphere is higher near the surface of the earth and becomes progressively lower at higher altitudes. A temperature inversion occurs when a pocket of warm air is trapped above cool air. Instead of a gradual changing density, abrupt changes occur, making complete refraction possible. The refracted wave does not travel back to earth; it is refracted back and forth by both boundaries on each side of the layer of warm air. The trapping medium is called a *tropospheric duct.* In order to use the duct, both transmitting and receiving antennas must be within the duct.

FADING

Fading can be broken down into two main categories: multiple path reception and selective fading. Multiple path reception occurs when signals arrive at the receiving antenna by more than one path. No path length is exactly the same from one moment to the next due to the dynamic nature of the ionosphere.

Multi-path signals may arrive in-phase one moment and produce signal enhancement and the next moment may arrive out-of-phase and produce, at times, total cancellation. Under such conditions, the received signal may fluctuate wildly. Selective fading occurs when individual frequency components undergo different amounts of refraction at the ionosphere to such an extent that some of the frequency components may not arrive at the receiving antenna.

Multiple path reception occurs when the sky wave returns to the earth within the ground wave region of the transmitter. Since ground wave coverage does not extend far, this type of fading only occurs when the skip distance is short. This effect is most noticeable in the 160-metre amateur band. AM broadcasts stations frequently employ 5/9 wavelength antennas. Antennas of this wavelength have a radiation angle which provides a skip distance that falls outside the ground wave coverage area, thus avoiding fade at the ground wave/sky wave boundary.

Fading due to multiple path reception can occur when the signals arriving at the receiving antenna travel by different ionospheric regions, or when signal reflection from a large object such as a man-made structure or a mountain range is involved.

Because a *single* radio transmission consists of a band of frequencies, then it is possible for a group of frequencies *within the signal bandwidth* to be refracted by a greater or lesser amount. This can result in some signal components not arriving at the receiving antenna - this phenomenon is called selective fading. Expectedly, the signals of wider bandwidths are more susceptible to selective fading. A double sideband signal (AM) is more likely to suffer selective fading than a single side band signal (SSB) due to its wider bandwidth. A telegraphy signal will not experience any selective fading. One classic example of this phenomenon is the well-known high school physics experiment using a light beam from a torch and a prism. The different colours (frequencies) are seen to be refracted by different amounts. Sunlight through water droplets undergoes selective refraction and creates a rainbow.

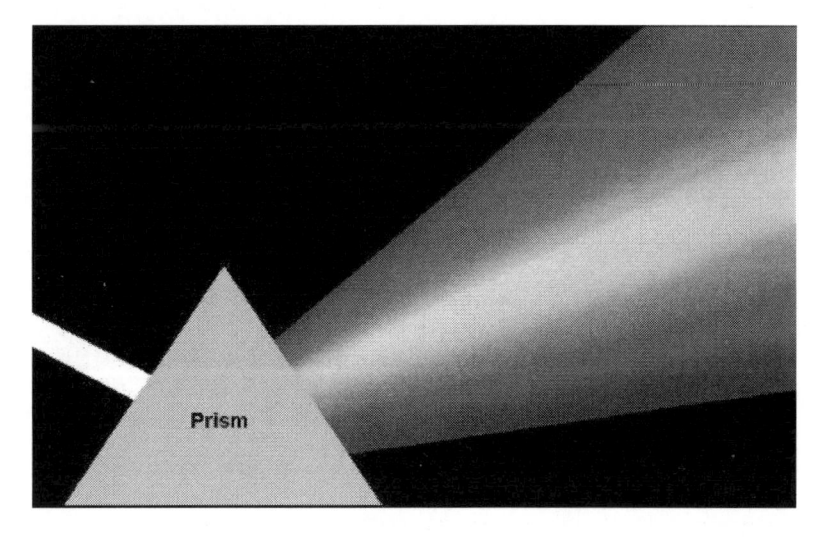

Figure 7-28. Radio waves are dispersed by the ionosphere just like light through a prism

There is more to propagation than has been covered in this chapter. Radio wave propagation is a fascinating subject and a science in itself.

HF (ionospheric) propagation is crucial and will always be so. In the event of war, communication satellites and submarine cables would be one of the first things to go, leaving nothing but HF communications.

This chapter has covered what is needed for Advanced Radio examinations. For more information, take a look at www.ips.gov.au. Or your local ionospheric prediction service.

29. AMPLITUDE MODULATION

We're going to be talking about amplitude modulation. This is the first time we will be discussing how radio is used to transfer information across a long-distance in virtually no time at all. Therefore, it is my intention to cover the concepts completely without the use of unnecessary mathematics. Please do not think that without all of the mathematics we are short changing the subject. I have worked with many trainees, who have come straight out of university and when I have asked them about amplitude modulation they reel off a whole set of mathematical equations in an attempt to explain it. However, when I have asked, "What really is AM"? I have often received a blank look. Mathematics alone does not *always* explain everything in such a way that you really get a *feel* for the subject. I hope that during this chapter you will gain an understanding of what radio transmission and amplitude modulation is about. This and the next two chapters may be verbose and sometimes repetitive, but this is done to help you understand the important *real* concepts of AM transmission and reception.

Firstly, let's have a look at what radio transmission is 'really all about'. It's fine to have a microphone to speak into and expect to be heard somewhere else in the world. What are the real principles involved? This does not apply just to amplitude modulation but to all types of modulation.

Modulation - in radio communications means combining signals, one of which is normally the information we want to transmit (which is usually a low frequency such as audio or voice), with a radio frequency.

Radio communication uses the ability of an electromagnetic wave to transfer information from one point to another. It is very easy to create an electromagnetic wave, send this wave to an antenna and have this wave propagate over a long distance at the speed of light. It's another matter to have the electromagnetic wave contain useful information such as voice, or some other form of intelligence, which after all is the main objective of radio communication.

Remember that electromagnetic radiation takes many forms. Light is electromagnetic radiation. The infrared signal from your TV remote control is electromagnetic radiation. The heat given off by a bar radiator is electromagnetic radiation. The whole secret of radio communication is to place information onto the signal, to convey intelligence or a message from one point to another.

Amplitude modulation (AM) is just one method of doing this and is the first method ever used to transfer voice information from one place to another. There are many misconceptions about how this is done and what AM really is. The first thing that should be understood is that voice frequencies range from about *50Hz to about 3000Hz*. Would it not be simpler if we could just talk into a microphone, convert the sounds from our voice to electrical energy, feed to an antenna and have it transmitted to anywhere we wanted. The problem is that voice frequencies are not high enough in frequency to be radiated by an antenna efficiently. You can put voice frequencies right into an antenna. However, when you work out the length of antenna needed from the wavelength equation, it is many thousands of kilometres long. We have a name for such an antenna it is called a telephone line! For a more practical antenna, the basic principle involves converting voice frequencies, without losing any intelligence, to a higher frequency, radiating them via an antenna and at the other end converting the electromagnetic radiation back to voice frequencies so that they can be heard.

Again, the basic principle is to *move voice frequencies to radio frequencies*, radiate them from an antenna, have them propagate at light speed and at the other end recover those voice frequencies.

The basic principle of amplitude modulation is to take voice frequencies and mix (or modulate) them with a radio frequency signal so that they are converted to radio frequencies which will radiate or propagate through free space.

It is very easy to convert low frequencies to high frequencies so that they will radiate. All we need to do is take our voice frequencies (say from a microphone), then *mix* them with a radio frequency carrier and convert them to radio frequencies. The term *carrier* is very misleading; it is simply a sinusoidal high-frequency radio signal which we use to convert voice frequencies to radio frequencies for the purpose of radiation. I have often heard and read that the carrier 'carry's the voice frequencies' - this is a misconception.

Rather than talk about the whole range of voice frequencies from 50Hz to 3000Hertz, we will look at the subject of amplitude modulation as if we were trying to transmit a single audio frequency of say, 1000 Hz, to a distant location. Using one audio frequency rather than the entire voice frequency range just simplifies our explanation. T

FREQUENCY CONVERSION

The method used to convert one frequency to another is quite simple. You take the signal that you want to convert (in our case a 1000Hz audio frequency) and a radio frequency signal and we are going to use a radio frequency of 1MHz to illustrate how 1000Hz can be converted to a radio frequency for transmission. All that is required to do this is to feed the 1000Hz signal and the 1MHz signal into a *non-linear device*. The output of that non-linear device will contain our converted frequencies, some of which we want and some of which we don't. In radio transmission, the non-linear device is called a *modulator*, in radio reception the non-linear device is referred to as a *de-modulator or detector.*

The apparatus we need is an oscillator for the radio frequency source, a mixer (modulator) and our audio to be frequency converted. The old name for this combination (mixer and oscillator) was called more clearly a *frequency changer*.

Remember, we are using a 1000Hz audio frequency for the purposes of simplification. In practice, we would be using the whole range of audio frequencies from 50Hz to 3000Hz and the same principles would apply.

Now then, we want to transmit a 1000Hz audio frequency using radio frequency transmission.

We take the 1000Hz audio frequency from the microphone and the radio frequency (poorly called the carrier) and inject them into a mixer or a modulator (which are just non-linear devices) - the output of the non-linear device will contain our wanted frequencies. The output of the non-linear device will contain the original frequency that is 1000Hz audio, the 1MHz radio frequency and also the sum and difference of these two frequencies. The output of the non-linear device will contain a 1MHz radio frequency signal, a 1.001 MHz signal (i.e. sum) and a 0.999MHz signal (i.e. difference). There will be other mixing products produced as well, but these will be eliminated in the following stages by filters. The 1.001MHz signal is called the upper side frequency. The 0.999MHz signal is known as the lower side frequency. If we were using the whole range of audio frequencies from 50 Hz to 3000 Hertz, then we would have an *upper sideband* (USB) and a *lower side band* (LSB) of frequencies.

WHAT IS AMPLITUDE MODULATION?

In amplitude modulated (AM) systems, the modulating audio is applied to the radio frequency carrier in such a way that the *total power* of the transmitted wave is made to vary in amplitude, in accordance or in sympathy with the power of the modulating audio.

It is a popular *misconception* that in an amplitude modulated system the carrier power (in our case the 1MHz signal) is made to vary with the application of the modulated audio. Contrary to this, it is, in fact, the *total wave power which varies in amplitude and* not the carrier power. The carrier power in an amplitude modulated system remains constant.

You might ask, "Why modulate the audio frequency at all" (yes, I know I am repeating myself). If audio signals are fed to an antenna, radiation will not occur, because the frequencies are too low to radiate. When an alternating current is made to flow in an antenna, a system of alternating electric and magnetic fields is developed in the vicinity of the antenna. With each 180 degrees of alternation, these fields will return the energy to the antenna by collapsing into it. At high frequencies, the fields do not have time to collapse into the antenna and are left stranded in the space about it. This is called a free space field. Successive free space fields exhibit a repulsive force on their predecessors causing the fields to move out from the antenna at the velocity of light. This mechanism of radiation begins to occur at around 10kHz.

The function of the carrier in amplitude modulation is simply to provide a signal to heterodyne (mix) with the modulated audio, to convert all the audio frequency components to a higher frequency so that the mechanism of radiation will occur. *The carrier contains no intelligence*. The carrier can be removed before transmission, as is the case with single side band (SSB).

One possibility for the misconception of the modulated audio riding on top of the carrier may have arisen from the pattern that an AM signal produces on the screen of a cathode-ray oscilloscope (the oscilloscope is to be discussed). The pattern produced does appear to support the fallacy. The oscilloscope is limited in its operation by the fact that it can only display the resultant *instantaneous* voltage amplitude of all the signals present on its deflection plates.

When discussing AM, it is important to realise that the term refers to several modes of transmission wherein the total wave power transmitted is made to vary in accordance with the applied modulating audio. Double sideband (DSB), single side band (SSB), vestigial side band (VSB), are common modes used on all amateur and commercial bands.

All of these are correctly referred to as amplitude modulation. The reference to double side band as AM and single side band as SSB on some amateur transceivers gives credence to the misconception that SSB is not an amplitude modulated wave when in fact it is.

HOW IS DOUBLE SIDE BAND PRODUCED?

Regardless of the type of AM system, the principle is the same. The radio frequency carrier is combined with the modulated audio in a non-linear device. The output of the non-linear device will contain the sum and difference frequencies of the original signals (with the exception of balanced mixers used for SSB).

WHAT DETERMINES THE BANDWIDTH OF AN AM (DSB) SIGNAL?

Bandwidth means the signal width in Hertz, kilohertz and in some cases megahertz.

The bandwidth of an AM signal is equal to twice the highest modulating audio frequency.

As the highest voice frequency that needs to the transmitted (for acceptable intelligibility) is 3kHz and the bandwidth of a double sideband signal is twice the highest modulating frequency, a double sideband transmission must, therefore, have a bandwidth of 2 x 3 kHz = 6kHz.

Many modern transmitters have audio filters that slice off some of the voice frequencies above 2.8kHz making the bandwidth even narrower. *The bandwidth is not dependent upon the power of the modulating audio* (provided the transmitter is not overmodulated). In commercial AM broadcast transmitters, the frequency of the modulating audio is permitted to be as high as 4.5kHz and since a double side band system is used, commercial stations, therefore, have a bandwidth of 9kHz.

HIGH AND LOW-LEVEL MODULATION

When the modulating audio is applied to the carrier in a low power stage (that is, before significant amplification), this is said to be low-level AM.

On the other hand, if the carrier is amplified to a high power level before the modulating audio is applied, then the system of modulation is called high-level AM.

With low-level modulation, the double sideband (DSB) signal is produced at low power levels. The advantage here is that the audio power required is low; no high power audio frequency amplifier is required. If it is necessary to raise the power of the DSB signal (and it usually is), *then all subsequent amplification must be done with linear amplifiers.* In a high power AM system, the use of linear power amplifiers is usually avoided because of their lower efficiency. Low-level modulation is usually restricted to transmission systems that operate at low power levels such as small hand-held transceivers and telemetry devices.

Note: once a signal has been modulated i.e. it contains intelligence, all amplification must be linear otherwise distortion (further modulation or inter-modulation) will occur.

It is interesting to note that SSB is, in fact, a low-level AM modulation system, since it is far more efficient (overall) to remove the carrier when it is in the milliwatt region, than to amplify the modulated signal and then have to remove a high power carrier which by then could be a hundred Watts or more.

When very high transmitter output power is required, it is usual to employ high-level modulation. In these systems (AM Broadcasters) the carrier is developed by an oscillator and amplified to the final transmit power, then the modulating audio is applied. Since the carrier has no intelligence before modulation, high-efficiency non-linear stages (Class C 60-80% efficiency) can be used for amplification. In a broadcast transmitter, modulating audio is frequently applied to the Power Amplifiers plate (anode) circuit (high-level plate modulation). If the final carrier power is 10kW, then the modulator must supply 5kW of audio power to the modulated wave for one hundred percent modulation.

The relationship between carrier power and sideband power in AM system is:

$P_t = P_c + 0.5m^2 P_c$
where:
P_t = the total power transmitted
P_c = carrier power
m = modulation index

Examination of the equation shows that since the *carrier power (P_c) remains constant,* then the sideband power is given by the equation as:

Power Sidebands = $0.5m^2 P_c$.

The modulation index (m) is simply the percentage of modulation divided by one hundred. As the percentage of modulation goes from 0% to 100%, the modulation index goes from 0 to 1.

A WORKED EXAMPLE

A 100-Watt double side band (AM) transmitter is modulated firstly to 50% and then to 100%. What is the total power transmitted in each case?

From the equation given, at 50% modulation:

$P_t = 100 + 0.5 \times (0.5)^2 \times 100 = 112.5$ Watts,

of which one hundred Watts is the carrier power and only 12.5 Watts is the total side band power. That is, 6.5 Watts in each side band. At 100% modulation:

$P_t = 100 + 0.5 \times (1)^2 \times 100 = 150$ Watts   i.e. 25 Watts in each sideband.

The only reason I've shown these calculations is to highlight the inefficiency of a double sideband AM system. At best (m=1) the power transmitted in the sidebands is only one-third of the total power being transmitted (e.g. in the example above for 100% modulation with m=1, 50 Watts total sideband is one-third of the total power of 150 Watts). The carrier does not convey any intelligence to the receiver. The carrier can be removed and all the information in the sidebands will remain.

AMPLITUDE MODULATION ON AN OSCILLOSCOPE

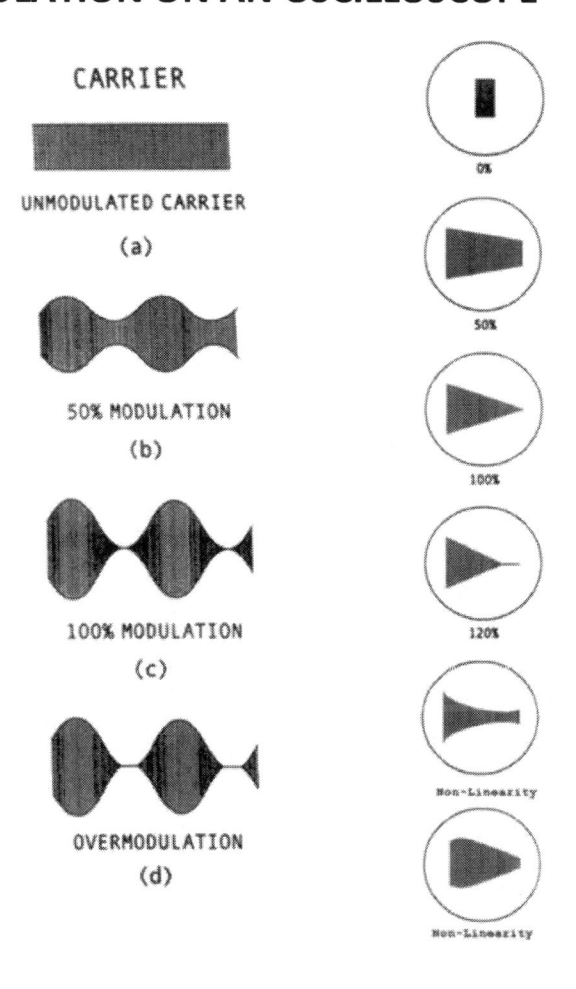

Figure 1-29

Figure 2-29

You do not need to know how these waveforms on an oscilloscope are produced. However, you may be asked in the exam to identify one of them. As you can see from the way the waveform of an amplitude modulated signal is displayed on an oscilloscope, it does appear at a glance that the power in the carrier varies.

An oscilloscope displays resultant voltage and not power. Carrier power remains constant in an AM system.

TRAPEZOID PATTERNS

The trapezoid patterns (figure 2-29) are another method of showing the percentage of modulation on the oscilloscope. Again, you do not need to know how this is done with the oscilloscope. However, such patterns have been shown in examinations and you may be presented with one pattern in the exam and asked a multiple choice question about what the display means.

I could explain to you how this is done. However, I think you have enough to remember without adding superfluous non-exam essential information. The trapezoid patterns have one advantage over the previous patterns - they show if the transmitter is linear. Look at the last trapezoid pattern and you will see that sides of the trapezoid are bowed inward (concave); this indicates the transmitter is operating in a non-linear way. The sides of the trapezoid could also be bowed outward (convex) this would also indicate non-linearity. You may be wondering what all the fuss about linearity is. Non-linearity causes interference, so it is highly undesirable. Whenever signals are fed to a non-linear device in an amplitude modulated system, some of the signals mix due to the nonlinearity and can create unwanted frequencies that can produce interference to other services.

In an amplitude modulated system no intelligence is transmitted by the carrier and the wanted intelligence is transmitted in both the upper and the lower sidebands. It is no wonder then that amplitude modulation is not a popular method of voice communication, though it is still used by AM commercial broadcasters (though many of these are converting to frequency modulation and digital radio).

On some transceivers, there is often a switch to change between AM (double sideband), upper sideband and lower sideband. Many radio operators have the impression that these are three different channels, when in fact, it is only one channel, with three distinct methods of transmission - all of them being amplitude modulation.

SINGLE SIDEBAND SUPPRESSED CARRIER (SSB)

SSB is a form of amplitude modulation (AM) where the carrier is suppressed (typically 50 to 70 dB) and one of the sidebands is removed.

For an interesting experiment on air particularly with amateur stations: Ask the other nearby station to key his or her transmitter say on upper side band and ask them not to say anything (don't modulate). Then you tune down about 1.5kHz and you will hear the suppressed but not eliminated carrier. You will hear a tone – this is the carrier.

Since all of the intelligence in a double sideband signal is present in each of the sidebands and none of the intelligence is in the carrier, it is possible to remove one sideband and the carrier before transmission. The advantages of this are:

❏ Conservation of spectrum space (an SSB signal occupies only half the band of a DSB signal);

❏ The useful power output of the transmitter is greater since the carrier is not amplified;

❏ The SSB receiver is quieter due to the narrower bandwidth (receiver noise is a function of bandwidth - this will be discussed later).

The major disadvantage of SSB is that the carrier must be reinserted in the receiver, to down-convert the sideband back to the original modulating audio. Another important fact that many textbooks on SSB fail to emphasise is that the carrier must be reinserted with the same frequency and phase it would have if it were still present. This is why clarifier control on a CB radio is so fiddly if the carrier is not inserted exactly correct you get the Donald Duck type voice. The amateur operator will use the VFO or RIT (receiver incremental tune) to do the same thing.

A single sideband signal on entering a receiver will go through many frequency conversions before being converted to the original audio. If the carrier was left with the sideband, it (the carrier), would experience the same conversions, so the carrier must be reinserted at a frequency and a phase relative to the sidebands current position in the spectrum for demodulation of the original audio (demodulation is the opposite of modulation).

Because of the necessity for the carrier to be reinserted, the SSB receiver is more complex than the DSB (AM) receiver and requires continuous attention from the operator to keep the reinserted carrier exactly on the correct frequency and phase. If the receiver wasn't so finicky, the public broadcasting service would have adopted the SSB system long ago.

TWO-TONE TEST SIGNAL

If you have an oscilloscope two-tone tests are easy to do and this will be explained in a coming chapter. You do need to know what a two-tone test is for examination purposes.

A two-tone test signal provides a standard whereby measurements made on an SSB transmitter can be duplicated. If an SSB transmitter is voice modulated and the output fed to the vertical plates of an oscilloscope, the pattern produced is so complex and dependent on the voice characteristics of the individual operator that it defies interpretation. A two-tone test signal consists of two sinusoidal modulating audio frequencies of equal amplitude which are *not* harmonically related (one is not a multiple of the other). The two-tone test signal produces the identifiable and repeatable patterns that I've shown in this chapter.

The reason the two tones of the two tone test signal must not be harmonically related is that the linearity of any system can be determined by injecting two tones in and looking for distortion products of these frequencies at the output.

Suppose tones of 1500Hz and 2300Hz were fed to an SSB transmitter in the upper sideband mode. The upper sideband output should then consist of two upper side frequencies: carrier + 1500Hz and carrier + 2300Hz. If any other upper side frequencies are present, this would indicate nonlinearity in what should be a linear system. If the tones were harmonically related, it would not be possible to distinguish easily between harmonics of the two tones. Unfortunately for the radio experimenter, such linearity measurements are best done with an expensive device called a spectrum analyser.

So in reality, two-tone measurements are rarely done by many radio experimenters. However, you need to know that a two tone (audio oscillator) is used to test an amateur SSB transmitter for non-linearity and produces the oscilloscope patterns shown earlier.

SINGLE TONE TEST

If an SSB transmitter were modulated with a single audio tone, what would you expect to appear on the oscilloscope?

Answer: an unmodulated carrier wave - see the earlier diagram 1-29(a).

In fact, if an SSB transmitter were modulated with a single audio tone it would be very hard to tell the difference between it and a CW (Morse code) transmitter.

If a single audio tone is fed to the microphone of an SSB transmitter, then the sideband output would consist of a single side frequency - the transmission would be indistinguishable from that of a telegraphy transmitter with the key down. In fact, if the audio tone could be keyed on and off, telegraphy signals could be sent. With the addition of a second tone and alternation between the two tones, code transmissions of teletype (RTTY) or facsimile (FAX) type are possible.

This method of modulating further SSB signals is called "audio frequency shift keying" or AFSK. This technique is extended further with slow scan television where a number of discrete audio tones are used sequentially to represent different shades of grey (these particular modes will be covered more fully later).

DOES MIXING SWAP THE SIDEBANDS?

In figure 3a-29 we have a DSB signal with a carrier frequency of 10MHz. The frequencies are shown on the diagram in kHz. So the carrier is 10000kHz. The upper and lower sidebands are 3kHz wide.

The lower frequency of the lower sideband (f1) is then 9997kHz.
The upper frequency of the lower sideband is the carrier frequency (fc) 10000kHz.
The lower frequency of the upper sideband is the carrier frequency (fc) 10000kHz
The upper frequency of the upper sideband is 10003kHz.

Now we are going to move the entire DSB signal on the left (Sig1) to a new carrier frequency of 27000kHz on the right (Sig2). One way to do this is to mix Sig1 in a mixer with an oscillator on 17000kHz.

There will be *many* mixing products at the output of the mixer, but the one we want is the sum of Sig1 and the Oscillator. We get rid of all the other unwanted frequencies by using a filter – not shown here.

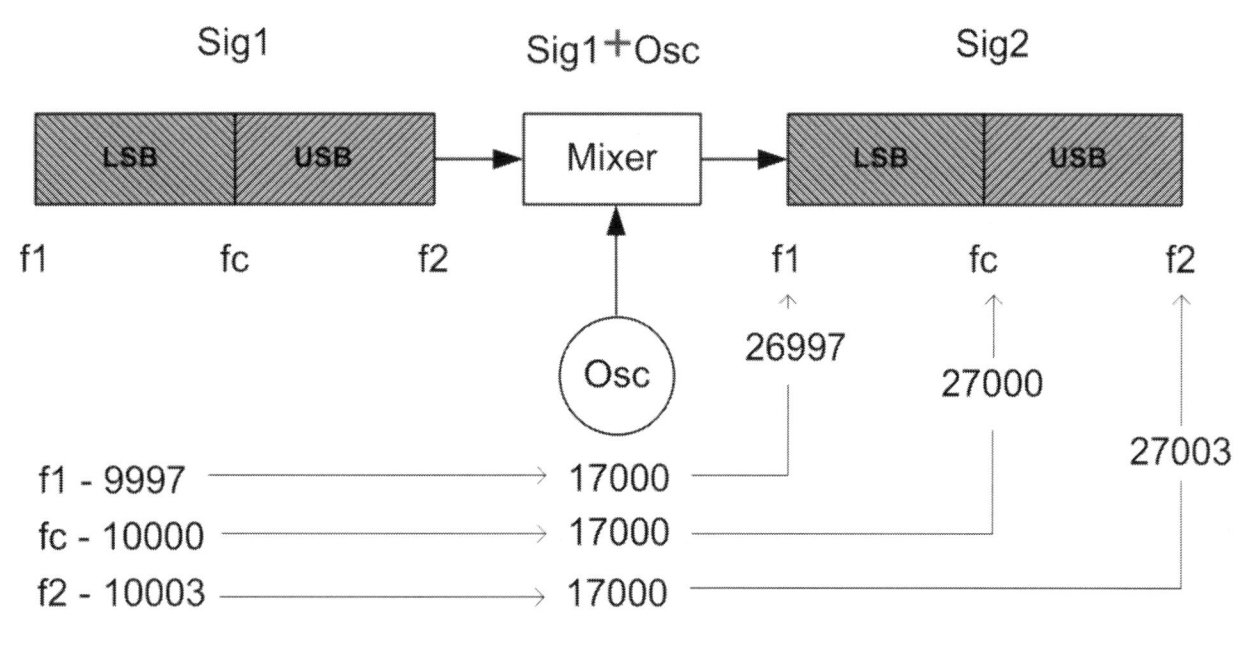

Figure 3a-29

As you see the DSB signal which ranges from 9997kHz to 10003kHz adds to the oscillator frequency and is converted to the new range 26997kHz to 27003kHz

This process, where we are summing the signal with the oscillator, does not change the *relative* positions of the frequency components in Sig1.

See that f1 is still the lower frequency of the lower sideband and f2 remains the upper frequency of the upper sideband.

When a single sideband signal is summed with an oscillator to move it in frequency, the sideband that is the upper or lower sideband remains an upper or lower sideband.

Figure 3b show the same process but this time, an oscillator frequency of 37000kHz is used. Our wanted signal is now the *difference* between the oscillator and Sig1. That is, we are now subtracting each of the frequencies in Sig1 from the oscillator.

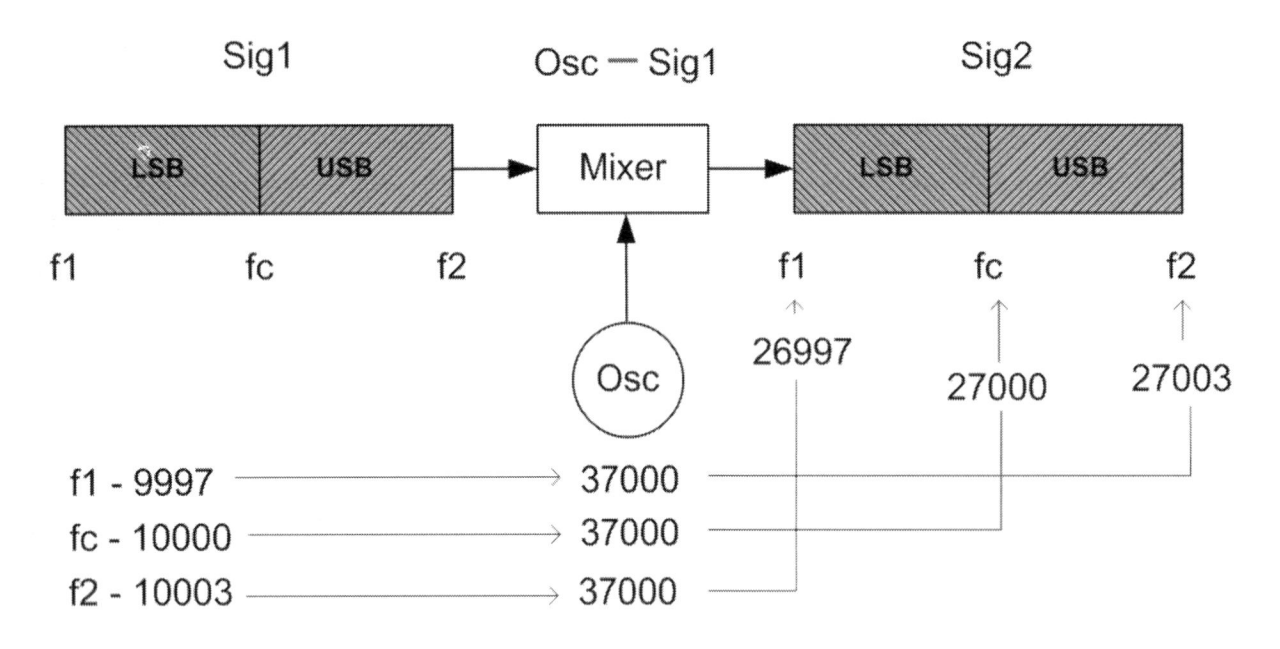

Figure 3b-29.

Notice now that f1 the lower frequency of the lower sideband comes out at the highest frequency of the upper sideband. f2 the highest frequency of the upper sideband is converted to the lowest frequency of the lower sideband.

So when we use an oscillator on the high side of the signal and convert using the difference between the signal and the oscillator the sidebands are swapped over. Not that is matters in the least as each sideband still represents all the frequencies in the modulating audio and can at any time be mixed with the carrier oscillator (BFO) in the receiver and converted to the original audio range so that they can be heard.

If you are asked a question about sideband conversion in the exam draw a sketch like the ones above and you will see straight away if a sidebands are reverse through mixing – or not.

In commercial and amateur radio, it is customary to specify the transmitter frequency for an SSB signal as that of the suppressed carrier. An amateur station transmitting on 28.5 MHz USB means that the suppressed carrier is on 28.5 MHz, on upper sideband the actual centre of the emission is 1.5 kHz higher than the carrier and on lower sideband 1.5 kHz lower.

CONVERTING AN SSB SIGNAL BACK TO THE ORIGINAL MODULATING AUDIO

After all, the whole idea of radio transmission is to take an audio frequency from the microphone, convert it to a radio frequency, transmit or radiate it, receive at a distant location and recover the original audio.

With a single sideband signal, the missing carrier is reinserted into the sideband at the receiver in the exact position it would be had it not been removed. The carrier re-insertion is done in the *product detector*. The difference frequency between the carrier and each individual side frequency component is the demodulated audio.

To put this in other words and demonstrate the whole aspect of radio communications regardless of the mode used, the problem is that audio frequencies cannot be radiated from an antenna. So, regardless of the mode of transmission, the audio frequency from the microphone must be converted to a radio frequency so that it can be transmitted and received at some distant location. The purpose of the receiver then is to convert that radio frequency back down to the original audio frequency that entered the microphone at the transmitter.

SCHEMATIC DIAGRAM OF A MIXER

The combination of a mixer and a local oscillator is often called a frequency converter. A mixer takes two input signals. Usually, a signal such as an AM, SSB or FM plus another signal from an oscillator is fed to the mixer. The mixer is just a *non-linear* device. The non-linearity will cause the mixing to take place. If you feed two signals into a *linear* device, you will get only those two signals out.

We can even just use a diode as a non-linear device. However, better mixers consist of active devices biased to operate on the non-linear portion of their operating curve.

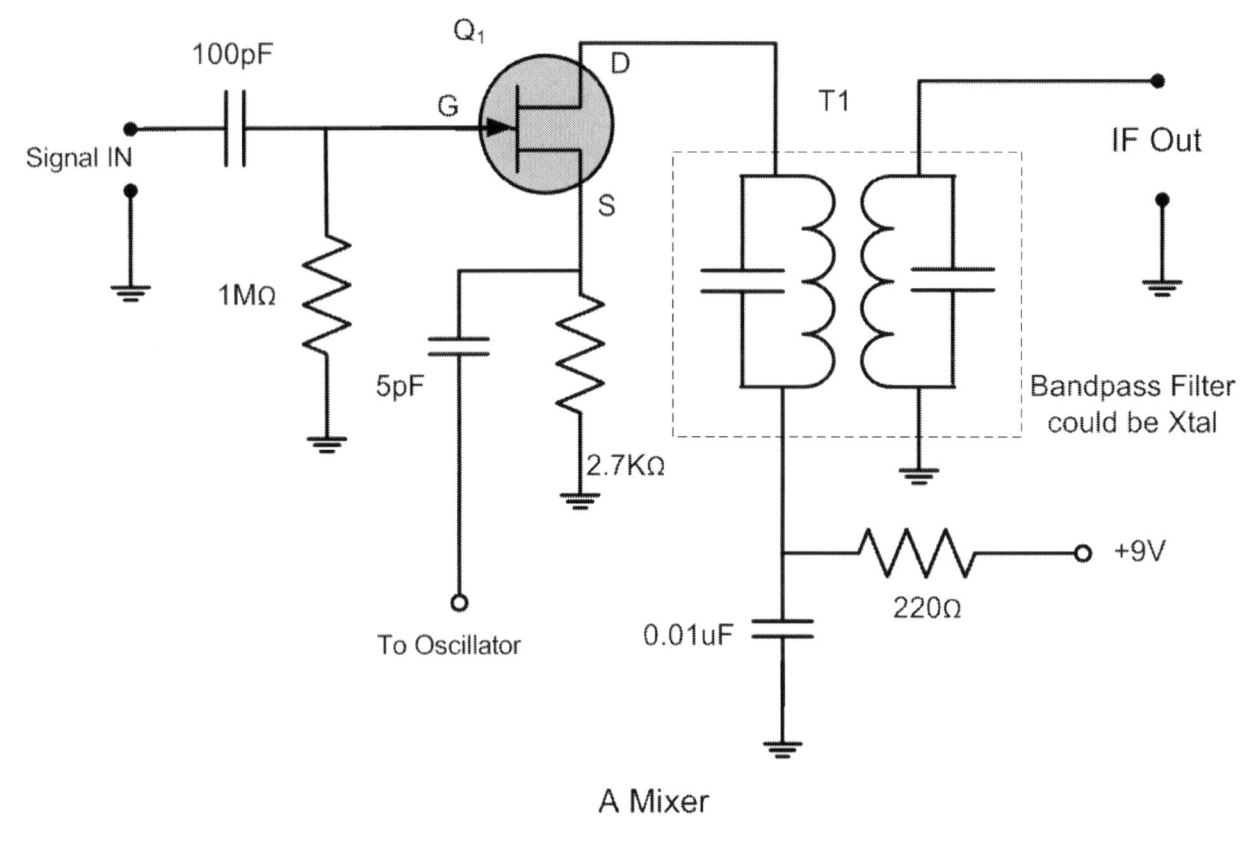

Figure 4-29

A Mixer

The schematic in figure 4-29 shows a JFET mixer. One signal is fed to the gate of the JFET while the other is injected into the source leg. Mixing will occur because the JFET will be operating as a non-linear device. The bias used is 'gate-leak' formed by the 100pF capacitor and 1MΩ resistor.

At the output of the mixer is an IF transformer with acts as a bandpass filter passing the mixing products we want and rejecting the others.

The 220-Ohm resistor and the 0.01μF capacitor act as a lowpass filter to prevent RF from entering the power supply.

Another popular method of mixing is to use a dual gate MOSFET. This is shown in the schematic of figure 5-29.

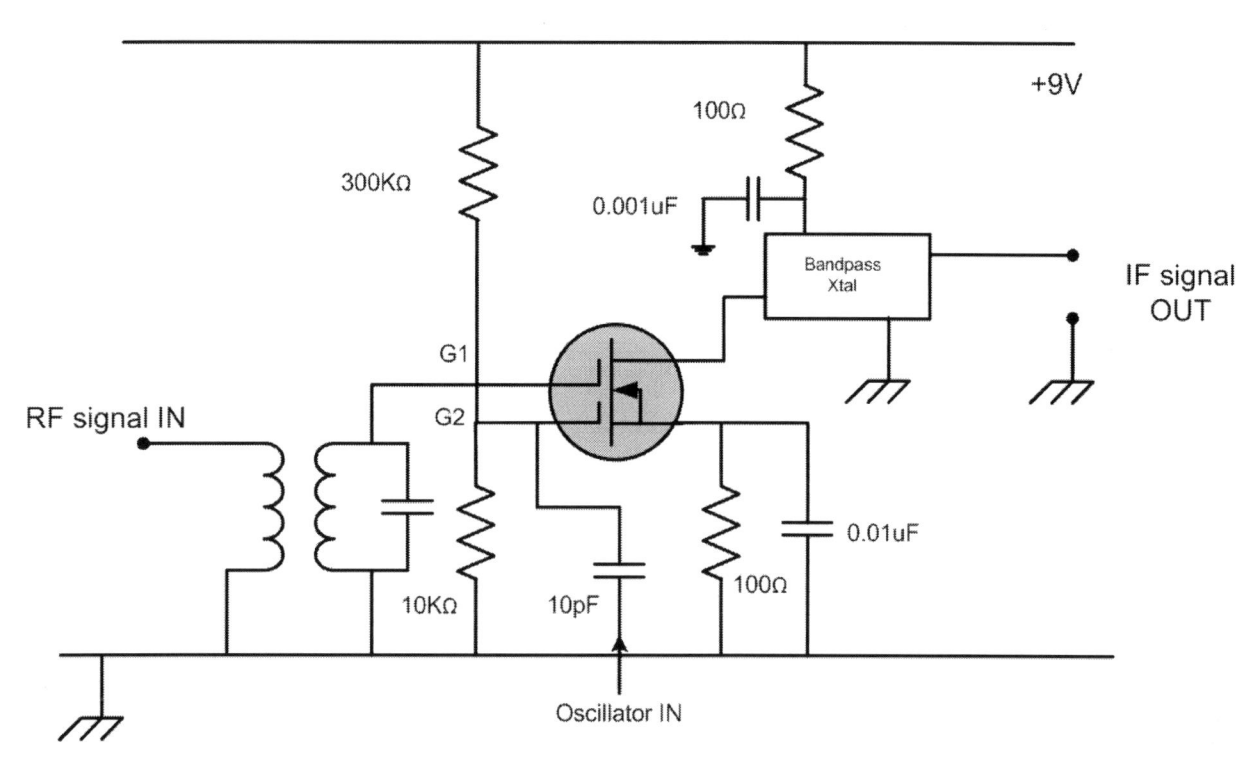

Figure 5-29

The signal that is to be moved in frequency is fed to gate 1 and the signal from an oscillator is fed to gate 2. This circuit uses voltage divider bias on gate 2.

30. AM TRANSMITTERS & RECEIVERS

Revision: Definition of amplitude modulation.

Amplitude modulation is when the modulating audio is combined with a radio frequency carrier in such a way that the total wave power is made to vary in accordance with the modulating audio.

The normal type of AM is double sideband or, to be more specific, double sideband full carrier.

Recall that the carrier does not convey any intelligence to the receiver. The information is in the sidebands. Either sideband contains all of the information.

The advantage of double sideband full carrier is simplicity in reception. The disadvantages are that it is not very efficient power wise and the bandwidth is greater than is necessary for other modes, particularly voice communications.

The carrier is used in the transmitter to heterodyne (mix) with the audio to convert the audio to radio frequencies (the sidebands) so that they can be radiated from an antenna.

The carrier is transmitted along with the sidebands. The advantage of this is that the carrier is available to the receiver to heterodyne once again with the sidebands, this time, to convert the sidebands back to the original audio frequencies, which can then be heard.

If the carrier were not transmitted (SSB), then the receiver would have to generate its own carrier.

Mixing or heterodyning in a transmitter is called modulating, whereas in a receiver it is called demodulating. The stage in a receiver which does the demodulating is referred to as the demodulator though it is sometimes referred to by an older name, the detector.

The most basic AM (DSB full carrier) receiver is the crystal set. Let's have a look at how a crystal set operates. Figure 1-30 is the schematic diagram of a crystal set, part of which we discussed in tuned circuits.

The AM (double sideband full carrier) electromagnetic wave and all the others in the air pass by the antenna and induce voltages into it. The parallel tuned circuit of the crystal set is resonant on one of these frequencies and signals on that frequency only will create a voltage across it. The parallel tuned circuit is selecting the desired radio channel and rejecting others.

The AM signal then causes a current to flow in the circuit, which includes a diode.

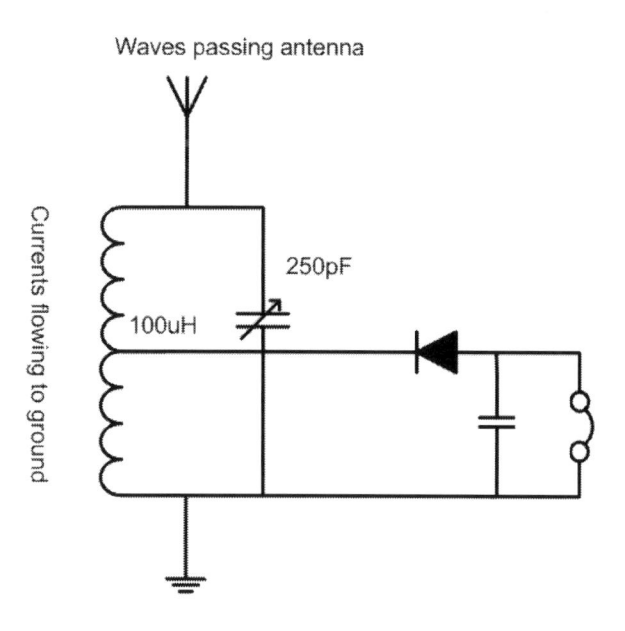

Figure 1-30

Now you can't get anything more non-linear than a diode. A diode is very non-linear as it conducts current in one direction only. Being non-linear, the AM signal will mix in the diode. The radio frequency carrier will mix with the sidebands, the resultant of which will be the original modulating audio. There will be all manner of other radio frequencies produced by the mixing in the diode as well - we don't care about them, we only care about the demodulated audio. So the output of the diode will contain the wanted demodulated audio and many other radio frequencies. The capacitor across the headphones has a low reactance to radio frequencies, so the radio frequencies do not go to the headphones (not that it would really matter if they did). The radio frequencies are bypassed to ground by the low reactance of the capacitor - so we are done with them. The low frequency demodulated audio passes through the headphones. The headphones, which will operate on the piezoelectric or the 'moving coil' principles, reproduce the sound to the listener's ear.

That was not so hard was it? Perhaps you may need to read through the paragraph several times till you are happy with the way the simple receiver operates.

If we wanted to, we could add an audio amplifier and loudspeaker to do away with the headphones and sit back and listen in comfort to the audio.

If there were lots of strong radio stations around and in close proximity frequency wise, more than one tuned circuit could be used to provide the receiver with greater selectivity. These earlier versions of AM receivers (with audio amplification) were called Tuned Radio Frequency (TRF) receivers because all tuned circuits were tuned to the incoming radio frequency.

Selectivity is the ability of any radio receiver to accept one particular radio signal and reject all others.

Sensitivity: Is the ability of any radio receiver to detect weak signals – measured in micro-Volts.

Just to get it firmly in mind, a linear device such as an amplifier is one which faithfully reproduces every signal at its input to the output without distortion. Signals *do not* mix together in a linear device.

A non-linear device, which can be as simple as a diode, or an amplifier operating in Class C, creates distortion of the incoming signals. What goes in is not what comes out. Yes, the nonlinearity causes all the individual frequencies entering the device to heterodyne or mix. In doing so, new frequencies which were not present at the input are produced in the output. The number of mixing products in a non-linear device are enormous, though if we know what we are after we can filter out the garbage and use the mixing products we want (the demodulated audio).

LOW LEVEL (AM) AMPLITUDE MODULATED TRANSMITTER

Figure 2-30 shows the block diagram of a low-level amplitude modulated transmitter.

AM Low level modulation

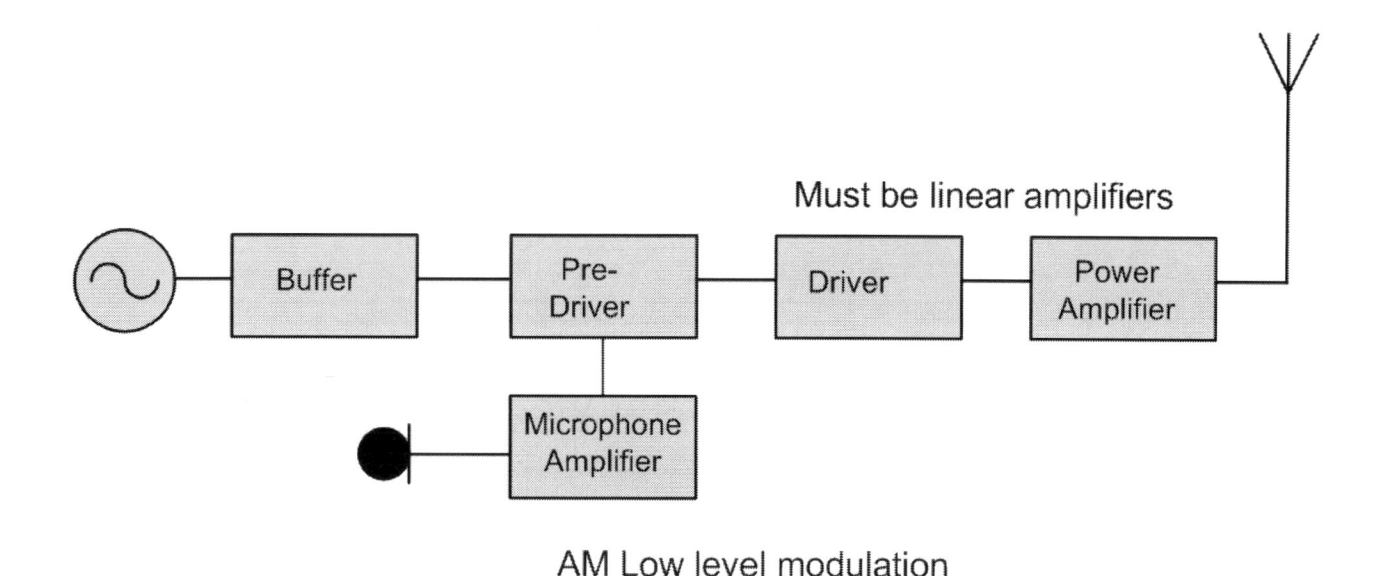

Figure 2-30

The carrier wave is produced by the oscillator. The buffer stage after the oscillator is just an amplifier which raises the power level coming from the oscillator to a useable level, without taking too much power (if any) from the oscillator. Trying to draw too much power from an oscillator could stop it from oscillating, or pull it off frequency, hence the buffer amplifier.

The signal from the microphone is amplified to a useable level and fed along with the carrier to the pre-driver. The pre-driver is just another amplifier - however the pre-driver is nonlinear, so it acts as a modulator. The output of the modulator will contain the double sideband full carrier signal, which is usually just referred to as AM.

After the pre-driver, which is acting as a modulator, we have created the signal that we want to transmit. All we need to do then is raise the power level to the desired level before feeding it to the antenna. All amplifier stages after modulation must be linear.

Whilst the low-level modulation was done in the pre-driver here; it can be done in any low-level transmitter amplifier – such as an IF amplifier.

The advantage of low-level AM modulation then is that not much audio power is required. The disadvantage is that all radio frequency amplifiers after the modulator (Driver in this case) *must be linear*. This is a disadvantage because linear amplifiers run at lower efficiency Class A or Class B push-pull.

HIGH LEVEL (AM) AMPLITUDE MODULATED TRANSMITTER

Figure 3-30 shows the block diagram of a high-level AM transmitter.

With high-level modulation, the carrier is created by the oscillator and amplified up to a high power level, in fact, the power level we want to transmit. The audio to be modulated is done in a high power stage. In this case, the power amplifier or otherwise called the final amplifier.

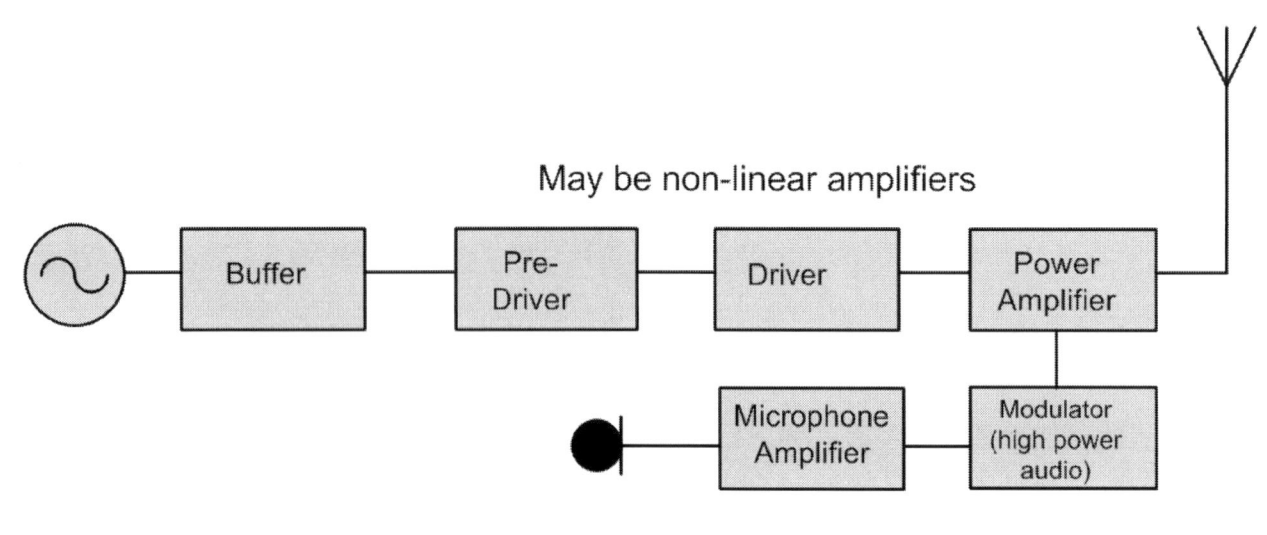

AM High Level modulation
Figure 3-30

Because the carrier power is high, we need a lot of audio power to produce 100% modulation. This is a bit misleading, but the stages, microphone amplifier and modulator are just audio amplifiers to bring the low-level audio from the microphone up to a high level.

Modulation is not occurring in the stage labelled the modulator. Modulation is taking place in the Power Amplifier. For modulation to take place in the power amplifier it must, of course, be nonlinear and is nearly always Class C.

The *advantage* of high-level modulation is that the carrier (the signal from the oscillator) can be amplified using high-efficiency Class C amplifiers. The *disadvantage* is that a large amount of audio amplification is required to modulate the high power carrier.

High-level AM modulation is typically used in high-power AM transmitters like broadcast stations and high power amateur transmitters which operate in AM mode only.

THE SUPERHETERODYNE RECEIVER

The Superheterodyne has been around for a long time and is still the receiver system of choice in modern receivers. Though we are talking about AM in this chapter, the Superheterodyne can and is used for other modes as well, but the principles are the same except the stage labelled detector will vary in its operation depending on the mode in use - we are going to stick with AM for now so the detector could just be a diode.

Earlier we mentioned that early receivers were called TRF (tuned radio frequency). Selectivity of a receiver is very important. We only want to hear the radio station to which we are tuned to. In the TRF receiver, there would be several tuned circuits like the one on our crystal set. The problem with the TRF is that every time you changed to a new radio channel or signal, you had to re-tune all of these circuits to the new frequency. This was a very fiddly affair, to say the least.

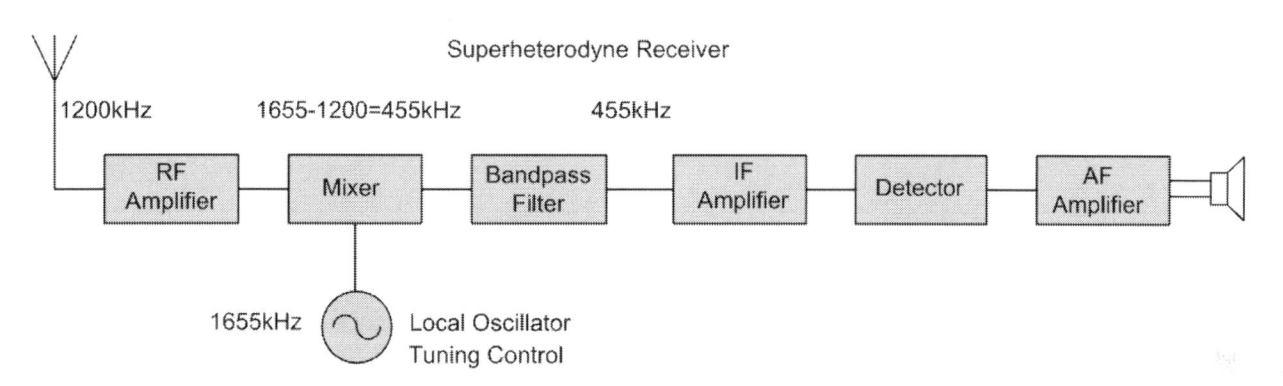

Figure 4-30

The idea of the Superheterodyne is to convert the signal (station) being received to a single standard frequency. This frequency is called the *intermediate frequency*. The great advantage of the Superheterodyne is that there is only one tuning control and the intermediate frequency amplifier, typically 455kHz, this never has to be adjusted and can contain many sharply tuned, high Q circuits, or even special crystal filters, tuned to 455kHz. This means the receiver has a very high selectivity. The I.F. amplifier also has very high gain giving the receiver very good sensitivity.

I think they must have thought up the name of this receiver from heterodyne for "mix" and "super" because it was so darn good compared to the TRF.

The problem is that to take advantage of the highly selective intermediate frequency amplifier, all signals to be listened to have to be converted to the intermediate frequency.

In the block diagram shown in figure 4-30, the first stage is an RF amplifier. It amplifies an entire band of frequencies coming down the antenna (such as the entire broadcast band, or an entire amateur band).

All of the radio frequency signals with all their information are fed to a mixer, along with a signal from a local oscillator. The mixer is a nonlinear device, so frequency conversion takes place.

In the diagram, I have shown a 1200kHz signal arriving at the antenna. Of course, there will be many other frequencies at the antenna as well. The 1200kHz AM signal is amplified by the RF amplifier and then fed to the mixer along with a signal from the local oscillator. The local oscillator is the *tuning knob* on the radio. I have shown the local oscillator operating on 1655kHz.

Now, in the mixer, all sorts of radio signals will be present. However, only the desired 1200 kHz signal will mix with a 1655 kHz local oscillator to produce a 455 kHz intermediate frequency.

So what we have effectively done with the mixer and the local oscillator is convert the desired signal on 1200kHz to 455kHz, by mixing it with a locally generated signal on 1655kHz.

This conversion of 1200kHz to 455kHz is just that, a frequency conversion. The 455kHz signal still has its carrier and sidebands intact and all with the same relationship to each other and contain all the information they ever had - we have just moved them from 1200 kHz to 455kHz.

The filter following the mixer is a bandpass filter tuned to 455kHz. (filters of this type are to be discussed) The filter will only pass signals on 455 kHz +/- the bandwidth of the signal we want to receive (a few kilohertz).

From here on the receiver never requires any adjustment by the user. The desired station on 455kHz passes through the bandpass filter and gets amplified by the 455kHz I.F. (intermediate frequency) amplifier. The I.F. amplifier is sharply tuned to amplify signals (our desired signal) on 455kHz +/- the small bandwidth the signal occupies.

The overall advantage of the Superheterodyne receiver is immensely improved selectivity, sensitivity and simplification of operation over the TRF.

The intermediate frequency amplifier is very aptly named, as the desired signal is received and converted to a standard middle (intermediate) frequency before being demodulated.

SUPERHETERODYNE RECEIVER - IMAGE INTERFERENCE

There are always two signal frequencies which will combine with a given local oscillator frequency to produce the same difference frequency. This means there are two signals that can enter a superheterodyne receiver and be converted to the IF.

The figure 5-30 block diagram shows a superheterodyne receiver tuned to the desired frequency of 1200kHz. This 1200kHz desired signal is converted to the IF by mixing with the local oscillator of 1655kHz as shown.

Now an unwanted signal on 2110kHz can also mix with the local oscillator and be converted to the same IF. If a signal were present on 2110kHz, then this receiver would most probably pick it up. The unwanted signal is called the image frequency and the type of interference is known as image interference.

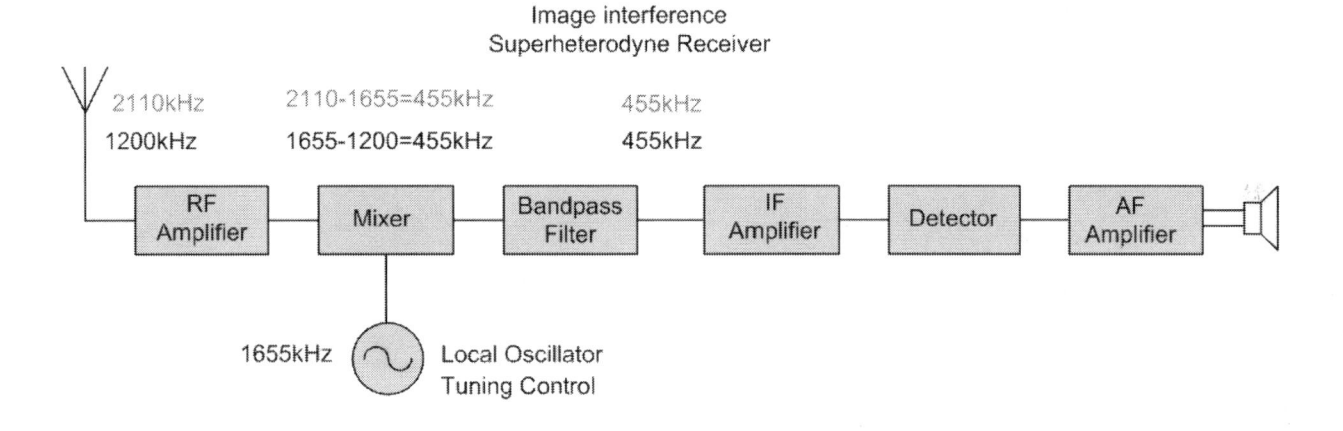

Figure 4-30

It is normal for most superheterodyne receivers to have the local oscillator on a frequency above the signal frequency as shown. The image frequency for any superheterodyne is twice the IF away from the wanted signal.

$$Image\ (frequency) = Signal\ Frequency\ +/-\ 2 \times IF$$

Equation 4-30

CURING IMAGE INTERFERENCE

Improving the selectivity of the RF amplifier will help with image interference. Also, use a high frequency first IF. Instead of a first IF of 455kHz we could use a first IF of say 10.7MHz.

Look at equation 3-30. Notice the image is twice the IF (2xIF) away from the wanted signal.

If the first IF is 455kHz, then the image is 910kHz away. In this close it is hard for the first RF amplifier to reject it. An IF of 10.7MHz was used this would place the image 21.4MHz away from the wanted signal. The RF amplifier is capable of rejecting a signal so far away.

If we use a high frequency first IF we will eliminate the image problem but unfortunately we get rid of one problem and create another. In IF amplifier on 10.7MHz does not have the high selectivity of a 455kHz IF.

So a high first IF will give us great image rejection but poor selectivity. The solution is to have two IF's the first one a high frequency and the second one low frequency for selectivity.

I have been using standard IF frequencies of 10.7MHz and 455kHZ – just remember this is just an example there are many other high and low-frequency IF's that can be used.

The problem is that we rely on the IF for selectivity and gain and at an IF of 10.7MHz, it's a lot harder to obtain the high selectivity and gain. Therefore, an ideal receiver will have two IF's - a high first IF for good image rejection and a further low IF for improved gain and selectivity. A superheterodyne receiver with two IF's is called a 'double conversion.' If we had three IF's it would be a 'triple conversion.'

Conversion Efficiency (MIXER)

Conversion efficiency is simply a fancy name for how efficient a frequency converter is, or more precisely, the efficiency of the mixer in the frequency converter. A frequency changer or converter is just a mixer and its associated oscillator. Mixers do not amplify. We lose amplitude when we do a frequency conversion. Efficiency percentage is a function of: (power out) / (power in) x 100. A good mixer would have a conversion efficiency of close to 100% or '1'.

A DOUBLE CONVERSION SUPERHETERODYNE.

Double conversion Superheterodyne Reciever

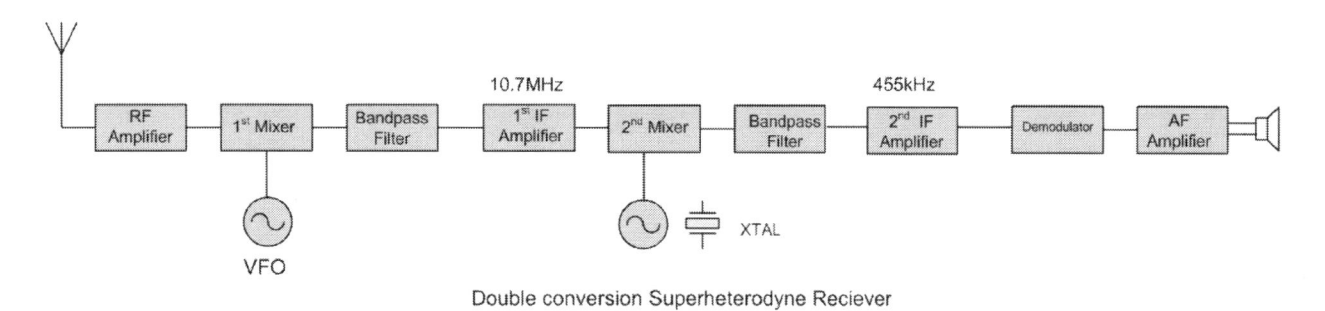

Figure 5-30

Figure 5-30 shows our truly good receiver. I have called the detector a demodulator. These are one and the same. The intermediate frequencies could be different from that shown. There could be another conversion stage and a third IF. Little modification would have to be done to this block diagram for any receiver in any mode. If this were an amateur radio DSB receiver, then the bandwidth would be 6kHz. In this case, the bandpass filters would have to have a bandwidth of 6kHz. That is the IF frequency +/- 3kHz. One of the bandpass filters would be a crystal filter. If we changed to a mode that uses a different bandwidth, then the appropriate bandwidth filter would have to be switched in. Many radio transceivers do not come with all the crystal filters pre-installed; they are optional extras.

For example, CW (Morse code) needs a very sharp crystal filter around 300Hz wide. FM needs a crystal filter around 16kHz wide. If you have a multimode radio and you do not have these filters fitted, you are not going to get high-quality reception on the modes with the missing filters.

Notice that the second oscillator is crystal controlled. The oscillator is always converting (along with its mixer) 10.7MHz to 455kHz so it can be a fixed frequency and never have to be adjusted.

31. AM TRANSMITTERS & RECEIVERS PART 2

SINGLE SIDEBAND (SSB)

Amplitude modulation occurs when the modulating audio is combined with a radio frequency carrier in such a way that the *total wave power* is made to vary in accordance with the modulating audio.

I would really like you to be very sure in your understanding of ordinary AM before we move to Single Sideband (which is another type of AM).

The basic principle of radio transmission is to convert the voice frequencies coming from the microphone to radio frequencies so that they can be radiated as an electromagnetic wave. At the receiver, the electromagnetic wave can be intercepted and the original voice frequencies recovered.

The audio frequencies from human voice range from about 100Hz to 3000Hz. Of course, the human voice can make sounds above this range, but for voice radio communications we do not need to use those higher voice frequencies. All of the intelligence (information) we need is contained in the voice frequencies below 3000Hz.

Because this is such an important topic, we are going to go over the process of amplitude modulation again with a slightly different approach. If we were to look at what was going on in an AM transmitter using an actual voice, there would be too many frequencies. Therefore, we are going to pretend that a sound is being made by a human voice, which only contains two frequencies. Our pretend human voice is making two audio sounds, one on 1000Hz and the other on 2000Hz.

The diagram in figure 1-31 shows the sound containing these two frequencies going into a microphone. The audio sounds are amplified by the microphone amplifier and fed to the modulator (a non-linear device). Also going into the modulator is a radio frequency signal of 1000 kHz (1MHz) from an oscillator, called the carrier oscillator.

The modulator is going to mix the audio and radio frequency signals together to produce new signals. This is a balanced modulator, so it suppresses the carrier by 40-70dB; other than that consider it a non-linear device. At the output of the modulator, there will appear the original carrier and the sum and difference of all the voice frequencies (which we have simplified to just two). The output of the modulator contains only radio frequencies and is called double sideband suppressed carrier, as shown. If it were not a balanced modulator, it would be a double sideband full carrier (DSB).

Look long and carefully at figure 1-31. The voice frequencies have been converted to two frequencies above the carrier and two below the carrier.

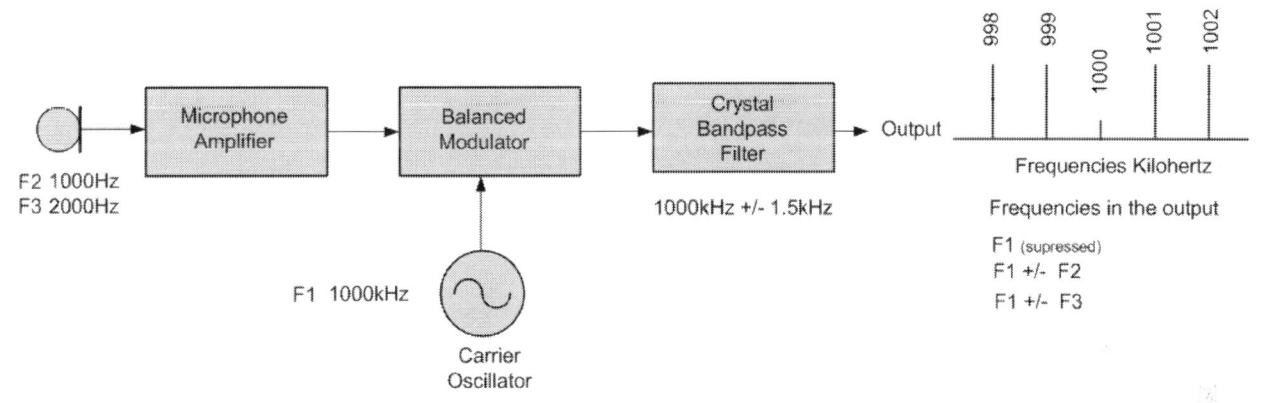

Figure. 1-31

The balanced modulator is just a nonlinear device like any other modulator the difference being that it suppresses the original carrier – in this example on 1000kHz or 1MHz.

Look at what goes into the microphone at the far left. Two audio tones. What is the resultant output signal? The set of two tones (1000kHz and 2000kHz) is frequency converted to two sets of single RF frequencies. One set on 998kHz and 999kHz. Another set on 1001kHz and 1002kHz.

The carrier is still there as well, but it is suppressed since this is a balanced modulator. How one such modulator does this will be covered shortly.

Can you see that really all we have done is convert two audio frequencies into two sets of radio frequencies? If we were to down convert those two sets of radio frequencies using the same carrier oscillator frequency, we would end up with the audio that went into the microphone.

SIDEBANDS

The frequencies above the carrier are the upper side frequencies. Those below the carrier are the lower side frequencies. Collectively, all of the upper side frequencies are called the upper sideband (of frequencies). Collectively, all of the frequencies below the carrier are known as the lower sideband.

There is no information in the carrier (1000kHz or 1MHz).

All of the *information* is in the upper sideband and the lower sideband. We really do not need both sidebands because all of the information is in each of them.

If we choose to amplify and transmit the lot, fine, we have what most people just refer to as AM. The highest voice frequency being 3000Hz means that the total bandwidth of the signal is 6kHz - twice the highest modulating audio signal.

We do not have to transmit this entire signal. We could just transmit either sideband and then convert those sideband frequencies back to the original audio in the receiver. This has obvious advantages, the main one being the bandwidth of one sideband would never be greater that 3KHz. Spectrum space is not an infinite resource and this would mean we could get twice as many channels into the same spectrum space.

There is another advantage of using just one sideband – since the bandwidth is only 3kHz – or in most transmitters 2.8kHz the receiver noise is less. The narrower than the bandwidth of any receiver the lower the noise is. I am sure you have heard the noise level decrease when you change from DSB to SSB. Imagine the low noise level if you used CW with a 300Hz filter.

Single Sideband

To transmit and receive single sideband has many advantages, but is technically more difficult. The principle of SSB transmission is simply to remove one of the sidebands (either one) completely and also get rid of the carrier. In practice, it is easy to get rid of one of the sidebands; a sharp bandpass filter will do this for us. We can reduce the carrier by using a balanced modulator. The correct full name for SSB is Single Sideband

Suppressed Carrier.

Now, if we were going to get rid of a sideband and suppress the carrier, it would make a lot of sense to do this down in the milliwatt range. Getting rid of a carrier of say 100 Watts (and a sideband) would be silly when it can be done more efficiently at low power levels.

The advantage of SSB is not just more efficient use of spectrum space and less receiver noise. The radio spectrum contains noise from many sources. If you turn on any receiver and do not tune to a signal, you will hear noise. This noise is from hundreds of sources, both man-made (QRM) and natural (QRN). Some noise is also generated inside the receiver by active devices and even resistors. The narrower the bandwidth of a receiver the lower its noise. Think of the bandwidth of a receiver like an open window. Think of the noise as rain. The more you close (narrow) the window, the less rain (noise) gets in. So yes, SSB is technically harder but well worth the benefits in power efficiency at the transmitter and noise reduction in the receiver.

SSB TRANSMISSION THE BALANCED RING MODULATOR

A Balanced Ring Modulator produces at its output a double sideband signal with the carrier suppressed by 40-70 dB.

There several types of balanced modulators. The balanced ring modulator is the most common. If there is a circuit in your assessment to identify it will be a balanced ring modulator.

Figure 2-31 shows the schematic diagram of a balanced ring modulator.

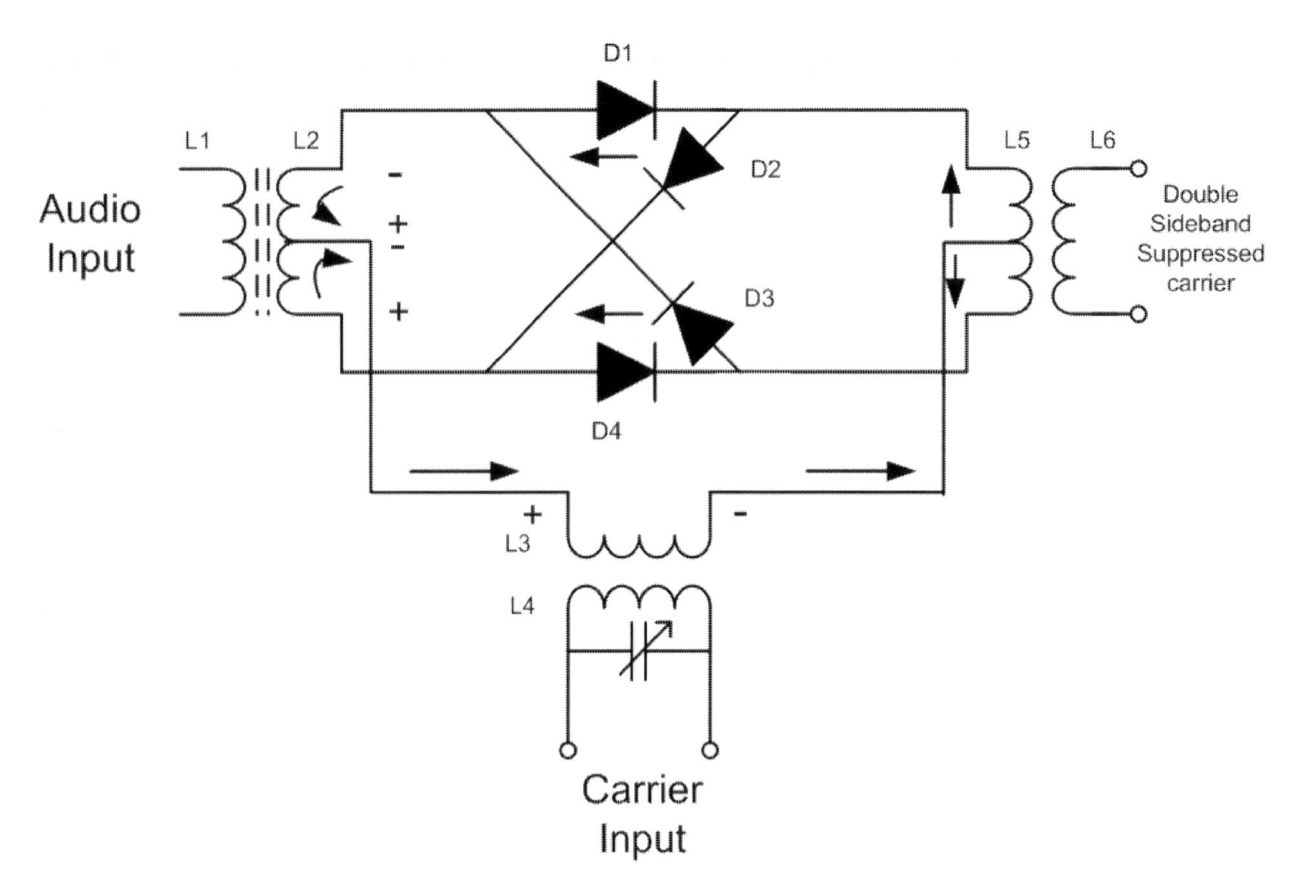

Balanced Modulator

FIGURE 2-31

You should learn to identify the circuit. I am going to provide a description of its operation. However, this is for interest sake only. Though you will need to know the operation as a block - that is inputs and outputs and what it does.

OPERATION OF THE BALANCED RING MODULATOR

With carrier input alone applied, RF currents will flow during one half-cycle through diodes D1 and D4 as shown. During the next half-cycle of RF, current will flow through D2 and D3. Notice that no matter which pair of diodes conduct, the current flows equally and in opposite directions through the upper and lower parts of the primary winding of L5. Because of these equal and opposite primary currents through L5, the voltage induced into the secondary L6 by the carrier will be zero.

This means that, provided the carrier controls the diode switching (conducting), no carrier will appear in the output.

With audio input only applied, current will flow through D3 and D4 or D1 and D2, depending on the polarity of the audio voltage, but no current will flow through L5 - so, for audio alone, the output voltage at L6 is again zero.

Suppose that both audio frequencies AF and carrier are applied with the instantaneous values as shown in the diagram. The carrier input level is 6 to 8 times higher than the AF and, as such, determines which diodes will conduct. The arrows show the current produced by the carrier. With the polarities as shown, the audio voltage is assisting current flow through the top half of L5 and opposing current flow through the bottom half. The current through L5 is now upset (unbalanced) and an output at the sideband frequencies will appear in L6.

In practice, some type of balancing of the circuit is required. This is usually a simple resistive voltage divider network, the resistance of which is adjusted to balance the modulator with carrier only. In an SSB transmitter, this control is inside the transmitter on the circuit board and labelled *carrier balance* on the schematic diagram.

A COMPLETE BLOCK DIAGRAM OF AN SSB TRANSMITTER

It's time we took a walk through the full block diagram of an SSB transmitter and discuss the operation of each stage.

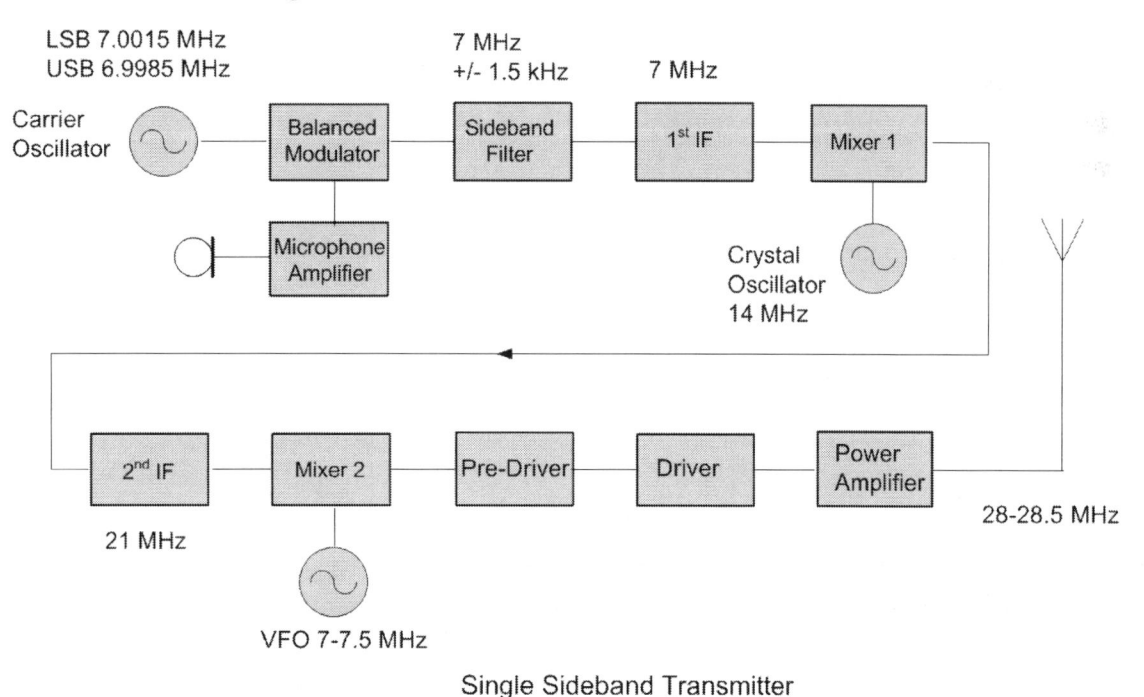

Single Sideband Transmitter

FIGURE 3-31

The block diagram of figure 3-31 is that of an SSB Transmitter (TX) which will operate between 28 to 28.5MHz. The carrier oscillator produces the radio-frequency voltage necessary to heterodyne with the modulating audio voltage in order to convert the audio signals to radio frequencies.

The carrier oscillator must be able to operate at two frequencies to enable switching between upper and lower sidebands if this is required. The carrier oscillator frequency will determine which sideband will pass through the crystal filter (to be discussed). The crystal filter is a high selectivity bandpass filter. The carrier oscillator frequency will place one sideband inside the crystal filters passband. It is the crystal filter then that passes one sideband and rejects the other.

Bandpass filter – definition: A filter that passes a band of frequencies with minimum attenuation and rejects (attenuates) all others.

Following the carrier oscillator and all other oscillators in the block diagram, there would be buffer amplifiers (not shown) to prevent the carrier oscillator from being too heavily loaded - such loading could pull the oscillator off frequency.

The microphone amplifier increases the signal power from the microphone to the level required to obtain a satisfactory modulation percentage with the signal from the carrier oscillator.

Adjusting the power output of the microphone amplifier controls the percentage of modulation and the transmitted power. This is a control on the front of the transmitter usually called 'mic gain.'

The RF and modulating AF signal voltages are fed to the balanced modulator. The fundamental requirement of the balanced modulator is to provide non-linearity to enable the process of modulation to take place. Balanced modulators have a second important characteristic that enables the original carrier frequency to be suppressed. The output of the balanced modulator then will consist of two sidebands and a suppressed carrier. The carrier suppression should be typically at least 40dB down from the carrier's full amplitude.

The bandwidth of the crystal filter (bandpass filter) will only permit the passage of one sideband. Which sideband will pass through the filter is determined by which sideband is positioned on the centre frequency of the crystal filter. A typical crystal filter for this use will have a bandpass of about 2.8kHz, which allows most of the sideband to pass through.

The purpose of both IF amplifiers is to simply amplify the required sideband as much as possible while maintaining the high selectivity in a narrow band system (SSB is narrow band).

In any communications system, the greatest amplification is provided by the IF amplifiers. Amplification gains of 40-60 dB are typical, whereas power amplifiers usually have a gain of around 10dB. The IF amplifier can produce such high amplification gains because it is a fixed and sharply tuned amplifier.

A combination of a mixer and an oscillator provide the function of a frequency converter. Both mixers and their oscillators serve to position the required sideband on the output frequency. Notice how the first mixer has an oscillator which is crystal controlled. This is possible since the conversion from the first IF frequency to the second IF frequency is always the same, requiring no adjustment of the oscillator. The crystal oscillator also provides high-frequency stability.

The variable frequency oscillator (VFO) connected to the second mixer enables adjustments to be made to the final transmit frequency. The VFO frequency should be low to obtain stability and is typically no more than 8 MHz. At higher frequencies, the LC components of the VFO would become so small as to make the frequency of oscillation very susceptible to external effects such as temperature variations.

In modern transmitters the VFO can be made to operate at higher frequencies while still maintaining stability - these designs employ the phase locked loop (PLL) system. With the PLL type VFO, the output of the transmitter is adjusted in discrete increments, typically as low as 10 Hz per step.

The chain consisting of the pre-driver, driver and final amplifier uses only linear amplification. Their sole purpose is to raise the power level of the selected sideband with minimum distortion. The best practical linear RF amplifier is Class AB.

As the carrier and modulating audio were combined at an early low power stage, the SSB transmitter is by definition a low-level modulation system - so all amplification after the balanced modulator must be done with linear amplifiers. A feedback path between the final amplifier and the pre-driver marked could be drawn in as a line and marked ALC. ALC stands for Automatic Level Control. Part of the RF signal arriving at the final amplifier is rectified and turned into a DC voltage, which is used to control the gain of an earlier stage, which could be a pre-driver or an IF amplifier.

The purpose of ALC is to prevent the operator from over-driving the transmitter. The higher the transmitter output, the higher the ALC voltage - this ensures lower gain in the preceding stage that the ALC voltage is applied to. ALC is a type of negative DC feedback that prevents the transmitter from being over-modulated.

If the transmitter were to be over modulated the linear amplifiers would become non-linear and further unwanted modulation and new frequencies would be created, causing out of channel interference. This out of channel interference is referred to as "splatter".

MORE ON THE CARRIER OSCILLATOR OF AN SSB TX

As mentioned earlier, it is the carrier oscillator which determines on what range of frequencies each sideband ends up. If we want to transmit on the upper sideband, the carrier oscillator must place the upper sideband on the band of frequencies that will pass through the crystal filter. Likewise, for lower sideband. So the carrier oscillator must be switchable between two frequencies and of course, crystal controlled for frequency stability.

The circuit of figure 4-31 is one of the best ways of switching the carrier oscillator and hence selecting the sideband to be used.

We have discussed oscillators earlier. Can you recognise what type of oscillator it is? The two capacitors C1 and C2 should give you a clue to the identification of this circuit. It's a Colpitts crystal oscillator.

Forward biasing the appropriate diode achieves sideband switching. A positive potential connected to the anode of D1 will cause D1 to be forward biased, connecting the sideband crystal (X1) to the rest of the circuit.

In this circuit, D1 and D2 are acting as electronic switches. When a diode is forward biased, the switch is closed. Without bias, the switch is open.

Nothing else has happened. Do not try and read anything complicated into this diagram. It is a frequency converter. We call it a carrier oscillator and balanced modulator only because of what we are doing with this frequency converter; that is, moving audio to radio frequencies in a DSB system. The two voice frequency tones remain the same distance apart in the radio spectrum as they were at audio. Their amplitude is the same. All we have done is move the two audio frequencies that is one set of voice frequencies (1000kHz & 2000kHz) to two sets of radio frequencies.

Carrier Oscillator
FIGURE 4-31

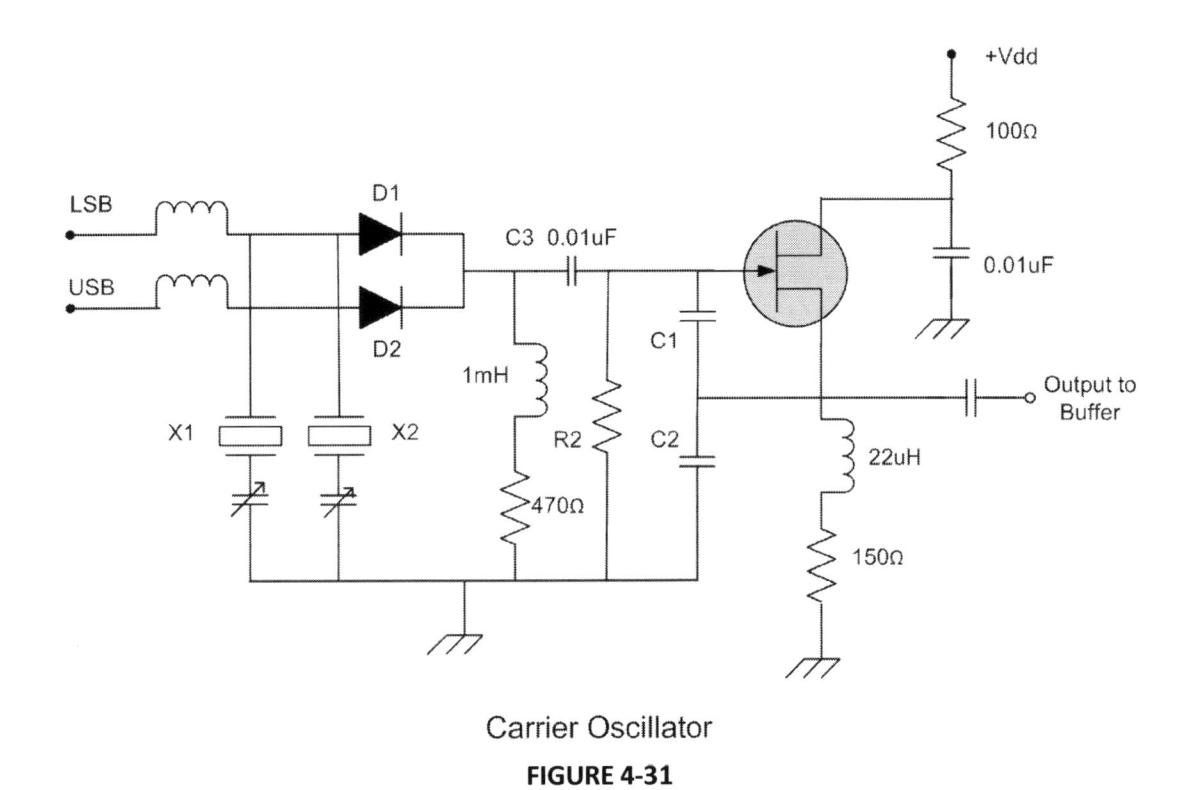

The bias applied to the diodes would upset the operation of the N-channel FET if it were not for the DC blocking capacitor C3.

You do not need to know this circuit for assessment purposes. However, you do need to be able to identify a Colpitts oscillator.

SSB RECEPTION

To finish off the topic of AM transmitters and receivers, we need to look at two final things.

An SSB receiver is no more than a Superheterodyne receiver. The difference is that the demodulator is called a product detector, along with another oscillator, called a beat frequency oscillator or BFO.

Imagine receiving a single sideband signal like the one we created at the beginning of this chapter. If the carrier were present as in a DSB full carrier signal, we would just have to feed the signal into a non-linear device. The carrier would mix with the sideband and one of the mixing products would be the demodulated audio.

When we just have a sideband in the receiver, we need to reproduce the carrier in the receiver. We need to create the carrier in the receiver, on the frequency, it would be on had it not been removed.

As you have seen, the sideband started as an audio frequency band of frequencies. It has been frequency converted. Moved through different IF stages, which involves changing frequencies. There may be many frequency conversions between transmission and reception. The important thing to realise is that all of the individual audio frequency components are still in the same place relative to each other – and if the carrier had been transmitted, it would be there for us to use. The carrier is at the bottom of the upper sideband and the top of the lower sideband.

In an SSB receiver, we don't have a carrier, so we need to make one. The oscillator that makes the carrier in an SSB receiver is called a BFO (Beat Frequency Oscillator) It's just an RF oscillator which reproduces the carrier. We feed the carrier and the sideband into a demodulator - and just to make life hard, they call the demodulator in an SSB receiver a product detector. The output of the product detector is the demodulated audio.

An SSB receiver would also contain a crystal filter (sharp bandpass filter) on one of its IF frequencies, to obtain high selectivity.

So, look at your Superheterodyne diagram in an earlier chapter. Add a crystal filter in the IF stage. Call the demodulator a product detector and feed the carrier into the product detector from a BFO and you have an SSB receiver.

The product detector and BFO are really just another mixer and oscillator forming a frequency conversion of the sideband from RF back to audio. This audio is the original transmitted audio.

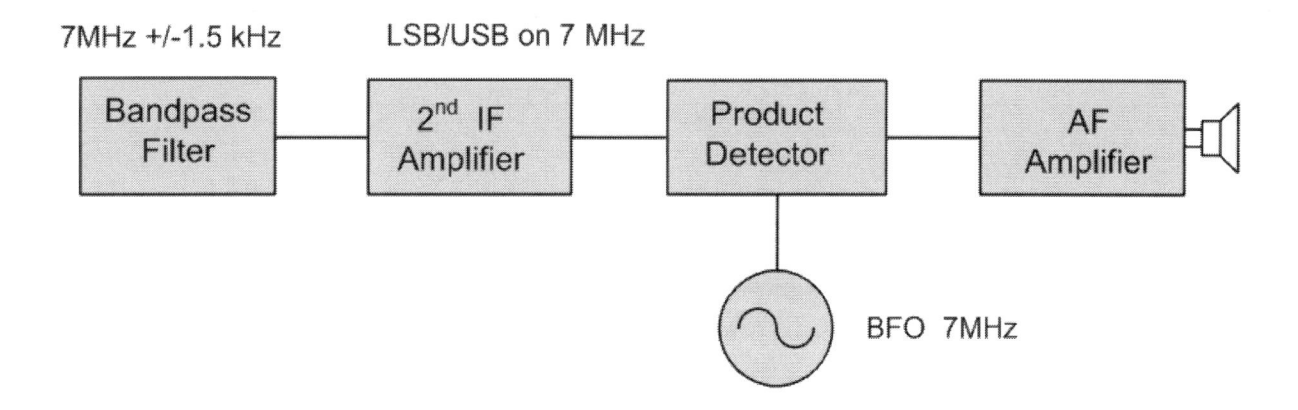

```
7MHz +/-1.5 kHz    LSB/USB on 7 MHz

[Bandpass Filter] — [2nd IF Amplifier] — [Product Detector] — [AF Amplifier] 🔊
                                              |
                                           BFO 7MHz
```

Portion SSB Receiver

FIGURE 4-32

THE DUAL ROLE OF BFO AND CARRIER OSCILLATOR IN SSB TRANSCEIVERS

In SSB transceivers (transmitter and receiver) it makes sense not to duplicate circuitry. In many transceivers, the same oscillator is used for the BFO and carrier oscillator. So on transmit the oscillator is the carrier oscillator and on receive it functions as a BFO.

AUTOMATIC GAIN CONTROL - IN ALL RECEIVERS

Automatic Gain Control (AGC) applies to all receivers and not just for SSB. You are required to know what AGC is.

What ALC is to a transmitter, AGC is to a receiver.

Almost without exception all radio receivers whether AM, SSB or FM, are superheterodyne. Review the block diagram of the superheterodyne if you have forgotten it. The problem with receivers is that the signals they are receiving can vary enormously in strength. For example, one signal from one transmitting station could arrive at the receiver at a level of -120dBm (about 0.2 microvolts) and the operator may be in communication with another station whose signal arrives at -20dBm. There is 100 dB difference between these two signals. Without AGC the stronger signal would either be too loud or completely overload the receiver causing distorted output. To receive the weaker signal, the operator would probably have to adjust the receivers gain control. AGC does all of this for you.

A small amount of received signal from an early receiver stage is rectified and converted to DC. This DC voltage is then used to determine the gain of an IF and/or RF stage in the receiver. Recall that most of the gain (amplification) in any receiver system occurs in the IF stages. Pretty simple eh! The AGC voltage determined by the strength of the received signal automatically adjusts the gain of the receiver.

That's about all you need to know about AGC for the exam. AGC can be a pain, though. For example, suppose an amateur station is in contact with two stations, one of which is very strong and the other which is very weak. After the strong one has finished transmitting the AGC has made the gain of the receiver low and if the weak station was to begin transmitting immediately, then he or she may not be heard for a while because the AGC has not acted fast enough to return the receiver to full gain.

For this reason, many transceivers (or receivers) have a switch for the AGC. This switch is usually marked Slow, Fast and Off.

In the 'Off' position the AGC is disabled. The 'fast' and 'slow' positions determine how fast the AGC returns the receiver to full gain. This gives the operator more control over the receiver. Slow is handy when talking to a single station on SSB, who pauses a lot during transmission. If the operator of an SSB transmitter pauses (to think perhaps), there is no sideband output. The receivers AGC could interpret this brief pause as an end of transmission, when it is not and raise the gain of your receiver. When the transmitting station transmits (speaks) again, your receiver may be overloaded. Therefore, the 'slow' AGC position is useful here. The 'fast' AGC position is great for controlling a signal which is fading up and down.

A Completer SSB Transceiver.

Figure 4-33 brings together everything we have learned about SSB transmission and reception.
This block diagram is not a blueprint for all SSB transceivers since many configurations are possible. However, the principles are the same in all transceivers. Notice the stages which are common to both the transmitter and the receiver. The two frequency carrier oscillator is the beat frequency oscillator in the receiver. The sideband filter, is a crystal bandpass filter, with a bandpass of 3kHz. It is used to on transmit to remove the unwanted sideband and again in the receiver to greatly improve selectivity.

Radio Theory Handbook

A Single Sideband Transceiver

FIGURE 4-33

Controls:
1 – Sideband switch. 2 – Carrier balance. 3- Mic gain/ Power out. 4 – Band switch. 5 – Main Tuning/RIT. 6 – Carrier Level. 7 – Pi-Coupler
8 – RF RX Gain. 9 – Notch – IF Shift. 10 – AF Gain. 11 – ALC On/Off/Slow/Fast.

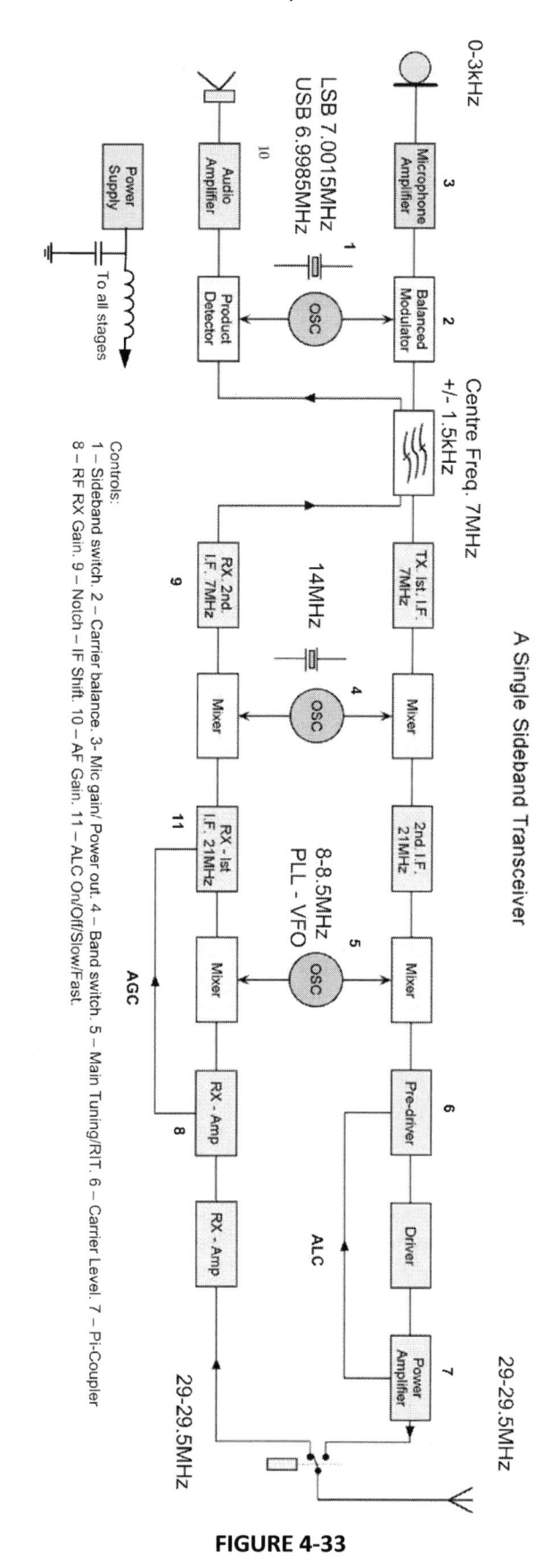

The 14MHz crystal oscillator is used to do an upward frequency conversion in the transmitter to the 2nd IF amplifier and a downward frequency conversion the 2nd IF of the receiver. Similarly, the PLL type VFO is used on transmit for an upward conversion to the transmit frequency and in the receiver as a downward conversion to the 1st IF.

The driver and power amplifiers would most likely be Class AB (linear). Instead of a changeover relay to a switch the antenna between the receiver and transmitter, electronic switching could be used.

The carrier oscillator can be switched to two frequencies for sideband selection. The chosen sideband will be positioned in frequency to pass through the sideband filter and the unwanted sideband will be stopped.

For reinforcement let us look at the output of the balanced modulator on each sideband. Remember for either sideband to get through the sideband filter the centre of the sideband must be on 7MHz or 7000kHz. On USB the carrier is on the lowest extremity of the sideband. For LSB the carrier is on the highest extremity of the sideband. With the carrier oscillator on 6998.5kHz the USB passes through the sideband filter. When the carrier oscillator is switched to 7001.5kHz the LSB passes through the sideband filter. In each case the unwanted sideband is blocked by the crystal filter.

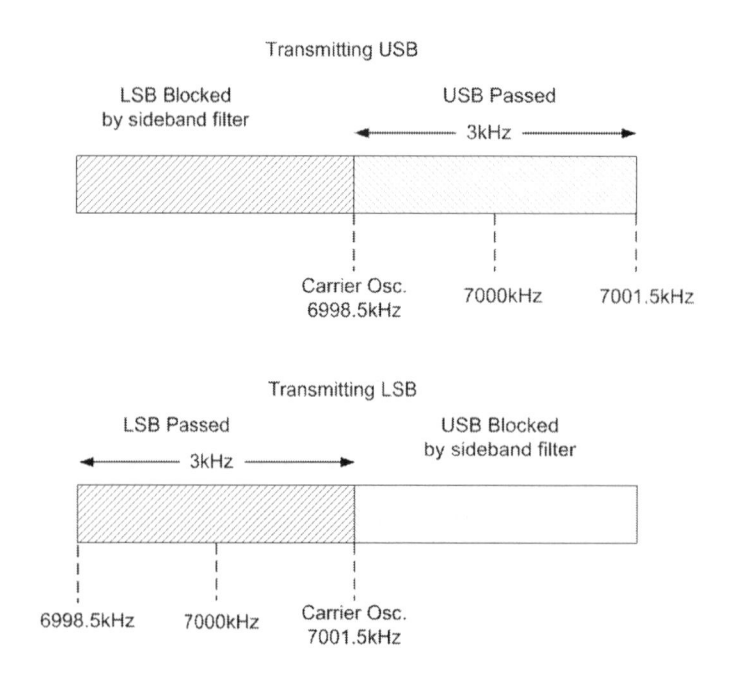

FIGURE 4-34

32. FILTERS - IN RADIO COMMUNICATIONS

RADIO SIGNALS

In radio communications, we talk a lot about radio signals. A radio signal is a very broad expression. The radio signal from a television transmitter is in the VHF, UHF or higher region of the spectrum and is about 7MHz wide. An SSB transmission from an amateur station on HF is about 3kHz wide. Therefore, when we talk about receiving, passing and rejecting 'signals' we need to keep in mind that the bandwidth of the signal can vary significantly. *The bandwidth of a signal is how much spectrum space it occupies.*

When radio communication was in its infancy, transmitters were very broad. Transmitters used a wide band of frequencies. Morse code was the only type of signal sent. Marconi, at first, had all sorts of problems selling his radio telegraphy because multiple transmitting and receiving stations would interfere with each other.

Selectivity is the ability of a receiver to receive only the wanted signal and reject all others.

With tuned circuits and crystal oscillators, we can now have a multitude of transmitters operating in a small space. Keeping them all from not being heard or interfering with each other is a mighty task, particularly in high-density metropolitan areas. I have worked on interference problems on Sydney tower. At that time, there were over 450 transmitters and receivers within a 600-metre radius of Sydney Tower. When transmitters do interfere with each other for whatever reason, many of which will be discussed, we often have to resort to additional filters other than the ones already built into the transmitter or receiver.

RADIO FILTERS

In electronics, radio and communications filters are used to select or pass desired signals and reject or block undesired signals. In other words, there may be frequencies or signals that we want to pass and others that we want to block. The only way of doing this is with a device that is frequency selective - it must behave differently towards different frequencies. Such a device is called a filter.

We have already encountered filtering. If you recall the chapter we did on power supplies, the object of the power supply is to convert alternating current to direct current. The first stage of the rectification process is to convert alternating current to pulsating direct current. We converted AC to pulsating DC using rectifier diodes. Immediately after the rectifier, we used a capacitive or inductive input filter.

The ripple frequency of a power supply is equal to, or twice that of the supply frequency. Typically, this is 50 or 100 Hertz. DC can be thought of as zero Hertz. So the aim of the filter of a power supply is to oppose frequencies above zero Hertz. A power supply filter is then a low pass filter. It passes DC (zero Hertz) and blocks or makes it harder for higher frequencies to pass.

We previously defined the property of *capacitance as that property of an electric circuit that opposes changes of voltage and inductance as that property of an electric circuit that opposes changes of current.* In a power supply filter, whether it is capacitive or inductive input, we are using the properties of capacitance and inductance to convert pulsating DC to smooth DC.

TUNED CIRCUITS

Capacitance and inductance combined, we have learnt, can form a tuned resonant circuit. The tuned circuits have a specific way of behaving at resonance and off-resonance. As a summary:

Series resonant LC circuits have low impedance at their resonant frequency and higher impedance at all other frequencies.
Parallel resonant LC circuits have high impedance at their resonant frequency and lower impedance at all other frequencies.

A tuned circuit *cannot* be used to block or pass just one single frequency. We would need an impossibly high 'Q' to do this. Tuned circuits can be designed to pass or block *signals* or a *range of frequencies*.

Using the properties of parallel tuned circuits and series tuned circuits and the fact that capacitive and inductive reactance change with frequency, we can build circuits to pass certain frequencies and stop others. Inductors and capacitors are ideal components for building filters. However, there are other specialist ways to build filters and this will be discussed.

TYPES OF FILTERS

Irrespective of the mechanism by which the filter does its job, filters can be broken down into four *basic* types.

a) low-pass filter
b) high-pass filter
c) band-pass filter
d) band-stop filter

A *Low-pass* filter passes all low frequencies up to a certain frequency; from that frequency and above significant attenuation begins to occur.

A *high-pass* filter passes all high frequencies down to a certain frequency; from that frequency and below significant attenuation begins to occur.

Band-pass filters pass a band or range of frequencies only; outside this band of frequencies significant attenuation occurs.

Band-stop filters stop or attenuate a band or range of frequencies only; outside this band of frequencies all frequencies are passed with minimal attenuation. Band-stop filters are sometimes referred to as *notch* filters.

SIGNIFICANT ATTENUATION

It is important for radio engineers to agree on the point where a filter is considered to have significant attenuation. The frequency at which significant attenuation begins to occur is called the cutoff frequency of the filter. The cutoff frequency is the point on the filter's characteristic curve where the desired frequency has been attenuated by 3 decibels. Take for example a typical low-pass filter with a cutoff frequency of 36MHz. This filter should have a very small insertion loss of say 0.2 decibels for all frequencies up to and around 34MHz. Beyond 34MHz, the filter will attenuate the signal more until at about 36MHz; the attenuation has reached 3dB. This 3dB cutoff point is considered to be the higher limit of the operating range of the low-pass filter. Above the cutoff frequency, the low-pass filter will begin to start attenuating signals dramatically.

Figures 1-32(a-d) shows the schematic diagram of the four basic types of filters.

The circuits shown are the basic features only of each type. It is very common for filters to use a combination of the four basic circuits to obtain the desired filtering effect. However, for your radio assessments, you should remember these basic types and the best way to remember to learn how they operate.

The low-pass filter is just like our power supply filter. The reactance of the inductor is low for low frequencies and increases as frequency increases. As frequency increases the reactance of the capacitor becomes lower and lower, effectively shorting out higher frequencies.

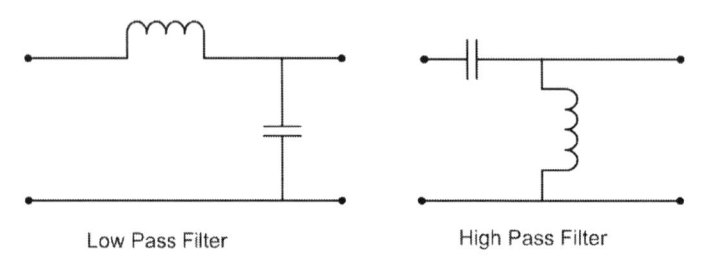

Low Pass Filter

High Pass Filter

Figure 1-32(a). **Figure 1-32(b).**

The basic high-pass filter just has the components reversed. X_L and X_C change with frequency causing higher frequencies to pass through this type of filter more easily than lower ones.

A band-pass filter could be made by combining the two filters above and this is sometimes done. However, a better method is to use tuned circuits. A series LC circuit will have low impedance to a narrow band of frequencies on or around its resonant frequency. Figure 2-32(a). Say we wanted the band-pass filter to pass 10.7MHz, +/- a bit of course. Then LC circuits would be tuned to 10.7 MHz. The series circuit would offer almost no impedance to a 10.7MHz signal since its impedance is very low at resonance. The impedance of the parallel LC circuit would be extremely high at 10.7MHz - in fact, so high it would be as if it were not there at all.

Not far away from 10.7MHz, the series LC circuits impedance would start increasing, making it harder for signals to flow through it. The impedance of the parallel LC circuit away from 10.7MHz would begin to drop causing it to begin shorting out (shunting) the signal.

The band-stop (notch) filter shown in figure 2-33(b) is exactly the same as the band-pass but with the tuned circuits swapped around.

You do not have to have *all* the elements shown. A lone series LC circuit 'in shunt' is a bandstop. A single inductor in series is a low pass. You can have more elements than I have shown. A high pass may have 5 series capacitors and 4 shunt inductors. A few get trapped with this in assessment so make sure you understand it or pester your trainer till you do.

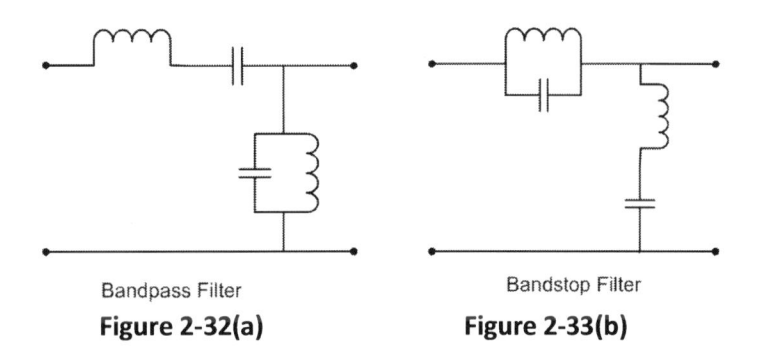

Bandpass Filter
Figure 2-32(a)

Bandstop Filter
Figure 2-33(b)

IDENTIFICATION of FILTERS

Identification of a filter is determined not only by its electrical function, that is, high-pass, low-pass, etc. but also by its physical shape when drawn in the schematic form. The components of a filter may form the letter 'L', 'T', or the shape of the Greek letter for Pi (π). Use of one of these letters is then made in the description of the filter. For example, low-pass filter 'L' type. For more complex designs letter designations are dropped. A further description of the filter may include the input reactance, such as low-pass, inductive input, L type.

The filters shown above will work as they are, however, more effective filters can be made by joining filter sections together. Each filter shown can be considered a filter section in a more complex filter design.

Combined filters.

The figure 3-32 circuit is a high quality combined high-pass-band stop filter that could be used on the back of a TV set to block the low HF transmission from an radio station and allow the higher VHF or UHF TV signals to pass. Although this is a more complex circuit, you should still be able to work out that it is a combined high-pass/bandstop filter. The series resonant circuits (L1,C1;L2,C2) provide for a bandstop on say 28 MHz. If C1 andC2 were to be removed, we would just have a high pass filter.

Combined Highpass - Bandstop

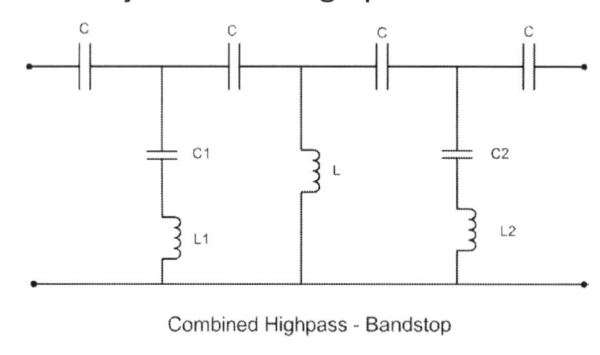

Figure 3-32

INSERTION LOSS

Ideally, on the band or range of frequencies that a filter is supposed to pass unimpeded, the loss should be zero dB. In practice, LC circuits have some insertion loss. An insertion loss of less than 0.5dB is good, though some more complex filters may have insertion losses as high as 2dB. Remember, insertion loss is undesirable attenuation to the *wanted* signal.

The best way to identify the function of a filter is to remember the behaviour of X_L and X_C and, impedance of series and parallel circuits. This is the method I much prefer you to use. Having said that, if you have trouble identifying the electrical function of a filter you may use the following straight memory method:

Try to remember this:
a) *Low-pass filters* - inductance in series and/or capacitance in shunt.
b) *High-pass filters* - capacitance in series and/or inductance in shunt.
c) *Band-pass filters* - series tuned circuits in series and/or parallel tuned circuits in shunt.
d) *Band-stop filters* - parallel tuned circuits in series and/or series tuned circuits in shunt.

CONSTANT K

If the values of inductance and capacitance are chosen correctly, any change of reactance of either one (due to a changing frequency) will be exactly offset by an equal and opposite change in the other. Therefore, the product of the reactances will always be constant. This product is known as the 'constant K.' This type of filter is called a Constant K.

CRYSTAL FILTER

We learnt in an earlier chapter that a quartz crystal has two resonant frequencies. A series resonant frequency and a parallel resonant frequency. A quartz crystal operating on its parallel resonant frequency behaves just like a parallel LC circuit; it has high impedance at this frequency and this frequency only. A quartz crystal operating on its series resonant frequency behaves just like a series LC circuit; it has low impedance at this frequency and this frequency only.

Figure 4-32 – ceramic filter

Figure 5-32.

The practical advantage of quartz crystal is that it has an extremely high Q and therefore makes an excellent high selectivity filter. They are far more expensive than LC circuits and cannot handle the same amount of power.

One of the most obvious applications of a crystal filter that we have discussed earlier is the crystal filter in an SSB receiver. This filter is a band-pass filter about 2.8kHz wide and very sharp (selective). It is asking a lot of a band-pass filter to remove a single sideband and reject the other - the best way to do this is with a quartz crystal filter. An arrangement of quartz crystals connected to form a filter is called a crystal lattice filter network.

The crystal lattice filter shown in figure 6-32 is a band-pass filter, suitable for use as a sideband filter in an SSB transmitter-receiver or transmitter. X1 and X2 are series resonant at the centre of the required sideband, while X3 and X4 are parallel resonant at the same frequency.

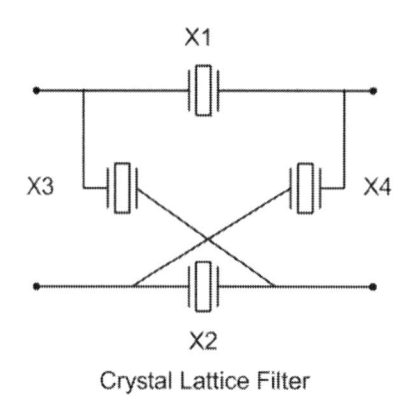

Figure 6-32. Crystal Lattice Filter

For a discussion of the two resonant frequencies of a quartz crystal, please refer to 'Chapter 27 – Oscillators' on this topic. In the diagram of the crystal filter (figure 6-32), X1 and X2 being series resonant, behave as a series LC circuit. They have low impedance to the desired signals and exhibit increasing opposition to signals removed from their series resonant frequency. Also, X3 and X4 being parallel resonant, behave like parallel tuned circuits in shunt with the signal path. A high impedance in shunt with the signal path will not affect signal frequencies, but will provide a low impedance path for signals away from the pass band centre frequency, effectively short-circuiting them.

Some ceramics also exhibit the piezoelectric effect (PE). A good example is the ceramic microphone which is a PE microphone just like the crystal microphone. (both are high impedance microphones). Ceramic filters are also be used in receivers along with crystal filters.

FILTER CURVES

A filter curve or selectivity curve is just a way of graphically describing how a filter works. These curves can actually be produced when testing a filter, using a spectrum analyser and a tracking generator. You will need to be able to identify each of the four basic filter curves.

The selectivity curve of a low-pass filter is shown in figure 7-32(a). This filter will pass all frequencies with minimum attenuation (the insertion loss of 0.5 dB) up to the cutoff frequency (fc), which for a typical HF transmitter would be around 35 MHz. Beyond the cutoff frequency, the filter begins to attenuate or block signals quickly. At about 50 MHz a low-pass filter of this type could attenuate signals by as much as 70dB.

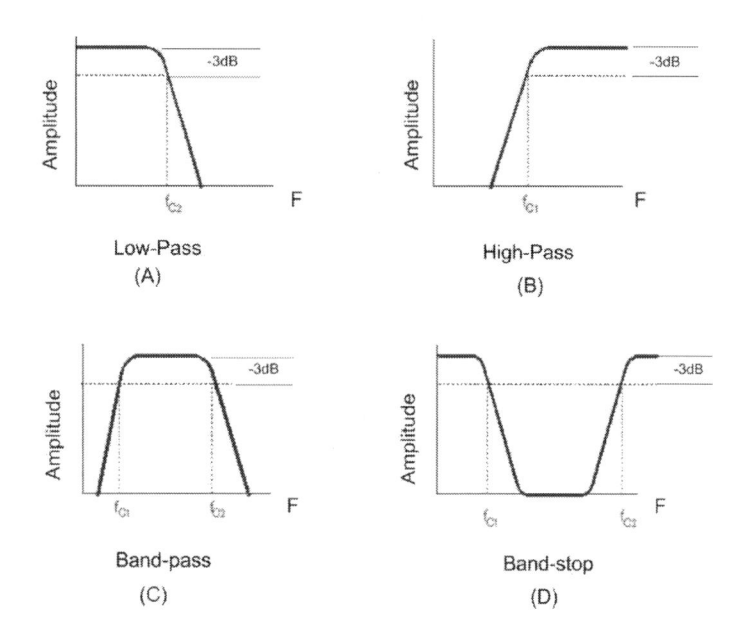

Figure 7-32.

It is a trap to assume that a low-pass filter is only used at low frequencies such as the HF radio band. A low-pass filter could have a cutoff frequency of 112 MHz for use with say an FM broadcast station. Such a filter would still pass all frequencies up to 112MHz.

The selectivity or characteristic curve of a high-pass filter is shown in figure 7-32(b). A typical high-pass filter may be used on the back of a TV set where the antenna leads connect to the set. The high-pass filter will pass the high TV signals (with the insertion loss of 0.5dB) and attenuate all frequencies below the cutoff (fc) to give the TV extra protection from say a nearby HF radio transmitter.

There are many other applications for a high-pass filter besides this one. The important thing is that you understand and can identify the curve. The output of the filter is high for frequencies above cutoff and low for frequencies below the cutoff (fc). I have not shown specific frequencies on the horizontal scale as the frequency range can be anything. In practice, the vertical axis would be marked in decibels and the horizontal axis in megahertz or kilohertz.

The band-pass curve in figure 7-32(c) shows that all frequencies between two cut-offs, f_{c1} and f_{c2}, will be passed. Signals outside that band will be significantly attenuated.

Most textbooks show band-pass filter curves nice and symmetrical. In practice, they often aren't, as my brilliant artwork illustrates!

The curve of a band-stop or notch filter in figure 7-32(d) looks a bit strange to begin with.

You may have a FM broadcast station nearby that is overloading one of your receivers. The FM broadcast band is from 88-108MHz. Therefore, a notch filter for FM broadcast would have an f1 of 88 MHz and an f2 of 108MHz - or at least somewhere in that band.

Again, it is important you can recognise the curve. I suppose I should point out that your examiner could get cunning and change the axis. For example, "Output" could be shown increasing downward, instead of as I have shown it.

PLACEMENT OF A LOW-PASS FILTER

Probably one of the most useful and most commonly found filters in HF radio stations is the low-pass HF filter. The idea of placing a low-pass filter at the output of an HF transmitter is to make doubly sure that all emissions (and they do exist at all stations) above the HF band are attenuated even further, to reduce greatly the potential for interference. From my time working in radiocommunications, I was often astonished at the controversy of where the low-pass filter should be placed.

There is an important rule with all filters, irrespective of their type and frequency range: filters are made with a specific input and output impedance which is usually the same. A filter could have an impedance of 50Ω or 300Ω to cite just two examples.

If you do not place a filter into a system that matches its impedance, then it will simply not work as it is intended. In a radio system, the degree of impedance match or mismatch is normally determined by measuring the VSWR, with a VSWR meter (to be discussed in full later). If you have 50-Ohm coaxial cable, it will only behave like 50Ω coaxial cable if the VSWR is 1:1. If you are going to insert a 50Ω low-pass filter into a 50Ω coaxial line, then there must not be significant VSWR on the line at the point of insertion. Otherwise, the filter may not work and can even be damaged by high voltages and cause interference.

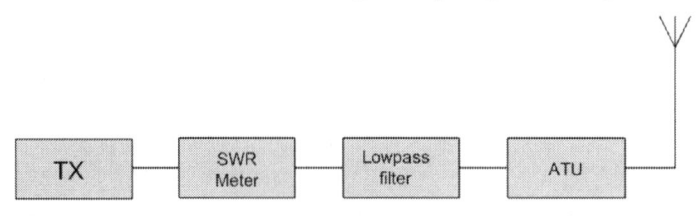

SWR meter and filter placement

Figure 8-32

In the diagram of an HF radio system in figure 8-32, the low-pass filter is placed last, but before the Antenna Tuning Unit. If there was no ATU, then the low-pass filter would be last, but the VSWR must be near to 1:1. In this arrangement, you should not swap the positions of the VSWR meter and the low-pass filter. A VSWR meter can actually produce harmonics (interference because of the diodes used) and these can only be stopped by a low-pass filter placed after it. So, a low-pass filter should be connected to a line that matches its impedance and it should be the last device in a transmission line with the exception of an ATU and antenna.

A MAINS FILTER

A 240-Volt household power line filter is more commonly called a mains filter. The purpose of a mains filter is usually to prevent radio frequency energy from going into the mains network and causing interference to other devices in your own or to nearby households. Alternatively, a mains filter can be used to protect a device from radio frequency interference coming in via the mains.

Mains Filter

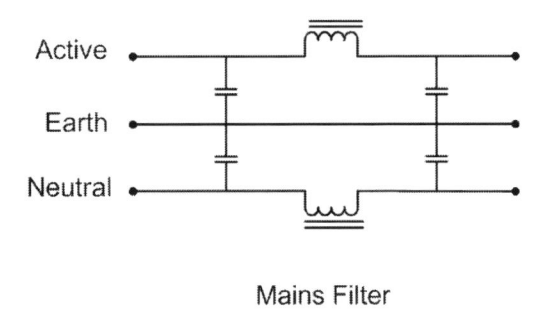

Figure 9-32.

If you can imagine your house being invisible with only the electrical wiring remaining in place, then you can start to imagine the large antenna system your household wiring is.

The diagram of the mains filter is a configuration called double Pi type. A mains filter is a low-pass filter, passing the mains frequency and blocking the higher radio frequencies. Notice the iron core inductors - these are a good indicator of large inductance and low-frequency operation. Ceramic capacitors are normally used to obtain the necessary high voltage rating, which is typically 2kV. The high voltage rating is necessary to prevent damage from possible high voltage line transients (voltage spikes). This type of filter could be placed in the mains lead of a TV receiver, to prevent RF energy entering the receiver after being picked up by the household wiring. The filter may also be useful in preventing power tools from interfering with radio and television reception, provided the interference is conducted from the power tool to the affected device via the mains lead. A mains filter will not be of assistance where the interference is being radiated and *not* conducted by household wiring.

CAVITY FILTERS

Cavity filters are used a lot in radiocommunications. Radio repeater stations nearly always use cavity filters. Cavity filters can be made to act as band-pass or band-stop (notch) filters.

The advantage of cavity filters is that they are very high Q, but not as high as a crystal filter. However, they have the advantage of being able to handle high power levels. You will find cavity filters, or combinations of them, at the output of receivers and transmitters, used as repeater station or in high RF environments.

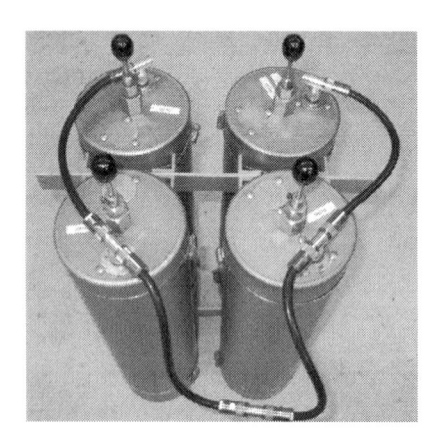

Figure 10-32

A repeater station is a receiver connected to a transmitter. Whatever is received by the receiver is retransmitted on the transmitter, relaying the radio operators message. The repeater is usually placed in a prime location to give extended coverage and may run at high power. The problem is keeping the repeaters transmitter from interfering with its own receiver. The transmit and receive frequencies may only be 600kHz apart. A combination of band-pass and notch filters made from cavities can easily achieve the amount of receiver-transmitter isolation (in decibels). By the way, the receiver and transmitter of a repeater often share the one antenna, though they do not have to.

Cavity filters are rarely used below 30 MHz as below this frequency they are physically too large. The arrangement of 4 cavity filters shown in figure 10-32 is called a duplexer. A duplexer is just a system of cavities comprising of notch and band-pass filters, to enable a transmitter and receiver on close frequencies to share a single antenna.

The cavities shown in figure 10-32 are about 300mm in diameter and about 700mm long and operate in the VHF band. The most common shape of a cavity filter is that of a cylinder as shown, however any shape can be used such as squares or rectangles.

Imagine for a moment a parallel tuned circuit with a conventional L and C. Figure 11)-32. Now remove the L and replace with a single wire. That single wire has some inductance – though not much and you would expect the resonant frequency to be very high. What if you paralleled more straight wire inductors across the capacitor? Well, inductances in parallel cause the inductance to decrease so the resonant frequency increases. Now get carried away and place some many straight wire L's in parallel with the capacitor that you form a solid cylinder around the capacitor. You now have a cavity resonator. All you need is coupling loops and a method to adjust the internal C to tune the cavity.

Cavity Filter

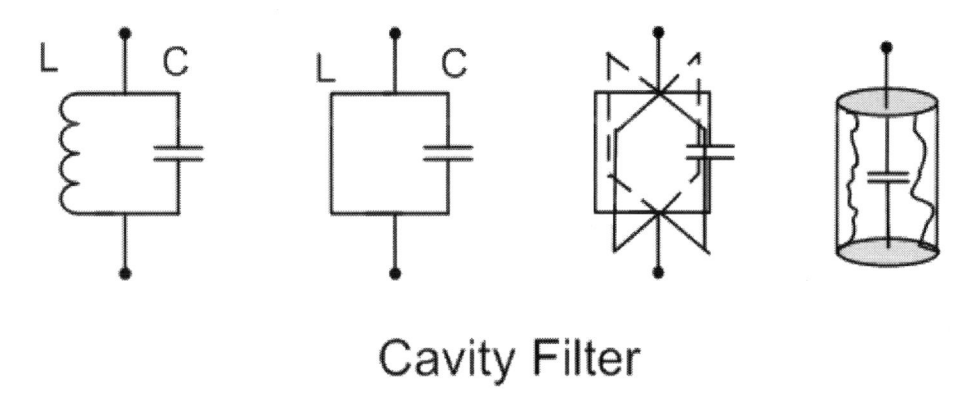

Figure 11-32

REPEATER DUPLEXER

A repeater duplexer is shown in figure 12-32. The purpose of the duplexer is to prevent the repeaters transmit signal from overloading, desensitising or blocking the repeaters receiver. This is called *transmitter-receiver isolation* and is measured in decibels. The transmitter has two cavities to allow the transmitter frequency signal to pass through with minimum attenuation and cause all other signals to be attenuated. Similarly, the receiver has two cavities that pass the received signal and block everything else.

This duplexer is a fairly simple one. How much filtering is required depends on the RF environment where the repeater is located. If it is a multi-transmitter site, then more cavity filters may be required and a device called and *isolator* may need to be installed to stop other transmitters getting into the repeaters transmitter and causing transmitter intermodulation. Likewise, on a multi-transmitter site the are other transmitters that may get through the duplex and cause receiver intermodulation interference.

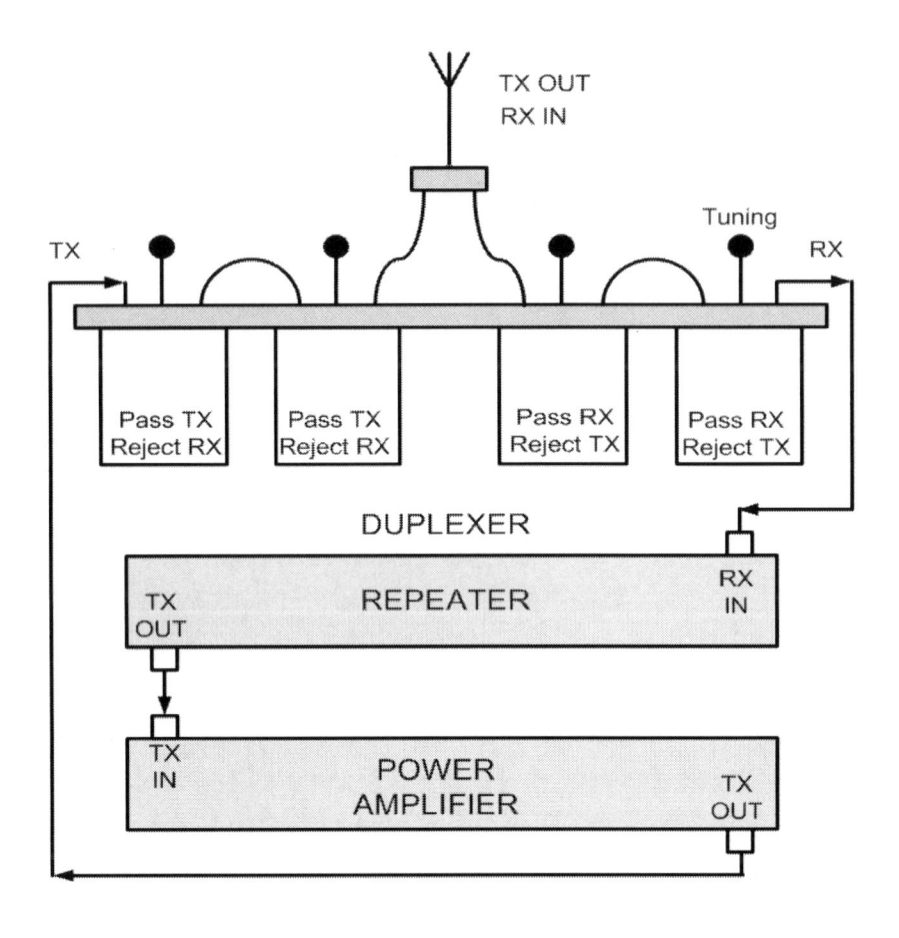

Figure 12-32

There are many ways to tune a duplexer using simple equipment. The best way to tune cavities in a duplexer is to use a spectrum analyser with a tracking generator. This produces an easy to understand graphical display of the bandpass and bandstop characteristics of the duplexer.

The duplexer will introduce an insertion loss of about 0.5dB. Increasing the insertion loss is sometimes done deliberately to achieve a sharper bandstop and bandpass characteristics. It is a bit of a trade-off between insertion loss and duplexer characteristics can be chosen considering the RF environment where the repeater is located.

Another method that help creates better isolation between the transmitter and receiver is to use two antennas rather the one. If this method is used an extra 20 or so dB isolation can be achieved with cautious mounting of the two antennas usually with vertical separation.

CIRCULATORS (Ferrite) – an introduction.

Figure 13-32

A circulator is a three terminal device. (Figure 13-32). It has three ports. Inside there is a ferrite material and magnets. The magnets are usually adjustable to tune the circulator. A signal going into port 1 (see photo) will exit on port 2. A signal going into port 2 will exit on port 3. A signal going into port 3 will exit on Port 1. That is why it is called a circulator. It works this way because the ferrite under the influence of a magnetic field changes the polarisation, velocity and phase of the electromagnetic wave at the junction of what is in effect three transmission lines.

It might be strange for you to think about transmitters receiving signals from other transmitters. Why not? Transmitters have antennas – those antennas receive signals from other transmitters and funnel their energy into the output stage of the transmitter. The transmitter you will recall, often has a non-linear output stage. The external signals from other transmitters will enter our transmitter and mix with our transmitter signal in the power amplifier and re-radiation will occur. The result it a great mess of unwanted signals *very likely* to cause interference. This interference is called *transmitter-intermodulation*.

Now if we are wanting more isolation between our repeaters transmitter and other transmitters we can use a circulator to gain an extra 40-70dB of isolation. That is a lot of isolation but it comes at the price of an expensive circulator.

To convert the *circulator* to an *isolator* all we need to is terminate port 3 into a small (10-20W) dummy load.

Our transmitter is connected to port 1. A 50Ω load is connected to port 3. Our antenna is connected to port 2. Our transmitted signal enters port 1 and exits port 2 to the antenna. Any signals coming down our antenna (from other transmitters) will enter port 2 and exit port 3 and be dissipated in the 50Ω load. What we have achieved is isolation between our transmitter and all other transmitters on the same or nearby radio site. Dual isolators, two singles isolators connected end to end; can achieve 70dB of isolation.

STRIPLINE FILTERS

At UHF frequencies and above, normal filter components like capacitors and inductors would become too small physically to be stable. From UHF and beyond a common filter used inside communications equipment is the stripline. I am not going to go into the operation of these filters here as their principle of operation requires some knowledge of transmission line theory, as this is how these filters operate. Suffice to know for now that ordinary capacitors and inductors cause problems in filter construction for UHF frequencies and above and an alternative is the stripline filter, with its operation to be covered in the chapter on transmission lines.

See chapter 38 (Transmission lines) for more information on how stripline filters work. Striplines are covered in transmission lines because they use transmission line principles for their operation.

Radio & TV broadcast filters

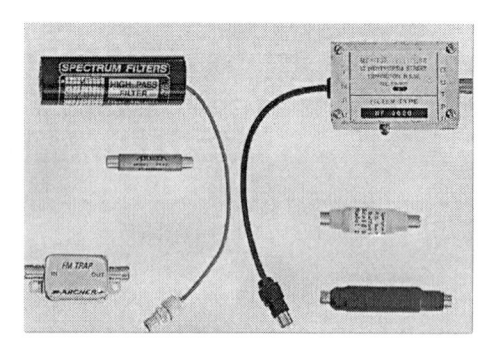

Figure 14-32

Figures 14-32 is a number of filters useful to help prevent or cure interference (mostly overload) to TV and broadcast band receivers.

A high pass filter suitable for a television receiver will cure the majority of interference problems between an HF (0-30MHz) radio station and a TV set which is being overloaded. Good filters of this type can be expensive – read later how a very cheap transmission line notch will often quickly, simply and cheaply do the same job.

33. INTERFERENCE

As a radio experimenter, you are expected to be able to identify interference, particularly to television, radio and domestic electronics. Most of the time when interference occurs it is due to some fault in a receiver or a deficiency in the electromagnetic immunity of an appliance. Though as a licensed radio operator you will no doubt get the blame for everything, because if you were not there with your radio, then the problem would not exist. I have also heard: "we didn't have the problem until you moved in." Your neighbour, being non-technical, will not appreciate being told by you that the problem is theirs even if it really is. **The very best thing that you can do for yourself is to ensure that you have no interference whatsoever in your own home**. You need to be very cautious in assisting a neighbour with their interference problem. If you do not know what to do or if you just do not want to risk touching expensive equipment in your neighbours home is to tell them to contact the radio licensing authority; just make sure your own house is in order first.

Since the introduction of digital TV much of this chapter is not applicable. The syllabus has not been updated for Digital TV. With analogue TV, the screen is a diagnostic tool; this is not the case with DTV. I will cover analogue TV interference since it is still in the syllabus and then we have to go into slightly more of an investigative approach for DTV.

TELEVISION AND OTHER INTERFERENCE

Listed below are the terms you should understand. Some are an interference type, while others are the results that interference produces.

Cross hatching
Sound bars
Herringbone pattern

Ghosting
Power line interference
Receiver overload
IF Interference
Harmonic interference
Audio rectification
Co-Channel interference
Alternator Whine

Several types of the most common analogue TV interference are shown in figure 1-33. Some of these images are taken from the archived and now publicly available in electronic version.

Figure 1-33(a) is a normal analogue television picture without interference.
Figure 1-33(b) is a normal picture with no interference and weak signal strength (snow).

Figure 1-33 (c) is interference from an AM radio transmitter. This interference pattern is referred to as a "herringbone weave" because there are alternate bands with diagonal lines in them sloping the opposite way. If you use your imagination a little it is a bit like a herringbone of a fish. This interference could be from any modulated AM transmitter and there would likely be some interference to the sound but this may not be intelligible.

Figure 1-33(d) Is interference from a modulated FM transmitter. The interference lines may "wiggle" with modulation. Sound may be affected and if it is it my very likely be intelligible. The width of the interference lines become finer as the frequency as the frequency of the source FM transmitters increase. With practice an interference investigator can make an estimate of the transmitter's frequency.

Figure 1-33(e) is electrical interference. This could be from almost any electrical device and usually close by. Perhaps in the home of the viewer. If the horizontal bands of interference are in two distinct bands the most likely source is powerline interference. The two bands of lines are caused by flash over of the insulators twice for each cycle of AC. If the interference comes and goes with some sort of routine, then suspect devices the is controlled by a thermostat. Most types of electrical interference (except powerlines) is easy to locate by turning the power circuits off and using a process of elimination.

Figure 1-33(f) is co-channel interference from another transmitter that could be far away. It is a bit like ghosting but the displaced image is not the same as the main image.

Radio Theory Handbook

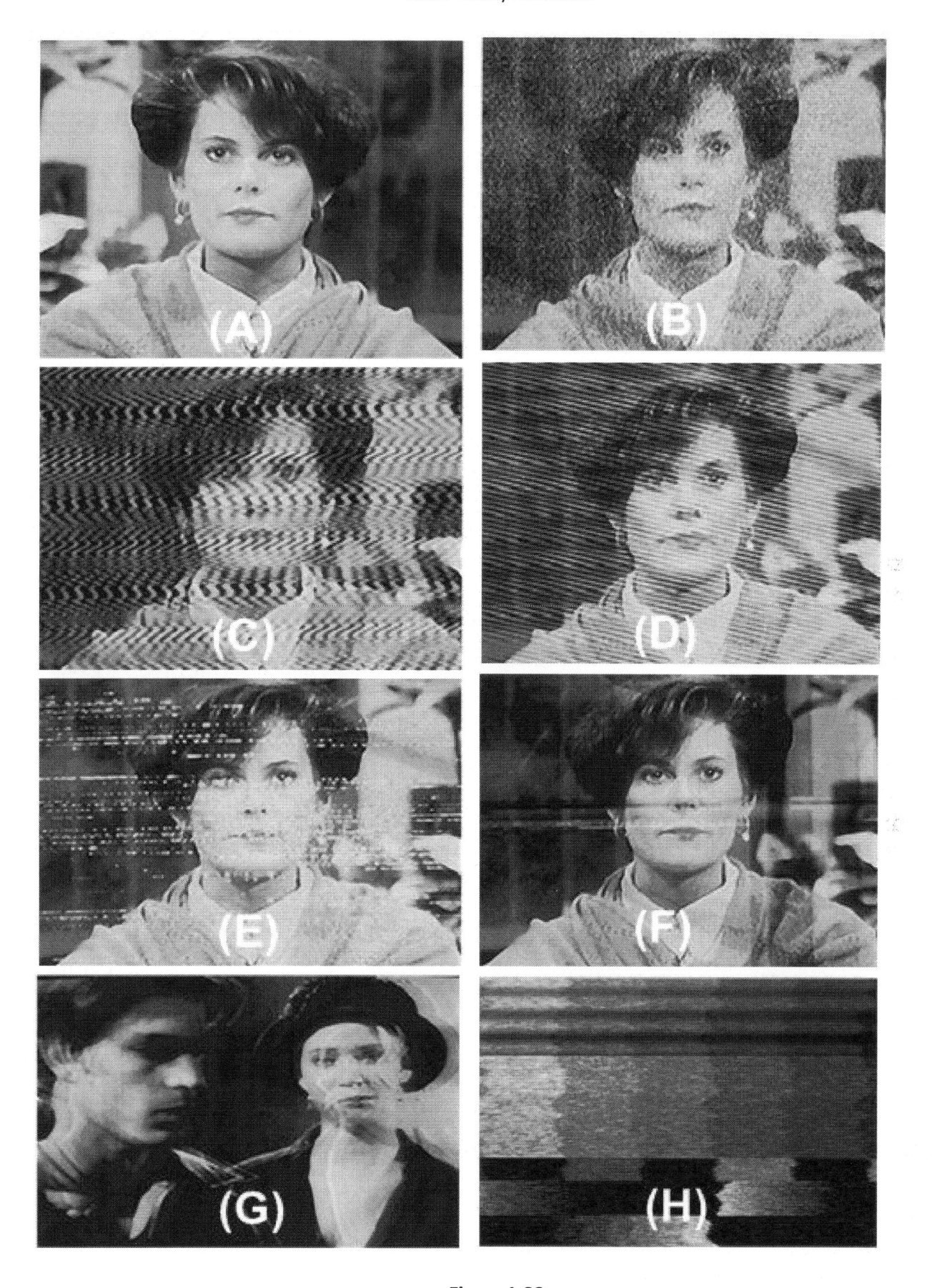

Figure 1-33

Figure 1-33(g) is called ghosting. The main image appears twice and shifted in distance. This is nearly always cause by multipath signals.

That is the same signal arriving at the TV receiver via more than one path. This can be reflection from buildings, aircraft or even mountain ranges. Aircraft in the vicinity can cause temporary ghosting. The best approach for this is to improve the directivity of the TV antenna and experiment with its direction. Ghosting can because by standing waves of a faulty TV antenna system.

Figure 1-33(h) is called "sound bars". This is usually interference from an RF transmitter that is very close and being overmodualted.

These types of interference are becoming far less with the move to DTV. They will still occur but you will not have the picture on a cathode ray tube to use as a diagnostic tool.

BE DIPLOMATIC AND CAUTIOUS

There was a time when I would have recommended that a radio experimenter intervene and do all they could to help a neighbour with the interference. Install filters and the like. These days' home entertainment equipment is far more complicated and expensive. I would be reluctant to touch it. What if you do install a filter and the appliance develops an unrelated fault a few days or weeks later? Well, it was fine before you touched it!

However, let's be reasonable about this. If you have an amateur station and are running something of the order of +50dBm to +56dBm of output power (100-400 Watts) and you place your transmitting antenna on the same property boundary and pointing at, or a few metres away from, your neighbour's TV antenna. Then, even if your neighbour's TV antenna is correctly installed and has filters connected, there is still going to be an overload problem and *morally* it is your problem, though *technically* the problem lies with the TV receiver being unable to reject such a strong nearby off-channel transmission.

If cooperation is not forthcoming, particularly from the amateur station, then the inspector can change the license conditions of the amateur on the spot so that the station cannot be used above a certain power level. If your antenna is rotatable and interference only occurs when the antenna is pointed at the TV antenna, then the inspector could give you instructions not to transmit in this direction. In extreme cases, the inspector could direct you to transmit during particular hours of the day.

Usually, sensible placement of antennas, consideration for neighbours and good communications skills eliminates all problems even if the help of an inspector is required. Keep in mind also that if your neighbour does call an inspector to investigate the problem and the problem is found to be the neighbour's TV antenna, feedline, or TV itself, then they will be paying for the inspector's advice. How the regulator will respond to you and interference depends on the regulations of your country.

SOME OTHER TYPES OF INTERFERENCE

HARMONIC INTERFERENCE

Every radio transmitter emits some energy on multiples of its operating frequency. These unwanted frequencies on multiples of the operating frequency are called Harmonics. It is not a matter of if and when harmonics occur; it's a matter of what level the harmonics are. It is no accident that most of the radio experimenter's bands are harmonically related to each other. This is so that harmonics generated by experimenters will most probably fall on other experimenter bands. An interesting experiment is to search deliberately for harmonics with a receiver, whether they are yours or that of another station or even a commercial broadcaster.

Harmonic interference is a transmitter problem and should be cured at the transmitter with filtering - usually a low-pass filter.

An example of harmonic interference:

An amateur transmitter is operating on 28MHz and only causes interference to a TV receiver when it is tuned to Channel 9 (196.25MHz Vision). What is the most likely cause of this interference? What cure would you suggest?

Since the interference is to Channel VHF 9 only, the most likely cause is in-band interference. That is, energy is being radiated from the amateur transmitter within the band of frequencies used by Channel 9. Channel 9's vision carrier is 196.25MHz, the seventh harmonic of an amateur transmitter operating in the 28MHz band of frequencies will fall within the channel allocation for Channel 9. Since harmonic radiation is the problem, the only solution is to install a low-pass filter at the transmitter.

The clue given in the question was that interference was to *one* Channel only. If fundamental overload, intermediate frequency (IF), or audio rectification were the cause, then interference would generally occur to *all* channels (including unoccupied channels). Let's have a look at these types of interference now.

FUNDAMENTAL OVERLOAD

Fundamental overload is not a transmitter problem. When a transmitter is producing fundamental frequency overload, that is, receiver overload on another receiver not tuned to the transmitters frequency, then it is a receiver problem. However, when speaking about the problem from the transmitting stations point of view, we call it fundamental overload. Provided you are operating within the prescribed power levels you are not at fault, though do read the advice given on this matter earlier in this chapter. The majority of interference, particularly to TV, is from fundamental overload. Fortunately fitting a high pass filter to the TV receiver usually cures this problem easily.

Figure 2-33

A high pass filter installed on the back of a TV receiver blocks (attenuates) the HF radio transmissions and allows the VHF/UHF TV signals to pass virtually unattenuated.

IF INTERFERENCE

When a transmitter causes interference to an IF stage of a radio or TV receiver the interference will occur without change no matter what channel or frequency the affected receiver is tuned to. The Interference is to an IF stage. One typical IF frequency is 10.7MHz. Generally, again this is a receiver problem and is normally due to lack of receiver shielding. Don't laugh! To prove IF overload, I have had to wrap receivers and TV sets in aluminium foil to prove to the owner or their TV technician that the IF is insufficiently shielded.

AUDIO RECTIFICATION

Many household appliances contain audio amplifiers. TV's, radios, sound systems and intercoms to name just a few. Bedside clock radios send a tremor down my spine! Now, audio amplifiers are not supposed to pick up radio signals are they? Well, if enough RF energy does get into any audio amplifier, the AF amplifier will become overloaded, become non-linear and start acting like a demodulator for radio frequency signals.

The problem is with the audio equipment for sure. Sound systems often use cheap figure eight cable for the speaker wires and there are usually bundles of it shoved behind the system. Fortunately, the fix is usually very easy for sound systems. Winding the excess speaker cable around a ferrite rod, say 10mm diameter and 250mm long, will usually get rid of the problem. You are just adding inductive reactance to the audio line, which hardly affects the audio performance at all, but offers a high reactance to high RF currents.

TV, radios and the like require a technician to install internal RF bypassing onto the audio amplifiers input. This is usually fairly simple. Clock radios are not worth the trouble, bin them. Of course, you should eliminate mains interference first.

TELEPHONE INTERFERENCE

Telephones are not supposed to pick up radio signals, but sometimes they do. Usually, a simple 0.5μF or higher capacitor across the line will fix the problem - the line is usually a blue/white and/or red/black cable - better still, call the Telephone company and complain to them.

INDUSTRIAL HEATING

Industrial machines, of which there are many, can cause severe interference. Such interference is non-selective and can happen to anyone. Report it to the radio licensing authority in your country or track it down and advise the owner of the machine about the problem.

This can be extremely costly interference to fix and the only way is to advise your licensing or regulatory authority who investigate large numbers of interference problems for the public, in most cases for no charge. If this type of interference occurs it will affect a large number of households and commercial radio services.

ALTERNATOR WHINE

A mobile radio station experiences a variable pitch whining noise in the receiver. What is the likely source?

Of course, I can hear you say, the alternator (AC generator) in the car. The diodes in the alternator can cause this interference. The diodes are needed to convert the AC to DC to charge the car battery. The varying pitch is due to the changing RPM of the car engine.

This interference is usually eliminated by installing a 1µF or larger coaxial capacitor at the alternator output terminal.

Alternator whine can also be transmitted and not heard. If other radio stations report to you a variable pitch whining noise on your signal, do something about it, it is most irritating to listen to.

INTERFERENCE TO DIGITAL SERVICES

Diagnosing the cause of interference for Digital TV, Radio and other things like modems is more difficult as the nature of the interference on the device does not tell you very much at all.

Suppose you or your neighbour have interference to Digital TV reception. Where do you start? The onscreen interference is pixilation or picture drop out.

The best place to start is to try and determine if the cause of interference is the DTV's receiver or antenna system. All DTV's have a user option to check the level of the TV signal to the receiver.

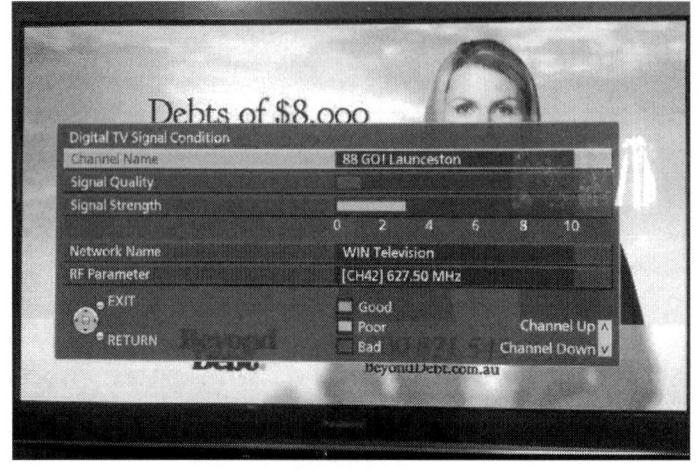

Figure 3-33

Access the menu system and check the signal strength. It won't give you and absolute reading in dBuV usually, but it will tell you if it is good, fair or bad. If it is not good, then go no further, make it good. Look at what could be wrong with the feedline, antenna, connectors or any amplifiers inline.

DTV WEAK SIGNAL

Make sure the antenna is of the right type and pointed in the correct direction. Many people install an antenna going by their neighbour's installation. This is usually okay but sometimes it is the blind leading the blind and they all end up pointing the wrong way. Better is to move the antenna while looking at the signal meter on the DTV.

Check the coax cable. All the connectors and inline amplifiers. Masthead amplifiers need a DC supply from a small plug pack inside the house. Check the DC output – if you can to make sure that DC is getting to the masthead amplifier. Make sure all the DTV connectors are properly 'home' in their sockets. The plugs and sockets oxidise over time clean them up with a bit of emery or light sandpaper.

Internal amplifiers need a DC supply as well. Check the signal strength with the amplifier in and out of the circuit. There should be a large difference if the amplifier is not working. A faulty amplifier will act as an attenuator, that is, the signal will be better without it.

One of the problems with mast head amplifiers is they can oscillate. If they do you will have pixelation or no picture and it is very likely, your neighbours will have it from your oscillating masthead amplifier. To determine is a masthead or another amplifier is oscillating switch off the DC power while watching the interference on another system.

So you have checked everything about the DTV system, antenna, feedline, connectors, amplifiers and signal strength are all good. So we can rule out a problem with the receiver.

NATURE OF THE INTERFERENCE

Now you have to become a bit more of a detective.

When does the interference occur? All the time? At night or during the day or at certain times of the day or night. Look for anything repetitious about the interference. There are many things in your home and neighbours homes that can cause radio interference in a cyclic manner.

Refrigerators switch ON and OFF related to the ambient temperature.
Butter conditioners switch ON and OFF.
Electric blankets are usually switched ON and OFF by the user at certain times of the day.

Hot water systems
Solar systems and their inverters
Alarm systems
Automatic lighting – in particular, some LED drivers can cause interference.
Anything with a thermostat.
Power tools.
Pool chlorinators.
The home and workshop are just full of devices that can produce radio frequency interference (RFI)

ELIMINATE POTENTIAL SOURCES

The best approach is to power down all devices. You can do this one by one, but my preferred approach is to power down all electrical circuits in the switchboard accept of course the circuit that the DTV is ON. If there are other appliances on the same electrical circuit as the DTV, then power them all down. Preferably switch OFF the mains power. Some devices still operate when their onboard power switch is turned OFF. Don't forget external systems such as Solar Power, Alarms, Pumps and Chlorinators, etc.

If the interference disappears when you do this, then you're almost home. Once you have identified the source of the interference, you can then take appropriate steps to fix or replace it. Only qualified persons should work on mains electrical appliances.

If it is your Radio Station, then everything that applies to analogue TV applies to DTV. If it is overload when operating on HF, look at placing high pass filters on any VHF/UHF device or amplifiers used by them.

I have no interference, but I have RF chokes wound on ferrite rod everywhere. They are cheap, quick and easy to do. I have RF chokes on my router telephone lead and Ethernet. Wind the leads at the router around a ferrite rod and tape or use cable ties to keep in place. Do the same on all power supply leads, external speakers, coaxial cables, sound systems. Ten such rods do not cost much and RF chokes are your best friend.

If the interference is external to your radio station. You may have to canvass neighbours and ask if they are getting it as well and work from there.

If the interference is to your Radio Station then if possible, run the affected receiver from battery power and switch off your mains. You then eliminate your entire property in one hit. If the interference is external, you may have to use a portable receiver and even a directional antenna to locate it. One thing to remember about external sources if they are broad bandwidth (which most are) is that the higher you go in frequency and can still receive it then closer you are to the source.

Ask for help from experienced radio users. If the external interference is to yours and your neighbour's domestic TV and Radio devices, then your regulatory authority would generally investigate the problem particularly if it is affecting a large number of people.

34. FREQUENCY MODULATION

WHAT IS FREQUENCY MODULATION?

Before defining Frequency Modulation, let's just go over again what we mean by 'modulation' as applied to radio communications. Modulation is the process of converting low-frequency signals into high-frequency signals so that they can be radiated and recovered at a remote location. Usually, the 'intelligence' we are trying to transmit is voice, but of course, can be other things such as encoded video or data. With amplitude modulation, we combine the modulating audio with the radio frequency carrier in such a way that the *total wave power* of the transmitted wave is made to vary. Remember, the *total* wave power varies. The RF *carrier power* in an AM system remains constant.

Now besides using the modulating audio to change the total power of the wave being transmitted, we can make the modulating audio change the frequency of the RF carrier.

FM is a system by which the modulating audio is made to change the instantaneous frequency of the carrier.

Popular belief is that in an AM signal the carrier power varies and in an FM signal the carrier power remains constant. This is incorrect. In an AM system the total wave power varies with modulation and the carrier power remains constant and in an FM system the total power transmitted remains constant and the power in the carrier varies.

The simplest FM transmitter is where the modulating audio is fed directly to an oscillator and the voltage level of the modulating audio causes the frequency of the oscillator to vary.

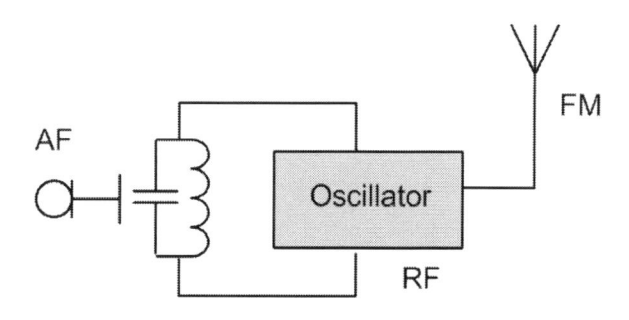

Figure 1-34

A SIMPLE FM TRANSMITTER

Modulating audio could be fed to a capacitor microphone that forms part of an LC circuit that in turn controls the operating frequency of an oscillator. Hence, the voltage of the modulating audio controls the frequency of the oscillator. This signal could then be amplified and transmitted to the receiver. At the receiver, we would need a method of converting RF frequency changes back into audio.

BACK TO THE VOLTAGE CONTROLLED OSCILLATOR (VCO)

In an earlier chapter, we discussed how voltage could be used to control the frequency of an oscillator. We did this by using a varactor diode. The reverse bias on the diode changed its capacitance and in turn that capacitance change was used to make small changes in the operating frequency of a quartz crystal. To jog your memory, have a look at the circuit of the VCO again, shown in figure 2-34.

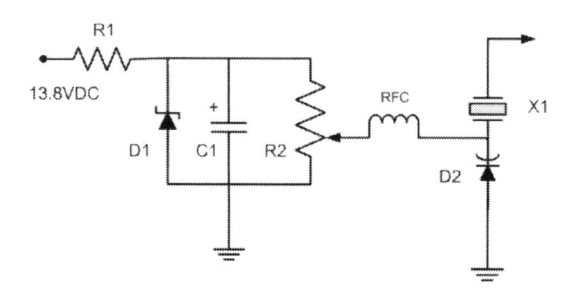

Figure 2-34

The potentiometer (a variable resistor with a centre tap) can be adjusted to bring about small changes in the frequency of the quartz crystal. Everything to the left of the radio frequency choke (RFC) in the circuit provides a regulated DC voltage using a Zener diode.

Now, what if we did away with the Zener diode and applied (modulating) audio to the varactor diode. We would have a method of producing FM.

THE REACTANCE MODULATOR

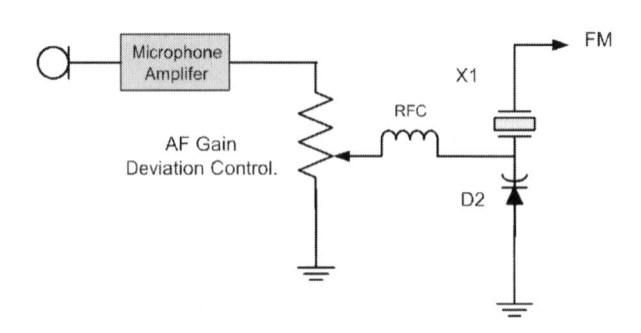

Figure 3-34

It is that easy to produce Frequency Modulation. Simply feed the modulating audio to a varactor diode and the diode will bring about frequency deviations of the quartz crystal proportional to the modulating audio amplitude.

I have not drawn the complete circuit of the oscillator, as it could be any type. As you can see, the modulating audio is used to change the reverse bias on the varactor diode, hence changing its capacitance and in turn the frequency of the oscillator.

The RFC (radio frequency choke) is just to keep RF out of the AF circuit.

Just to make sure you have got exactly what we are on about here, let's go over it. Without any modulating audio, the reactance modulator (which is just a fancy voltage controlled oscillator) will produce RF energy on one single frequency only. This is shown below in as the 'carrier' figure 4.34 - the frequency is irrelevant, but it would be high, at least in the order of a few megahertz. This is called the *centre frequency*.

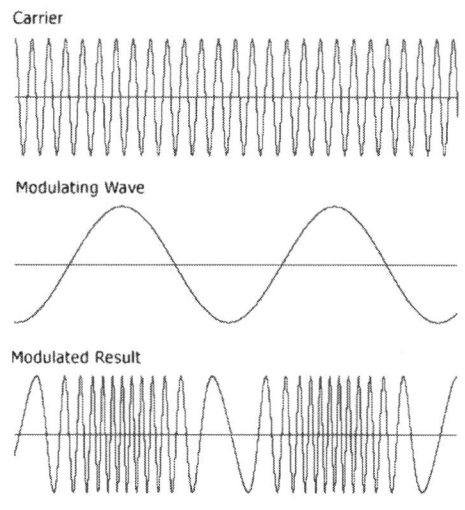

Figure 4-34

Then we have the audio frequency coming from the microphone, which in the case of radio voice communication would be somewhere between 100Hz and 3kHz maximum.

Now, this modulating audio is applied to the varactor diode in the reactance modulator causing the frequency of the quartz crystal to change up and down from the centre frequency. So we end up with a frequency-modulated wave as depicted at the bottom of figure 4.34. *The total wave power does not change, the frequency does.*

These graphs could be shown on a device called an oscilloscope. The oscilloscope clearly shows the amplitude (power) of the FM wave does not change, but it is evident that the average power of the centre frequency does change as it is not there all the time.

Frequency modulation is a method of modulating a carrier wave whereby the modulating audio causes the frequency of the carrier to change.

Major E.H. Armstrong developed the first working FM system in 1936 and in July 1939 began the first regular FM broadcasts from Alpine, New Jersey.

The capacitance of the diode in the reactance modulator has a direct effect on the oscillators frequency. The following important conclusions can be made concerning the output frequency:

1. The frequency of the sound waves arriving at the diaphragm of the microphone determines the *rate* of frequency change.
2. The amplitude of sound waves arriving at the diaphragm determines the *amount* of frequency change.

The previous diagrams illustrate the fundamental differences between an FM and AM wave. If the sound waves striking the diaphragm was to be doubled in frequency from 1kHz to 2kHz with the amplitude kept constant, the *rate* at which the FM wave swings above and below the carrier would change from 1 kHz to 2kHz. However, since the intelligence (amplitude and frequency) of the sound wave has not changed, the *amount* of frequency swings above and below the carrier frequency will remain the same.

On the other hand, if the frequency of the modulating audio was constant but the amplitude was doubled, then the rate of frequency swing above and below the carrier would remain at 1kHz and the amount of *frequency deviation* (movement from the centre frequency) would double.

DEVIATION OF AN FM WAVE

Without modulation, an FM transmitter produces a single carrier frequency (fc). When modulation is applied, the output frequency will swing above and below the unmodulated carrier frequency (fc). The amount of decrease or increase around *fc* is called the *frequency deviation*.

MODULATION INDEX OF AN FM WAVE

The modulation index is the ratio of the amount of frequency deviation to the audio modulating frequency. Suppose an FM *broadcast* transmitter has a deviation of 75kHz when a modulating audio frequency of 15kHz is applied to the microphone input, then:

modulation index (μ) = deviation / modulating frequency
modulation index (μ) = 75/15=5.

The modulation index has no unit. In this text, the modulation index will be denoted by the Greek letter (μ). Sometimes the modulation index is denoted by the letters (mf).

There is a *special case* concerning modulation index and this is when the frequency of the modulating audio is at its *highest*. In such a case the modulation index is called the *deviation ratio*. For narrow band voice FM transmissions, the deviation is restricted by regulation to 5kHz. However, this may differ in some countries. Also some regulatory authorities are loosening up the rules on occupied bandwidth. Since the highest modulating audio used for voice communications is 3kHz, the deviation ratio for an amateur FM transmitter is then:

Deviation / highest modulating frequency = 5/3 =1.66.

The modulation index is known as the deviation ratio when the highest modulating audio frequency is used. **Radio experimenters do not generally use 'deviation ratio' they would simple refer the the 1.66 as the modulation index.**

SIDE FREQUENCIES OF AN FM WAVE AND MODULATION INDEX

When a carrier (fc) is frequency modulated by a signal (fs), then theoretically an infinite number of side frequencies will be produced on each side of the carrier.

The *distance* between each of the side frequencies is equal to the frequency of the modulating signal (fs).

The *amplitude* of each side frequency does not follow any simple pattern and is dependent upon the modulation index (μ) or deviation ratio.

At a certain distance from the carrier (fc) the power in the side frequencies will decrease to a level where they can be considered insignificant. A side frequency is considered insignificant if its voltage amplitude is less than 10% of the unmodulated carrier amplitude. The number of significant side frequencies is directly proportional to the modulation index and can be found from significant side frequencies = 2 (μ + 1) where μ = modulation index. To determine the relative amplitude of each of the side frequencies, it is necessary to consult a table of *Bessel functions of the first kind*.

Mf	J0	J1	J2	J3	J4	J5	J6	J7	J8	J9	J10	J11	J12	J13	J14
0.000	1.00														
0.250	0.98	0.12	0.01												
0.500	0.94	0.24	0.03												
1.000	0.77	0.44	0.11	0.02											
1.500	0.51	0.56	0.23	0.06	0.01										
2.000	0.22	0.58	0.35	0.13	0.03	0.01									
2.405	0.00	0.52	0.43	0.20	0.06	0.02									
3.000	-0.26	0.34	0.49	0.31	0.13	0.04	0.01								
4.000	-0.40	-0.07	0.36	0.43	0.28	0.13	0.05	0.02							
5.000	-0.18	-0.33	0.05	0.36	0.39	0.26	0.13	0.05	0.02	0.01					
7.000	0.30	0.00	-0.30	-0.17	0.16	0.35	0..34	0.23	0.13	0.06	0.02	0.01			
10.00	-0.25	0.04	0.25	0.06	-0.22	-0.23	-0.01	0.22	0.32	0.29	0.21	0.12	0.06	0.03	0.01

BESSEL FUNCTIONS OF THE FIRST KIND – R. Bertrand. 1997

Table 1-34.

Example 1. A commercial FM broadcast transmitter has a deviation ratio of 5 and the frequency of the modulating audio is 15 kHz. Determine (a) the amount of frequency deviation and (b) the occupied bandwidth (BW) of the transmission.

Answer: the deviation is 75kHz and the occupied bandwidth is 180kHz.

Explanation:

We learnt that modulation index = deviation / frequency of the modulating audio. This is more often written as $\mu = \Delta f / f_s$. This equation can be transposed to give:
Δf (deviation) = $\mu \times f_s$ and for our problem this gives
Δf (deviation) = 5 x 15kHz = 75kHz.

The significant number of side frequencies present in the transmitted wave is given by:

significant side frequencies = 2 (μ+1) = 2(5+1) =12.

Since each of these 12 significant side frequencies are spaced apart by an amount equal to the frequency of the modulating audio (15kHz), then the bandwidth is:

BW = 12 x 15 kHz = 180kHz.

A slightly easier method for determining the bandwidth of an FM wave can be derived. In the last example we found the bandwidth from:

BW = 2 (μ+1) x f_s

By multiplying each term within the brackets by f_s we get:

BW = 2($\mu \times f_s$ + 1 x f_s).

Since $\mu \times f_s$ equals the deviation, the equation can be simplified to:

BW= 2(Δf + f_s).   **Equation 1.34**

The bandwidth (how much spectrum space it occupies) is equal to 2(Δf + f_s). Where Δf is the deviation and f_s is the frequency of the modulating audio.

Example 2.

Assume a transmitter has a peak deviation of 3 kHz (which is typical for narrow band voice) and that the frequency of the applied modulating audio is 3kHz). With the aid of Bessel functions, construct the spectrum of this wave as would be seen on a spectrum analyser.

The modulation index 'μ' of this wave is 1. There are 2(μ+1) or 4 significant side frequencies (two each side of the carrier) each spaced apart by (fs), or 3kHz. By consulting a table of Bessel functions, we see that the voltage amplitude of the carrier for a modulation index of 1 is 0.77 and the first two pairs of side frequencies have amplitudes of 0.44 and 0.11. The third pair of side frequencies has a relative amplitude of 0.02 and since this is less than 10% of the unmodulated carrier value of 1, the third pair of side frequencies can be considered insignificant.

We can now construct the graph from the above information as shown in figure 5-34.

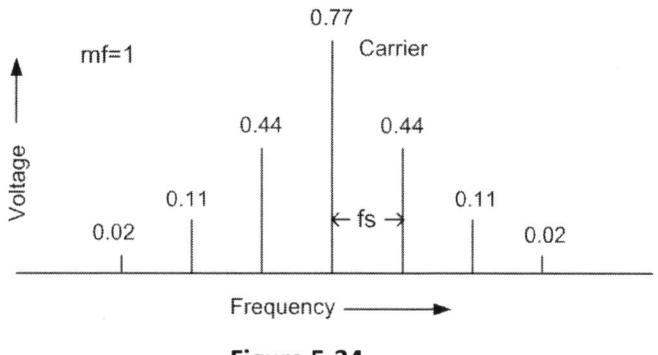

Figure 5-34

Show using a table of Bessel functions that the total power of an FM wave remains constant.

A table of Bessel functions gives the relative voltage amplitude for the carrier and each pair of side frequencies. Since the power is directly proportional to the square of the voltage, the relative power amplitude of the carrier and its side frequencies can be found by squaring each value listed in the table. In addition, since the total power in an FM wave remains constant, the sum of the squares of the values in the Bessel table should equal 1 for any particular modulation index.

The figures given on tables of Bessel functions are for practical purposes rounded, usually to two decimal places. So the sum of the squares will be almost, but not quite, 1. If we used Bessel functions taken to say six significant places, we would find the result is extremely close to 1.

Modulation Index = 1

Relative	E	E²	Multiply by	Power
J(0)	0.765198	0.585528	x1	0.585528
J(1)-	0.440051	0.193645	x2	0.387290
J(2)-	0.114903	0.013203	x2	0.026406
J(3)-	0.019563	0.000383	x2	0.000766
J(4)-	0.002477	0.000006	x2	0.000012
			Total Power:	1.000002

Table 2-34.

Here we have demonstrated that the total power in an FM wave remains constant. If we did the same for any other modulation index, we would find the same result. The reason it may seem I have carried on about this a bit is because there is a popular false belief that the power of the carrier of an AM wave varies with modulation and the power of an FM carrier remains constant - when precisely the opposite is the case.

SUMMARY SO FAR

AM - a system of modulation whereby the modulating audio is combined with the carrier in such a way as to create new frequencies (sidebands) and where the total wave power transmitted varies with modulation.

FM - a system of modulation whereby the modulating audio causes the frequency of the carrier to vary in proportion to the amplitude of the modulating audio. The total power transmitted does not vary. The power in the carrier does vary and can fall to zero (for example, look at mf = 2.405 in Table 1.34).

THE ADVANTAGES OF FM WHEN COMPARED TO AM

An FM system provides a better signal-to-noise ratio than an AM system. Put simply; this means it has less noise.

During its transmission (propagation), a frequency modulated wave will be subject to noise and interference voltages. The effect of these unwanted voltages is to vary the amplitude and phase of the FM wave. The noise amplitude variations have no effect on the performance of the system. Intelligence, information or modulation if you like, are not carried in the amplitude of an FM wave. *Amplitude variations are removed in the FM receiver stage called the limiter.* The phase deviation caused by noise means the carrier is effectively frequency modulated and some noise will appear at the output of the receiver.

The improvement in the signal-to-noise ratio that FM has over AM is primarily dependent upon the system deviation ratio (**D**). This is why FM broadcast stations sound so good. The degree of improvement can be determined by the equation:

Signal-to-noise ratio improvement = 20Log(**D**sqrt3) dB

The deviation ratio used in mobile FM voice communications (narrow band FM) is 5/3 or 1.66. The signal-to-noise ratio advantage over AM, by using the previous equation, is found to be 9.21 dB. Nothing to get excited about, but it is none-the-less a worthwhile improvement. Significant advantages in signal to noise ratio are only realised by commercial broadcasters. FM broadcast stations have a system deviation ratio of 5, giving a signal-to-noise ratio advantage of 18.75 decibels compared to AM.

THE DISADVANTAGES OF FM WHEN COMPARED TO AM

FM has a greater bandwidth requirement than AM. Narrow band FM occupies 2(5+3) =16kHz of spectrum space compared to AM's 6kHz (or in the case of single sideband 3kHz).

FM systems generally have a much wider bandwidth than AM systems. This makes FM (or any wider bandwidth signal) *more prone to selective fading.* Selective fading applies primarily to, but is not limited to, HF propagation – see the chapter on propagation for more information.

CAPTURE EFFECT

Another disadvantage of FM is that of *capture effect*. Capture effect is the ability of an FM receiver to lock onto and capture one transmitting station only. This can be a good thing, but it can lead to weaker signals not being heard at all. Typically, if a signal is about 3dB weaker than the desired signal, then it will not produce an output from the receiver. The receiver's limiter stage has "captured" the stronger signal. This explains the reluctance of airport authorities to switch to FM. Using an AM system allows an aircraft to break into a working channel should an emergency situation develop.

PRE-EMPHASIS AND DE-EMPHASIS

Pre-emphasis is used at the transmitter to boost the modulating audio frequencies above 1kHz. De-emphasis is employed in the receiver to restore the modulated audio back to its original power distribution.

Pre-emphasis and de-emphasis have the effect of improving the signal-to-noise ratio at the receiver for higher audio frequencies.

Pre-emphasis circuit

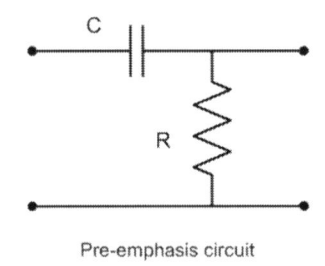

Figure 6-34

Pre-emphasis - boosting of audio frequencies above 1kHz in the FM transmitter.
De-emphasis - attenuation of frequencies above 1kHz in the FM receiver

De-emphasis circuit

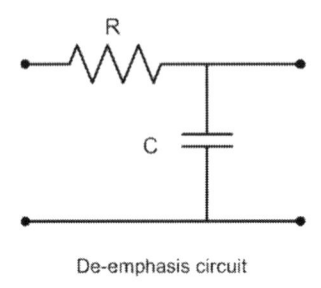

Figure 7-34.

Frequencies contained in human speech mostly occupy the region from 100 to 10,000Hz, but most of the power is contained in the region of 500Hz for men and 800Hz for women. The problem is that in an FM system the noise output of the receiver increases linearly with frequency, which means that the signal-to-noise ratio is poorer at the higher audio frequencies.

Also, noise can make radio reception less readable and unpleasant. This noise is greatest in frequencies above 3kHz. The high-frequency noise causes interference to the already weak high-frequency voice. To reduce the effect of this noise and ensure an even power spread of audio frequencies, pre-emphasis is used in the speech amplifier of the transmitter.

A pre-emphasis network in the transmitter accentuates (boosts) the audio frequencies above 1kHz, so providing a higher average deviation across the voice spectrum, thus improving the signal-to-noise ratio at the receiver. Without corresponding de-emphasis at the receiver, the signal would sound unnatural.

Pre-emphasis and de-emphasis are obtained by using simple audio filters. Normally we use LC filters for high and low pass filtering, however for simple low power audio, where we are not concerned about some resistive loss, simple RC filters can be used. You may be required to identify one of the circuits in an Advanced Radio assessment. The pre-emphasis circuit produces higher output at higher frequencies because the capacitive reactance of 'C' is decreased as frequency increases. As the resistor and the capacitor form a voltage divider and the reactance of the capacitor has decreased, more voltage must appear at the output across the resistor.

A similar reasoning, but reverse in effect, can be applied to the de-emphasis circuit. The standard time constant (T=CR) for these circuits in narrow band transceivers is 75 microseconds (uS).

Summary.

Pre-emphasis is used in the transmitter to boost (emphasise) the higher audio frequencies in speech (above 1000Hz).

Pre-emphasis increases the average deviation of a transmitter resulting in better signal-to-noise ratio. In addition, the effects of higher frequency noise is less noticeable.

De-emphasis in the receiver attenuates the high audio frequencies in the receiver to restore the original audio.

THE BLOCK DIAGRAM OF AN FM RECEIVER

Figure 8-34 below shows the partial block diagram of an FM receiver. All the stages preceding the last IF amplifier would be the same as that of any superheterodyne receiver. The bandwidth of the IF amplifier would be approximately 16 kHz in an amateur FM receiver. Compared to an AM receiver, the IF amplifiers would have sufficient gain to allow for full limiting action to occur at the limiter. We will discuss the limiter stage shortly. By the time the FM signal reaches the limiter, it has been amplified sufficiently for the limiter to be able to remove any amplitude modulation. Any amplitude modulation would be noise in an FM system. The stage following the limiter is called the *discriminator.* The purpose of a discriminator is that of an FM demodulator. An FM demodulator's job is to convert frequency variations in the signal back to the original modulating audio (the audio that went into the microphone at the transmitter). The discriminator can work on one of several principles. You will not have to know how these work, but you may have to know what they are called. The basic type of discriminator are the Foster-Seeley, Ratio Detector and the PLL type. The PLL type being the most commonly used FM demodulator today.

All these discriminators perform the same function in that they convert frequency variations into audio amplitude variations. The greater the frequency deviation, the higher (louder) the audio output voltage. The discriminators job is to convert frequency variations back to amplitude variations – which is the demodulated audio.

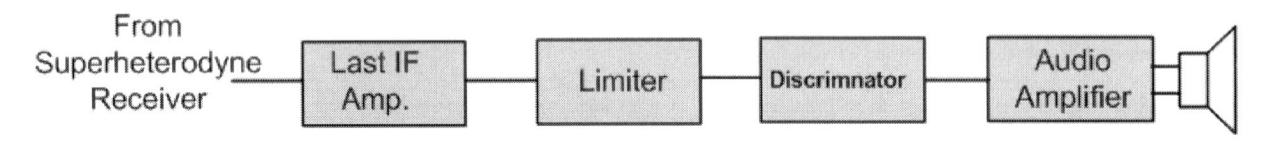

Figure 8-34.

The de-emphasis circuit (usually part of the audio amplifier) restores the audio to its original power distribution by attenuating frequencies above 1kHz. All FM receivers contain a *mute* or *squelch* circuit to cut off the input signal to the audio amplifier in the absence of an FM input signal.

OPERATION OF THE LIMITER STAGE

Figure 9-34 shows what happens to the input signal after being fed to a limiter. Positive and negative amplitude peaks drive the FET (though other active devices could be used) into cutoff or saturation, therefore *slicing off the amplitude variations which represent noise. The limiter is actually an over-driven RF amplifier.* If the signal entering the limiter was not of sufficient amplitude to drive the FET between cutoff and saturation, then limiting action would not occur. This happens on weak signals and considerable noise is heard at the receiver output. In an attempt to ensure full limiter action, most FM receivers have more IF amplification than AM types.

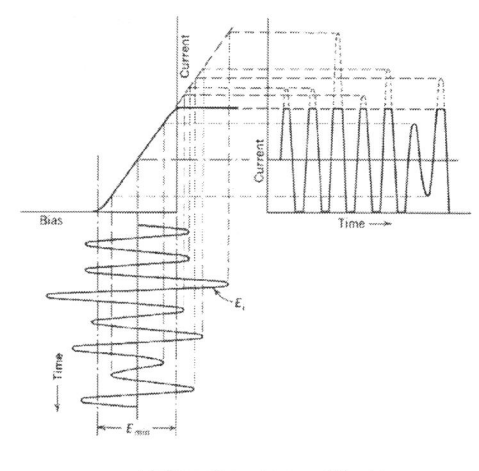

Limiter action – removes AM noise

Figure 9-34

The limiter shown in figure 9-34 consists of a sharp cutoff FET. You can see that the input to the gate has amplitude variations (noise). The output from the drain has all of the amplitude variations removed. The intelligence is in the frequency that is left unchanged by the limiter.

THE FOSTER-SEELEY DISCRIMINATOR

Foster-Seeley Discriminator

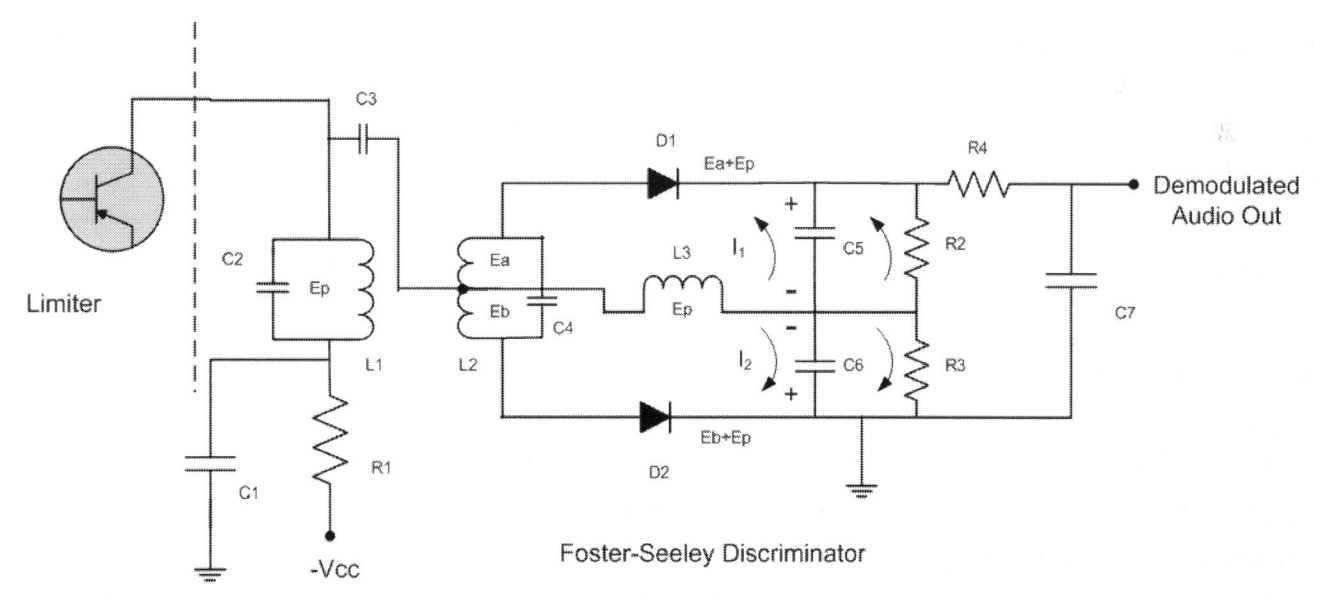

Figure 10-34

The Foster-Seeley circuit (Figure 10-34) and its operation are explained in detail. You do not need to explain the operation for assessment. You may need to know the schematic's name and to be able to identify it.

Operation:

Remember first - the purpose of any FM demodulator is to convert frequency variations back to audio frequency changes.

In order to understand the operation of the circuit, it is important to realise that the capacitances C1, C3 and C6 have very low reactances at the operating frequency. The left side of L3 is connected to the top of L1 via C3 and because of the low reactance of C3, there is no appreciable voltage across C3. Therefore, the left end of L3 and the top of L1 are at the same potential. Similarly, the right end of L3 is connected to C6 and C1 to the bottom end of L1 and is the same potential as the bottom of L1.

L3 is electrically in parallel with L1. The FM signal from the last IF stage produces a voltage Ep across L1 and since L1 is in parallel with L3, then the voltage Ep will also appear across L3. There will be a voltage induced into L2 from L1. Because of the centre tap, half of the induced voltage will appear as Ea on the top half of L2 and as Eb on the bottom half. The total voltage applied to D1 is then Ep + Ea and the voltage applied to D2 is Ep + Eb.

At the centre frequency (no deviation) Ep + Ea = Ep + Eb. Also, the currents I2 and I3 through R2 and R3 will be equal and opposite. Likewise, the voltages appearing across R2 and R3 will be equal and there will be no net output voltage. The output voltage is the net or resultant voltage across R2 and R3. This is as it should be with no deviation. At the centre frequency, the tuned circuit consisting of L2 and C4 is resonant and as such the current and voltage are in phase.

At any frequency other than the centre frequency (when modulation is present) there is a change of phase between Ea, Eb and Ep. This means that Ep + Ea will not equal Ep + Eb. How much difference exists will depend on the deviation from the carrier frequency. Whether Ep + Ea is greater or less than Ep + Eb will depend on which way the frequency swings. The overall result is that for any frequency other than the centre frequency, currents I2 and I3 will not be equal. The voltages R2 and R3 will not be equal and opposite and a net voltage will appear at the output with an amplitude directly proportional to the deviation. By converting the deviation into amplitude variations, we have recovered the original modulating audio.

A de-emphasis circuit is formed by R4 and C7. R4 and C7 form a simple low-pass filter.

THE RATIO DETECTOR - another FM discriminator

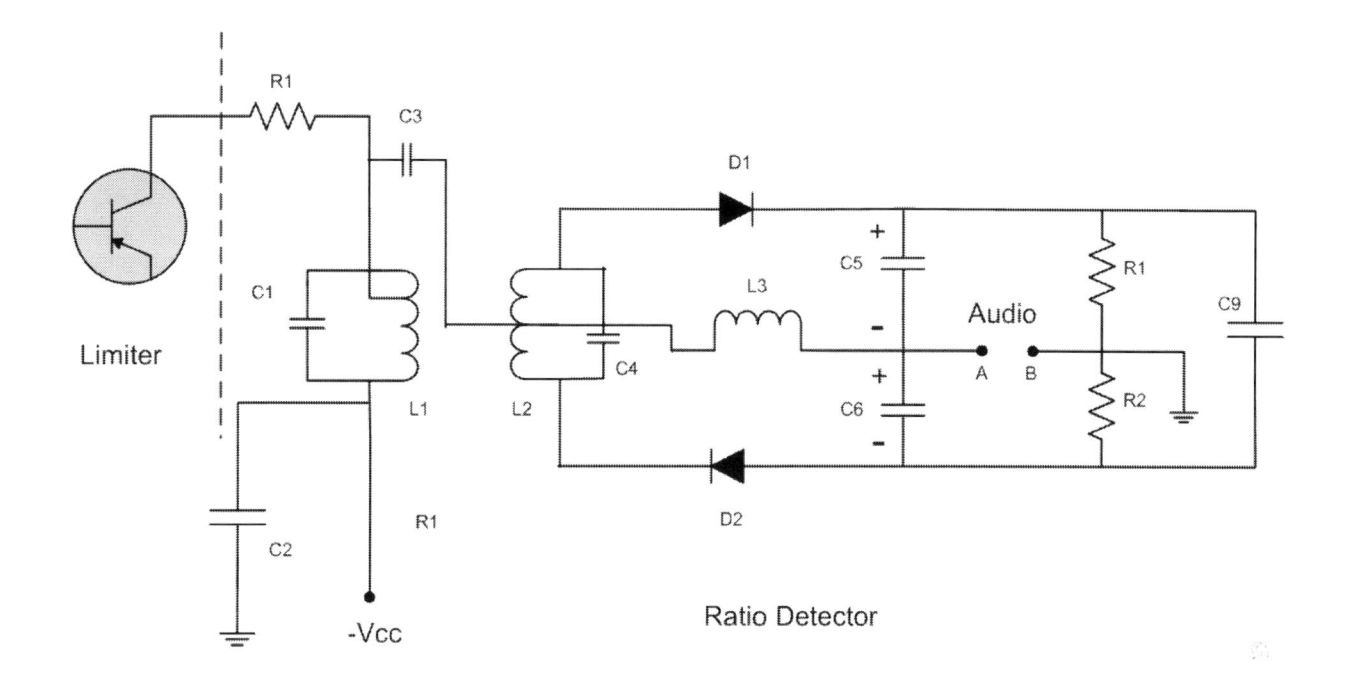

Figure 11-34

In a ratio detector, one of the diodes is reversed so that current can flow in the overall circuit. The circuit operation is similar to the Foster-Seeley discriminator, but the output voltage is now taken from the points (A) and (B) as shown in the schematic diagram. Because C9 is connected in parallel with R2 and R3 and also C5 and C6, the voltages C5 + C6 and R2 + R3 are held at a constant value by the charge on C9.

At the centre frequency, there will be no potential difference between points (A) and (B). When deviation is present, one of the diodes will conduct harder and the voltage across its associated capacitor will increase; the voltage on the other capacitor will decrease (the total voltage C5 + C6 remaining constant); the circuit is unbalanced and a potential difference must now exist between points (A) and (B). This net voltage is the demodulated audio output. Because amplitude variations do not unbalance the circuit, a ratio detector does not *theoretically* require a limiter stage. In practice, however, most manufacturers provide a limiter stage, as large amplitude variations will produce some noise output.

PHASED LOCKED LOOP (PLL) FM DISCRIMINATOR

The PLL discriminator is one of the common methods of FM demodulation today. It is not necessary to have a broad knowledge of PLL systems for Advanced exam purposes. The block diagram of a PLL FM detector is shown in Figure 12-34.

PLL - Discriminator

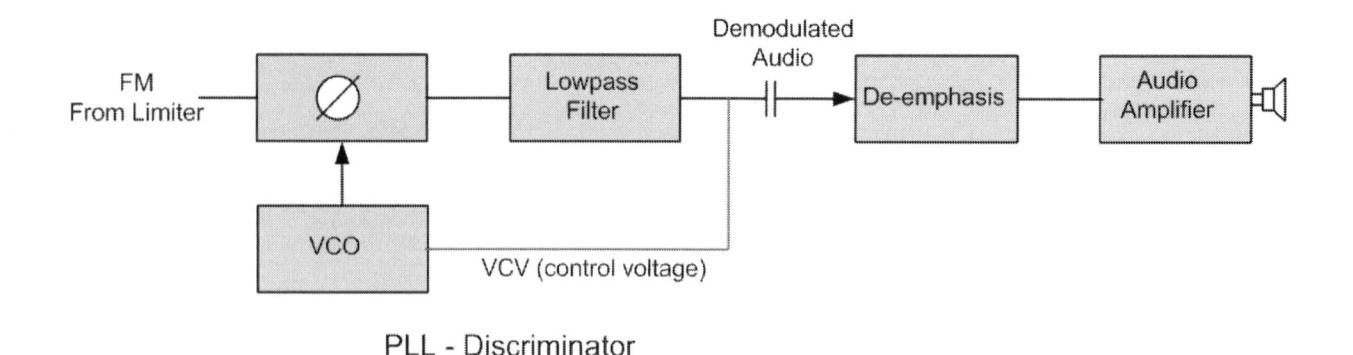

Figure 12-34

The FM signal from the limiter is fed to the phase detector. At the centre frequency, *only the carrier will be present and the PLL circuit will lock onto the carrier and is held there by the correction voltage produced by the phase detector (VCV)*. If the IF was 10.7MHz, then the VCO would now be oscillating at 10.7MHz. When deviation is present, the carrier swings above and below the centre frequency. *The PLL circuit will attempt to stay in lock by producing a correction voltage (VCV)* which makes the VCO track the carrier frequency swing. As the correction voltage is proportional to the amount of deviation from the centre frequency, the *correction voltage is the demodulated audio.*

FM transmitters.

Let's look at the block diagram of the FM transmitters shown in figures 13(a)-34 and 13(b)-34. The transmitter in Figure 13(a)-34 is the most common type used in amateur and general radiocommunications. The audio voltage from the microphone is increased in level by the microphone amplifier. The amplified audio is then applied to the reactance modulator which, in most cases, consists of little more than a reverse biased varactor diode. The audio voltage will cause the effective capacitance of the diode to vary. The varactor diode is connected across the crystal in the 10.695 MHz oscillator; thus, the applied modulating audio makes the crystal oscillator vary above and below its centre frequency. The microphone amplifier will have an internal gain control that can be adjusted to obtain the required amount of deviation from the crystal. The *output of the crystal oscillator has the required deviation for transmission*. The signal is then fed to the mixer stage along with a very high-frequency carrier from the heterodyne oscillator (VFO). For stability, the heterodyne oscillator (VFO) would consist of a complete PLL circuit.

At the output of the mixer will be the required transmit signal which is passed through a band-pass filter in order to remove other unwanted mixing products. The FM signal is then amplified by the power amplifiers that typically consist of 3-4 active stages. RF amplification can be Class C for maximum efficiency.

The block diagram of two types of FM transmitter

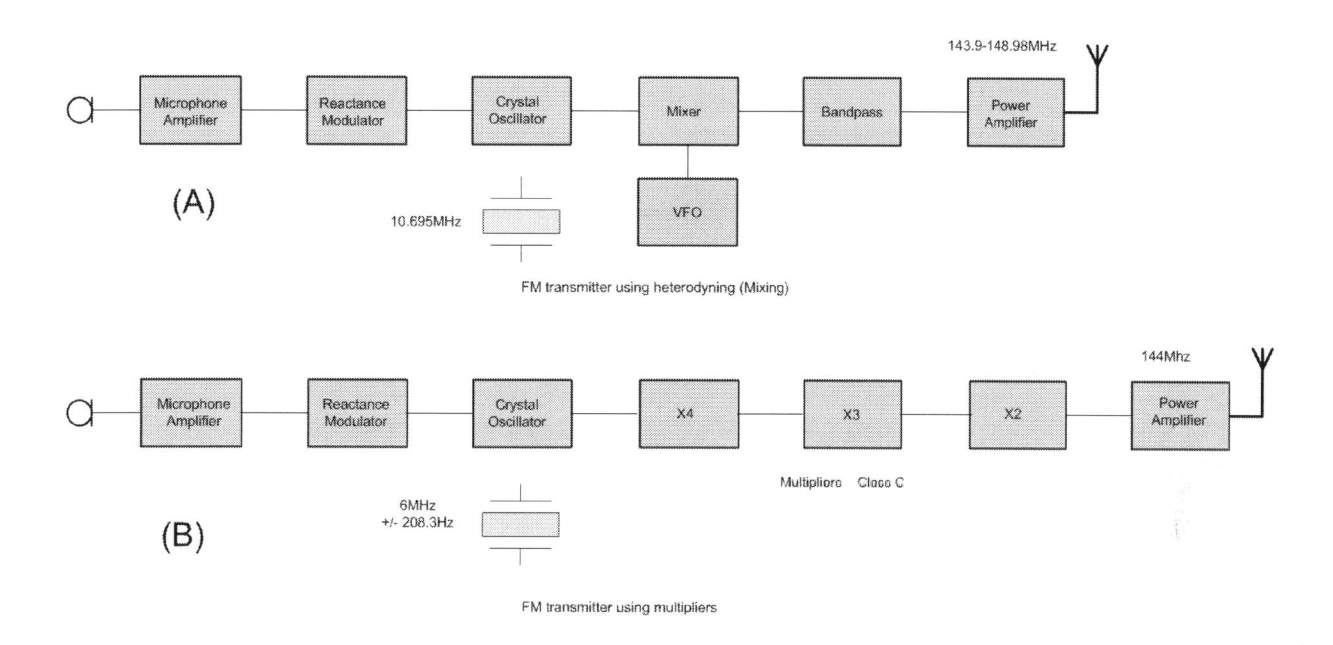

Figure 13(A)(B)-34

Can you see what is not shown on these block diagrams? There would be Buffers after every oscillator. The power chain would contain 1-2 driver amplifiers before the power amplifier. There is no power supply. There would generally be in 13(a)-34 a microprocessor controlling the VFO (through PLL) and driving a digital readout for the transmit frequency and repeater offsets.

THE FREQUENCY MULTIPLIER METHOD

The earlier experimental FM transmitters more often relied upon *frequency multiplication* to obtain the final transmit frequency. In these transmitters, the frequency deviation produced by the reactance modulator was much less than the required +/- 5kHz *because frequency multiplication has the effect of multiplying the deviation as well*. A block diagram of an FM transmitter that uses frequency multiplication is shown in figure 13(B)-34. With integrated PLL circuitry now commonplace, the heterodyne technique described is now universally used to obtain the required transmit frequency. Frequency heterodyning (frequency conversion by mixing), however, leaves the deviation unchanged. In FM broadcast transmitters where a deviation of 75 kHz is required, it is usual to use frequency multiplication to obtain the required deviation, then frequency heterodyning to place the FM signal on the correct transmit frequency. Another advantage of multiplication for broadcasting is the crystal can be of much lower frequency. Low-frequency crystals a larger and more stable in frequency then high-frequency thin crystals. These days with PLL and DDS achieve high stability.

Frequency heterodyning is just another name for a mixer and oscillator combination that forms a frequency converter.

35. A COMPLETE FM TRANSCEIVER

This is a consolidation chapter.

The purpose of this chapter is the consolidation of transmitters and receivers by way of looking at the complete block diagram of a typical (and real) 2-metre band transceiver.

The block diagram (Figure 1-35) is that of an amateur transceiver capable of operation on the 2-metre amateur band. I have not shown buffer stages as they should be assumed.

Although there are two antennas shown in the diagram, there is in reality only one and a small changeover relay is used to switch the antenna between the transmitter and the receiver. This transceiver is PLL controlled and the PLL's programmable divider is controlled by a microprocessor. The microprocessor also controls the LCD display and keypad. The operator only has to key in the transmit and receive frequencies on the keypad and the microprocessor will set the programmable divider to the correct value (300-1100). The microprocessor will also take care of displaying the appropriate transmit and receive frequencies. It is common to work through repeaters on the 2-metre band and the 'standard' transmitter/receiver offset is 600kHz. The offset can be set via the keypad and the microprocessor looks after the rest. Notice first that the receiver section is drawn in 'blue,' the transmitter in 'green' and the PLL section is 'orange.' If you are reading this in greyscale the receiver is at the top, transmitter in the middle and PLL at the bottom.

Assume the transceiver is set to transmit and receive at *147.995MHz.*

This frequency would have been entered on the keypad by the operator and the microprocessor will display the TX and RX frequency on the LCD display as 147.995MHz. To operate on this (147.995) frequency, the microprocessor sets the programmable divider to divide-by-1100.

The reference oscillator is part of the PLL and provided by OSC3 on 5.76MHz. This frequency is divided by the fixed divider (divide-by-1152) down to the *reference frequency of 5kHz.*

The 5 kHz reference frequency signal is fed to the Phase Detector. This means two things:

1. The PLL will be in lock on 5kHz and
2. The channel spacing for this radio (the increments between the individual channels) is 5 kHz.

When I say the PLL is in lock on 5kHz it means that for "lock" to occur the signal coming from the programmable divider is 5kHz. This is shown in the block diagram as (B) 5kHz.

If the signal coming from the programmable divider is not 5kHz, then the PLL will not be in lock and the VCO frequency will change until the PLL comes into lock. The VCV is the DC control voltage from the phase detector to the VCO. The VCV voltage will make the VCO move in frequency very quickly to 131.095 MHz. When the VCO hits 131.095MHz, the PLL comes into lock and the VCO is held there.

When the VCO is on 131.095MHz, the PLL will come into lock. The 131.095MHz signal is fed from the VCO to MIXER4 along with a signal of 125.595MHz from OSC2. This produces a frequency difference F1-F2 of 5.5MHz which, when divided by 1100, equals the reference frequency of 5kHz, which is fed to the right-hand side of the phase detector.

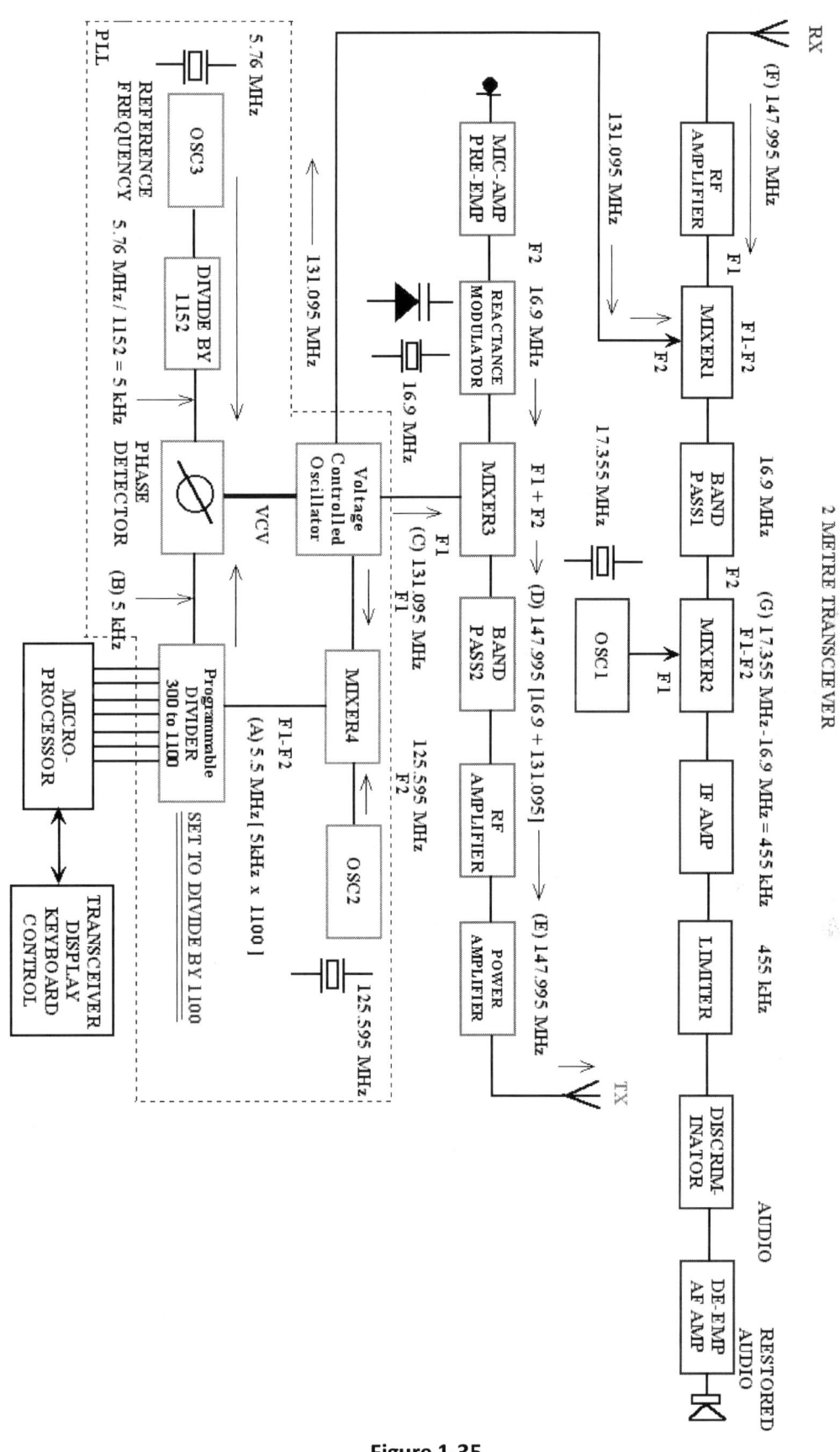

Figure 1-35

The display says the transmitter and receiver are on 147.995MHz and the PLL is producing a very stable signal on 131.095MHz, for use by both the transmitter and the receiver.

TRANSMISSION

The microphone most likely works on the dynamic or capacitor microphone principle. The audio signal from the microphone is amplified and pre-emphasis is applied. The audio signal is then fed to the *Reactance Modulator,* which operates on *16.9MHz*. Since there is no frequency multiplication in this system, an FM signal with full deviation must leave the reactance modulator (F2). The FM signal F2 is frequency converted to *147.995MHz* by summing it with the VCO signal F1 in MIXER3. The output of MIXER3 also has other unwanted mixing products and these are eliminated by BANDPASS2. The FM signal is then amplified using efficient *Class C* amplifiers and radiated from the antenna.

RECEPTION

The operator releases the PTT (push-to-talk) button and the antenna is switched quickly to the receiver circuit. The received FM signal is fed from the antenna to a wideband RF amplifier and then into MIXER1 where it is mixed with the signal from the VCO on 131.095MHz. The product F1-F2 of MIXER1 is 16.9MHz. BANDPASS1 at the output of MIXER1 eliminates unwanted mixing products from MIXER1. The wanted FM signal (now on 16.9 MHz) is fed to MIXER2 and frequency converted to the IF frequency of 455kHz.

The 455kHz FM signal is then highly amplified by the IF amplifier before being fed to the LIMITER to remove any amplitude modulation (which would be noise). Upon leaving the LIMITER, the FM signal is demodulated by the DISCRIMINATOR. The recovered audio from the discriminator undergoes DE-EMPHASIS and further audio amplification before being fed to the speaker.

That's it! The full operation of the transceiver in brief. You should now have a good idea how an FM transceiver works and what all the stages do. If you print this diagram, you should be able to work out what frequency the transceiver would be operated on if the programmable divider is set to say divide-by-300. I have not shown ALC (Automatic Level Control) or AGC (Automatic Gain Control) and lines representing these could be drawn onto the diagram. There are a number of options for ALC and AGC. As an exercise you may wish to draw these for yourself.

36. TRANSMITTER FAULTS

As a radio experimenter you have a responsibility to ensure that your transmissions do not interfere with other radio users, domestic electronics, radio and non-radio devices. You may have no interest in building your own transmitting equipment, but your qualification along with a License enables you to do this if you so choose. Even if you do not build your own equipment, you will most likely make internal adjustments and or modifications. This chapter will cover some material already covered and some new material, all relating to what can go wrong with a transmitter and cause interference.

CHIRPING

Chirping only applies to CW (Morse) transmitters. Keying chirps are *quick* changes in the frequency of a transmitter and occur each time the telegraphy key is closed or opened. Keying chirps are usually caused by an oscillator stage in the transmitter being pulled off frequency each time the transmitter is keyed. Chirping was more of a problem when many radio experimenters used to build their own equipment and when CW was more popular. CW is still very popular today in spite of it not being a licence requirement.

In CW mode, if the stage being keyed is too close to the oscillator, then the oscillator may shift slightly off frequency each time the key is closed. These quick changes in frequency sound very much like a canary chirping, hence the term *chirping*. Chirping can also occur if the voltage regulation to the oscillator stage is inadequate. A well-regulated power supply and good buffer amplification will prevent Chirping.

KEY CLICKS

Key clicks also only apply to a CW transmitter.

Key clicks occur when the CW transmitter is being turned ON and OFF by the Telegraphy Key (Morse Key if you like) too quickly. When the telegraphy key is closed, the transmitter begins to transmit, when the key is opened it stops. The clicks occur when the telegraphy key is opened and closed. If the transmitter is keyed on and off too sharply, then the leading and trailing edges of the Morse signal will have a rapid rise and short decay time, like the leading and trailing edges of a square wave.

This results in sidebands being produced on each side of the carrier frequency. The diagram in figure 1-36 shows part of a telegraphy character on an oscilloscope. The key clicks are identified by an almost vertical attack and/or decay time of each element in the character.

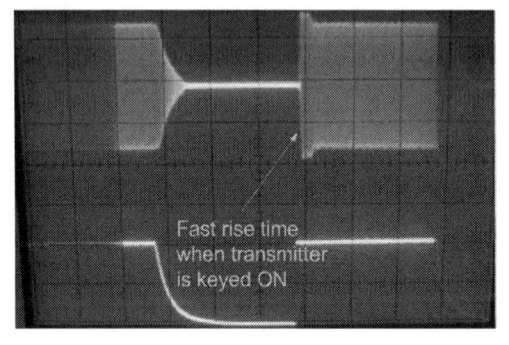

Figure 1-36

Take any sinewave and add to it all of its harmonics and you produce a square wave. The more harmonics that are added, the more perfect the resulting square wave. If a fundamental sinewave plus it harmonics can produce a square wave, then a perfect square wave must be made up of its fundamental plus all of its harmonics. If a series of square wave dots is transmitted (dots with key clicks), then the signal must be made up of the dot frequency plus the many harmonics of the dot frequency. If the dot frequency is 10Hz (10 dots per second) and the wave shape of the dots is square (it has key clicks like in the oscilloscope diagram of figure 1-36), then there may be 50 or more strong harmonics present that will produce sidebands on both sides of the carrier, extending the bandwith the telegraphy signal by 1 kHz (2 x 10 x 50). The resultant received signal sounds broad and a clicking sound can be heard.

Installing a "key-click filter" in the keyed stage will prevent transmitter key clicks. The filter normally consists of a simple RC time constant that prevents the bias of the keyed stage from turning ON and OFF too quickly. The waveform at the beginning and end of some Morse elements is shown in figure 2-36. Take note of the slower decay and attack time.

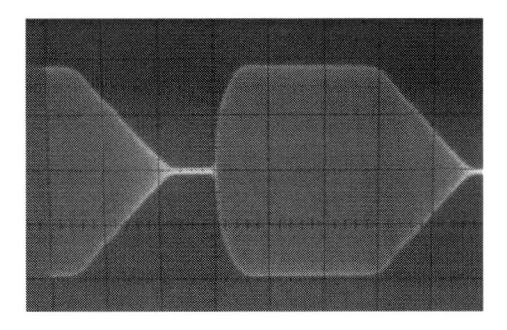

Figure 2-36

It is a fallacy that key clicks are caused by dirty telegraphy key contacts. Dirty contacts may cause the key to stick closed or become sloppy. This is called squawking. Some examiners are aware of this fallacy and may use distractors (wrong answers) that target this to try and really test you. Also, clicks have nothing to do with the speed of the Morse key. When the Morse key is closed the transmitter should take a few milliseconds to come to full power. When the key is opened the transmitter power needs to drop to zero power over a few milliseconds. If this does not happen key clicks will occur. Your bandwidth is increased and it sounds awful.

SELF-OSCILLATION

Self-oscillation is caused by unwanted positive feedback (regeneration) in an amplifier. If the positive feedback is sufficient, the amplifier will break into oscillation on the operating frequency. The amplifier is no longer operating effectively as an amplifier. Self-oscillation in an RF stage of a transmitter could cause that transmitter to come 'on air' by itself. To prevent self-oscillation, an equal amount of negative feedback (degeneration) is applied to the amplifier, cancelling out the unwanted positive feedback. This process is called *neutralisation*.

All active devices have some unwanted capacitance between the input and output terminals. In a bipolar transistor it is the capacitance between the collector and the base; in the FET it is the capacitance between the drain and the gate; in the electron tube, it is the plate to control-grid capacitance. At low frequencies, this capacitance is not a problem as the capacitive reactance is high. At higher frequencies this reactance decreases and more signal is coupled from the output to the input. This feedback is regenerative and if it becomes too high, it will cause the amplifier to oscillate.

The partial schematic diagram of a power amplifier in figure 3-36 illustrates one method of applying an equal and opposite amount of negative feedback to prevent self-oscillation.

The capacitance shown dotted as 'C_{bc}' is the internal capacitance of the BJT - it is not external as shown.

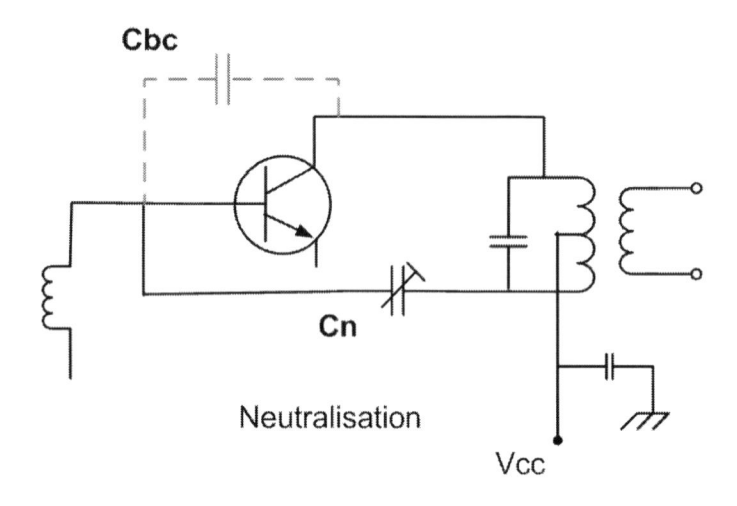

Figure 3-36

Negative feedback is applied by the variable capacitor C_n (a trimmer capacitor). Notice that the collector supply is connected to the centre tapping of the inductor in the collector tank circuit. Since the supply is bypassed to ground, the centre tap is at RF ground potential. When the top of the tank is positive, the bottom must be negative and vice versa. Hence, negative feedback is available at the bottom of the tank circuit.

Figure 4-36 shows another method of obtaining neutralisation. In this circuit, the negative feedback is taken from the secondary winding of the RF transformer. Since transformer action induces an 180-degree phase change between primary and secondary, the secondary voltage provides the necessary negative feedback.

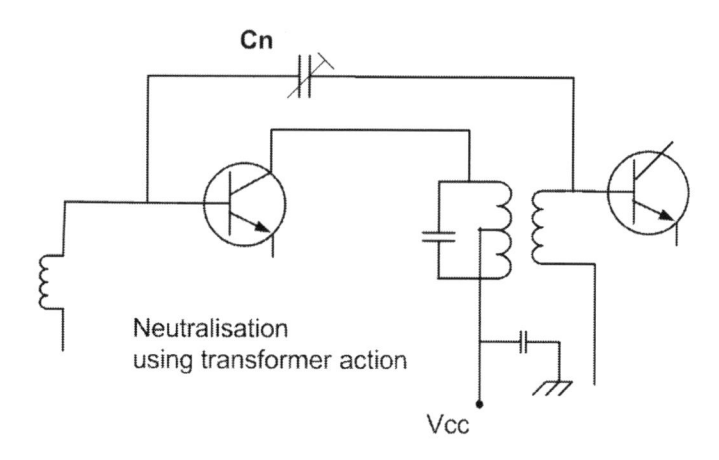

Figure 4-36

PARASITIC OSCILLATIONS

A parasitic oscillation is an unwanted oscillation on any frequency other than the operating frequency. A parasitic oscillation is caused by stray resonant circuits in an amplifier causing the amplifier to act as a tuned-input tuned-output oscillator.

The tuned-input tuned-output oscillator has not been mentioned before and it is not in the syllabus. Basically, a tuned-input tuned-output oscillator is an RF amplifier with a tuned resonant circuit at its input and at its output. Both these tuned circuits are resonant on the same frequency. Add some positive feedback and we have a pretty good oscillator.

Parasitic (unwanted) resonant circuits could be created by bypass capacitors and radio frequency chokes. If the stray reactances are high, then *low-frequency parasitics* can be created. On the other hand, if the stray parasitic resonant circuits are created by wiring loom capacitance and inductance, the parasitic oscillation will be near to or even above the operating frequency and in this case, they are called *high-frequency parasitics*.

Parasitic oscillations generally produce a large number of harmonics. If a transmitter operating on 28MHz has a low-frequency parasitic on 500kHz, spurious signals may be heard from the transmitter across the spectrum every 500kHz up to and beyond the operating frequency. You will only know you have a parasitic if other operators tell you or you look at the spectral purity of your transmitter on a spectrum analyser. The fundamental frequency of the parasitic can be determined by using a remote receiver.

When you know the fundmental frequency of a parasitic, a dip oscillator may be used to find the *physical location* of the parasitic.(to be discussed further). With the dip oscillator tuned to the parasitic frequency, it can be moved around the circuit (inside the transmitter which is turned OFF). Tuned to the parasitic frequency a dip will indicate the *physical location* of the parasitic circuit.

PARASITIC STOPPER

In electron tube amplifiers that only carry a low current, the easiest method of eliminating parasitic oscillations is to install a low-value resistor (10-100 Ohms) in the grid and plate circuits as close to the electron tube as possible. In high current circuits, 10-20 turns of wire can be wound over a resistor and soldered to each end. This, in effect, creates a low-Q RF choke. Such a device is called a 'parasitic stopper.'

For many assessments, you need to know:
(a) what are parasitic oscillations,
(b) the difference between high and low-frequency parasitics,
(c) what is a parasitic stopper,
(d) how to physically locate the parasitic circuit causing the problem and fix it.

In transistor amplifier stages, small ferrite beads can be threaded over the leads of the active device to form a parasitic stopper.

How parasitic stoppers work

All types of parasitic suppressors work by increasing the resistance in the parasitic circuit to a value equal to or higher than the *critical resistance*. When any resonant circuit (in this case a parasitic) has a resistance equal to 2xsqrt(L/C), the circuit is said to be *critically damped* and will not oscillate. By critically damping the parasitic circuit, oscillation is prevented. Critically damped means the Q is so low that oscillation cannot be sustained.

$$Critical\ Resistance = 2\sqrt{\frac{L}{C}}$$

Equation 1-36

OVER DRIVING A LINEAR RF AMPLIFIER

Over driving a linear RF amplifier stage will cause 'splatter' and therefore a widening of the occupied bandwidth and possible interference. Remember that you may over drive linear amplifiers in an SSB transmitter simply by turning the microphone gain up too high.

When a complex waveform such as an SSB signal is fed to a linear RF amplifier, it is increased in amplitude by the gain of the amplifier. All linear amplifiers have some non-linearity that results in some distortion. Non-linearity causes the individual frequency components within the signal to mix with each other and produce new signals. New signals (frequencies) are distortion. Normally the amount of non-linearity is so low as to be insignificant. When a linear amplifier is over-driven it ceases to behave in a linear fashion and severe distortion may result. Excessive harmonics and spurious side frequencies are created, which almost certainly produce interference to other services such as television reception and other nearby operators will report 'splatter.'

FLAT TOPPING

Flat topping results from over driving an amplifier stage. The input signal becomes so high that the amplifier is driven into saturation and/or beyond cutoff. The resultant waveform on an oscilloscope appears as though the top and bottom peaks have been flattened. The term 'Flat Topping' is applied to SSB signals only.

Pi-Coupling at the transmitter output

The schematic diagram of a pi-coupler is shown in figure 5-36 and redrawn to emphasise that a Pi-coupling network is just an extension of the output 'tank' circuit.

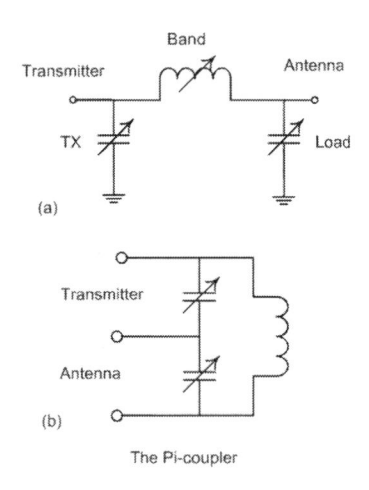

Figure 5-36

The Pi-coupler does a few important things, namely:

a) Matches the impedance of the power amplifier of the transmitter to the transmission line (feedline).
b) Acts as a low pass filter for suppression of harmonics.
c) It can if necessary, supply the flywheel effect (FM or CW).

It is evident from the configuration that the circuit will function as a low-pass filter. If this is not clear to you, then I suggest you revise the chapter on filters.

To see how the pi-coupler is used to obtain impedance matching, it is best to visualise the redrawn lower part of figure 5-36. The Pi-coupler is simply a parallel tuned circuit with a tapped capacitance across the inductor. The impedance of the parallel tuned circuit at resonance is very high. However, the impedance seen across A-B and B-C will depend on the ratio of C1 and C2.

When the transmitter is tuned properly, that is, the Pi-coupler is tuned to match the transmitters impedance to the transmission line, the output impedance of the transmitter will be equal to the impedance seen between A and B and the impedance between B and C will equal that of the transmission line.

When the output of the amplifier is matched to the input impedance of the transmission line, maximum power transfer will be obtained. In practice, the operator alternately adjusts the three controls on the coupler for maximum power or minimum output current. I have labelled the controls if figure 5-36 as, Transmitter, Band and Antenna. They can be called other names for example, Tune, Inductance and Load. It really does not matter. A good trick is to adjust all three controls for greatest receiver noise. Then check if the frequency is clear, identify in a mode that can be understood by listeners, announce a *test* or *tuning*, then switch to a full carrier mode set to low power and adjust the controls alternately for maximum power while transmitting.

It is possible to adjust the Pi-coupler by monitoring the circuit current drawn in the output of the final amplifier. At resonance, the circuit impedance will be high and anode (or equivalent) current drawn will 'dip' to a minimum. The operator will see the 'dip' by monitoring the anode current or collector current.

Figure 5-36

Either method (dip or maximum RF output) can be used alone, or in conjunction, to tune any transmitter output stage. All transmitters with tuneable pi-coupling in the output, or some variation of it, work this way.

INTERMODULATION INTERFERENCE

Two or more signals can, upon entering a non-linear device, mix together to produce new signals. The new signals are called intermodulation products and they may be on frequencies that can result in interference. The difference between modulation and intermodulation is that modulation (or mixing) is deliberate.

The most common place for this mixing to occur is in a receiver. However, mixing can happen in the power amplifier stages of two or more transmitters operating in close proximity.

Transmitted signals from two or more transmitters can be received by some metallic structure such as a tower. Rusted or bad metallic connections can cause the connection to act as a rectifier (non-linear device) causing intermodulation and re-radiation of the new unwanted signals.

Intermodulation can also occur in the output stage of a transmitter (since it is often non-linear). This is called *transmitter intermodulation*. Now, many radio operators do not understand transmitter intermodulation. Imagine two or more transmitters in close proximity to each other. Do you really think that if two transmitters were transmitting on different frequencies and perhaps even different modes, that signals from each transmitter would not enter the other transmitter? They do! Transmitters transmit to other transmitters - they are not smart enough to know that the antenna is a transmission antenna.

Signals are received and mixed in the output stages of transmitters and are sometimes re-radiated causing interference to other services.

Receiver intermodulation occurs when two or more signals enter a receiver and none of the signals are on the receiver frequency, but they mix in the receiver to produce an on frequency signal and interference occurs.

For example, say the receiver was tuned to frequency C. Two signals enter the receiver on A and B. Neither A or B alone cause any problem, however when they are both present A and B might mix (in the receiver) so that C=2A-B; now there in interference heard in the receiver. The combination is called a second order intermodulation (because there are two orders A and B) is the most common. A third order intermodulation might be D=A+B-C; that one is fairly common also; and you have three RF sources mixing to produce interference on D.

You do not need to go any deeper into intermodulation interference for examination purposes.

CARRIER SUPPRESSION

Operating on SSB means transmitting one sideband, either upper or lower, by getting rid of the unwanted sideband using a band-pass filter and suppressing the carrier. Carrier suppression in experimenter's SSB transmitter should be 40dB or more. The stage looking after carrier suppression is the balanced modulator.

The carrier balance is one of the 'inside controls' you will most probably adjust. If you are 'on air' talking to another station on say USB, transmit but do not send any audio to the SSB transmitter (just be quiet). The only thing being transmitted now will be your suppressed carrier. Have the nearby station you were just talking to tune down about – 1-2kHz and they will hear and report back to you that they can see your suppressed carrier on their *S-meter*. They will report a tone as well; it is their receiver that produces this tone when it picks up your suppressed carrier.

A better way is to use another receiver at your station and have your transmitter going into a dummy load. You can then do the same thing and adjust the 'carrier balance' control in the balanced modulator for maximum carrier suppression. Modern transmitters have no trouble achieving 40dB or more of carrier suppression.

I have heard SSB stations on the air with such badly adjusted carrier balance that they were transmitting single sideband full carrier, or at least close to it.

37. ANTENNAS

The purpose of an antenna is to transmit and receive electromagnetic radiation. When the antenna is not connected directly to the transceiver, we need a transmission line (feedline) to transfer the received or transmitted signal to and from the antenna.

I will mostly be talking about transmission. Whatever is said about transmission applies equally to reception. This is known as reciprocity. An antennas characteristics are the same on transmission and reception.

In a perfect transmitting system, the transmission line would transfer all of the power from the transmitter to the antenna without any losses. The antenna should then radiate all of the power it receives as electromagnetic radiation.

We have discussed the mechanism by which an electromagnetic wave is radiated by an antenna. If you have forgotten this, refer back to the chapter on electromagnetic radiation.

THE HALF WAVE DIPOLE

If you take a basic balanced transmission line such as 300Ω TV ribbon, split it and pull it apart, you will form a dipole antenna (refer to figure 1-37). Each side of the half wave dipole will be ¼ wavelength. We learnt earlier that the free space wavelength of an electromagnetic wave is found from:

$$\lambda = \frac{300}{frequency\ (MHz)}$$

Equation 1-37

This is derived from:

$$\lambda = \frac{c}{f}$$

where c = 300,000,000 metres/second.

The constant in this equation is 'c', the velocity of any electromagnetic wave in free space.

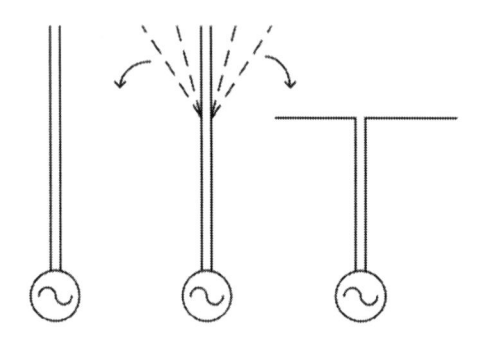

A dipole is a fanned out transmission line

Figure 1-37

A halfwave dipole is physically 5% shorter than its free space wavelength because the velocity of propagation of the wave along each leg of the dipole is slower than 'c'.
The capacitive end effect, and the dielectric constant of the antenna conductor, whilst very low, are enough to slow a wave down.

$\lambda/2 \times 0.95$

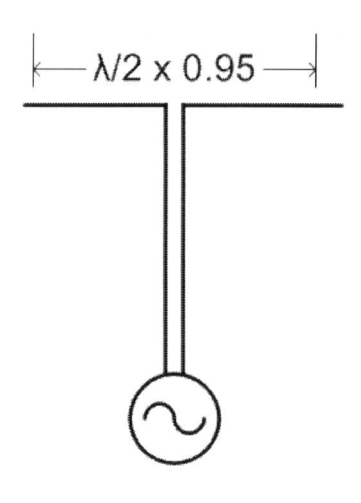

Figure 2-37

Example: If we wanted to make a halfwave dipole for 28MHz then:

Dipole length in metres = 300/28 x 0.95 x 0.5 = 5.09 metres.

We calculated the free space wavelength with 300/28, took 5% off it by multiplying by 0.95, then found a half wavelength by multiplying by 0.5. I strongly recommend you do not simply memorise the equations for antennas as some textbooks provide, but rather understand the calculations.

CURRENT AND VOLTAGE DISTRIBUTION ON A HALFWAVE ANTENNA

When a wave leaves the transmission line and enters a halfwave antenna, electromagnetic radiation is radiated. Any antenna which is a ¼ wavelength or any multiple thereof that antenna is resonant; you will see why shortly.

Currents and voltages will occur along the length of a dipole antenna. We do not care about the individual values of voltage and current. What we are most interested in is the *ratio of the voltage to current* at different points on the antenna.

Why? Do you remember earlier we talked about 'impedance' not actually being a physical thing? Any E/I ratio is, in fact, impedance. If we were actually to measure the current flowing in a dipole antenna, we would have to move an RF ammeter along the dipole and record the current at the centre, move out a bit and record the current again and so on many times until we got to the end of the antenna.

If we did this, we would be able to plot the current and voltage distribution of a dipole antenna. Figure 3-37 shows what we would get:

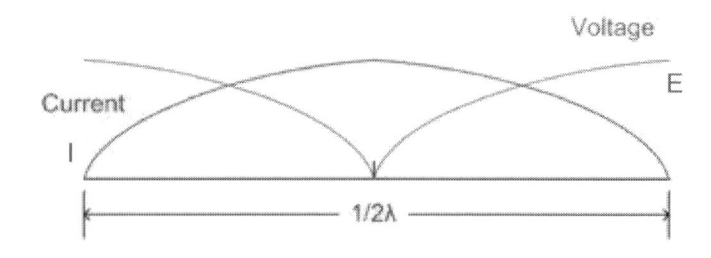

Figure 3-37

This is an important diagram. It is a halfwave dipole, but has many other uses as well, so we must learn what this diagram means.

The voltage distribution is labelled 'E' and the current 'I'. What sort of voltage-current ratio do we have at the centre of the dipole? Well, current is high and voltage is low. If I told you that I had a high current for a low voltage anywhere, what could you tell me about the impedance? The impedance must be low, as that's what a low impedance is, namely lots of current with little voltage. So the impedance at the centre of a halfwave dipole is low. The impedance is lowest at the centre than anywhere else on the dipole.

Look at the ends of the dipole. voltage (E) is high and current (I) is low. This must mean that the ends of a dipole have a high impedance, as that's what a high impedance is, a lot of 'E' for only a little bit of 'I.'

Why can we say that the halfwave dipole antenna is resonant? Simply because it is the correct physical length to accommodate the complete halfwave of current and voltage distribution.

Let's look at this idea of resonance from a different perspective. On the dipole antenna in figure 3-37, the length of the antenna is long enough to fit in a full halfwave of current and voltage distribution. A full halfwave is simply two-quarter waves.

At the centre of the antenna the current and voltage represents a low and resistive impedance. The impedance at the centre is approximately 72 Ohms. At the ends of the antenna, the current and voltage are representative of high and resistive impedance. The impedance at the end is 2000-3000 Ohms. In theory, the centre is zero Ohms and the ends infinite Ohms.

You also know now what the current and voltage distribution on a λ/4 antenna is, it's just half of the distribution shown on the antenna in figure 3-37. That is, from the centre to one end.

We can say that a quarter wave antenna is resonant simply because it is the correct physical length to accommodate a complete quarter wave of current and voltage distribution. At either end of a quarter wave, the impedance is resistive only. Being resistive implies resonance. With antennas, like tuned circuits, being resistive is synonymous with being resonant. Any antenna which is λ/4 or a multiple of λ/4 is a resonant antenna.

Why are we concerned so much about impedance? Well, we have learnt that for a load to dissipate (or radiate in the case of an antenna) all of the power it must be resistive and the resistance must match the source resistance; remember Jacobi's Theorem. In the case of an antenna system, the source is usually, though not always, the end of the transmission line. For a halfwave dipole, we know (irrespective of what frequency it operates on) that the feed point impedance at its centre is about 72Ω and resistive, provided the antenna is resonant (the right length). So *ideally* we should feed a dipole with 72Ω transmission line and if we don't we need to do an impedance transformation, say from 50 to 72Ω using a transformer (balun). In practice using 50Ω line to a centre fed dipole is close enough. However, a 1:1 balun *must* be used on any coaxial fed dipole, or you will not have a dipole and the feedline *will radiate*.

In practice, the actual feedpoint impedance, radiation pattern and many other properties of a dipole or any other antenna are affected by the environment. Some factors to consider include the height above ground, the proximity to metal roofs and other conductive objects and other antennas.

PHYSICAL and ELECTRICAL LENGTH

Because the velocity of propagation of a wave along an antenna is slower than through free space, the electrical and physical antenna lengths are not the same.

The physical length of an antenna is the material length in metres you would get with a measuring tape.

The electrical length is not a physical measurement. It is the electrical wavelength and this cannot be physically measured. The electrical wavelength must be calculated. The electrical wavelength is typically 5% longer than the physical length of the antenna that has open elements such as a dipole. Loop antennas, do not have the same end effects as open ended antennas.

ISOTROPIC ANTENNA

An isotropic antenna is a theoretical antenna. An isotropic antenna is one which radiates equally in all directions. If you imagine an isotropic antenna at the centre of a large balloon and the isotropic antenna is radiating visible light, then the surface of the balloon would be equally illuminated.

The isotropic antenna is used as a mathematical model in order to evaluate the directional properties of practical antennas. Almost all antennas have a gain greater than an isotropic. If an antenna with gain was placed at the centre of a large imaginary balloon and the radiation was visible light, the illumination of the balloon would not be even. There would be brighter regions because the antenna has gain in the direction of those regions.

A dipole has a gain of 2.14dB above an isotropic. Hence, a dipole antenna is said to have a gain of 2.14dBi, i.e. dB relative to an isotropic. Electrically very short rubber helical antennas, often referred to as a 'rubber duck', have a loss of several dB when compared to an isotropic. Such a lossy antenna would be described as having a gain of $-n$dBi. A negative gain means less gain than an isotropic antenna.

Some antenna manufacturers will use the isotropic antenna as their reference point when quoting antenna gains; others will use the dipole as the reference. The purchaser should be aware which reference is used when gain figures are quoted, as there is a 2.14dB difference between the two reference levels. A gain of 3dBd – means 3dB gain above a dipole. The same antenna would be 3dBd + 2.14 – 5.14dBi The last gain figure looks better. This is often done to promote the sale of the antenna.

ANTENNA GAIN

Most radio communications are point to point or at least confined to an area. An antenna which directs most of its energy toward the receiving station can be said to have a power advantage over an omnidirectional antenna. The power gain of a transmitting antenna is the ratio of the power radiated in its maximum direction of radiation compared to that radiated by a standard antenna, usually a dipole or the theoretical isotropic.

Suppose radio stations (A) and (B) are in contact with station (C). Station (A) is using a dipole antenna and (B) a Yagi with a gain of 6 dB above a dipole.

How much would station (A) need to increase their transmitter power to the same power advantage of station (B)?

Station (A) would need to increase their transmitter power by 6dB or, in other words, quadruple it.

A transmitter power increase by a factor of four or an antenna gain of 6dB should increase the received signal by one 'S' point. That is, provided the S-meter is correctly calibrated.

There are many traps when attempting to evaluate antenna performance on air, as in the latter example. Sometimes a dipole antenna will out-perform an antenna with higher gain because it has a more favourable angle of radiation for the propagation conditions in existence at the time of the comparison. Improved communications when using a directional antenna may at times be due to enhanced received signals and reduction in noise at the receiver, rather than transmit power gain.

The great advantage of antenna gain over increasing transmitter power is that you get the gain advantage twice. A gain antenna gives you gain on both transmit and receive.

Radiation Resistance.

Radiation resistance is an imaginary resistance which, when put in place of an antenna, dissipates as much power as the antenna radiates.

The higher the radiation resistance, the more efficient the antenna is as a radiator. With some antennas (e.g. a dipole or a quarter wave antenna) the feedpoint impedance and the radiation resistance are the same. If the feedpoint impedance of an antenna were to be changed by some impedance conversion device, this would not alter the radiation resistance.

Loss resistance

Loss resistance is an imaginary resistance that represents all the power losses of the antenna. For example, some power is lost due to the electrical resistance of the antenna, ground losses and nearby conductors.

Feedpoint impedance

Feedpoint Impedance is that impedance seen by the feedline where it connects to the antenna. Feedpoint impedance is determined by the E over I ratio at the point of connection.

MORE ON RESONANCE AND RESISTANCE

Any tuned circuit is resonant when both the inductive and capacitive reactances are equal.

At resonance the reactances completely cancel, leaving only the resistive part of the impedance. The shortest length of wire that can be resonant without any artificial loading is a quarter wavelength. A quarter wavelength of wire will have zero net reactance and a resistive feedpoint impedance of about 36Ω (half of the dipole).

If an antenna is less than a quarter wave, it will appear capacitive and will require the addition of some inductance, known as a loading coil. The added inductance cancels out the capacitive reactance and resonates the antenna.

An antenna which is longer than a quarter wave but less than a half wave will have a net inductive reactance and will require the addition of some capacitive loading to resonate it.

Antennas that are too short are capacitive and need inductance to be added. Antennas that are too long are inductive and need capacitance to be added.

Example 1.

Say you are going to use an end-fed antenna (just a piece of wire from the antenna tuner) on 80 Metres and the length of wire you have is 25 metres. Is this antenna too short or too long and if it is not the correct length will you need to add capacitance or inductance to it?

There is no need to run to the calculator. This antenna is to be used on the 80-metre band; that's a broadsword but close enough to tell us a quarter wave is roughly 20 metres. We have 25 metres of wire and we only need 20 metres. If you use all the wire, then the antenna will be too long. All long antennas are inductive; therefore, they need capacitance added to cancel out that inductance. This antenna will also require impedance matching. This is all done by the ATU. The ATU is doing the impedance match and also adding capacitance to bring the antenna to resonance.

Example 2.

An end fed wire antenna is 2.2λ long. Is it too long or too short and what do you have to do to bring it to resonance? We will assume that you do not want to cut the wire. Okay first is too long or is it short? What is the nearest multiple of a 1/4λ? Well, it is definitely 2.25λ, so the antenna is a bit short. Short antennas are capacitive; therefore, they need inductance to be added and also an impedance match. The ATU will do this for us by adding inductance and impedance match to make the antenna electrically 2.25λ.

FIVE-EIGHTH WAVELENGTH

One popular antenna is the 5/8λ vertical (figure 4-37). Now a 5/8λ antenna cannot be resonant since it is not a multiple of a quarter wave. The whip part of the antenna is in fact 5/8th of a wavelength and has a net *capacitive reactance*.

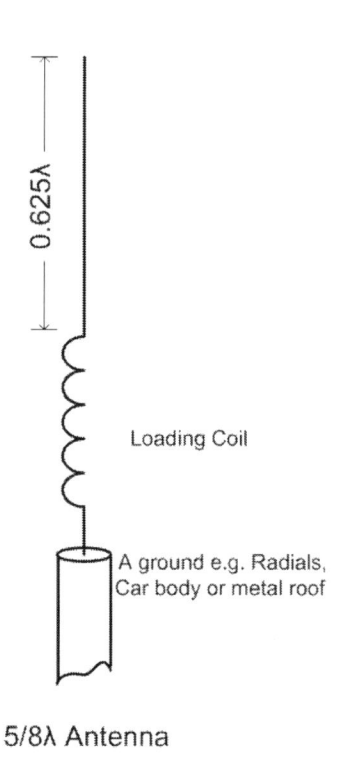

Figure 4-37

All 5/8λ antennas have an inductance (loading coil) fitted somewhere along the length of the antenna to cancel out the capacitive reactance and resonate the antenna. The 5/8λ antenna is tuned by the loading coil to resonate as a 3/4λ antenna.

So really; a 5/8λ antenna is electrically really a 3/4λ.

THE QUARTER WAVE ANTENNA

A quarter wave antenna (figure 5-37) is normally used with radials or installed on a flat conducting ground plane, such as the roof of a car or house/shed, or just a good conducting ground.

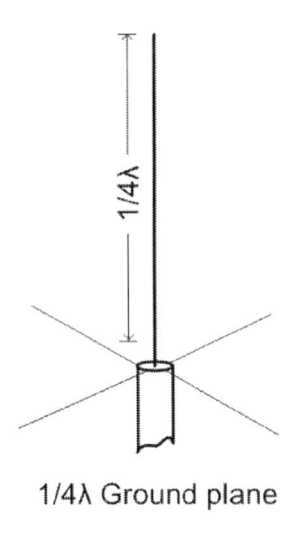

1/4λ Ground plane

Figure 5.37

A ground plane is simply a reflector. Some of the radiation from the antenna will be reflected by the ground plane and this reflected wave will interact with the incident wave from the antenna in such a way as to modify the radiation pattern of the antenna. In the case of a quarter wave antenna, the 'ground' should be as uniform as possible around the antenna so as to provide an omnidirectional pattern. This antenna has a low angle of radiation which makes it a good DX antenna.

Where it is not possible to mount the antenna on a conducting surface such as the earth or a car body, the ground plane effect can be simulated by attaching at least three; quarter wavelength radials to the feedpoint on the earth side of the feeder. Radials have another advantage, in that by lowering them to an angle of about 45 degrees to the horizontal, the feed point impedance can be raised from 36Ω to 50Ω to provide a closer match to a 50Ω coaxial cable.

I have seen it asked in exams why the radials of a ground plane antenna are drooped downwards at 45 degrees. One of the amusing answers, which I have often seen ticked as correct is, "to prevent birds from standing on them"! - Wrong.

ANTENNA LOADING

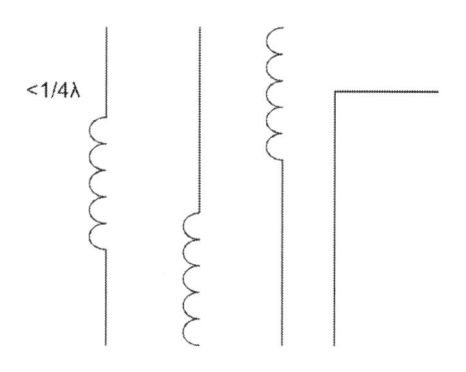

Figure 6-37

Loading, as discussed earlier, is adding inductance or capacitance to an antenna in order to make it resonant.

Top loading is a name given to various methods of adding capacitance or inductance to the top of the physically short antenna. Top loading not only adds the desired reactance to resonate the antenna but shifts the current distribution pattern higher up the antenna, increasing the radiation resistance and hence the radiation efficiency. What I am saying is that top loading is preferable to bottom or even centre loading. However, for mechanical reasons, particularly with mobile antennas, bottom loading is often used. The higher the loading coil, the better. An alternative is to distribute the inductance along the length of the antenna, perhaps even winding turns closer as you approach the top. This can be done on a fibreglass rod and the whole antenna coated with paint or sealant, or heat shrink tubing can be used. This is the method used for so-called helical mobile antennas.

On the low HF bands, the length of a grounded quarter wave can become impractical. An antenna that is physically short will not be resonant and its input impedance has a high capacitive reactance. Consequently, antenna current decreases and radiation decreases.

The radiation resistance decreases from 36Ωs for a quarter wave antenna towards zero as the length is reduced. Because of the short length, the current maximum will not occur on the antenna, but on the transmission line. The loss of the high current peak on the antenna drastically reduces the radiator power. Adding a loading coil at the base of a short vertical would resonate the antenna and increase the radiation resistance by about five Ohms. However, the current loop would still be low on the antenna and high I²R losses in the coil would reduce the radiation efficiency.

These difficulties are overcome by using top loading. A cylinder, sphere or disc placed at the top of the antenna acts to increase the shunt capacitance to ground which, in effect, is the same as adding inductance in series with the antenna, thereby bringing the antenna to resonance. The current distribution on the antenna is "pulled up" higher on the antenna resulting in higher radiation resistance and efficiency. There is no real significance in the shape of the metallic top loading device. The shapes mentioned earlier are normally used for the ease of capacitance calculation and appearance sake. A piece of wire can be used for top loading making the antenna appear as an inverted "L" or a "T".

Figure 7-37 – A capacitive hat creates shunt capacitance between the top of the antenna and ground. This shunt capacitance is in series with the antennas capacitance. When you connect capacitors in series the capacitance is reduced.

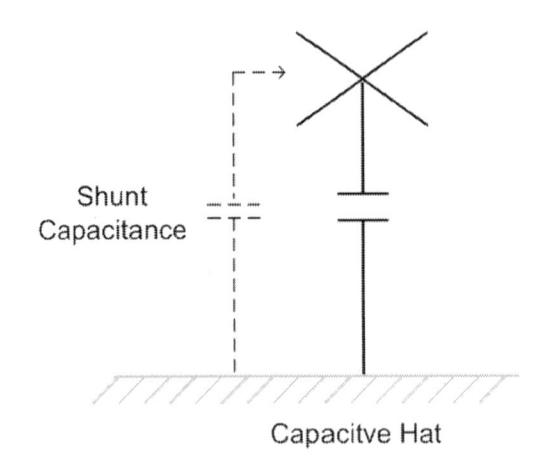

Figure 7-37

The price paid for loading any antenna

All antenna loading comes at a cost. The biggest cost is that the more an antenna is loaded the narrower its operating bandwidth. For example, a full-size quarter wave antenna on low HF may operate across an entire amateur band (200-300kHz) whereas linear loaded two-metre-long whip antenna may only operate across 20-30kHz before adjustment has to be made.

Loaded antennas have:

1. less operational bandwidth
2. lower feedpoint impedance sometimes making matching more difficult
3. greater losses – the loss resistance is greater.

YAGI Antenna

The parasitic beam, or Yagi array (named after Dr. Hidetsugu Yagi of Tokyo University), was invented in 1926 and was first used in the amateur service about 1935.

A Yagi antenna must have one driven element, a dipole and at least one other parasitic element (refer to figure 8-37). In our earlier discussion of a dipole, I mentioned that the radiation pattern, input impedance and many other characteristics, are determined by the height above ground and nearby antennas, among other things.

Well, a Yagi is just a dipole with other parasitic dipoles nearby (on the same boom). The parasitic dipoles receive radiation from the dipole and then re-radiate. So, we have energy being radiated from the main driven element and then after a time delay, the energy is picked up and re-radiated by the parasitic elements. By controlling the spacing and length of the parasitic elements, the antenna array can be made to have a main lobe in one direction, providing substantial gain.

THREE ELEMENT YAGI

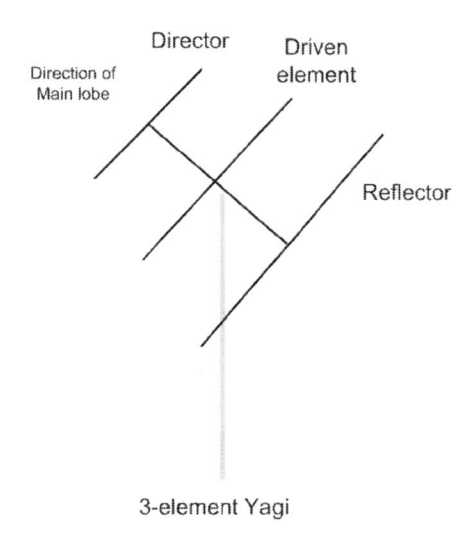

Figure 8-37

This antenna has three elements: driven element, reflector and director. The main lobe of radiation is in the direction of driven element to the director. The physical length of the driven element is found by the same equation as that given for a dipole. The reflector is 5% longer than the driven element and the director 5% shorter. The spacing between the elements can vary from 0.15 to 0.2 wavelengths. Maximum forward gain is obtained at a spacing of 0.18 wavelengths.

The director and reflector are called parasitic elements as they intercept energy radiated from the driven element and then re-radiate it. When the parasitic elements absorb power, they re-radiate it with a wave pattern like that of a dipole. However, because of the propagation delays introduced by the spacing of the elements, wave cancellation occurs toward the rear of the antenna and wave reinforcement in the forward direction. This wave reinforcement results in a gain of about 6-8.5dB over a dipole. The feedpoint impedance is about 18-25Ω. Some type of matching device is required to convert the feed point impedance to 50Ω if coaxial cable is used.

Additional parasitic elements can be added to a 3 element Yagi to increase the power gain further. Additional elements are always directors, placed in front of the driven element. Adding more reflectors provides no appreciable gain advantage at HF frequencies. Doubling the number of directors will increase the power gain by approximately 3dB. Adding parasitic elements to a Yagi array decreases the antenna bandwidth.

The front-to-back ratio of a three element Yagi varies from about 15dB to a maximum of 25dB. The front-to-back ratio is the ratio of the power radiated in the forward direction to that radiated to the rear of the antenna, expressed in decibels.

Each element of a Yagi antenna has a current and voltage distribution like that of a simple dipole. As the impedance at the centre of each element is the same no current will flow between points of equal impedance on an antenna, a metallic boom may be used.

If two identical Yagi antennas are stacked in phase and the gain of each individual Yagi is 6dB, the stacked array will have a total gain of 9dB, that is, 3 dB more.

The additional second antenna would theoretically double the amount of power radiated in the forward direction. This is an effective power increase of 3dB, giving a total system gain of 9dB. In practice, the additional gain would be slightly less than the theoretical maximum. Stacking a third antenna would result in an improvement of an additional 33% or 1.24dB at best.

The impedance of a Yagi depends on the number of elements. The elements are just dipoles in parallel. Think of 70Ω resistors in parallel, if you had 7 elements you can expect the impedance to be close to 70/7=10Ω.

FOLDED DIPOLE

A folded dipole antenna (refer to figure 9-37) can best be described as two dipoles connected in parallel. Folding a dipole has the effect of increasing the feed point impedance. With two parallel dipoles, the feedpoint impedance is increased by a factor of $2^2 = 4$, giving a total impedance of 300Ω for a standard dipole. With three parallel dipoles, the feed point impedance becomes $2^3=8$ times the original feed point impedance. The dipole forming the driven element of a Yagi antenna is often folded to increase the feed point impedance by a factor of 4.

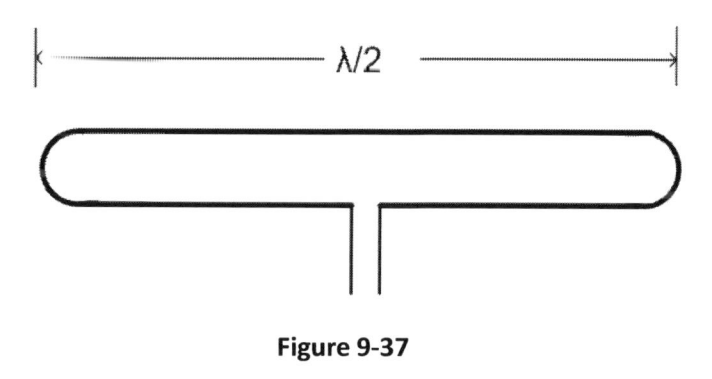

Figure 9-37

Folding a dipole also increases its bandwidth; the more parallel dipoles you have with their ends shorted, the greater the bandwidth. This technique to obtain very wide bandwidth is used by commercial stations on HF.

SINGLE QUAD LOOP ANTENNA

All loop antennas are just one wavelength loops. As there is no end effect like this say for a dipole, the electrical length of a one wavelength loop is a little longer than 1λ. A good place to start is about 5% longer. There are many loop antennas that simply take their name from the shape. Delta; shaped like a triangle, Quad; shaped like a square. On VHF and above (due to the ease of mounting small size loops) you could use a circle. The loops can form the elements of an array like a Yagi. So you could have a 3-element delta loop beam for an example. A cubical quad is just a two-element beam made from two quad loops one for the driven element and the other for the reflector. If you are only using two element beams, then you would choose a reflector over a director for a small gain advantage.

A quad antenna is really a "pulled out" folded dipole. Just imagine pulling out a folded dipole to form a square or diamond. A folded dipole has a radiation resistance of 288Ω (4 x 72). When half pulled out to form a quad loop, the antenna is half way between being a folded dipole and a shorted transmission line (see next chapter). A shorted half wavelength of transmission line would have an input impedance of zero Ohms one-half wavelength back from the termination. It is not surprising that the feedpoint impedance of a single quad loop is the mean (average) of these two values: (288+0) / 2 = 144Ω. When fed at the centre of one side, the impedance decreases slightly to 125Ω.

With this antenna, there is no "end" to the wire elements as the quad forms a continuous loop. The shortening effect which occurs with other antennas is not applicable since there is no "end effect". In fact, for a quad loop to be resonant, it has to be slightly longer than a quarter wave on each side.

Summarising the important characteristics of a quad loop:

Radiation resistance is 144Ω when fed at a corner.
Radiation resistance is 125Ω when fed at the centre of one side.
Power gain is 1.4dB above a dipole.
Bi-directional radiation pattern.
The physical length of each side is 0.257 λ.

TRAPPED DIPOLES AND VERTICALS

Traps used on HF antennas are parallel tuned circuits. Recall that at resonance; a parallel tuned circuit has very high impedance. The higher its "Q", the higher the impedance. In fact, if the "Q" was high enough, a parallel tuned circuit impedance could be likened to an open circuit switch.

Trapped antennas use the high impedance of a parallel tuned circuit to switch in and out different sections of the antenna. Have a look at the three band trapped dipole of figure 10-37.

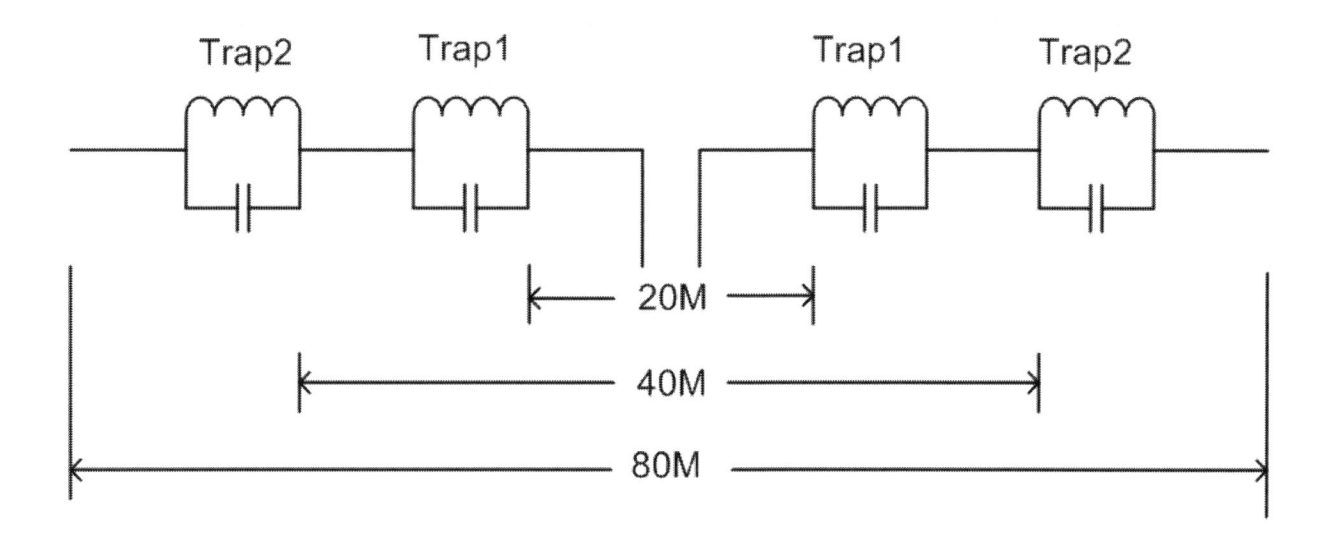

Figure 10-37

The highest frequency band is 20 metres. The Trap1's will be parallel resonant on a 20 metre frequency and the transmitter will only "see" the smaller antenna in the centre. On 40 metre the Trap1's are no longer resonant. The 40-metre band is the middle section of the antenna shown. The Trap2's are parallel resonant on 40 metres, isolating the ends of the antenna. On the 80 metres band (the lowest frequency) the whole antenna is used. When the traps are not used at their resonant frequencies, they act like loading coils and effectively shorten the length of antenna wire needed.

Half of this antenna could be stood up vertically and you would have a trapped vertical. You can even make a trapped multiband Yagi using this same principle for the driven element, directors and the reflector.

The traps may be made from an actual high "Q" inductor and capacitor. However, there are easier methods on HF. On VHF and above the options for traps and other matching techniques becomes more difficult.

MULTIBAND DOUBLET

The simplest multiband dipole is just an ordinary dipole fed with good quality balanced transmission line. To permit operation on multiple bands an antenna tuner (ATU) must be used. This dipole will be resonant on any band where it is a halfwave long or a multiple of a halfwave. If a radio experimenter does not have sufficient room to put up a resonant dipole then, if a good ATU is available, the operator can put up a dipole of any length. This is usually the maximum length that the available space will permit. Such an antenna is often referred to as a doublet rather than a dipole as a dipole infers a halfwave antenna.

Figure 11-37 shows a doublet antenna fed with parallel line. It is best if each side of the doublet is the same length as you want it to be a balanced antenna. If you can't get exactly the same length do not be overly concerned. You can configure the doublet as an inverted 'V'. This antenna will operate well on any band that you can achieve an impedance match to your transmitter using the ATU.

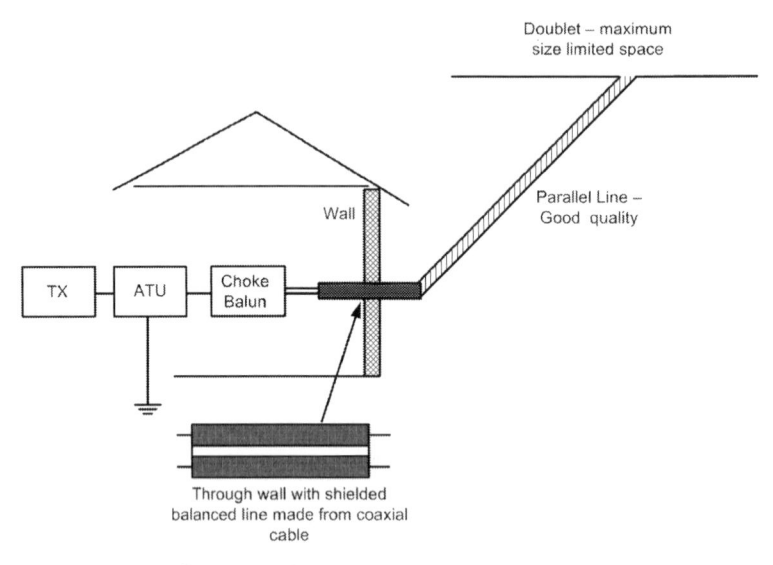

Figure 11-37

Entry to your station can be a problem because you have to maintain the *balance* of the transmission line. You can make a short shielded balance line using to two short lengths of coaxial line configured as a length of shielded balanced line. You must use balanced line to the antenna as this antenna will have high SWR. With reasonable quality balanced line, you do not care what the SWR is, as long as your tuner can provide a match to the transmitter. This antenna system will be efficient due primarily to the use of balanced line. High SWR on balanced line will not substantially increase the line loss as it would if coaxial cable were used.

Is this antenna resonant on all bands used despite its short length? Well it might seem odd but, yes it is resonant. The antenna system is brought to resonance by the ATU. That is the ATU provides the reactance necessary to make the entire antenna system (feedline and antenna) resonant. Since the feedline is balanced and does not radiate, the radiation is being done by a resonant antenna. Suppose on a particular band the antenna had a feedpoint impedance of 100-j20Ω. The 20Ω of capacitive reactance indicates that the antenna is not resonant right? No. When we match this antenna with an ATU in the station, the ATU well reflect a conjugate impedance to the antenna. That is, impedance reflections created by the ATU will reflect to the antenna the conjugate of 100-j20Ω which is 100j20Ω. So the antenna is reflected a reactance which makes it resonant. It does not matter what the 'R' is, it just matters that there is no reactance. Such a doublet will not behave exactly like a resonant dipole because it is the wrong length. However, it will behave as a resonant antenna.

If we tried to do the same thing with coaxial cable it would just not work as well. Why? This antenna system will have high standing waves on some bands, probably most of them. If coaxial cable were used the line losses would be much greater due to the electrical stress the standing wave produces on coaxial line. This means that the ATU will provide the correct conjugate impedance; but the coaxial line, due to is losses, will not reflect this conjugate to the antenna and the antenna will not be brought to resonance. It will still work, but not nearly as well and the losses will be much greater.

This doublet fed by parallel line is arguably the best multiband antenna you could use, if you have space restrictions.

THE FAN DIPOLE

The Fan dipole is just a number of resonant dipoles connect at the same feed point. This is shown in figure 12-37.

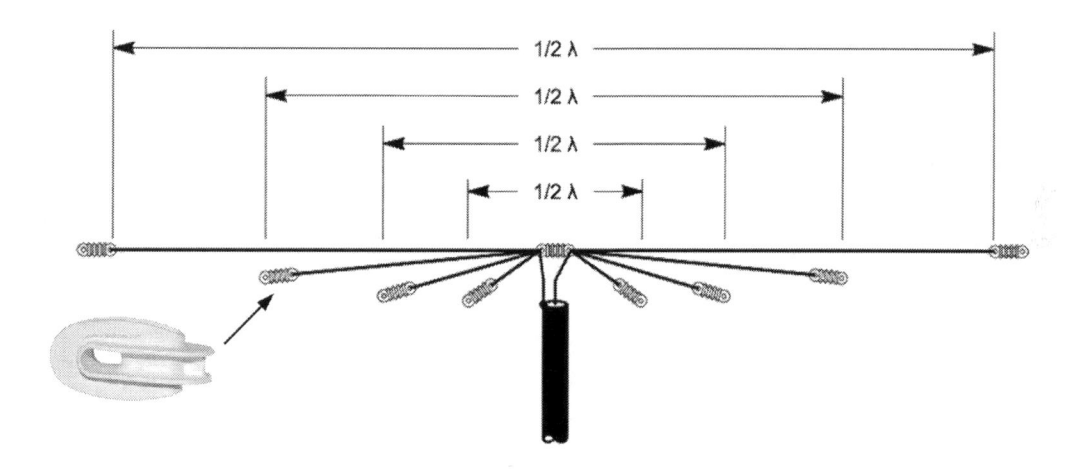

Figure 12-37

This antenna works very well on each band where there is a resonant dipole. The only downside to the Fan dipole compared to a single dipole is that the dipoles not in use do have some affect on the active dipole. This means they are fiddly to tune. If you're willing to do this then once you have got it right, or nearly right, you will have a good and efficient multiband antenna. This antenna is called a 'Fan' because of the multiple centre fed dipoles look a bit like a hand 'fan'. The dipoles can be parallel with each other and supported by spacers. The less bands you build it on the easier it will be to tune. A great antenna for restricted space. If you only have enough room for a 40 metre antenna then you can add dipoles for 20, 15 and 10 metres. If you wish you could feed the antenna with parallel line which would make it even more efficient. If you use coaxial cable, which is unbalanced, to feed any balanced antenna you **must** use a choke balun at the feedpoint. The Fan dipole shown in figure 12-37 must have a 1:1 balun at the feedpoint. This balun could just be the coaxial feedline wound around a ferrite rod and held in place by PVC tape or potted in a short length of PVC pipe using encapsulating resin.

OFF-CENTRE FED (OCF) DIPOLE

Any dipole not fed at the centre is an off-centre-fed dipole. The OCF dipole normally refers to a dipole antenna that is fed at a point 1/3rd from one end. Sometimes this dipole is referred to a 1/3rd – 2/3rd dipole. The reason for feeding the antenna at this point is that the impedance at that point is in the vicinity of 200Ω.

If the OCF designed for the 80 metre band it will be a ½λ on long on 80 metres. Use the standard half wavelength equation to work out this length, not forgetting to reduce the physical length by 5%. If this OCF is used on 40 metres it will be 1λ. However the feedpoint impedance will still be in the vicinity of 200Ω. This is the same with 20 metres and 10 metres. Figure 13-38 shows the dimensions of an OCF dipole.

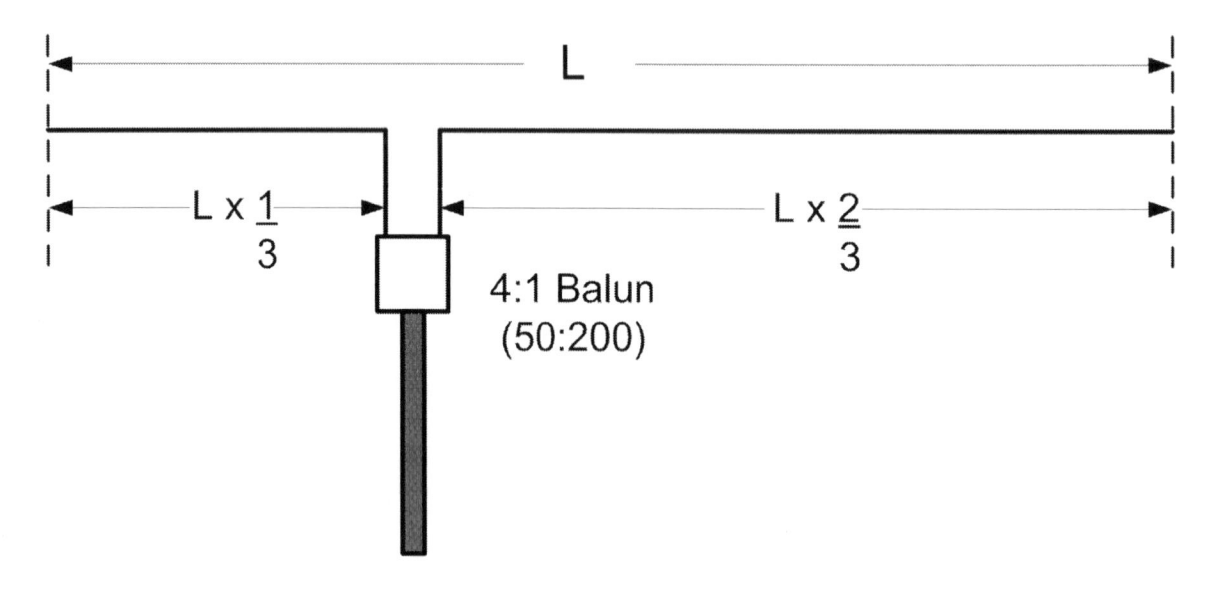

Figure 13-37

The balun is a 1:4 (200Ω:50Ω). The balun should be placed right at the feedpoint. Since the impedance at the feed point is around 200Ω the balun will be around 50Ω on the coaxial cable side. The impedance of the feedpoint will not be 200Ω on all bands but the balun will give us close to 50Ω for connection to 50Ω coaxial line. The impedance may vary at the feedpoint for 100Ω-300Ω. On the transmitter side this will translate to around 25-100Ω for a reasonable SWR, at no more the a little over 2:1. So it works well with a bit of a compromise. We are breaking a rule here which I will talk about shortly.

THE OCF – Why it works

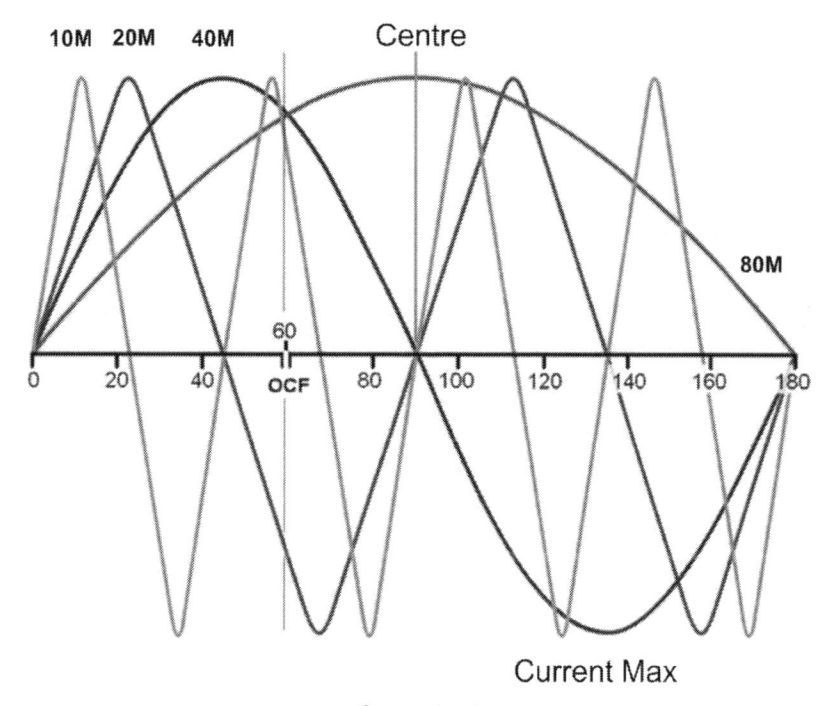

Figure 14-37

Figure 14-37 shows the current distribution on four radio bands 80, 40, 20 and 10. Notice how the current at 60° is much the same for all of these bands. The current is almost maximum for each band. If the current is about the same then the impedance at that point is about the same. This makes it possible to operate on the four bands shown and get an impedance we can easily cope with. You may have to tweak the SWR a little with an ATU.

The OCF is not, strictly speaking, a balanced antenna. There is a degree of unbalance due the different lengths of antenna on each side of the feedpoint. We have neglected this rule. We get away with it because the amount of unbalance is not great. Secondly, baluns have two main purposes. They allow us to connect two unequal impedances and they choke common mode currents. See Chapter 38 Transmission lines. For details of how transmission line baluns work.

For some, having space to put up an 80 metre halfwave dipole is difficult. You can build the OCF just the same as described on 40, 20 and 10 metres. You can droop the sides of the OCF within reason – just be careful not to unbalance it too much. If you are pruning the length of an OCF remove antenna wire proportionately from both sides to maintain the one third two thirds ratio. For all wire antenna adjustment, it is better to fold the wire back along the antenna rather than cut it off.

The OCF performance is close to that of a standard full size dipole on each band. The difference being that as the length of the antenna increases in wavelengths new lobes are produced towards the long dimension of the antenna. Figure 15-37 shows increasing radiation towards one end on the higher HF bands.

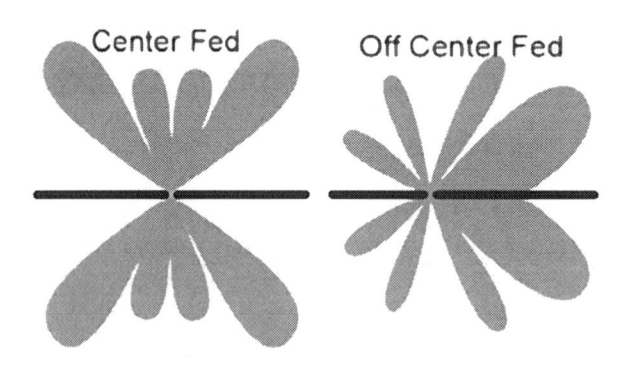

Figure 15-37

This is usually not a problem with the lower HF bands. All long wire antennas behave this way. The longer an antenna is in wavelengths, the more directional they become, in the line of the antenna conductor.

CO-LINEAR ANTENNA

In radiocommunications, a collinear antenna array is an array of dipole antennas mounted in such a manner that the corresponding elements of each antenna are parallel and collinear, that is they are located along a common line or axis.

Collinear arrays of dipoles are high gain omnidirectional antennas. A dipole has an omnidirectional radiation pattern when in free space and not influenced by any other conductors in that it radiates equal radio power in all azimuthal directions perpendicular to the antenna, with the signal strength dropping to zero on the antenna axis.

The purpose of stacking multiple dipoles in a vertical collinear array is to increase the power radiated in horizontal directions and reduce the power radiated into the sky or down toward the earth, where it is wasted.

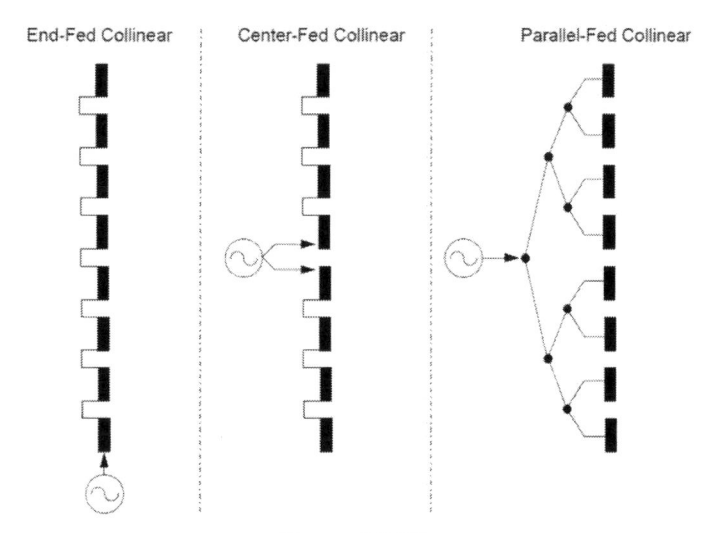

Figure 16-37

Theoretically, when stacking idealised lossless dipole antennas in such a fashion, doubling their number will produce double the gain, with an increase of 3.01 dB. In practice, the gain realized will be below this due to imperfect radiation spread and losses. The approximate gain of a collinear is 10Log(n) where n is number of elements. So for a 4 element collinear array of dipoles the gain is 10Log(4) = 6dB. Collinear arrays do not have to be made of dipoles, they could be Yagi's or many other types of antennas.

Collinear dipole arrays are often used as the antenna for repeaters. They are also sometimes used for broadcasting. Multiple directional antennas mounted vertically separated are referred to as "stacked" and if alongside each other as "bayed".

You will find many methods of building a collinear antenna with a Net search.

THE CORNER REFLECTOR

The corner reflector and related antennas with a reflector sheet or screen behind a dipole can achieve an antenna gain approaching 10dB. Corner reflectors become very practical at UHF frequencies. The corner reflector is related to the parabolic reflector shown in figure 17-37(a). The parabolic reflector is analogous to the parabolic mirror found in optical systems. If a parabolic reflector is sufficiently large, so the that the distance from the focus to the reflector is many wavelengths then optical conditions are approached.

If the reflector is close to the same dimensions as the operating wavelength, or less, the analogy to optics is not complete, since the radiator is coupled to the reflecting sheet.

An alternative reflector consists of two flat conducting sheets which intersect at an angle so as to form a corner. This is called a corner reflector antenna, 18-37(b) and (c) show two views of a corner reflector. An advantage of such a sheet reflector. as long as the sheet is sufficiently large, the dimensions are not critical and there are no tuning adjustments. There is no focus point as there is with a parabolic reflector.

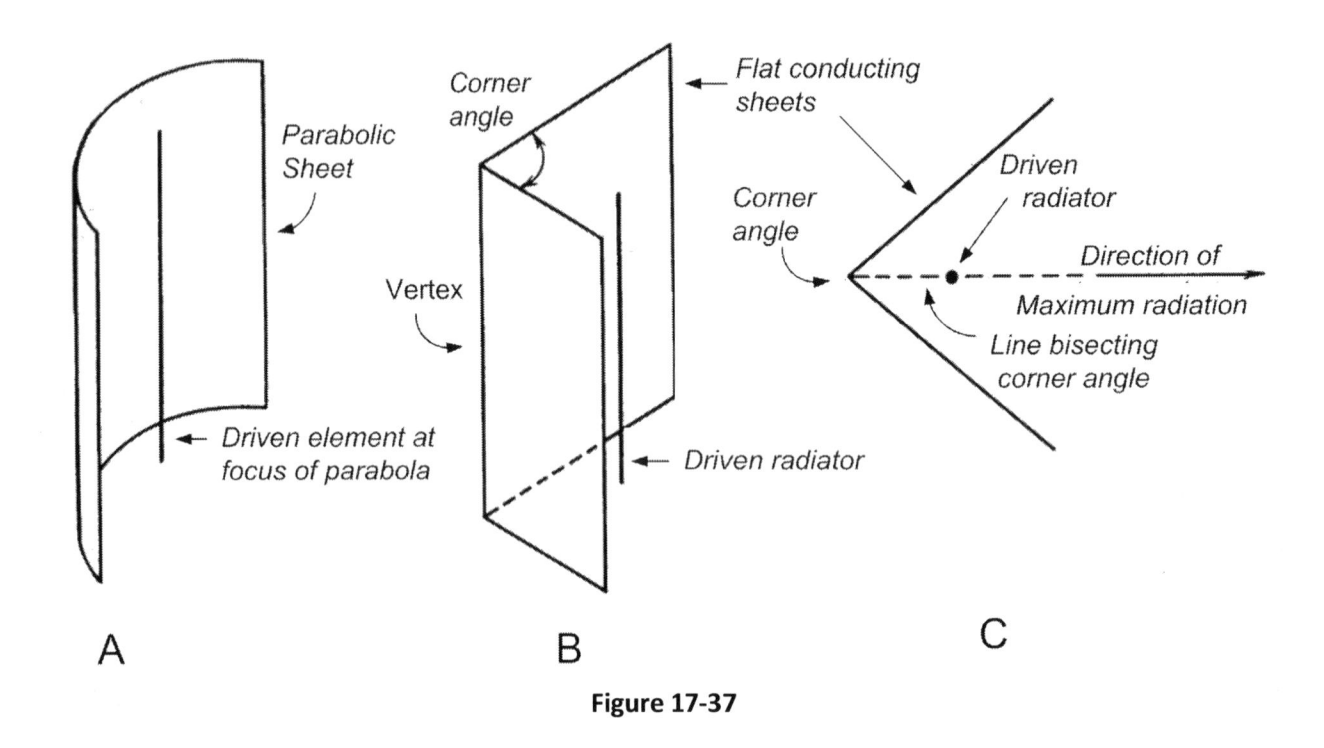

Figure 17-37

I have shown a vertical end fed antenna in figure 18-37. A centre fed dipole is more often used. The corner does not have to be exactly 90°. At 180° we would have a flat sheet reflector with some loss of gain. The distance from the corner to the radiator is not critical and be varied over wide limits with little change in the directivity of the antenna. Especially suitable are values of 0.35 and 0.5λ. The antenna can be rotated for vertical or horizontal radiation.

You may have seen a so called coke-can antenna to improve mobile telephone reception. These are just a coke-can unfolded to make either a parabolic or corner reflector held behind the phones antenna. They do work; though you do not have to use a coke-can. Try it if you have poor mobile phone reception.

The V-Beam

The unterminated V-Beam is a long dipole parallel to the ground drawn into a V and at least 1 wavelength on each leg. The radiation pattern of the unterminated version is *bidirectional* much the same as if you had two Yagi's back to back.

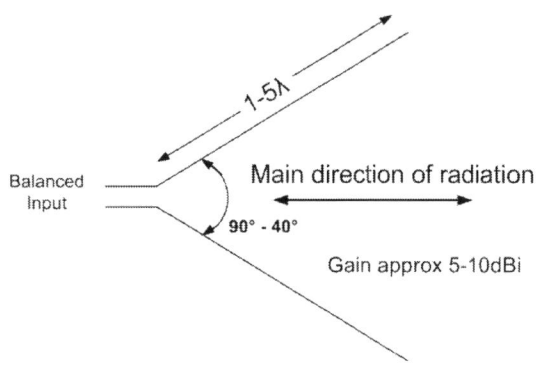

Figure 18-37

The exact gain depends on the angle between the antenna conductors and their length. At a length of 1λ and an angle of 90° the gain is approximately 5dB. At 5λ and 40° this increases to 10dBi.

The V-beam can be converted to a *unidirectional* antenna by terminating ends of the antenna though a 500-600Ω non-inductive resistor to ground. The terminating resistors prevent reflected waves on the antenna. The direction of radiation is then mainly from left to right in figure 18-37.

BASIC RADIATION PATTERNS

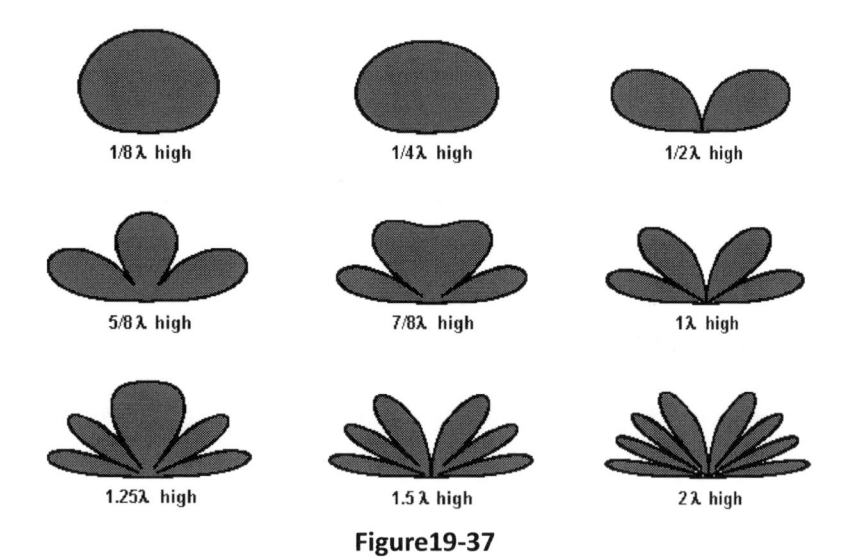

Figure 19-37

Figure 19-37 shows the radiation patterns for a horizontal dipole at different heights above ground. The dipole has a theoretical gain of 2.14dBi. The thing to notice is that the closer the dipole is to ground the higher the radiation pattern. At 1/8λ most of the radiation is straight up! Suppose you were on the 80M band using a half wave dipole 10M above the ground – then you can expect to have a cloud warmer as shown above. However how many of us can get our antennas up higher than 10-12M. So we just put up with the fact that much of our radiation is going upward and our skip distance is short.

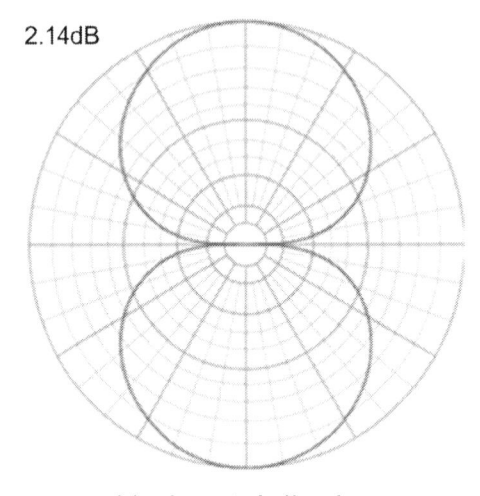

Horizontal dipole

Figure 20-37

The typical radiation pattern of a horizontal dipole is shown in figure 20-37. This is the radiation pattern as seen from above. A horizontal dipole run East-West will radiate best North-South. The horizontal dipole is bidirectional. If the dipole is made longer than one-half wavelength, radiation moves towards the ends of the antenna conductor. Such is the case if you have an OCF (off centre fed) dipole for 80 Metres and used it on 10 metres. A standard dipole running East-West will radiate the best in a North-South direction. Inverted V's and G5RV etc. are just different versions of a dipole with different mounting shapes and strategies to overcome impedance matching and/or multiband operation. None will work as well as a dedicated dipole for that band, but the G5RV and the OCF offer greater band coverage with a single antenna with a just a small performance trade-off.

The Yagi is a unidirectional antenna (one-way). Typical elevation and azimuth patterns are shown in figure 21-37.

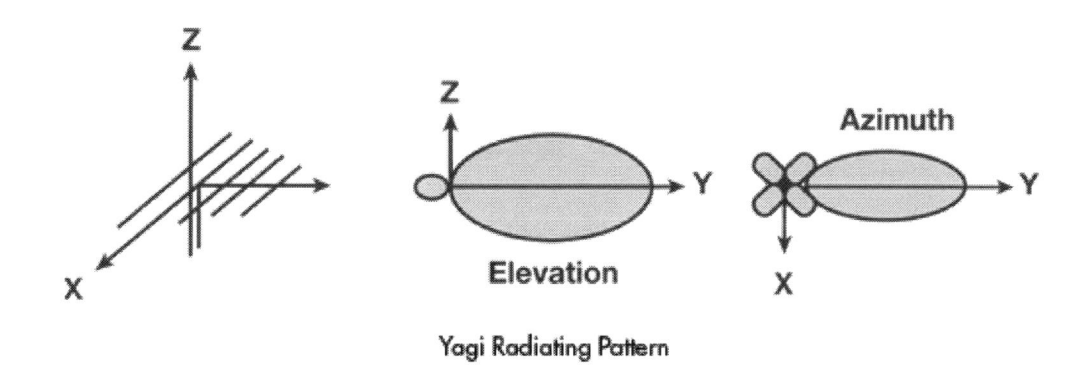

Yagi Radiating Pattern

Figure 21-37

Vertical antennas have an "omnidirectional" radiation pattern. That is not strictly the definition of "Omni" which means all-around – suggesting an isotropic. Omnidirectional with regard to vertical antenna means all-around in the horizontal plane.

ANTENNA MATCHING

Gamma Match

The Gamma match is most often used for matching a Yagi. The impedance of a Yagi depends on its number of elements. The elements are just shorter or longer dipoles than the driven element and act as if they were all connected in parallel. The Gamma match is really a stub match. The centre of feedpoint impedance of a Yagi with several elements can be very low 10-20Ω. As you move out from the centre of the driven element the impedance increases. At the end of the driven element the impedance is very high. It is logically then somewhere between the centre and the end of the driven element that impedance will be 50Ω. The Gamma match extends the centre of the coaxial cable out along the driven element to where the impedance is 50Ω. This position is adjusted by the shorting bar. The Gamma Rod has a small amount of inductive reactance. A series capacitor exists between the Gamma Rod and the sliding capacitor rod to tune out any inductive reactance with an equal amount of capacitive reactance.

The shorting bar and Capacitor Rod are adjusted for minimum SWR or using an impedance meter; the correct impedance, usually 50Ω, and no reactance.

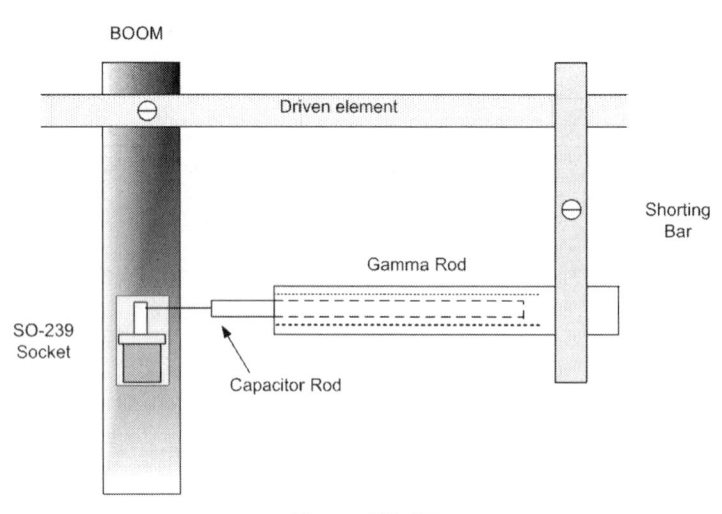

Figure 22-37

This type of matching arrangement can be used on many other antenna types. It is not difficult to adjust. When you are adjusting most matching devices you are not bringing the antenna to resonance. You are just adjusting for the best impedance match between your feedline and the load, your antenna. The Gamma and other stub type match is so effective you could get a match to one or two soft drink cans in series. I have also heard of people "loading up" tree trunks! A Gamma and other stub type matching will allow you to match to almost anything. This match will have little effect on the efficiency or performance of your antenna.

Stub Match - J-pole

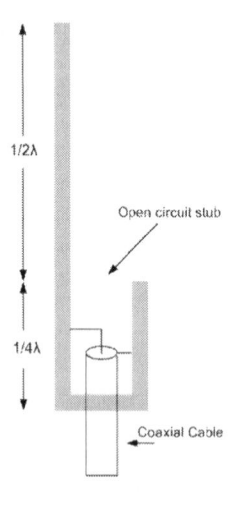

Figure 23-37

The standard J-pole is just an end fed halfwave antenna as shown in figure 23-37. The end of a halfwave dipole is a high impedance. The low section of the J-pole is a quarter wave short circuit stub at the bottom and an open circuit at the top. The end of the half wave antenna 'sees' a high impedance at the open end of the stub. Between the open and short circuited ends will be a 50Ω impedance were you can connect your cable for 1:1 SWR. Usually some mechanical sliding arrangement its used to move the feedpoint on the stub up and down. The J-pole is made from one piece of aluminium or copper pipe. The stub can be clamped to a metal pipe to raise the antenna height if needed. It is a good idea to place a simple 1:1 choke balun (Guanella) on the feedline close to the antenna. Just wind the feedline around some ferrite rod and tape it in place. You can pot the balun in resin if you want but there is no need if the coaxial cable is designed to be in the weather. If you are going to put connectors on the balun, then you will need to box or pot it. Once again a stub balun will match almost anything. The transmitter does not know or care about what sits on top of the stub. Adjusting the stub is not tuning the antenna. The stub just provides the correct match for the feedline. A good SWR does not mean anything about the performance or efficiency of the antenna as a radiator.

Matching – Mutually coupled transformers

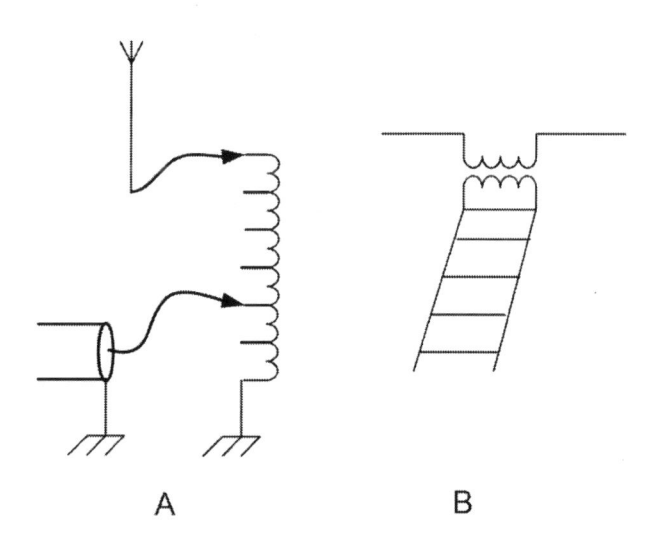

Figure 24-37

Figure 24-37(a) shows a popular method of matching a mobile antenna installation using an auto-transformer. The transformer is placed usually in the boot of the vehicle and may be just 15-20 windings tapped on a toroid core. Both the antenna and the position of the coaxial centre tapped can be moved around to get the best match. When you are actually adjusting the amount of inductance in the antenna you are also adjusting its resonant frequency. Mutually coupled transformers are narrow band and therefore not popular at a base station. In a HF mobile station though the antenna is going to be heavily loaded and as a consequence of this the bandwidth will be narrow. No matter what type of matching you use you may only get 20-30kHz of operation bandwidth. For most HF mobile communications this limited bandwidth is easy to cope with. Once you know the tap positions for several frequencies and bands it is not hard to stop, open the boot and adjust the tap positions. If you travelling, you may not bother with this and choose to just operate on or near your favourite HF frequency.

Figure 24-37(b) shows how you could use a mutually coupled transformer at the feedpoint of a dipole. The transformer is not a 'transmission line balun'. It is a transformer operating on same the principles as transformer in linear power supply. Such transformers have a narrow bandwidth and are not the method of choice. A better choice for impedance for transformation would be a transmission line balun (TLT). See Transmission lines chapter 38.

There are a number of other matching methods but most a variation on the ones covered. A Delta match is similar to a gamma match. The Delta match is just a double sided gamma match. The stub match used on the J-pole could be used on just about anything for example, the centre of a dipole.

38. TRANSMISSION LINES

A transmission line connects between a transmitter and an antenna and its purpose is to deliver all the signal power to the antenna. A perfect transmission line does not radiate or receive any energy and does not have any losses.

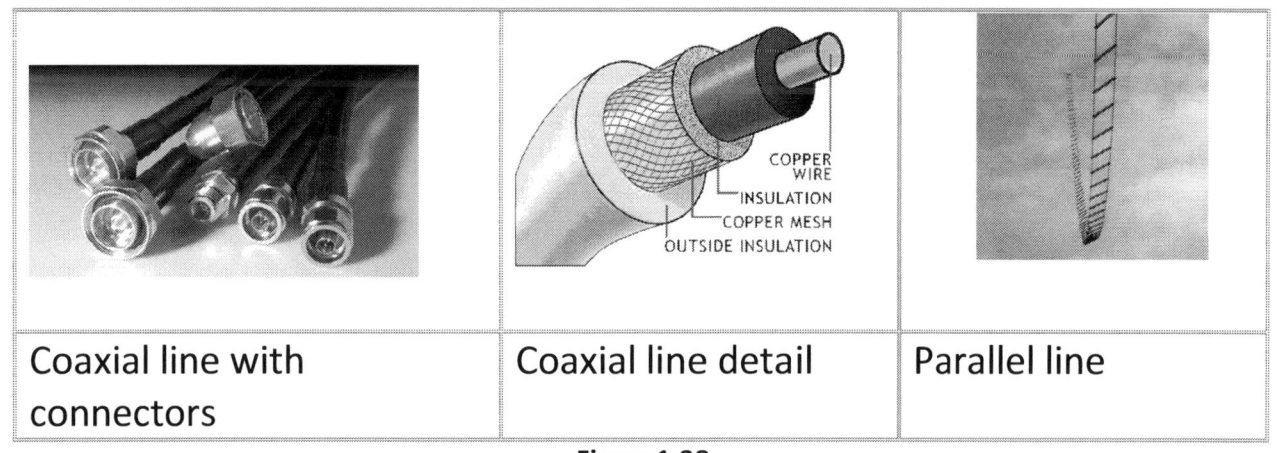

Coaxial line with connectors	Coaxial line detail	Parallel line

Figure 1-38

There are many other types of transmission line other than that shown in figure 1-38. However, for radio work below microwave frequencies, coaxial and balanced line are the only two you are likely to encounter.

CHARACTERISTIC IMPEDANCE

When an electromagnetic wave travels through free space, the current and voltage distribution of the wave settles into a particular ratio. In free space, the current and voltage distribution of an electromagnetic wave settles into the ratio E/I equal to 120Π or 377Ω. So we say the characteristic impedance (Z_o) of free space is 377Ω.

As we learned in the chapter on electromagnetic radiation, it is the properties of free space that determines the velocity of light and all other electromagnetic radiation.

Similarly, when a wave travels along an antenna or in a transmission line of infinite length, the current and voltage distribution of the wave will settle to a particular ratio of E/I and this is called the characteristic impedance (Z_o) of that line. Now the reason why I said a line of infinite length is to eliminate what is connected to the end of the line, that is, the load. We will come back to this later and connect a load.

Take a length of transmission line **without losses** and of infinite length. Transmit a wave into this transmission line. If we were to measure the current of voltage at any two points on this transmission line, we would get the same E and I. The ratio E/I will be constant and equal to the characteristic impedance of the line in Ohms.

WHAT IS A TRANSMITTER'S LOAD?

When you connect an antenna directly to a transmitter, the load for the transmitter is the antenna. When we use a transmission line to connect a transmitter to an antenna located elsewhere, the load for the transmitter is no longer the antenna. It is the transmission lines input impedance which, under most circumstances, will be the same as the characteristic impedance (Z_o).

We have discussed before the importance of matching a source (transmitter) impedance to a load (transmission line) impedance. Do you recall when we did the exercise on connecting different resistors to a battery? Only when the resistance connected to the battery was equal to the internal resistance of the battery did we get maximum power dissipated in the load resistor. Likewise, for maximum power to be transferred from the transmitter to the transmission line, the output impedance of the transmitter must match the input impedance of the transmission line. Keep in mind again that although we are talking about transmitters, the same applies to the reception of radio waves.

WHEN IS A 50 OHM TRANSMISSION LINE 50 OHMS?

At first, this may seem like a very silly question. After all, if you purchase 50Ω transmission line you expect it to act like 50Ω line. Unfortunately, whether a 50Ω transmission line behaves like 50Ω transmission line depends on how we use it. Just because it has 50Ω written on its side is no guarantee that it will behave as 50Ω. That's up to us to ensure! A transmission line will only exhibit its characteristic impedance when it is terminated in a load equal to its characteristic impedance. A 50Ω cable is 50Ω when it is connected to a load consisting of 50Ω of pure resistance. If a transmission line is terminated in a load not equal to its characteristic impedance, then the impedance of that line will vary from one point to the next along its length due to the presence of *standing waves*.

FACTORS THAT DETERMINE CHARACTERISTIC IMPEDANCE

The characteristic impedance of any transmission line is a function of the size and spacing of the conductors and the type of insulating material (dielectric) between them.

If the distributed inductance and capacitance per unit length of a line is known, then the characteristic impedance can be found from:

$$Zo = \sqrt{\frac{L}{C}}$$

Equation 1-38

To find the distributed inductance and capacitance per unit length, you can either measure it or look it up from a cable data book. A data book will tell you distributed L and C of a cable is for example a one-metre length. It does not matter what the length is as long as it is the same length for both L and C. Using this L and C, you can calculate the Z_o.

For example, looking up engineering tables for the inductance and capacitance per unit length for **RG58** we find: -

Inductance=L=333x10^{-9} Henry/meter or **333nH/metre**
Capacitance=C=1.333x10^{-10} Farad/meter or **133.3pF/metre**

If we insert this L and C into Equation 1-38 we get: -

$$Zo = \sqrt{(333\text{x}10^{-9}/(133.3\text{x}10^{-12})}$$

Do the division first then find the square root of that answer.

$$Zo = 49.98Ω$$

If we know the physical dimensions of the coaxial line, we can calculate the Z_o using equation 2-38. You very likely will not have to do this in radio assessments however you will probably need to know what determines the characteristic impedance of any cable. The answer is its dimensions and the materials, particularly the dielectric, used to make the cable. Equation 2-38 shows how to calculate Z_o from the construction of a coaxial cable. ε in these equations is the permittivity of the dielectric.

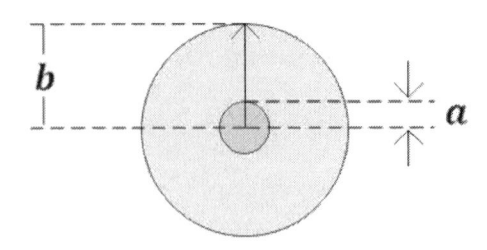

$$Zo = \frac{138}{\sqrt{\varepsilon}} \log \frac{b}{a} \ \Omega$$

Equation 2-38

Similarly, the characteristic impedance (Z_o) of a parallel line can be determined by primarily the dielectric used and its physical construction.

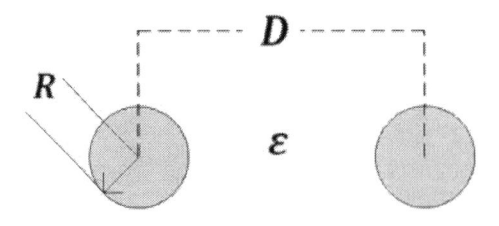

$$Zo = \frac{276}{\sqrt{\varepsilon}} \log \frac{D}{R} \ \Omega$$

Equation 3-38

RANGES OF Z_o

In the design of transmission lines, there are certain constraints which restrict the range of practical impedances that can be achieved. For two wire parallel lines, the Z_o is usually restricted to a range of 100Ω to 600Ω, while for coaxial lines the practical range of characteristic impedance is typically 30Ω to 100Ω.

Interestingly, it is no accident that in radiocommunications a Z_o of 50Ω is most common for transmission. However, for receive-only systems 75Ω cable is the most common. The reason? It can be proven that 50Ω is the best compromise between power handling ability and losses, whereas 75Ω cable is optimised for having low losses with no regard to the power handling ability.

BALANCED AND UNBALANCED LINE

On a balanced line, such as a parallel wire line, the impedance between each leg of the line above the earth is the same. This line is said to be "balanced." On the other hand, a coaxial line has a larger outer concentric conductor with a smaller diameter solid conductor through the centre. Because of this construction, it is impossible for each leg of the line to have the same impedance above the earth. A coaxial line is said to be "unbalanced."

If you find this concept hard to understand, imagine placing an Ohmmeter between each side of a parallel line and ground (as in dirt, earth). You will measure a very high resistance (impedance) in the megohm range. However, you will measure the same value between each leg and ground. This line is balanced. This is why we have twisted pair cable and why 300Ω ribbon is often twisted - in order to maintain the balance. A coaxial line, on the other hand, has its centre conductor at megohms above ground while the sheath (outer conductor) is at ground - definitely an unbalanced line.

A two wire parallel line such as 300Ω TV ribbon will "behave" as balanced line only if it is installed correctly. In a TV installation, this line must be held away from metal structures such as the antenna mast, by using stand-offs. If this were not done, the metal structure would unbalance the line and alter its characteristic impedance. A correctly installed 300Ω TV feeder is twisted at least once every 150mm. The purpose of this is to ensure that each side of the line is "influenced" to the same degree by nearby objects such as metal stormwater down pipes. Attaching 300Ω ribbon to a wall using thumbtacks driven into the centre of the dielectric is absolutely out for the same reasons (don't laugh, I have seen it done). The latter practice is commonly found in domestic TV installations and frequently leads to poor reception and interference. A balanced line is difficult to install in order to maintain the balance between each leg. Having said that, parallel balanced lines have much lower losses than coaxial lines.

WHY BE CONCERNED?

If a balanced line is installed correctly, then induced currents from your radio transmissions will flow in opposite directions on each leg of the TV line and be equal in amplitude, thus completely cancelling out and greatly reducing the possibility of interference (for the same reason, a parallel line does not radiate when used on a transmitter). If, however, the line is unbalanced, it will function more like a long wire antenna and funnel your signal into the TV set, greatly increasing the chances of interference.

A coaxial line is unbalanced by virtue of its non-symmetrical construction. At the transmitter, it is usual practice to connect the outer conductor to ground. The cable can be run any way you like and can even be buried in the ground (preferably in conduit). The induced voltages in the shield are conducted to earth and do not affect the shielded, inner conductor circuit.

VELOCITY OF PROPAGATION

The velocity of propagation is the speed with which an electromagnetic wave travels through a transmission line. The velocity of propagation within a line depends on the construction of the line. In particular, the dielectric used can significantly alter the velocity of propagation. Manufacturers of transmission line describe the velocity of propagation by stating the velocity relative to the velocity of light (or any other electromagnetic wave) in free space, commonly referred to as the velocity factor. The velocity factor can range from 0.56 to 0.95 depending on the type of cable. A line with a velocity factor of 0.66 means that the wave can travel along this line at 66% of light velocity (light velocity = 300,000 km per second).

Figure 2-38.

Some typical velocity factors are:

1. Parallel line, air dielectric, 0.95 - 0.975
2. Parallel line, plastic dielectric, 0.80 - 0.95
3. Coaxial, air dielectric, 0.85
4. Coaxial line, polyethene dielectric, 0.66

Higher quality coaxial cable (unbalanced line) has excellent shielding and has more air as the dielectric. However, care must be taken with bending, or the cable will be crushed. Crushing upset the physical dimensions and therefore the impedance.

The most important ones to remember for everyday use are 0.66 for coax and 0.80 for 300Ω parallel line. If you use something else look up the velocity of propagation on the manufacturers data sheet.

There is an interesting approximation for determining the velocity factor of coaxial lines. The reciprocal of the square root of the dielectric constant (Permittivity) is a close approximation to the velocity factor - $\frac{1}{\sqrt{\varepsilon}}$ Polythene has a dielectric constant of 2.3. So a coaxial line with a polythene dielectric has a velocity factor of: $\frac{1}{\sqrt{2.3}}$ = 0.659 or 0.66 rounded.

A LINE TERMINATED IN ITS Z_o

An electromagnetic wave is travelling down a transmission line but has not yet reached the load. Remember, a wave travels at a finite velocity, a fraction of the speed of light. As it travels its current and voltage distribution, or ratio E/I, will be equal to the Z_o of the line. The wave has no idea how the line is terminated until it reaches the termination. The wave has not reached the load – it is on its way down the line. The current and voltage of the wave must obey Ohm's law. When the wave reaches the load, which is equal in impedance to the Z_o, it will be totally dissipated in the load or radiated if the load is an antenna. Such a line is called a *flat line* as it has no standing waves.

REFLECTED WAVES

An electromagnetic wave upon reaching a mismatched termination must conform to Ohm's law. You need to imagine a wave travelling in a lossless transmission line of infinite length. The voltage and current ratio (E/I) of the wave will be that of the characteristic impedance. Now let's do away with the infinite line and place a load at the end of the line. The wave has not reached the end of the line yet, so its E/I distribution is still representing the characteristic impedance (Z_o). If the load is not equal to the Z_o, the wave upon reaching the load must go through a current and voltage redistribution so that E/I now represents the load impedance. To do this, the wave goes through a sudden redistribution of the energies contained in its magnetic and electric fields, so that the current and voltage across the load represents the Ohms law load impedance.

In going through a redistribution of current and voltage, an induced current and voltage wave is created (Faraday's law of induction) and this new wave opposes the wave that created it (Lenz's law). The induced wave will now begin to propagate through the line back towards the generator or transmitter. This is called a reflected wave. How much of the incident wave is reflected and how much is dissipated (or radiated) in the load is determined by the amount of mismatch between Z_o and the load impedance (Z_L).

Again - a wave, before reaching the termination (load), has no knowledge of the termination conditions. The wave's current and voltage distribution will be representative of the characteristic impedance of the line. As close as one micron (a millionth of a metre) away from a load, the wave is still unaware of the conditions at the load. Suddenly, upon reaching the load, an instantaneous change in impedance occurs. The voltage and current must now redistribute themselves to conform to the Ohm's law value established by the load. This rapid redistribution (of electric and magnetic fields) causes an induced reflected wave that travels back down the line from the load to the transmitter.

A line **not** terminated in its Z_o will have an *incident* or forward wave and a *reflected* wave travelling in the opposite directions.

STANDING WAVE

Standing waves are produced when reflected waves travel from the load back towards the transmitter and interact with the incident (forward moving) waves from the transmitter. The result of this interaction is called a standing wave.

Standing waves are an *interference pattern* caused by the interaction of incident and reflected waves. At certain points along the line, incident and reflected waves will be additive; at other points, they will be subtractive. If you the throw two stones in a pond a distance apart concentric waves will radiate outward. When the two waves collide they will interfere with each other and a distinct interference pattern will be seen on the pond. This is the same effect with incident and reflected waves on a transmission line.

When incident and reflected waves interact and form a standing wave, the impedance of the line is no longer its Z_o. The impedance at any point along the line is equal to the resultant and measurable E/I at that point.

Imagine placing a Voltmeter at some fixed point on a transmission line which has reflected and incident waves present. At the point of attachment, the voltmeter will measure the resultant standing wave voltage. *Moving the Voltmeter along the line will reveal that the interference pattern exhibits a periodic pattern of maxima and minima.* Common terminology refers to the resultant voltage minima as *nodes* and to the maxima as *anti-nodes*. Nodes are produced on a line where the incident and reflected waves are equal in amplitude and 180° out of phase. Anti-nodes occur when both waves are equal in phase and therefore additive.

The term 'standing wave' comes from the fact that the position of the nodes and anti-nodes do not move - they are stationary. The distance between adjacent nodes or anti-nodes is a half wavelength.

The effect of standing waves is most dramatic when the line is terminated in an open or a short circuit. In such cases, all of the power arriving at the termination is reflected.

STANDING WAVE PATTERN ON A SHORT CIRCUITED LINE

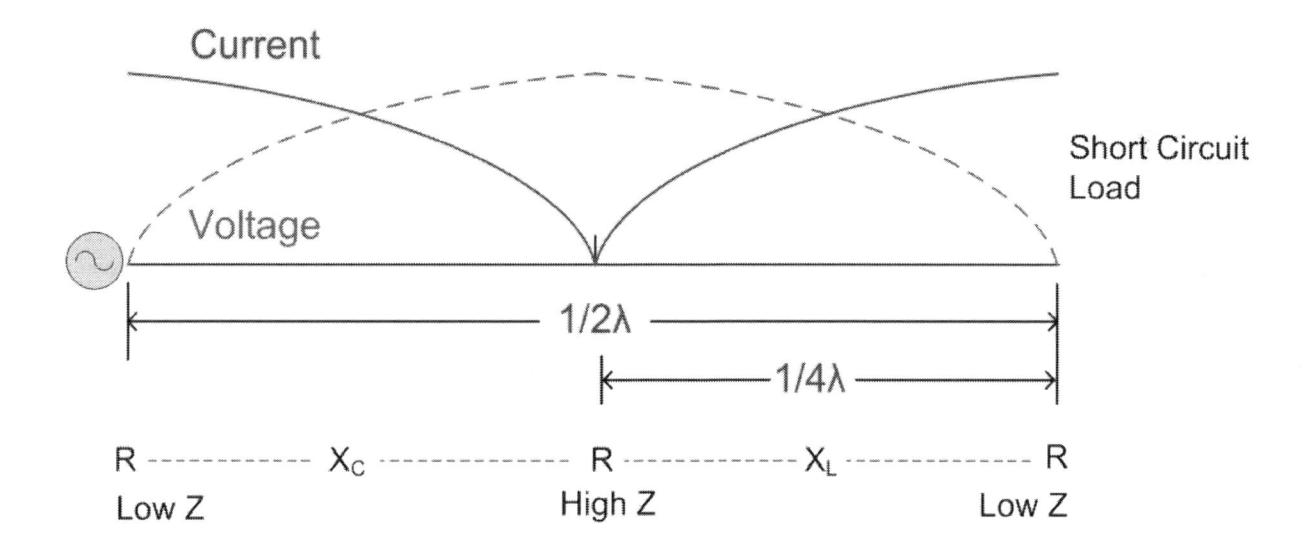

Current and voltage distribution with short circuit load

Figure 3-38

Figure 3-38 shows the actual resultant standing wave of current and voltage on a short-circuited line. If you have trouble working this pattern out for yourself, just remember this: The voltage-current ratio E/I must represent the impedance at that point on the line. Now there is one place where you definitely know the impedance; at the load. This line is short-circuited, so the load impedance is close to zero. Can you see that at the load (the right-hand side) the current is high and the voltage is low? Low voltage causing a high current is representative of a low impedance. If the impedance of the load was zero (short circuit), then the impedance 1/4λ back from the load will be infinite. Half a wavelength away from the load the impedance will be equal to the load impedance. I have only drawn a half wavelength of the line as this pattern just keeps repeating.

STANDING WAVE PATTERN ON AN OPEN CIRCUIT LINE

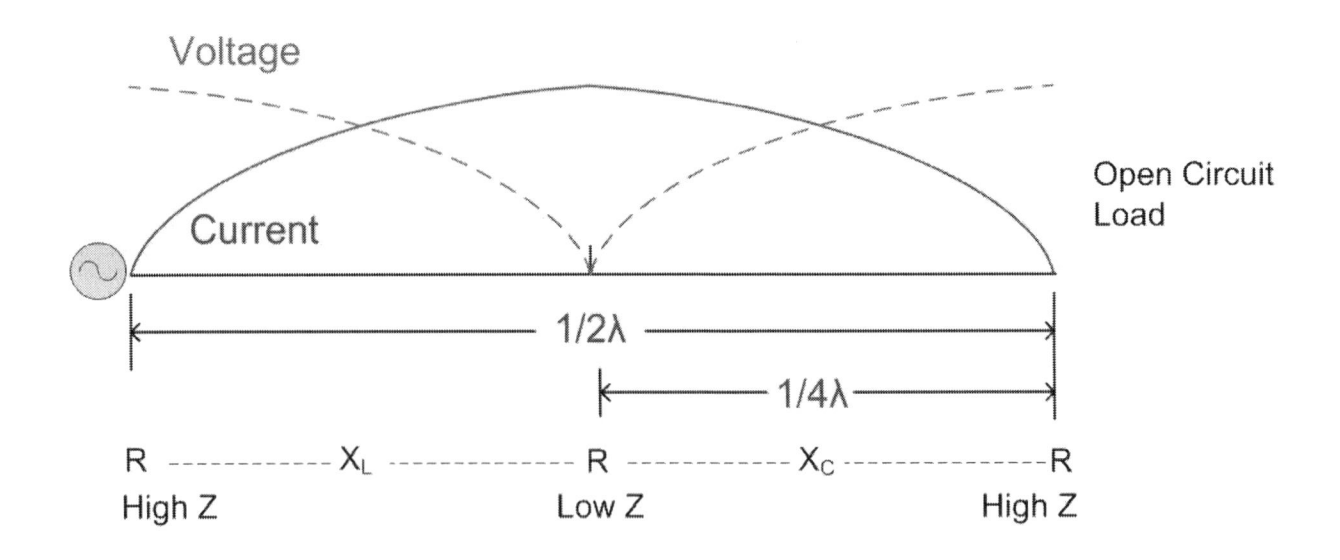

Figure 4-38

Figure 4-38 shows the standing waves of current and voltage on an open circuit line. If the pattern looks familiar to you, it should, as this is exactly what happens on a dipole antenna. Again, please note how the E/I ratio represents the impedance at that point. The load is open circuit (infinite impedance). At the load, the diagram shows a very high voltage with little or no current. Is not high voltage and low current the same as a high impedance? It is!

VOLTAGE STANDING WAVE RATIO (VSWR)

The ratio of the voltage maximum (anti-node) to the voltage minimum (node) is called the VSWR. The VSWR is an indication of the degree of match or mismatch between the line's Z_o and the load impedance. VSWR means the SWR is just obtained by voltage measurement. You could, in fact, determine the VSWR by current measurement – this is called ISWR – all three are the same thing.

VSWR (OR SWR) FROM FORWARD AND REFLECTED POWER

Many wattmeters allow the measurement of the forward and reflected power on a transmission line. Equation 4-38 can be used to convert these forward and reflected power measurements to VSWR.

$$SWR = \frac{\sqrt{FWD\ PWR} + \sqrt{REF\ PWR}}{\sqrt{FWD\ PWR} - \sqrt{REF\ PWR}}$$

Equation 4-38

Example:
Forward power = 100 Watts
Reflected power = 10 Watts

$$SWR = \frac{\sqrt{100} + \sqrt{10}}{\sqrt{100} - \sqrt{10}} \qquad SWR = \frac{13.6}{6.84} \qquad SWR = 1.9 : 1$$

We could have used any 1 to 10 ratio and obtained the same answer.

VSWR METER

Radio stations usually have an instrument which measures the amount of *reflected voltage* relative to the amount of *forward voltage* on a transmission line. Such a device is called a VSWR meter. The forward voltage is first 'SET' to full scale and then a reading of the reflected voltage is taken. The scale on the metre is calibrated to read VSWR directly. This type of VSWR meter is, in fact, measuring the *coefficient of reflection,* which is just the ratio of the forward and reflected voltage. Mid-scale on these type of VSWR meters corresponds to a coefficient of reflection of 0.5 (VSWR=3) and full scale is 1.0

(VSWR = infinite). There is a simple mathematical relationship between VSWR and the coefficient of reflection (ρ - Rho), so the manufacturer is able to calibrate the scale in terms of VSWR:

$$VSWR = \frac{1+\rho}{1-\rho}$$

Equation 5-38

If you want to calibrate your VSWR meter in terms of coefficient of reflection, then just put a linear scale on the meter face from 0 to 1. Quarter scale would be ρ = 0.25, third scale ρ = 0.33, half scale ρ = 0.5 etc.

IMPEDANCE MATCHING

For the transmitter to develop its full power and also to obtain maximum possible power transfer to the load, all components in the transmission system must be matched.

While the above statement is true, it often leads to the false conclusion that reflected power is lost power. Reflected power is *not* lost since it will be re-reflected at the transmitter. Reflected power does *not* go back into the transmitter and burn out the finals! The power amplifier may be damaged if the VSWR is high, however, this is due to the power amplifier itself creating high voltages across its components due to impedance mismatch. Reflected waves actually travel back and forth from the transmitter to the antenna – each time they arrive at the antenna a little more power is radiated. They bounce back and forth, until most of the power is eventually radiated along with some power dissipated in the transmission line due to the multiple reflections.

Generally speaking, a matched system is better because:

1. If VSWR is low, then line losses are low.
2. The transmitter will develop its full output power.
3. A flat line can carry more power than one with a standing wave present.
4. There will not be multiple reflections (and multiple attenuation) on the transmission line

The above is particularly true of coaxial lines. *Parallel open wire lines have virtually no losses even with very high VSWR.*

Standing waves place the transmission line under unnecessary electric stress. Anti-nodes (maxima) of voltage can break down the dielectric while the anti-nodes (maxima) of current cause increased copper losses (heat loss).

The impedance of a transmission line varies along its length if it has standing waves. Look at how the impedance varies along the length of a transmission line which is terminated in a short circuit or an open circuit. By a line terminated in a short circuit, we mean there is no antenna or other load, the two sides of the transmission line are just connected together. By a line terminated in an open circuit, we mean a transmitter connected to a transmission line that goes nowhere and is just left unconnected.

The line terminated in a short circuit will be *inductive* for the first quarter wave, then *capacitive* for the next quarter wave, then inductive again and so on. Whether the line is inductive or capacitive depends on the phase of the current and voltage at that point. On short circuited lines less than $1/4\lambda$ the current is increasing towards maximum at the short circuit and the voltage has already been at maximum and is decreasing. The current is lagging the voltage. Hence, the impedance is inductive.

Let's get this clear even though you may not be examined about the input impedance of a transmission line which is short or open circuited. The principles are the same for antennas.

Let's make the statement again:

A line terminated in a short or open circuit will have an input impedance which is either inductive or capacitive. Your problem is to work out what the input impedance type is, that is, inductive or capacitive. Don't memorise it, but work it out by looking at the current and voltage on the line.

Before we start, recall this:

Current LAGS voltage in an inductive circuit (think of L for inductance and L for Lag).

Current LEADS voltage in a capacitive circuit.

Figure 5-38 shows a transmission line terminated in a short circuit. The short circuit is the load and the line is less than ¼ wavelength.

Ron. Bertrand

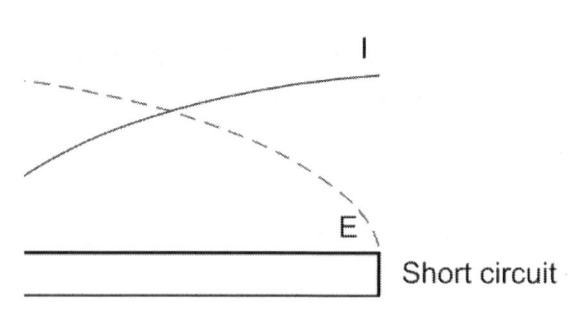

Figure 5-38

First, look at the load and work out what the voltage and current would be at the load. The load is a short circuit. A short circuit means **high current** and **low voltage**. That's what a short is. As we move back from the load, from right to left, the current must begin to fall since it is maximum at the load. Voltage is minimum at the load and as we move back from the load, from right to left, the voltage must rise. This is the voltage and current distribution on a short circuit line *less than* ¼ wavelength long.

Now the important bit. Looking into the transmission line from the left-hand side – is it inductive or capacitive? Looking into the line from the left-hand side we see current increasing and voltage decreasing. Current is on the increase and is lagging. Hence, the input impedance of this transmission line is *inductive*.

A length of transmission line terminated in a short circuit and less than ¼ wavelength is inductive. In fact, by changing the length, you can obtain any value of inductance you like. So if you need to add inductance to something (perhaps an antenna) you could use a shorted length of transmission line and adjust its length until you got the correct inductance.

For interest sake, I have calculated what the inductance would be for a shorted length of transmission line at various wavelengths. These calculations are for 50Ω coaxial cable. Remember, this is the inductance you would get looking into the end opposite to the short circuit:

0.1λ = 36.327Ω (inductive reactance)
0.125λ = 50 Ω (notice anything?)
0.2λ = 153.884Ω
0.22λ = 262.109Ω
0.24λ = 794.727Ω
0.245λ = 1591Ω
0.248λ = 3978Ω
0.25λ = $2.50197630472884 \times 10^{15}$ Ω (infinite impedance – open circuit)

As you can see from the values I calculated, if the length of a shorted transmission line (less than ¼ wave) is varied, its input impedance is inductive and varies with length. At 0.125 wavelengths the inductive reactance will always be equal to the characteristic impedance of the line (50 Ohms in our example). A line with a Z_o of 600Ω terminated in a short circuit and 0.125λ long would have an input impedance of 0j600Ω.

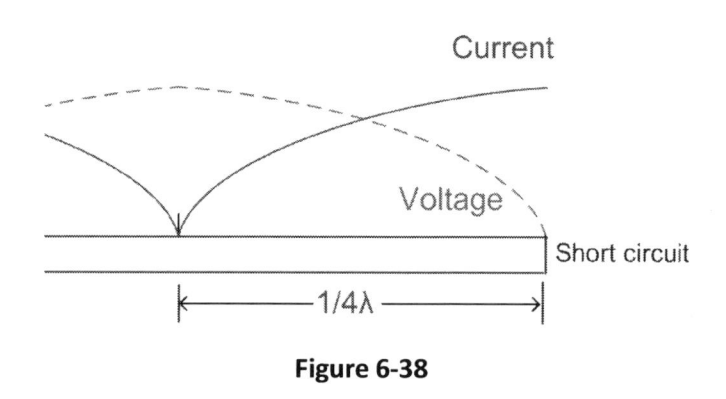

Figure 6-38

What happened at ¼ wavelength? Well, that number is huge and it was the best my calculator could do to define an open circuit. A ¼ wavelength of line terminated in a short circuit (zero impedance) will have an open circuit (infinite impedance) at its input.

Let's increase the length of the line beyond ¼ λ but less than ½ λ (refer to figure 6-38).

Looking into the line now (from the input) the current is high and falling and the voltage is rising. Current is ahead of the voltage, that is, current is leading. The input impedance is *capacitive*.

Let's increase the length of the line now to exactly ½ λ (refer to figure 7-38).

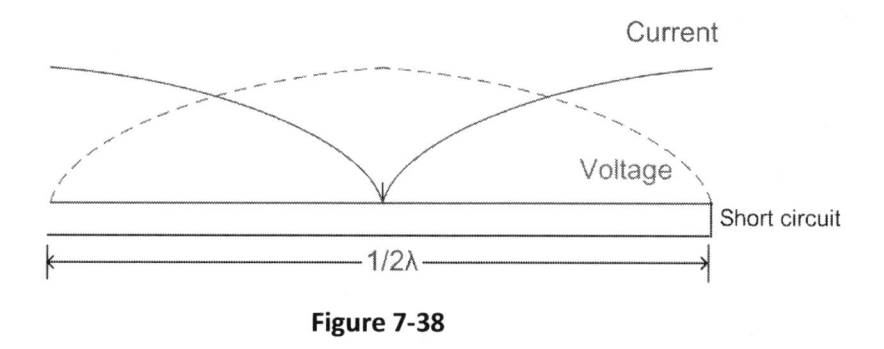

Figure 7-38

Can you see that the current and voltage at the input (the left-hand side) is now exactly the same as it is at the load? At ½ λ back from a short-circuited transmission line we again have a short circuit, or if you prefer, zero impedance.

If the load were not a short circuit. Say the load was 88-j20Ω. That is the rectangular format for 88 Ohms or R and 20 Ohms of capacitive reactance. Then on a half wavelength of line (irrespective of the characteristic impedance) the input impedance would be the same as the load. That is 88-j20Ω.

This pattern continues for any length of short circuit transmission line.

Exactly the reverse happens on a line which is terminated in an open circuit. Try drawing the voltage and current distribution on an open circuit length of transmission line one wavelength long for yourself.

The current and voltage distribution on an open circuit length of transmission line is the same as that which appears on an antenna.

What we have learnt is that standing waves on open and short circuited lengths of transmission line can be used to simulate any value of inductance or capacitance we want. We can also convert a short circuit to an open circuit or vice versa. All of this has many applications - from tuning antennas, impedance matching and even making filters. Who said standing waves weren't useful?

THE STUB

A stub is a length of open or short-circuited transmission line. The input impedance of such a line is reactive and the amount of reactance and its type (X_L or X_C) can easily be determined by equation or measurement.

Stubs are frequently used to bring an antenna to resonance. If an antenna is not resonant, it will be reactive. The reactance of a stub can be used to cancel the unwanted reactance of the antenna, thus bringing it to resonance. Short circuited stubs are favoured for this application; as open circuit stubs tend to radiate somewhat.

Figure 8-38 shows a vertical and dipole antenna which are electrically short and should, therefore, have inductance added to it in the form of a loading coil. Here have shown both ways – adding inductance with a literal inductor and adding inductance with a short circuit transmission line less than ¼ wave long. Sometimes the antenna and its stub are made from the one bent piece of aluminium tube.

A short vertical and a short dipole

Adding inductance using stubs
<1/4λ

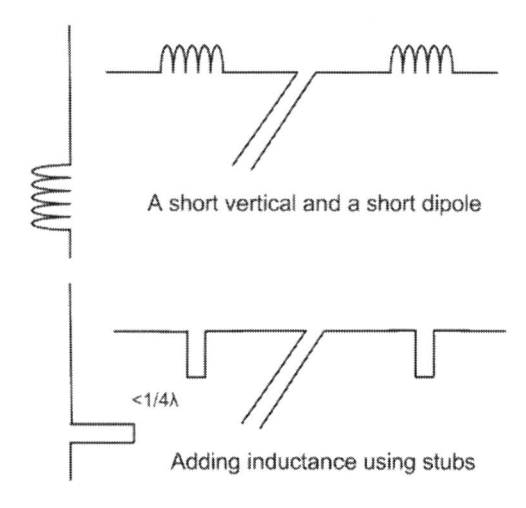

Figure 8-38

THE QUARTER WAVE TRANSFORMER

A length of transmission line called a "quarter wave transformer" can be used to match an antenna's feedpoint impedance to that of the transmission line. The characteristic impedance of a quarter wave transformer is given by:

$$Z_o = \left[\frac{\lambda}{4}\right] = \sqrt{Z_1 \times Z_2}$$

Equation 6-38

Where Z_o is the impedance of the transmission line used for the quarter wave transformer. Z_1 and Z_2 are the two impedances to be matched.

As a practical example, suppose it was necessary to connect 300Ω transmission line to a dipole antenna which has a feedpoint impedance of 75Ω (figure 9-38).

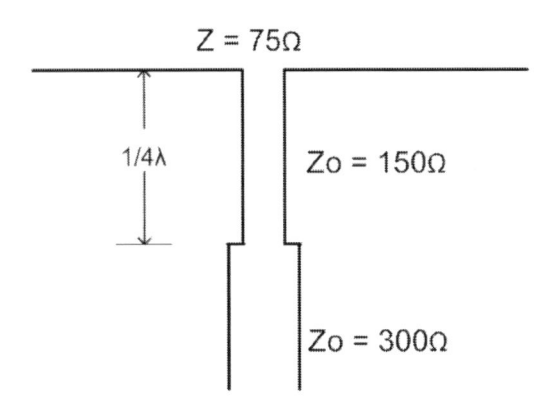

[Figure: Quarter wave transformer — Z = 75Ω feeding a 1/4λ section with Zo = 150Ω, connected to a line with Zo = 300Ω]

Figure 9-38

$$Z_o = \left[\frac{\lambda}{4}\right] = \sqrt{300 \times 75} = 150\,\Omega$$

As Figure 9-38 shows, a quarter wavelength of transmission line which has a characteristic impedance of 150Ω will match the 75Ω and dipole impedance to the 300Ω impedance of the transmission line. Remember that when calculating the length of a quarter wave transformer, the velocity factor (V_f) of the transmission line must be taken into account as follows:

$$Length(m) = \left[\frac{\lambda}{4}\right] = \frac{300}{f(MHz)} \times 0.25 \times V_f$$

Coaxial cable can be considered to have a velocity factor of 0.66 and parallel transmission line 0.80 unless otherwise told. The problem with quarter wave transformers is that they only work correctly on the frequency where the quarter wave of the line is indeed a quarter wave.

A ONE HALF WAVELENGTH 1:1 TRANSFORMER

In figure 10-38, we see a *half wavelength* of transmission line of unknown impedance terminated to a load of any impedance.

Figure 10-38

Whatever the impedance of the load, this impedance will be reflected to the input of the transmission line. Suppose the load was 72j0Ω then the input impedance will also be 72j0Ω.

A half wavelength of transmission line has the property of reflecting its load impedance to its input. The characteristic impedance of the line makes no difference - whatever the load impedance is, this impedance will be 'reflected' to the input of the line. So a half wavelength of transmission line acts as if it were a 1:1 transformer. Of course, the same would apply to one full wavelength of the transmission line, or in fact any multiple of half wavelength of the transmission line. This can be very useful when making antenna measurements. Suppose you are using a device called an Impedance Bridge to measure the feedpoint impedance of an antenna. Suppose it is not practical to work at the antenna. Provided you use a half wavelength or multiple thereof, of transmission line, the measurement taken at the input of that line will be as if it were taken at the antenna (neglecting the effects of losses).

Some impedance bridges allow you to measure the impedance of an antenna with any line length. To do this, you have to tell the impedance bridge the length of line in wavelengths and its loss. The bridge measures the input impedance and then computes the impedance at the load.

QUARTER WAVE NOTCH (Bandstop filter)

We have learnt that a half wavelength of transmission line 'reflects' the identical impedance to its input. If you look at the diagrams showing the current and voltage distribution of an open circuit line, you will see that on a line terminated in an open circuit the impedance 1/4λ back will be a short circuit. Remember that the line is only a quarter wavelength for a particular narrow band of frequencies.

AN EXAMPLE OF USING A STUB NOTCH

Suppose a TV viewer was getting overload by a nearby radio operator on 29 MHz.

To fix the problem, we need to connect a filter on the back of the TV. The purpose of the filter will be to block the 29MHz and pass all other frequencies - most importantly the VHF/UHF TV band.

A quarter wavelength of transmission line on 29MHz, terminated in an open circuit, will act as a notch filter.

Say we are using 300Ω TV cable to make our notch (we could use anything), then the length of the line is: The length of the open circuit notch is quarter of a wavelength times the velocity fact of 0.8 for 300Ω line

$$Zo \left[\frac{\lambda}{4}\right] notch = \frac{300}{29} \times 0.25 \times 0.8 = 2.0689 \ metres$$

Equation 7-38

In centimetres, that rounded up to 207cm. I would cut the notch a bit longer say 209cm. Remember it is open circuit and ribbon so it is going to be very easy to prune bits off the end to get it just right.

Connect 207cm of open circuit 300 Ω line to the back of the TV set in parallel with the existing feeder which goes to the TV antenna. You will effectively be placing a notch of around 20-25dB on the 29 MHz (+/-400kHz) band for the cost of 2 metres of ribbon.

Please do not take this method too lightly. It is extremely useful. Notches can be made easily and cheaply for all radio bands and they really do work very well to stop overload. I would say nine times out of ten this has worked for me.

A trick if you are using 300Ω line – tape the open end up and make a sleeve of alfoil which you can slide on and off the end if the notch to tune. You could solder a smaller trimmer capacitor onto the open end. The only problem with using 300Ω balanced line is your hand capacitance will alter it a bit – so adjust let go; rinse and repeat.

Your notch can just as easily be made with a coaxial line and connected to the back of the effected receiver through a "T" connector.

Something to try

If you have a radio receiver on any band, one that you can get to the antenna terminals, tune the receiver to a constant signal and make a notch for that frequency to eliminate or reduce that signal. If you try this, work out the stub length from the above equation and cut it a little long. Connect it to your receiver and then trim 2-3 mm off at a time using pliers. Watch the 'S' meter dip as you cut (tune) the stub. This is a practical way to tune any transmission line notch filter.

THE STRIPLINE

We mentioned earlier in this book that filters for UHF and above are often too difficult to make from standard capacitors and inductors. We mentioned, that filters could be made using transmission line principles. One type of filter and/or transmission line is called Stripline. At UHF and higher frequencies, transmission lines become so short in terms of wavelength that half and quarter wavelength lines can easily be made of copper printed circuit board track. These printed circuit board transmission line filters are called Stripline filters. They can be used to create an inductive or capacitive reactance, act as resonant tank circuits, or filters.

The Stripline for Advanced Theory is examined only to the point that you need to know stripline can be used for filters and transmission lines at UHF and above. Microstrip is not in the Australian syllabus and there is often confusion between the two, I will cover both.

Microstrip vs. Stripline

The importance of PCB interconnect techniques increases with frequency of operation. At higher frequencies, transmission line wiring provides superior performance by minimizing crosstalk, signal distortion and radiation (as compared to ordinary point-to-point wiring). When implementing transmission lines on a printed circuit board (PCB), there are two options: Microstrip and Stripline.

Microstrip.

A Microstrip transmission line consists of a copper trace separated from a ground plane by an insulating substrate. This configuration is depicted in the image below.

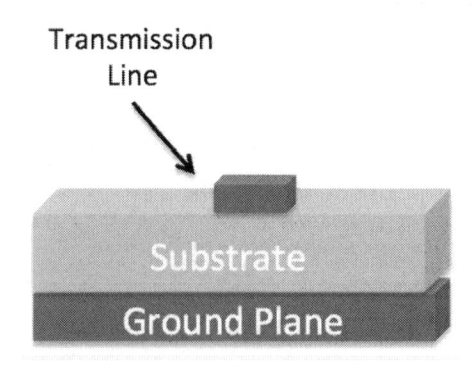

Figure 11-38. Microstrip Transmission line

Since one side of the conductor is exposed to air, these transmission lines can only exist on top and bottom PCB layers. The trace impedance is influenced by both the dielectric of the substrate material and the air above it.

Stripline.

Unlike Microstrip, stripline transmission lines are fully contained within a substrate – which is sandwiched between two ground planes.

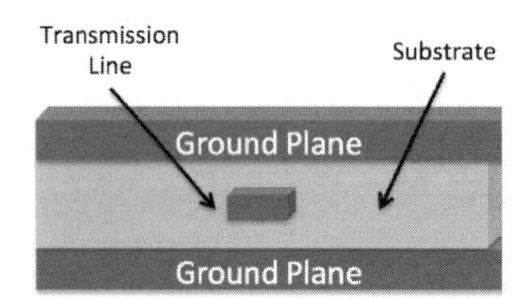

Figure 12-38. Stripline Transmission line

Due to the substrate impregnated nature of stripline, these transmission lines can only exist on internal routing layers and require a minimum of 3 board layers (2 ground planes and a routing layer).

There is more to learn about transmission lines than I have covered in this chapter. You will find that you can do all sorts of 'magic' things using transmission lines.

Like all transmission lines, the characteristic impedance depends on the physical dimensions and the dielectric used. Also like all transmission lines, Stripline and Microstrip can be used to make filters and stubs.

TRANSMISSION LINE BALUNS

The most popular type of Balun and the type most often used by Radio experimenters is the Transmission line type. These are called "Transmission line transformers" (TLT) because they work on transmission line principles hence their inclusion here and not in the chapter on "transformers". Ordinary transformers that have a primary and a secondary are more correctly called "mutually coupled transformers." This is because they are really two or more inductors magnetically coupled and operate on the principle of Faraday's Law.

TLT's are not mutually coupled transformers

These Baluns are also widely misunderstood. It does not help when amateur radio literature going back to the 1950's contains much conflicting information or unsubstantiated claims. Often when fallacies survive for a long time, they become entrenched and more or less accepted as fact.

An exhaustive treatment would require a book. We will try to cover the essentials here and perhaps dispel some of the worse fallacies.

I think one of the major misconceptions comes from the fact that a transmission line Balun looks like a transformer. That is, it looks similar to a conventional mutually coupled transformer which also can be used as a Balun. Mutually coupled transformer baluns are not broadband – TLT's baluns are the preferred choice for radio use.

The name is also confusing. A "Transmission line Balun" does **not** mean a balun used in a transmission line. Though it is used in a line, this is *not* what the name means. Such a Balun is one that utilises "transmission line principles" for its operation.

Parallel wire (or coaxial cable) lines can be wound on to ferrite cores or even air cores to create a 1:1 balun. When a toroid or rod is wound with a parallel wire transmission line it may appear on casual inspection that it is a conventional mutually coupled transformer – it is not – it is actually a 1:1 Guanella balun though many just refer to this balun as an RF choke.

The two most common baluns used in radio are the *"Ruthroff"*, often referred to as a *"voltage balun"* and the *"Guanella"* often referred to as a *"current balun"*.

Figure 13-38 shows a 4:1 transmission line (Ruthroff) 4:1 balun. This is one type of TLT balun. When we look closely at TLT baluns, we find some very interesting features. Firstly, the magnetic core is nearly passive when it comes to transferring energy between input and output. Ordinary mutually coupled transformers are affected by a degradation in balun action at high frequencies due to the transformer acting more like a capacitor as frequency is increased together with increased core losses. Transmission line baluns, on the other hand, have an astonishingly wide operating bandwidth. They don't appear to have enough windings on them and it seems hard to believe that such small devices can transfer so much power. The power transfer efficiency of a good TLT balun is in the order or 95-98%.

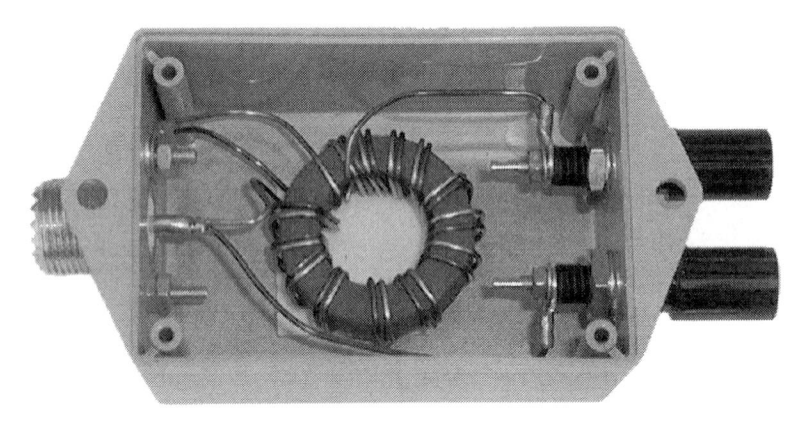

Figure 13-38

Figure 13-38 shows a 4:1 Ruthroff Balun also sometimes called a voltage balun. Wound on a ferrite toroid with a Permeability (μ) of about 140.

Note: *having excessive μ causes too much choking reactance and high voltages to be produced across a balun. Not having enough μ will mean that you do not have enough choking reactance.*

The reason for this is that the "windings" of a transmission line balun are not windings at all! What we have wound around the toroid is a very short "transmission line". There is no reliance on inductive (magnetic) effects between what "appears" to be windings. These short transmission lines are less than $1/8^{th}$ λ. The coiling of the very short line around a toroid has no direct electrical significance for balun action. However, if there are leakage currents down one side of the antenna through the balun, there will be a choking reactance or RF Choke effect. These leakage currents are often referred to as "common mode currents". Common mode currents would cause the transmission line to radiate and significantly increase the probability of interference.

The use of a toroid also enables the transmission line to fit into a convenient small space and brings the "opposite" ends together for easy interconnection. Remember we can coil these transformers from parallel or coaxial line. It makes no difference provided the correct impedance of line (Z₀) is used. If we can wind them from coaxial line, then it should be obvious that we are not relying on mutual inductance – (the linking of magnetic fields) for the operation of these devices.

The pictorial diagram of the Ruthroff 4:1 balun of figure 13-38 is shown as a schematic diagram in figure 14-38.

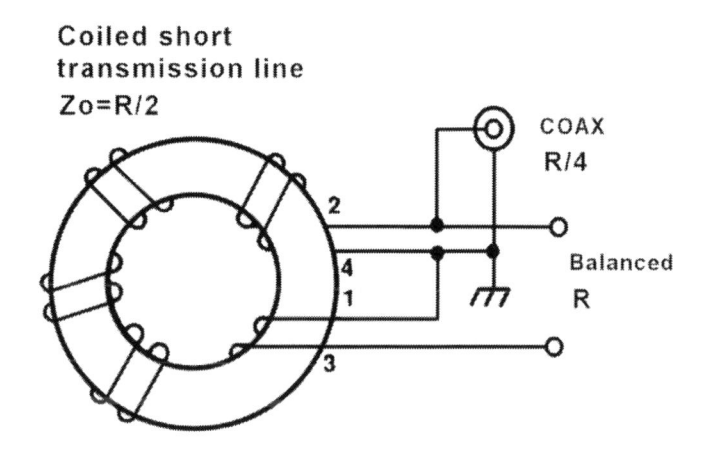

Figure 14-38

The characteristic impedance of the coiled transmission line should be the geometric mean of the input and output impedance. This is a 4:1 balun so it could be used to match 200Ω (balanced) to 50Ω (unbalanced). If we did this the characteristic impedance of the short length of transmission line wound around the toroid should be:

$$Z_o = \sqrt{R_{in} \times R_{out}} = \sqrt{200 \times 50} = 100\Omega$$

Equation 8-38

What does 4:1 mean?

Now it is often wrongly thought that a 4:1 Transmission line Balun can be used to match any two impedances with a ratio of 4 to 1. *Not so.* A 4:1 balun designed to match 200 to 50Ω cannot be used to match say 100 to 25Ω. The characteristic impedance of the line wound on to the toroid is an intrinsic parameter of the design. The Z₀ of the line wound onto the toroid is a determining factor of the input and output impedances to which the balun can be connected.

Hopefully, you are beginning to see how different these devices are from conventional transformers – though this critical difference is not always stated in much of the popular literature including some major handbooks.

In practice, the three impedances (R_{in}, R_{out} and Z_o of the coiled line) from the equation given above can vary somewhat and the balun will still work. For example, the Z_o of the coiled line in the 4:1 balun shown in figure 14-38 should be exactly 100Ω. If you think about what determines the Z_o of a transmission line you should realise you can't just wind a balun with any two wires of any diameter and with just any old insulation as it is the size of the wires and their distance apart and the type of insulation between them that determines the Z_o of the line that is created.

There are some that go to great lengths to get this just right. In practice, you don't have to be too fussy.

If you look at the balun in figure 14-38, you will notice how the spacing of the line around the toroid varies slightly in a couple of spots. Well, this would mean that there is a variation in impedance at those spots. It would be better if the two wires making up the line were bound together every 10-15 mm with a narrow strip of tape 3-4 mm wide. Having said how it should be ideally, small variations in spacing (and consequent aberrations in Z_o) don't seem to matter that much. The SWR of the balun figure 14-38, when terminated in 200Ω, should have an SWR of 1.5:1 or better from 2-30MHz.

I mentioned that the type of transmission line used to make the balun is not important as long as its characteristic impedance is correct. Figure 16-38 shows the same (as in figure 15-38) 4:1 Ruthroff Balun made using coaxial cable and coiled around a ferrite rod. Here it is even more obvious that the magnetic core does not influence the transmission line. The core does not participate in the transfer of energy as in a conventional transformer. This is why high power baluns can be such small devices.

R → $R/4$ →

$Z_o = R/2$

Figure 15-38

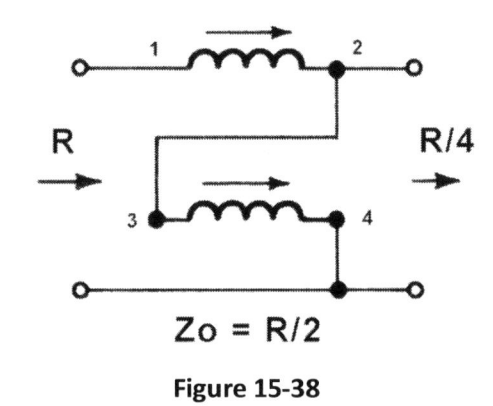

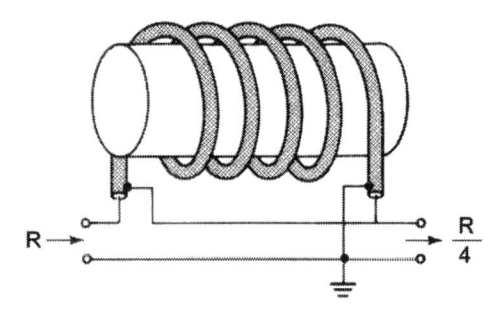

R → $\dfrac{R}{4}$ →

Figure 16-38

Figure 17-38 shows the schematic diagram of the Ruthroff balun. The terminals 1 and 3 represent one end of a short parallel or coaxial transmission line. Terminals 2 and 4 the other end. I think it is the common use of this type of schematic which leads to the misconception that this type of balun relies on inductive transformer action. The coil symbols are not meant to imply inductive coupling. However, we do have the property of inductance from end to end. Looking at the schematic you can see that coil (3-4) is in shunt across the output. It appears to be short-circuiting the output. There really is no short circuit because the line element (3-4) does present an ordinary reactance in shunt with the output terminals. The important thing to understand is that this shunting reactance comes from the line elements behaviour as a single conductor. The shunting reactance has nothing to do with the (3-4) element working in conjunction with the (1-2) element to form a transmission line.

This undesirable low frequency shunting action of element (3-4) is reduced if the inductance is increased. A higher inductance in element (3-4) means a greater low-frequency response from the balun. How can we increase the inductance of element (3-4) without changing the characteristic impedance of the line? Well, we would wind the transmission line into a coil. If this is done on a toroid, it also brings the ends close together for easier interconnection. The effect of a ferrite core is to make element (3-4) a more effective RF Choke. So we use a core really just to extend the low-frequency response of the balun. As we go up in frequency, the length of the coiled transmission line gets longer in wavelengths. As the length of the line approaches 1/8th wavelength the high-frequency response rolls off and the balun becomes ineffective.

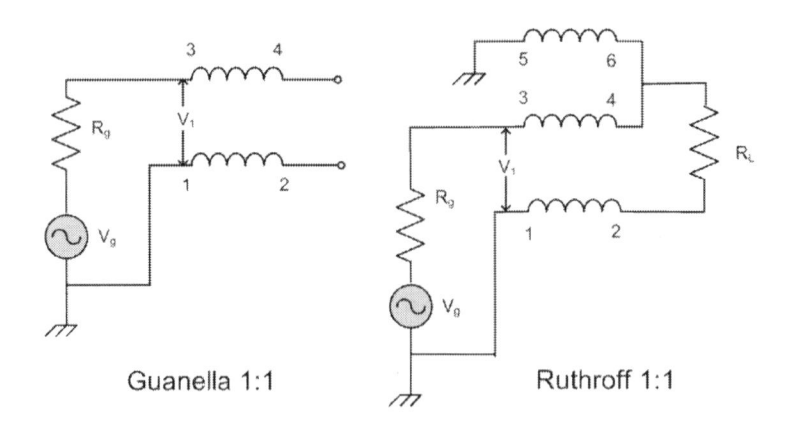

Figure 17-38

GUANELLA BALUN

This type is commonly referred to as a "current" balun. Whatever can be done with a Guanella can be done with a Ruthroff.

I find the Guanella (current) Baluns easier to construct and a little more tolerant to variations in impedance.

Guanella and Ruthroff 1:1 Baluns

We use a 1:1 balun when we don't need an impedance transformation. Figure 17-38 shows two popular and effective 1:1 transmission line baluns. I prefer the Guanella type for ease of construction and I think it is the better of the two, but you experiment for yourself. This Balun is just what it looks like. It comprises of a transmission line of characteristic impedance *equal to the line in which it is to be placed* coiled around a ferrite rod or toroid. The coiled transmission line of the Guanella forms a choking reactance for any unwanted common mode currents. The choking reactance isolates the output from the input and only flux canceling transmission line currents are able to flow. This is the reason why this type of balun has been in recent years referred to as a "current" or "choke balun".

Making a 1:1 Guanella Balun

The Guanella 1:1 balun is a short length of transmission line, of any type, usually wound on a low Permeability (50-250 µ) toroid or rod.

The characteristic impedance of the short line should be equal to the impedance of the system. If you're using the balun in a 50-Ohm system, then the line used to wind the balun should be 50 Ohms. The balun should be used in its impedance. It should not be used in a system mismatched by more than 2:1 (VSWR) for best results. As the SWR increases, there will be a higher voltage gradient along the length of the coil transmission line resulting in increased dielectric type losses.

Since the line is short, it is easy to wind the toroid with parallel line or small physical size coaxial cable.

There are two common methods of winding a Guanella 1:1 Balun. These are both shown in figure 18-38. I have used coloured wires and a red (powdered iron core) here for just for photographic purposes.

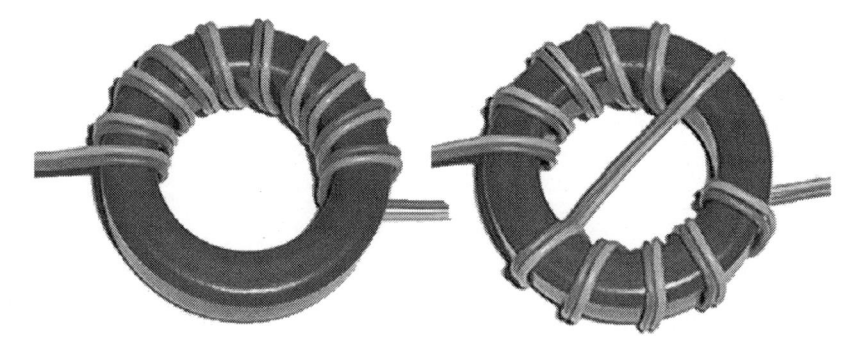

Figure 18-38

The winding method on the right of the photo is often referred to as a "magic crossover". Its proper name is a Reisert's 1:1 Balun. It really is just a Guanella with a crossover on the toroid. It is not a "magic" anything! Whilst it looks a bit more formidable than the previous photo, electrically these devices will work exactly the same. This method may have an advantage when using coaxial cable and you need to distribute the turns around the toroid to get your 10 turns.

Using Ferrite Rod

Ferrite rod works well, but the impedance may vary toward the upper end of HF. Ferrite rod does work well and it's even less expensive. About $3-4 at most popular parts stores and has a Permeability of 150. You can wind it with RG58 or similar and have connectors on each end. A popular mounting method is to place the entire assembly into a length of PVC pipe with end caps. Fill the pipe with encapsulating resin if you want to make it water and bug proof.

The 1:1 Guanella Balun is a basic building block for a range of baluns. It is easy to use two 1:1 Guanella Baluns to produce a 4:1 (or 1:4) impedance transformation. You can coil two 1:1 baluns on to each side of a toroid. At the low impedance end you connect these two transmission lines in parallel and at the high impedance end (four times higher) the lines are connected in series. This will be obvious if you look at the schematic diagram of the 4:1 Guanella balun in figure 19-38.

In figure 19-38, we simply have two short transmission lines. One line is formed by elements 1-2 and 3-4. A second line is formed by elements 5-6 and 7-8. Each of these lines taken in isolation form a 1:1 Guanella balun.

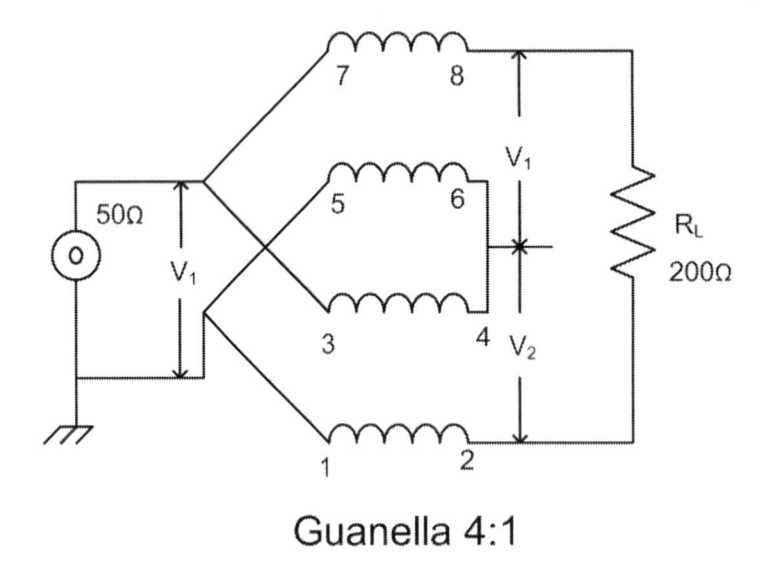

Guanella 4:1

Figure 19-38

On the low impedance side, they are connected in parallel and on the high impedance side they are connected in series. The phasing of this connection produces a 4:1 impedance transformation. The two 1:1 baluns can be wound on separate cores for very high power use or for most purposes on each side of a single toroidal core will be fine.

Figure 20-38 is a pictorial diagram using coloured wire to illustrate further how two 1:1 baluns are configured to produce a 4:1 transformation. The optimum impedance of the line used to wind this balun is as before the geometric mean of the input and output impedances. So if our 4:1 balun is matching 50 to 200Ω the line impedance making up the balun should be; $Z_o = \sqrt{(200 \times 50)} = 100Ω$.

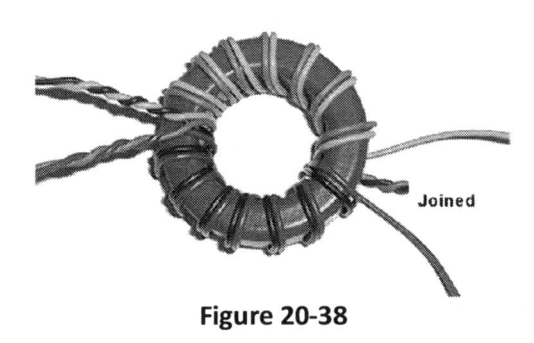

Figure 20-38

We have seen how two simple 1:1 Guanella Baluns parallel connected on one side and series connected on the other side can make a 4:1 impedance ratio. By using more parallel series connections other impedance ratios are possible.

With a Guanella balun we are not limited to using two transmission lines. Use your imagination. We could use 3 x 100Ω lines. On one side we series connect to get 300Ω. On the other side we parallel connect to get 100/3 = 33Ω. When then have a 300:33 or for practical purposes a 10:1 Balun.

If we had two 200:50 (4:1) Guanella baluns. We could series then one side to get 250Ω and parallel connect them to get 40Ω as shown in figure 21-38. This would give us a balun of 250:40 or 6.25:1.

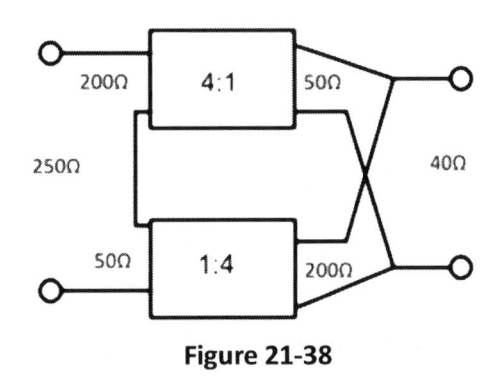

Figure 21-38

Many other impedance ratios are possible just be using different series parallel combinations.

39. TEST EQUIPMENT AND MEASUREMENTS

SSB MEASUREMENTS

The two-tone test signal comprises of two non-harmonically related sinusoidal audio tones of equal amplitude.
These tones are used to modulate an SSB transmitter and conduct linearity and power measurements.

Why use a two-tone test signal?

The two-tone test signal is the standard modulating signal used when conducting tests on an SSB transmitter. If an SSB transmitter is voice modulated then the output waveform, as seen on an oscilloscope, would be unpredictable and dependent on the voice characteristics of the operator. The peak power measured in this way would vary from one operator to the next. The two-tone signal overcomes this problem by providing a standard test modulating source which produces a predictable wave shape on an oscilloscope, enabling tests to be repeated.

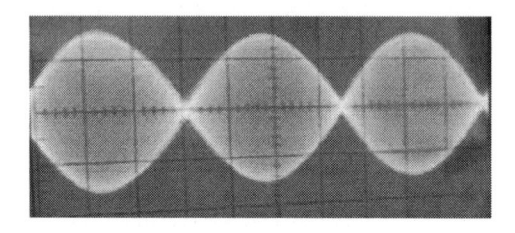

Figure 1-39

You should be able to identify the waveform in figure 1-39 as an SSB signal modulated with the two-tone test signal and displayed on an oscilloscope.

In addition, the two-tone signal enables the linearity of an SSB transmitter to be checked. If there were nonlinearity the pattern on the oscilloscope would not be nice and sinusoidal like that shown in figure 1-39 the crossover where the signal falls to zero would not be a sharp 'X'.

A nonlinear signal would look like this:

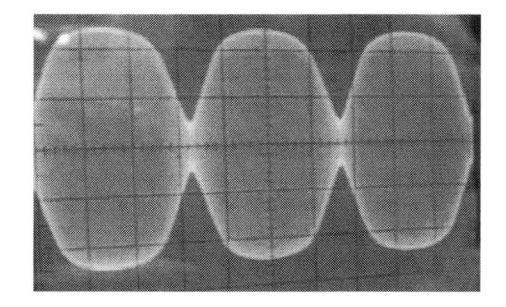

Figure 2-39

You can also see flat-topping on what should be a sinusoidal shape. Also, the crossover is not a sharp 'X'. This SSB signal would be producing splatter.

If the two-tone signal is used to modulate an SSB transmitter and the output of the transmitter checked with a spectrum analyser, then only those radio frequencies representing the two tones should be present if the transmitter is perfectly linear. In practice other signals will be present in the output - these will be the result of the two discrete signals mixing due to some non-linearity within the transmitter. The level of these mixing products is shown on specifications as intermodulation distortion products (IMD). The frequencies of the two-tone signal are not important provided they are sinusoidal, equal amplitude, not harmonically related and fit within the audio passband of the transmitter under test.

MEASUREMENT OF PEAK ENVELOPE POWER

To measure the peak envelope power (PEP) of an SSB transmitter:

1. Connect a dummy load to the transmitter output.
2. Connect a modulation monitor (usually an oscilloscope).
3. Connect an output level indicating device (an RF Voltmeter/ammeter or average/peak power meter).

4. Apply a two-tone test signal to the microphone input.
5. Adjust the two-tone signal or microphone gain for 100% modulation.
6. Read output level.

If a peak Voltmeter or ammeter is used, then the RMS value must be found by multiplying the result by 0.707. The average power can then be found from:

$P_{average} = E^2/R$

The peak envelope power, using the two-tone test, is the average power multiplied by two. In other words:

$PEP = 2 \times E^2_{rms}/R_{load}$

Keep in mind that the PEP of an SSB transmitter must be measured at 100% modulation. This creates problems for amateur operators who may reduce the microphone gain to control their power output while on SSB. It may be necessary to reduce power output because a large external linear amplifier might transmit over the legal limit at 100% modulation. Sounds fine (that is, reducing the mic gain to reduce the power out), however, it is not possible to measure the power at the reduced modulation percentage. This method is, therefore, unacceptable to the Australian Regulator. This potentially awkward situation does not usually come to a head unless serious interference problems exist and then the operator may be directed to make modifications to the transmitter in order to comply strictly with regulations.

CATHODE RAY TUBES

Figure 3-39 shows a cathode-ray tube (CRT) showing all the important parts.

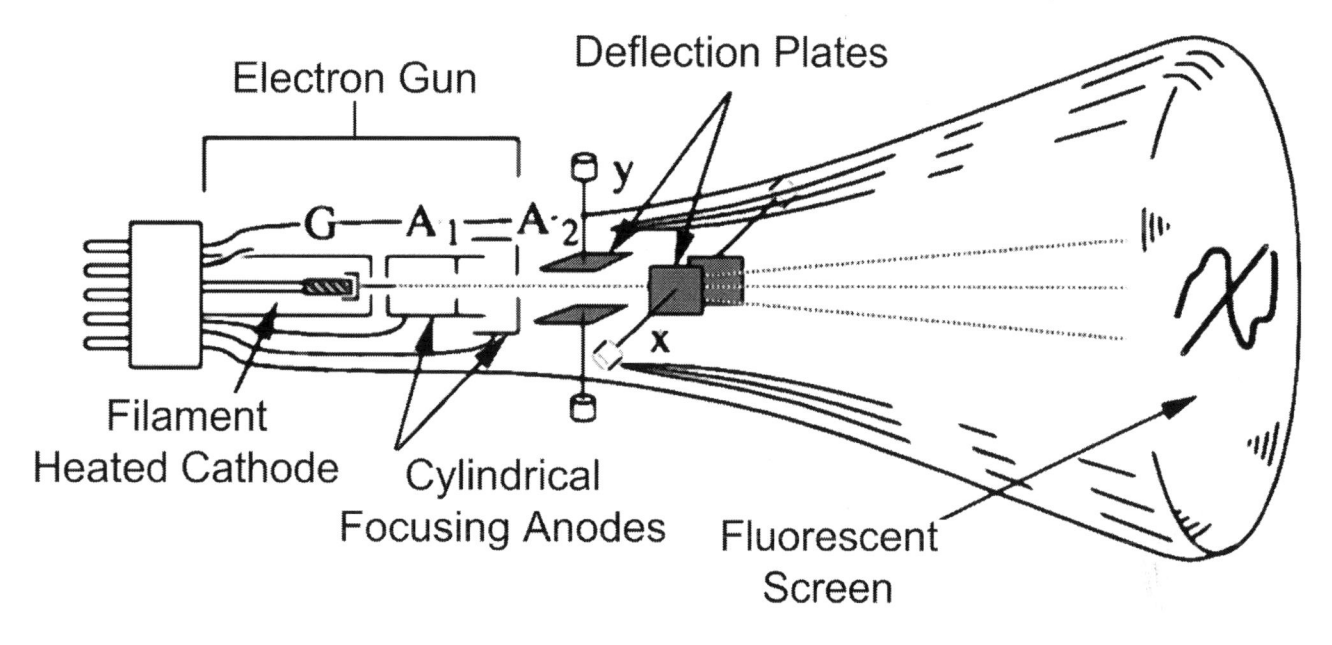

Cathode Ray Tube

Figure 3-39

The CRT is in many respects similar to the ordinary electron tube. The differences arise from the different uses which are made of the electron beam that is attracted to the anode. A controlled beam of electrons was originally called a cathode ray, hence the name "cathode-ray tube." Like the ordinary electron tube, there is a heater and cathode to create a space charge, an anode to attract electrons and a control grid to adjust the flow of these electrons. The anode current of a CRT is quite small, typically no more than 1 milliampere. The anode voltage is typically between 10 and 25kV. The anodes, as shown in figure 3-39, have a small hole, in the end, to allow the electron beam to pass through to the fluorescent screen.

It may seem that the electron beam should be attracted to the anode instead of passing through it. The reason this doesn't happen is that throughout most of its journey the electron beam is surrounded by the anode, so there is no difference of potential sideways across the tube to deflect the beam. The flanged sides of the CRT are coated with a conductive Aquadag coating which is at the same potential as the anode. This serves to keep the beam on track and drain off electrons once they have struck the screen and emitted a photon of light.

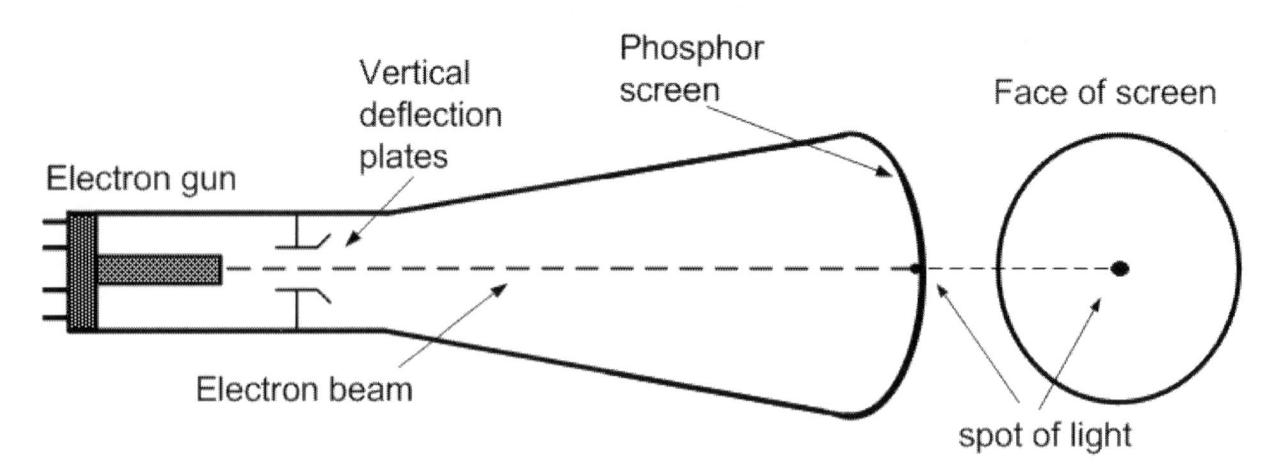

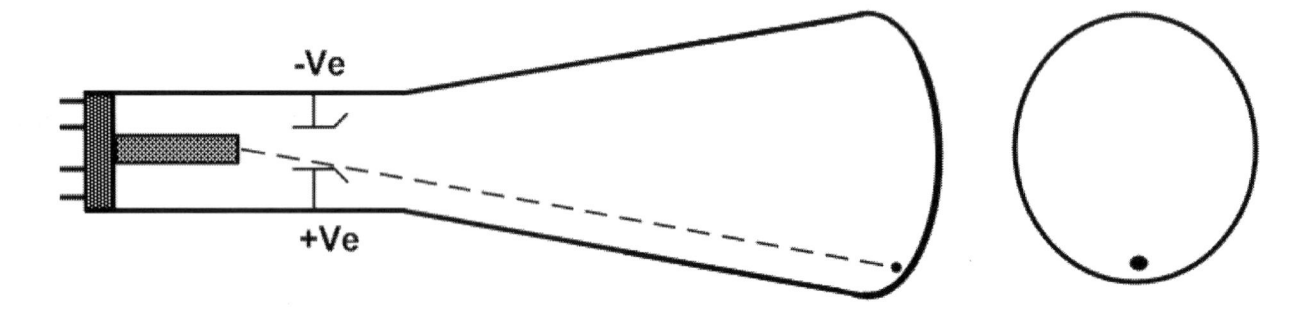

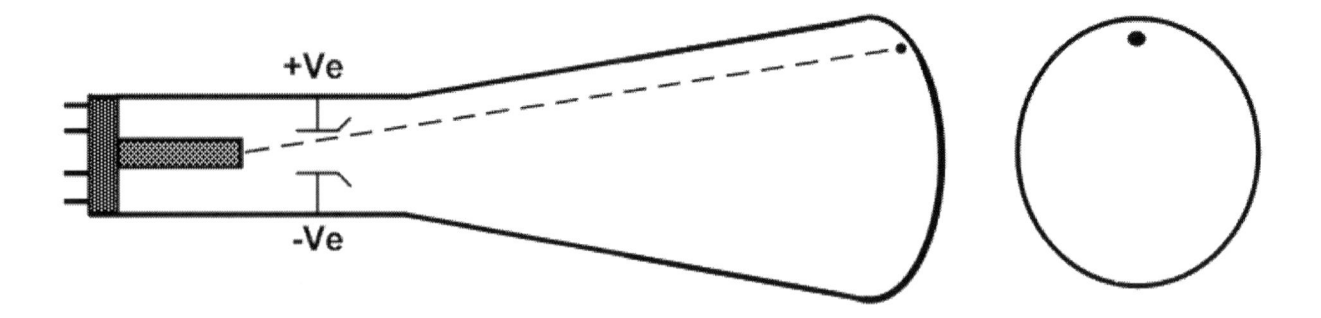

CRT – Cathode Ray Tube – how the deflection plates control the image.

Figure 4-39

The focusing electrode enables the beam to be concentrated into a narrow beam by creating an electrostatic lens (magnetic fields are used in older CRT TV receivers). Two pairs of deflection plates enable the beam to be deflected to any position on the screen by the electric field created between them from an applied deflection voltage (refer to figure 4-39).

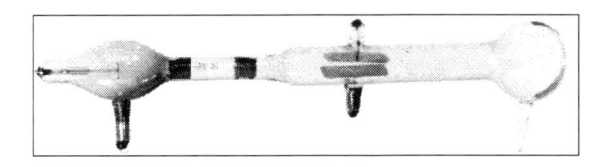

Figure 5-39

Figure 5-39 shows a picture of the first cathode-ray tube made by J.J. Thompson.

The process by which photons of light are emitted when an electron strikes the atoms within the phosphorus coating is called fluorescence. However, there is also another effect called phosphorescence, which has a continuance of the glow after the collision. This characteristic, together with the persistence of human vision, enables waveforms and TV images to be displayed without flicker on a cathode-ray tube.

THE OSCILLOSCOPE

The purpose of an oscilloscope is to provide a graphical representation of a voltage(s) amplitudes varying over time. In a sense, it is a graphical voltmeter.
A block diagram of a free-running cathode-ray oscilloscope is shown in figure 6-39. A sawtooth (sweep) oscillator is connected to the horizontal deflection plates via a sweep amplifier and a phase inverter.

Oscilloscope

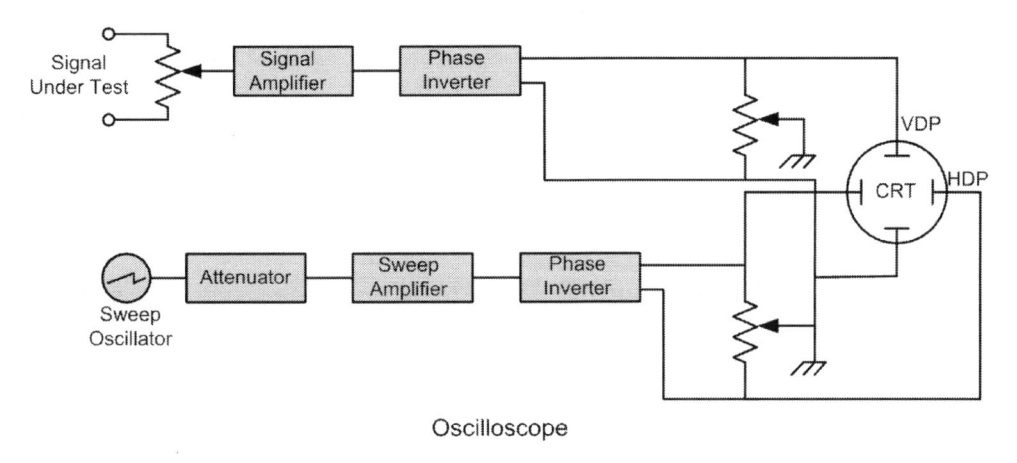

Figure 6-39

If the frequency of the sawtooth is set to 1Hz then, the electron beam will be swept across the screen once each second. The purpose of the sweep oscillator is to move the electron beam continuously horizontally across the screen.

A sawtooth waveform from the sweep oscillator is connected to the horizontal deflection plates. This sweeps the beam across the screen at the desired rate and the almost vertical time decay of the sawtooth prevents a retrace line from appearing. The signal under test is connected to the vertical deflection plates.

The sweep oscillator control on the front panel will be marked in time per division. The display is usually divided into ten divisions on each axis, each division equal to one centimetre.

If a household mains signal of 240 Volts and 50Hz is applied to the vertical deflection plates and the sweep oscillator frequency adjusted so that the electron beam moves entirely across the screen in a time equal to the period of the sine wave, then a full single sine wave would be displayed.

The signal under test is connected to the vertical deflection plates via a signal amplifier and a phase inverter. The phase inverter modifies a varying DC input signal to AC, permitting an AC signal to be positioned correctly on the centre graticule (marking) on the display. That is, positive above the centre graticule and negative below.

The brightness of the trace is controlled by the potential on the control grid of the CRT. The 'X' or horizontal axis represents time and the 'Y' or vertical axis the amplitude of the signal under test.

Multipliers - Shunts - Voltmeters - Ammeters (refresher)

Multipliers are very high-value resistances connected in series with a moving coil meter to enable the meter to measure high voltages. The excess voltage that would otherwise overload the meter is dropped across the multiplier.

Shunts are low-value resistances connected in parallel with a moving coil meter to enable the meter to measure high currents. Excess current is bypassed around the meter by the low resistance shunt.

Voltmeters must have a very high value of resistance relative to the circuit in which they are placed.

Ammeters must have a very low resistance relative to the circuit in which they are placed.

To measure the potential difference (voltage) across a resistance, the Voltmeter must be connected in parallel with the resistance. If the Voltmeter resistance is not at least 10 times higher than the value of the resistance across which it is placed, then the resistance of the meter will substantially alter the resistance in that part of the circuit and cause a false or inaccurate reading.

An ammeter measures current flow in much the same way that a liquid flow meter measures the flow of water or fuel through a pipe. If the flow meter does not have a low opposition to the flow of fluid through it, then inserting the flow meter into a line will reduce the flow of liquid and make accurate measurement impossible. Likewise, if an ammeter is to be connected in series with a conductor to measure the current flow through it, then its resistance must be so low as to have a negligible effect on the current being measured.

The Sensitivity of a Voltmeter

The total resistance of a Voltmeter can be determined by multiplying the sensitivity in Ohms-Per-Volt by the voltage scale to which the moving coil meter is switched.

If a Voltmeter has a sensitivity of 10,000 Ohms-Per-Volt and is switched to the 10-Volt scale, then the total resistance of the meter will be 10 x 10,000 = 100kΩ. If this meter was used to measure the voltage across a 10,000Ω resistor or less, then the meter would have a negligible effect on the circuit conditions and acceptable accuracy would be obtained. However, imagine if this Voltmeter was connected across a 500,000Ω resistor. The meter resistance, being only 100,000Ω, would totally upset conditions in that part of the circuit and an unacceptable voltage reading would be obtained.

A common mistake made by some students is that the resistance of the meter is determined by the voltage reading multiplied by the sensitivity - this, of course, is not correct. The total resistance of the meter is the sensitivity in Ohms-Per-Volt multiplied by the voltage range or scale to which the meter is switched.

Electronic digital Multimeters are not given an Ohms-Per-Volt rating. The input of these meters is constant regardless of the range to which they are switched. Their input resistance is typically in the order of 10 megohms or more.

Calculating the Sensitivity in Ohms-Per-Volt

The voltage drop across a 100-microampere meter movement is 100 millivolts at full-scale deflection. Calculate the sensitivity of the meter in Ohms-Per-Volt?

This question has more information than is needed to arrive at the answer. The sensitivity of a moving coil meter is directly related to the current required to deflect the meter movement to full scale, in this case, 100 microamperes. The sensitivity is easily found by finding the reciprocal of the full-scale deflection current:

Sensitivity of the meter = $1/(100 \times 10^{-6})$ = 10,000 Ohms-Per-Volt.

Calculating the Multiplier Resistance

Calculate the value of a multiplier resistance needed to enable a moving coil meter with a full-scale deflection (FSD) of 50 microAmperes and an internal resistance of 2000 Ohms, to be used as a Voltmeter to measure up to 10 Volts (refer to figure 7-39).

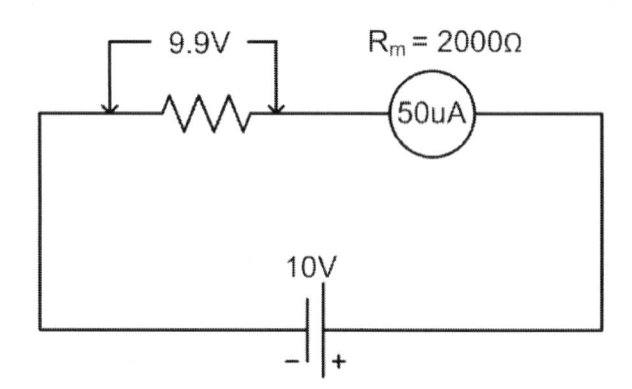

Figure 7-39

The sensitivity of the meter is $1/(50 \times 10^{-6})$ = 20,000 Ohms-Per-Volt. Since the Voltmeter is to be used to measure 10 Volts FSD, then the total resistance of the Voltmeter is 10 x 20,000 = 200,000 Ohms. Subtracting the meter movement's resistance from the total resistance leaves the value of the multiplier resistance as 198,000Ω. Putting this method into an equation we get:

$R_{multiplier} = (1 / I_{FSD}) \times \text{range} - R_{meter}$ **(Equation 1-39)**

Where:
- I_{FSD} = full-scale deflection current and
- R_{meter} = meter resistance.

The latter method can be proved by using Ohm's law to resolve the same problem. When 10 Volts is applied to the meter, the circuit current must be 50 microamperes for full-scale deflection. The voltage across the meter can be calculated from:

$E = IR = 50 \times 10^{-6} \times 2000 = 0.1$ Volts.

The remaining 9.9 Volts must be across the multiplier. Since the current through the multiplier and the voltage across the multiplier are both known, the multiplier resistance can be calculated from:

$R = E / I = 9.9 / (50 \times 10^{-6}) = 198,000$ Ohms.

THE RF PROBE

To enable a DC Voltmeter to measure RF voltages, an RF probe must be connected to the Voltmeter. The schematic diagram of an RF probe is shown in figure 8-39.

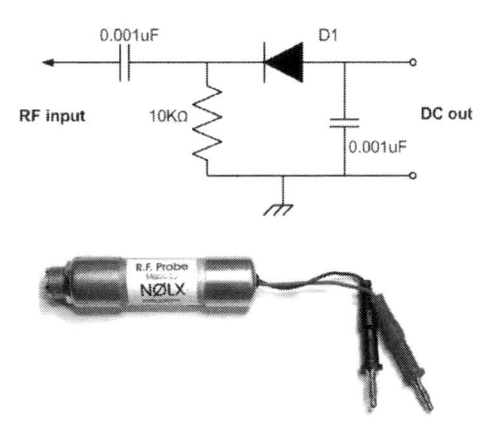

Figure 8-39

The RF probe consists of a simple half wave rectifier and filter. For the best frequency response, a point contact diode should be used (IN82A or similar). The probe must be housed in a shielded case and the lead to the meter must be shielded.

By measuring the voltage developed across a known load (say a dummy load of 50 Ohms), the output power can then be calculated from P = E²/R, where E is the RMS voltage and P the average power. Power levels as low as a few milliwatts at frequencies up to several hundred megahertz can be measured using an RF probe.

THE VSWR METER

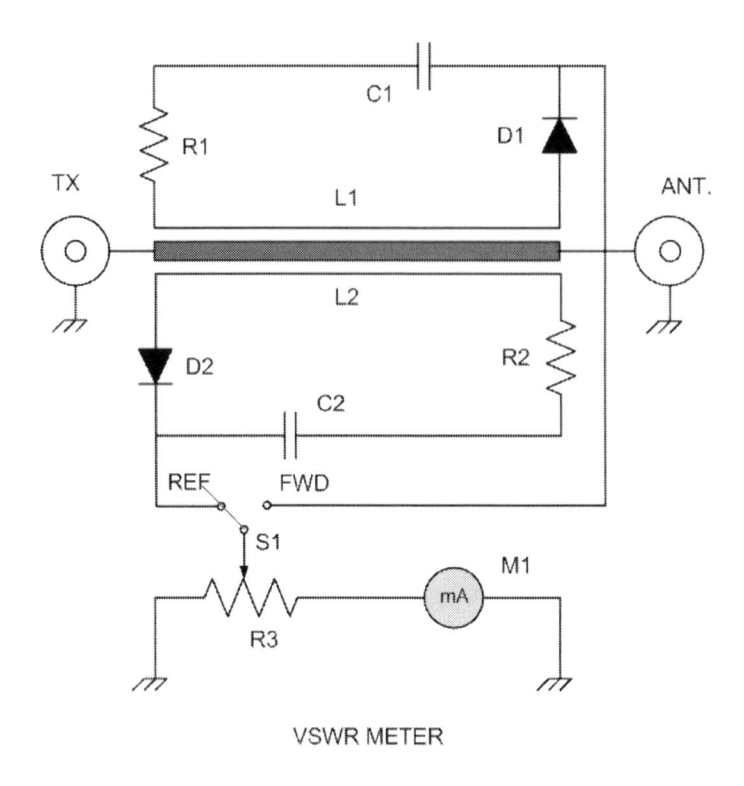

Figure 9-39

The following description of the operation of a VSWR meter is given for completeness sake and is not required for examination purposes. However, you may be required to identify the schematic diagram.

Refer to figure 9-39. A forward moving RF wave travelling through the VSWR meter from the transmitter to antenna will induce small forward-moving waves into L1 and L2.

The forward wave induced into L2 will arrive at R2 and be totally dissipated, since R2 is equal to the characteristic impedance formed by the centre conductor in the VSWR meter and L2. However, the forward moving wave induced into L1 will be rectified by D1 and this rectified current will produce a DC voltage across C1. The level of the voltage across C1 is directly proportional to the forward voltage.

If the transmission line is not terminated in the correct impedance, then a reflected wave will be present which will travel through the VSWR meter from antenna to transmitter and induce currents into L1 and L2.

The reflected wave induced into L1 is absorbed by R1, but the reflected wave induced into L2 is rectified by D2 and a DC voltage is developed across C2, which is directly proportional to the voltage of the reflected wave.

So it can be seen that the upper circuit consisting of L1, C1 and D1, is only responsive to forward waves while the lower circuit consisting of L2, C2 and D2, will only respond to reflected waves.

The switch S1 enables the operator to switch between the forward and reflected voltage readings. Meter M1 will read the voltage across C1 and C2, being the forward and reflected voltages respectively.

Though this meter could be calibrated to measure forward and reflected voltage, it is more commonly used to only measure VSWR and the operator is frequently unaware that it is, in fact, a forward and reflected voltage meter.

For VSWR measurements, the operator switches S1 to the forward position and, while transmitting, adjusts R3 for full-scale deflection on M1. The forward voltage is now referenced to full scale. S1 is then switched to the reflected position and the pointer on M1 will directly read the coefficient of reflection (K). If the forward voltage and reflected voltage are equal (open or short circuit termination), then the reflected reading will be full scale, indicating a K of 1. If there is no reflected voltage (perfectly matched system), then there will be no deflection of M1, indicating a K of zero. If 50% of the forward voltage wave is reflected, then M1 will indicate half scale deflection or a K of 0.5.

Therefore, you see, the most common type of VSWR meter really measures the coefficient of reflection. However, there is a simple relationship between K and VSWR:

VSWR = (1+K) / (1-K)   **(Equation 2-39)**

For example, suppose the reflected voltage came to half scale or K=0.5. What is the VSWR?

VSWR = (1+0.5) / (1-0.5) = 3:1

This explains why all (non-amplified) VSWR meters have '3' calibrated at centre scale. If you want to make your VSWR meter more useful, just calibrate the scale from 0 to 1 using rub on lettering and it will now double as a coefficient of reflection meter.

Reflectometer

In the schematic diagram of the VSWR meter (figure 8-39), the section of the circuit consisting of the two pickup inductors (L1 and L2) and their associated terminating resistors, is, in fact, a dual reflectometer since it has two loops for simultaneous measurement of forward and reflected voltage.

A reflectometer can be made or purchased as a stand-alone test device. To use a reflectometer, a 'level indicating device' such as a power meter or RF Voltmeter must be added. Some instruments have a moving coil meter calibrated in Watts connected to each port of a dual reflectometer.

THE DIP OSCILLATOR

A dip oscillator is just a portable hand held oscillator. The operating principle works as follows. The dip oscillator is set to oscillate at a known frequency, say 14MHz. Now suppose the dip oscillator is moved close to a circuit resonant on 14MHz. Remember when we talked about bringing a vibrating tuning fork close to another non-vibrating tuning fork on the same resonant frequency? Energy is transferred from one tuning fork to the other. A dip oscillator running on 14 MHz and brought close to a tuned circuit on 14MHz will result in RF power being transferred from the dip oscillator to the resonant circuit under test. The resonant circuit under test could be an antenna or a literal LC circuit. So, the dip oscillator then is ideal for finding the resonant frequency of an antenna or an LC circuit. Energy is not transferred from the dip oscillator to an external circuit unless the external circuit is mutually resonant with the dip oscillator. The dip oscillator has a meter - a current dip on the meter indicates mutual resonance.

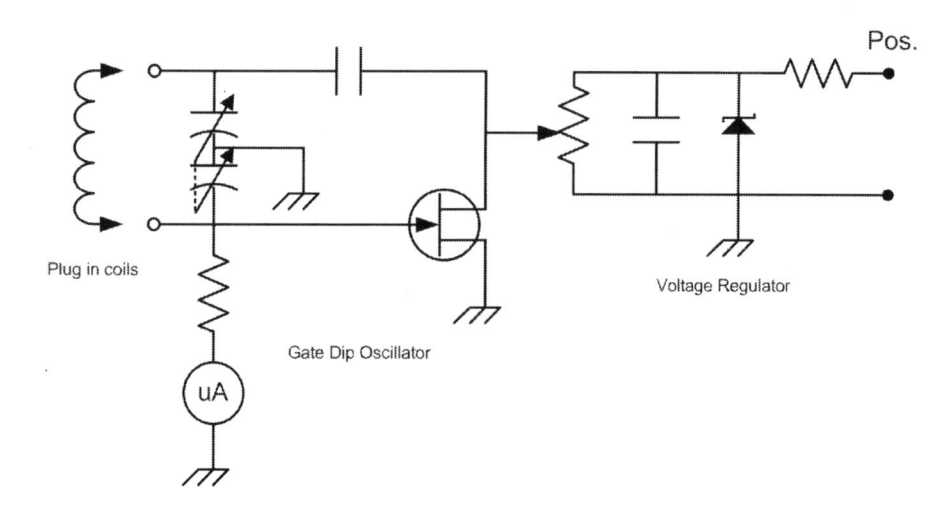

Figure 10-39

Refer to figure 10-39. When the circuit is oscillating, an RF voltage will appear across the resistor between the gate and ground. Because of the rectifying action of the gate-source junction, a DC current will flow through this resistor and indicate on the meter. In a free running oscillator, the gate voltage is high and therefore the current is high. When the coil of the resonant circuit is coupled to an external resonant circuit, power will be transferred when the two circuits are mutually resonant. This will be indicated by a dip in the gate current meter. The calibrated dial will indicate the frequency of the external circuit.

A dip oscillator only measures resonant frequency. However, if one component of a resonant circuit is known (L or C) then the resonant frequency can be measured and the value of the unknown component worked out mathematically.

I have often seen an exam question which tests for the function of a dip oscillator. One of the answer options might be to 'measure capacitance' or 'measure inductance'. Don't get caught here; a dip oscillator only measures resonant frequency. Capacitance or inductance may be calculated from knowing the resonant frequency, but the dip oscillator does not measure L or C directly.

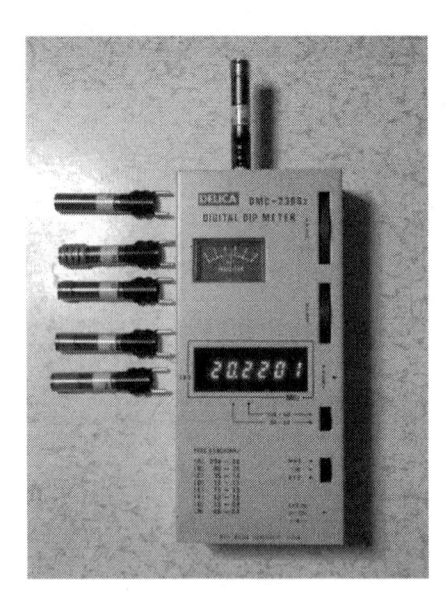

Figure 11-39

Figure 12-39 shows a practical dip oscillator.

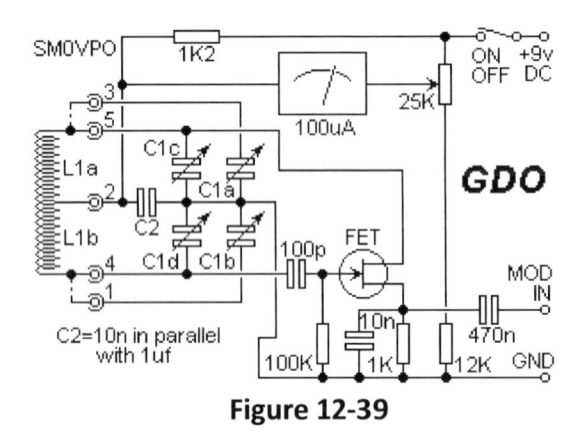

Figure 12-39

USING THE DIP OSCILLATOR AS A WAVEMETER

A wavemeter is just a simple selective well-calibrated receiver. By reducing the drain voltage to zero, oscillation stops and the gate and the source act as a diode to indicate energy is being picked up from an external 'live' resonant circuit and so the dip oscillator is used here as an absorption wavemeter. Such a wave meter would be useful for checking harmonic radiation from a transmitter, provided you do not overload the wavemeter with the transmitters fundamental. The oscillator can also be picked up on a receiver enabling it to be used as a simple signal generator.

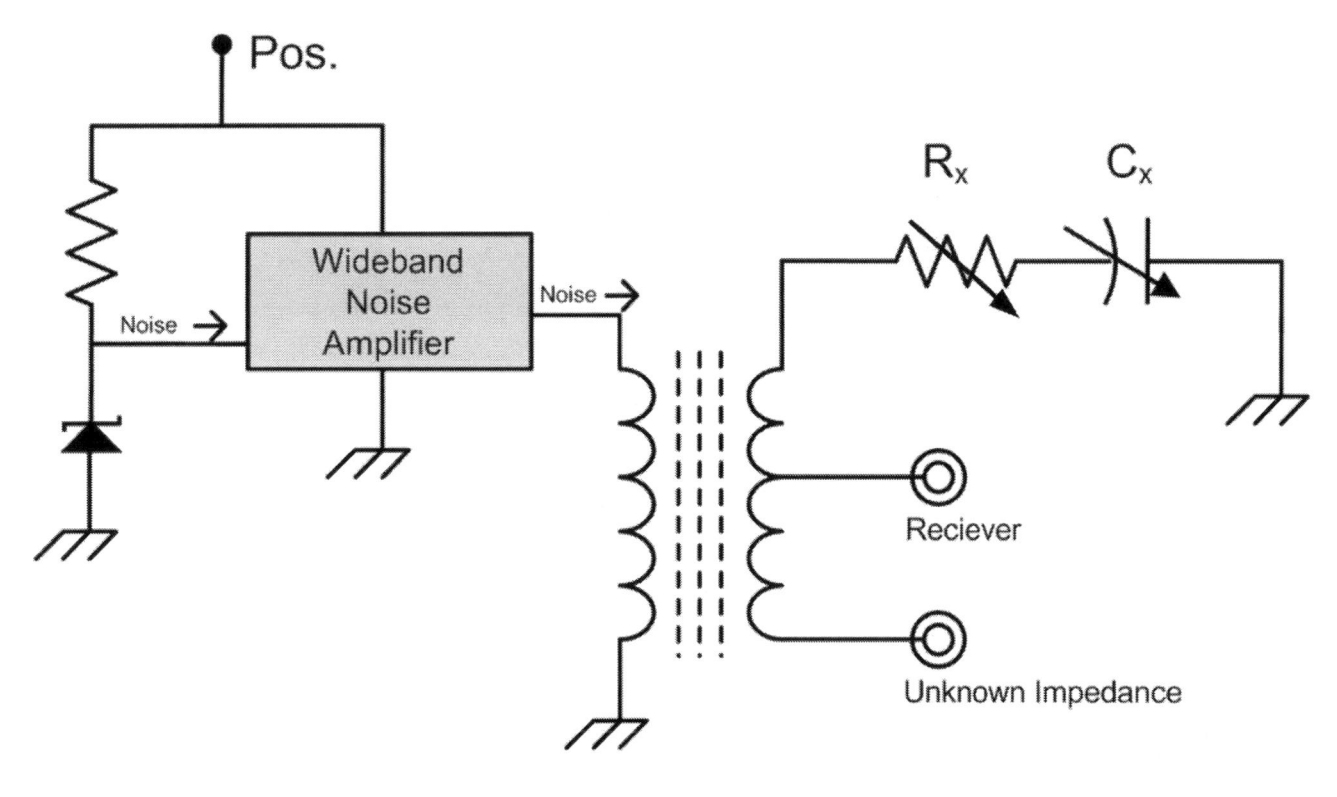

Noise (impedance) Bridge

Figure 13-39

THE NOISE BRIDGE

A noise bridge contains a noise source and an LC bridge. It is used to determine the resistive and reactive parts of an unknown impedance.

The noise source is typically a reverse biased Zener diode, which produces usable noise voltage output from about a few hundred kilohertz to around 100 MHz. The noise is amplified by a wide-band amplifier and then applied to a bridge network. An unknown impedance (usually an antenna) is connected to one arm of the bridge and the bridge adjusted for balance. A receiver is connected to the bridge to act as an RF indicator for balance. There are two balance controls, 'R' and 'X', which are calibrated for the resistive and reactive components of the unknown impedance respectively.

You simply connect the noise bridge to the impedance to be measured, usually an antenna. Using your receiver as an RF noise indicator. Then adjust the 'R' and 'X' knobs until the noise 'nulls' (when the bridge is balanced). You then read the antennas impedance as indicated by the 'R' and 'X' knobs.

If you use a noise bridge to measure the input impedance of an antenna you are trying to build or tune, then any reactance present would indicate that the antenna is not resonant.

Noise bridges are very easy to build and calibrate - a very inexpensive project. perhaps the only disadvantage of a noise bridge is the upper-frequency limit of around 100MHz.

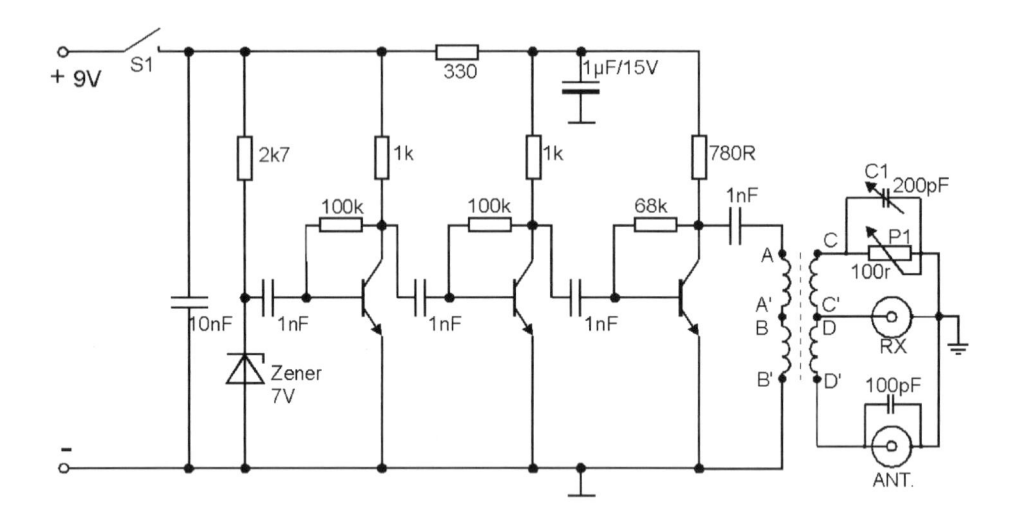

Figure 14-39

Antenna Impedance Analysers

Antenna analysers will measure impedance across a wide range of frequencies. Some go up to UHF. They do not rely on wide band noise for the source of RF. These devices, whilst extremely useful are costlier than a noise bridge though their price has been dropping dramatically. An antenna analyser will usually cover a wider frequency range than noise bridges. The noise source of a noise bridge (the Zener) drops off as frequency increases. Antenna analysers today are extremely useful instruments. They read the impedance in either rectangular or polar form. They can if given the length and loss of your feedline extrapolate measurements taken in the station to the measurement taken at the antenna. An analyser can also give you SWR. Some antenna analysers will sweep the antenna with a band of frequencies and give you a graphical result.

Figure 15-39 – a modern graphing impedance analyser.

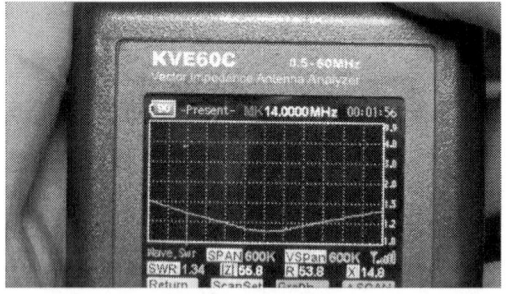

Figure 15-39

40. OTHER OPERATING MODES

Apart from the usual voice modes of AM, FM and SSB, amateur radio operators can experiment with dozens of different modes of operation. Different digital modes are introduced every year. There are literally dozens of special modes and protocols. The change is so fast that regulatory authorities in some countries just have difficulty keeping up with them all.

This chapter then, out of necessity, will be a *brief overview of some* of the many 'extra' modes that an amateur station may use. Amateur radio communication has progressed in many ways since its beginning in the early 1900's. General communications has progressed from spark to CW and voice from AM to FM and SSB. Similarly, data communications as a mode of experimenter communications have progressed from CW to FSK and from RTTY to more 'modern' modes of data communications utilising packets, etc.

Amateurs now communicate using both fast and slow scan television. An exciting developing area is high-speed wide bandwidth point to point links on the microwave bands.

I would like to encourage you to broaden your baseline in amateur radio once you have acquired your licence. Thinking that it is all too hard often stops the beginner. It is not. Have a go and you will be surprised. There is a wealth of information available for free from special interest groups and the Internet.

BINARY

Binary is a method of representing numbers with only two states - ON/OFF or High-Volts/Low-Volts.

Using the characters 0-9 forms all decimal numbers. If electrical circuits were to use 10 states like this, they would be very complex indeed. In the binary system, all numbers can be represented by a combination of 1's and 0's. Thus, it is very easy to represent any number or characters as a string of high and low voltage states. Table 1 shows the binary representation of the decimal numbers from 0 to 15.

2^3	2^2	2^1	2^0	
8	4	2	1	Decimal
0	0	0	0	0
0	0	0	1	1
0	0	1	0	2
0	0	1	1	3
0	1	0	0	4
0	1	0	1	5
0	1	1	0	6
0	1	1	1	7
1	0	0	0	8
1	0	0	1	9
1	0	1	0	10
1	0	1	1	11
1	1	0	0	12
1	1	0	1	13
1	1	1	0	14
1	1	1	1	15

Table 1

For example, a square wave digital data stream consisting of 1010 could represent decimal 10.

Binary data can be represented with many different types of modulation. For example, FSK (frequency shift keying). The secret to the lower bandwidth and higher quality of digital modes is that only two states a "1" and a "0" have to be transmitted. Analogue and Digital signals are subject to the same noise. The difference is that a noisy 1 or 0 is still a 1 or 0. Digital signals can be restored to noiseless provided the signal level is good enough. Algorithms can also be used for error checking. In sophisticated systems, much like an

Internet modem, full handshaking and error correction is implemented.

RADIO TELETYPE - RTTY

Baudot code is still the common denominator of digital modes but is far from being standardised itself. While many use the term Baudot code, others use the term Murray code. These two names acknowledge two important telegraph pioneers, both of whom made major contributions in this field. However, the correct term is CCITT International Alphabet No. 2. The differences between code types in practical systems are apparent in punctuation characters and would not be noticed between two operators unless they used punctuation such as ! : % @ etc. Baudot is sent at different speeds in different areas. Early transmission speed in Australia was mostly much standardised at 45 baud. These days, computers have almost completely taken over from the mechanical teleprinter, so baud rate changes are no longer the problem they once were.

ASCII

ASCII stands for *American Standard Code for Information Interchange.*

This is used to a very small degree on amateur bands. The proliferation of personal computers in the ham shack, often with inbuilt communications facilities designed to work on telephone networks, encourages this mode. However, there are much more practical ways to use a computer on the air. Transmission speed is generally standardised at 110 baud, but faster rates such as 200 and 300 baud are used, with varying degrees of success.

ASCII has become the standard code for communication between computers. The code was authorised for amateur transmissions in the United States in September 1980 and authorisation was given to Australian amateur stations shortly after. The ASCII code contains seven bits which are used to represent numeric values, the alphabet, punctuation marks, symbols and special control characters for teleprinters and terminals. An extra eighth bit is used as a parity check. If even parity is used then, the parity bit is set so that the sum of all the bits making up the character is even.

AMTOR - AMATEUR TELEPRINTING OVER RADIO

One common problem with direct printing RTTY (such as Baudot and ASCII) are errors in reception. If there is any slight disturbance to the received signal, it is quite likely that an incorrect character will be printed. This is not a major problem in the average amateur conversation, as the rest of the sentence will usually fill in the gaps. However, on all but the best contacts, important details (e.g. frequencies) must be repeated a couple of times to ensure that the other station gets what you want to say. This form of operation may be satisfactory for amateurs, but it is not satisfactory for commercial operations such as ship to shore teletype.

AMTOR can be thought of as a more advanced form of RTTY, which includes automatic acknowledgment of each group of characters sent, or a request for a repeat. This results in error free communication (at the expense of speed, especially in poor conditions). The protocol is very specific, so there is no variation in the transmission speed of 100 baud. As this mode requires more sophisticated equipment able to handle the error checking functions and a relatively fast receive/transmit changeover in the radio, the number of operators is slightly less than on the simpler modes.

The main feature of AMTOR is that it uses the Moore code, which is made up of seven unit characters (bits), but utilising a select few of the possible combinations, thus allowing a receiving station to tell if a received code is correct or not. AMTOR operation involves the transmitting station sending three characters, then waiting for a response from the receiving station. The response will either be "roger, go ahead", or "no go" (or perhaps nothing at all if the path has faded out). The transmitting station then either repeats the last three characters or sends the next three. This operation is a form of ARQ (Automatic ReQuest) known as AMTOR Mode A.

By sending text in bursts of three characters, awaiting a one-character reply, then sending three more, AMTOR requires radio equipment which can change from transmit to receive in a very short time. On-air AMTOR has a characteristically fast "chirp-chirp" sound.

There is another version of AMTOR (Mode B) which also uses the Moore code (i.e. has the ability to detect errors), but is an FEC (Forward Error Correction) system, rather than an ARQ system. Instead of a transmit and wait acknowledgment system, Mode B simply transmits every character twice, three characters apart. This allows the receiving station two goes at getting it right. As a result, Mode B AMTOR is more reliable than RTTY, but not as reliable as Mode A AMTOR. Mode B, as it does not require acknowledgment responses from the receiving station, is used for news broadcasts and CQ calls. Mode A is also limited by propagation timing to short-path HF contacts, due to radio propagation time intervals for around the world contacts - so Mode B is necessary for some contacts.

PACTOR

Pactor is a hybrid between packet and AMTOR techniques to try and provide a faster and more robust protocol for HF data links.

PACTOR transmits either 12 or 24 characters, depending on the baud rate which is either 100 or 200 respectively. Four characters are used for control including two checksum bytes. Errors are detected at the receiver by comparing the checksum with the accompanying data. PACTOR uses the AX-25 checksum (CRC-16). The receiver requests new data, a retransmission of data, or a change in system baud rate.

PACTOR has become a popular HF mode, as it is more efficient than AMTOR or Packet in most situations. It is readily available in commercial data controllers and is also available in software form for simple modem designs.

GTOR

The protocol that brought back those good photos of Saturn and Jupiter from the Voyager space shots was devised by M. Golay in the USA. His protocol has now been adapted by Kantronics for ham radio use.

GTOR is a protocol about four times as fast as the next fastest and has good reliability, which it gets from:

1. 16 bit CRC error detection.
2. Golay encoding with ARQ for error detection.
3. Interleaved data for noise disbursement.
4. Huffman compression and run length encoding for improved throughput.
5. 300, 200 or 100 baud to suit varying conditions.

It can transmit a full ASCII set and callsigns of up to 10 characters. GTOR is a linked mode between one station and another using a 2.4 seconds cycle, a data frame 1.92 seconds and acknowledgment of 0.16 seconds.

GTOR transmits either 24, 48 or 72 characters depending on baud rate of 100, 200 or 300 respectively. Errors are detected at the receiver using a CRC-16 checksum. The receiving station requests new data, a repeat of the last data or parity, or a change in baud rate.

GTOR is gaining some ground in HF data communications, but as it is only available from one manufacturer, it has not taken off as quickly as its benefits would suggest.

CLOVER

Clover is another high-performance HF data protocol, but like GTOR is proprietary and is also quite expensive. It can achieve quite high data throughput on a HF channel using various techniques, including data compression.

PACKET RADIO

Packet Radio is another error-free mode which has the added advantages of higher speed (if good radio signal is available) and increased economy of spectrum space by time-sharing a channel with other users.

Transmitted text is collected into "packets" before sending. Within this packet is the callsign of the packet destination, the source (sender's) callsign, information on the type of packet being sent (control, acknowledgment, etc.), data (where applicable) and a CRC (Cyclic Redundancy Check) which enables the receiver to determine whether there are any errors in the received packet. The exact format of the packet is laid down by an agreed protocol (a full explanation of packet radio can be found in recent editions of the ARRL Handbook and other such publications).

The first protocol widely used by amateurs was known as the VADCG V1 protocol. While this protocol facilitated early experimentation, it had a number of limitations. These limitations were overcome by a new protocol called AX.25. AX.25 is based on X.25, a commercial computer communications protocol.

Packet radio requires the use of a computer to handle the protocol and the exchange of data. This may be done by either programming a personal computer, or by using a dedicated computer called a TNC (Terminal Node Controller) connected between a computer/terminal and the radio. The TNC approach leaves the operator's computer free to handle application activities. A hybrid system that uses software and a computer I/O card is another alternative that has the benefits of both the programming and TNC approach.

Packet Radio has traditionally been used at data rates of 1200bps using AFSK modulation. With advances in applications and technology, this has become a rather limiting factor and 9600bps operation is becoming more common. Higher speeds are also becoming used on higher bands for networking applications and speeds of 19200bps through to

2Mbps are not uncommon.

Software

There is a wealth of software available for the different digital modes, but in particular for packet radio. Most software for amateur radio use is available free or very cheap through the shareware principle, but there is also quite an amount of commercial software available as well. The software that you use will depend on the hardware that you use, the computer system that you use and your own preferences. For most users, the main piece of software is a packet terminal program. To find out about available software ask around to see what others are using for the same equipment that you have.

Packet – more on frequency sharing

Packet radio, like the commercial X.25 packet switched systems, allows several simultaneous point to point connections to share the same frequency.

In packet radio, whenever we send a message to another station, our TNC builds a packet (like an envelope) around our message, adding details such as the callsign of the station this message is addressed to. Then, if the frequency is clear, the packet is transmitted into the airwaves.

Our TNC listens to the audio coming from the radio receiver and, if there are any other packet radio stations using the frequency, waits until there is a lull in the transmissions before transmitting our packet. At first, you might wonder how long it has to wait, thinking the other stations might be going for hours! However, in practice, a packet might take only a fraction of a second to transmit - a relatively short burst of time compared to the time it takes us to type a message. Unless the frequency is really busy, you might not even notice the delay!

The packet radio frequency in use carries messages for many people at the same time. When receiving, your TNC looks at every packet received, checks the address of each packet and identifies any packets addressed to you.

Any packets not intended for you may be ignored by your TNC, so you see on your screen only those messages addressed to you. This way you need not concern yourself with all the other information flowing around on the frequency and are free to concentrate on your contact with the other station.

If you wish, you can ask your TNC to display everything it receives. This is called monitoring. Every packet contains an AX.25 header ("envelope") which identifies the station who sent it and the station it is addressed to. The header contains other information too, but we won't get into that just now. When monitoring, your TNC will send to the computer the header and content of all the messages, so you can see who sent what to whom. Then you will see how the AX.25 systems are able to handle lots of different conversations on the same frequency at the same time. The other modes do not offer this capability.

FSK and AFSK

FSK is used on HF bands for RTTY (and other data modes). The transmitter frequency is taken as the mark condition. The space condition causes the transmitter to shift down in frequency, normally by 170 Hertz. The frequency shift is usually produced by switching a varactor diode across the carrier oscillator crystal.

AFSK is used on bands above 50MHz. In this mode, the RF carrier stays on all the time and a modulating audio tone is shifted in frequency. The standard frequencies are 2295Hz for 'space' and 2125Hz for 'mark', a separation of 170 Hz.

Many RTTY stations on HF use SSB transmitters and feed AFSK tones to the microphone socket. When an SSB transmitter is modulated with a single audio tone, it produces at its output a single upper or lower side frequency. The tone can change back and forth between the two audio frequencies, but the transmitter will only produce a single carrier frequency output for each tone. On air, this type of emission cannot be distinguished from FSK.

An SSB transmitter is not meant to operate at 100% duty cycle as is the case when modulated with AFSK tones. Consequently, the transmitter must be operated at reduced power.

DEMODULATING FSK AND AFSK

An RTTY (or other data) transmission consists of FSK or AFSK. The RTTY printer (or computer monitor) will not make any sense of such signals. The purpose of the RTTY demodulator is to convert the frequency shift signals to current pulses to operate the teletype machine. If a computer is used as the teletype decoder, then the FSK still needs to be converted into a digital pulse train acceptable to the logic circuits of the computer - the demodulator takes care of this.

SLOW SCAN TELEVISION (SSTV)

Slow scan television produces a very slow scan frame rate (no moving pictures with SSTV). Slow scan TV is popular because its narrow bandwidth requirement means that it can be used on the HF bands, enabling long distance communication. The pictures sent back from many space probes are in slow scan. For slow scan, the camera outputs a variable frequency audio tone. High tone is for bright picture information, and low tone is for dark areas. SSTV is not difficult to produce with computer software. All that is needed is some way of getting the image you want to transmit into the computer from an external camera. This can be done with a video capture card and a simple web camera. It is easy for computer software to scan this image and produce the audio tones which are simply used to modulate the transmitter. At the receiver the demodulated audio tones can be fed into the computer's sound card and software will reconstruct the image on the remote computer monitor. It is a simple matter to transmit colour pictures in SSTV. With colour SSTV the frame rate is slower.

With computers, SSTV is one of the easiest 'extra' modes to use with very limited cost. In fact, even if you do not have a camera you can transmit and receive images just using free software and your computer's sound card (and a radio transmitter of course). SSTV can be received just by audio-coupling your transceivers speaker to your computer sound card.

You can audio-couple for reception of many digital modes. Download software and you can be receiving in a few minutes. A low noise audio coupler between your sound card and the transceiver is recommended for transmission. Radio sound card interfaces are easy and inexpensive to construct yourself, or you can purchase a manufactured interface including cables for your transceiver.

Figure 1-40 shows a colour slow-scan TV picture received from the International Space Station (ISS) by radio amateurs. The amateur radio station on the ISS often communicates with radio amateurs using voice and SSTV. Many of the astronauts have an Amateur Radio Licence.

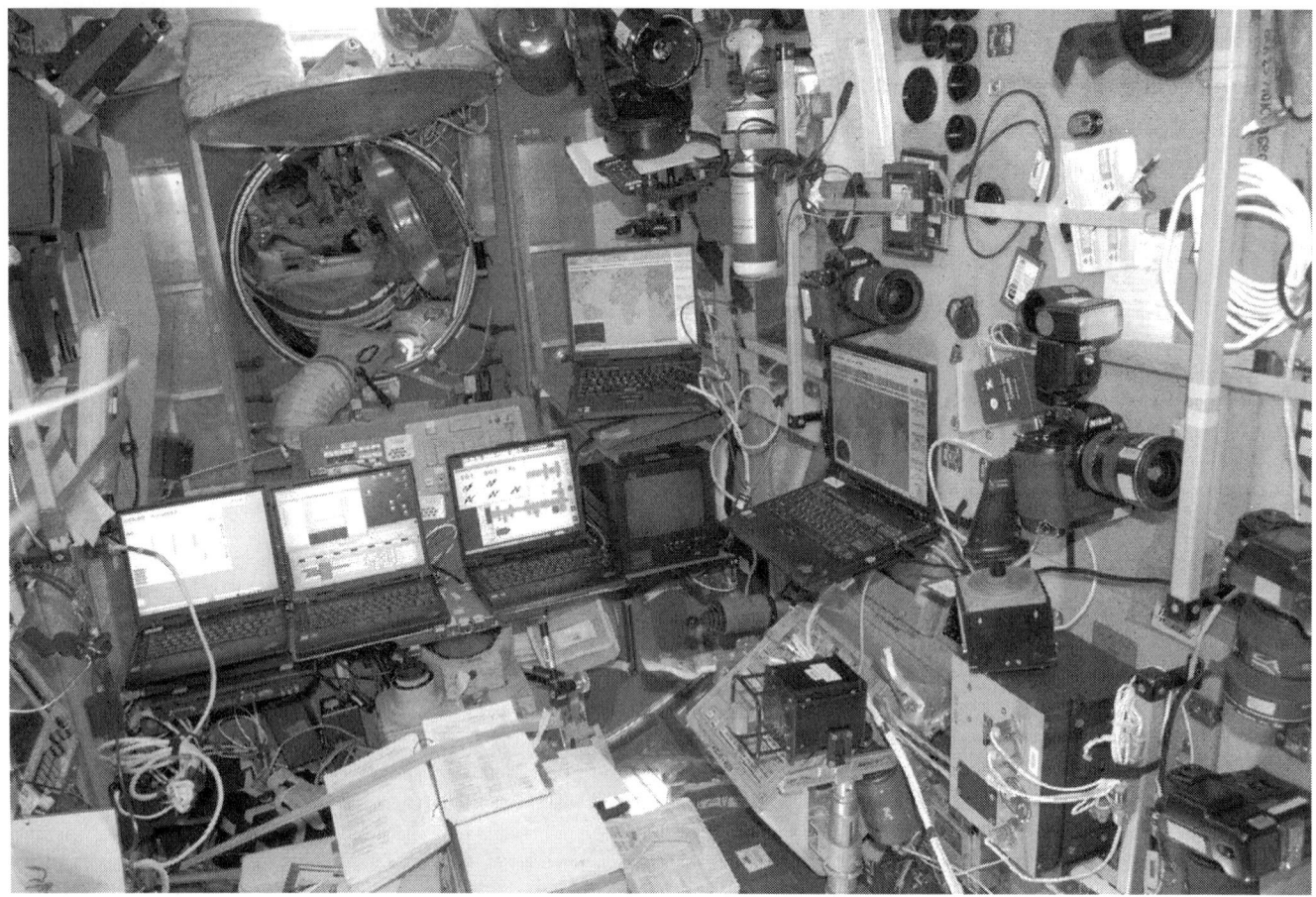

Figure 1-40

PSK31

PSK or Phase Shift Keying has become the most popular of the newer digital modes. There is a wealth of information on the web regarding BPSK (Binary PSK) and QPSK (Quadrature PSK). Because PSK31 has a bandwidth of only 31Hz, many signals can fit into the same bandwidth that would be occupied by an SSB signal (2.8 kHz approx.).

MFSK

MFSK is similar to the commercial Piccolo system. MFSK is very good under poor propagation conditions. The usual variant of MFSK is MFSK 16, but other types such as MFSK 8 are in development, along with other similar modes to MFSK (such as Domino). MFSK is sideband dependant, so you must have your receiver set to the correct sideband in order to decode it properly. Tuning is quite critical, although AFC helps somewhat.

MT63

MT63 is very robust and offers 100% copy when other modes fail. The trade-offs, however, are bandwidth and speed. MT63 is quite slow and occupies anything from

500Hz to a full 2kHz. Still less than an SSB voice channel. MT63 is usually confined to 14MHz and above.

HAM DRM

Similar in principle to the broadcast DRM signals heard on the SW broadcast bands. DRM is a very experimental mode at the moment, with the main exponents being found on 80m around 3733kHz. I have not had much success with this mode as yet, despite having good signal levels. The signals need to be very clean and strong in order to decode. Pictures can be sent using DRM, but time will tell as to how/if this mode grows in popularity.

DIGITAL / HD SSTV

If you tune to 14.233MHz you may well hear a strange signal that sounds very similar to the HAM DRM signals mentioned above. This will be one of the new Digital SSTV modes. Like all DRM modes, Digital SSTV produces excellent, noise/distortion free pictures which can be in high definition. However, for this to occur, the received signal needs to be very strong and relatively free from noise. If the program loses any part of the signal, due to a noise spike or a brief fade, the whole picture is lost.

WSPR

This introduction is taken from Joe Taylor (K1JT)'s WSPR 2.0 online user guide. WSPR (pronounced "whisper") stands for "Weak Signal Propagation Reporter." The WSPR software is designed for probing potential radio propagation paths using low-power beacon-like transmissions. WSPR signals convey a callsign, Maidenhead grid locator, and power level using a compressed data format with strong forward error correction and narrow-band 4-FSK modulation. The protocol is effective at signal-to-noise ratios as low as −28dB in a 2500Hz bandwidth. Receiving stations with internet access may automatically upload reception reports to a central database.

JT65

JT65 is intended for extremely weak but slowly-varying signals, such as those found on troposcatter or Earth-Moon-Earth (EME, or "moon bounce") paths. It can decode signals many decibels below the noise floor, and often allows amateurs to exchange contact information successfully without signals being audible to the human ear. Like the other digital modes, multiple-frequency shift keying is employed. However, unlike the other digital modes, messages are transmitted as atomic units after being compressed and then encoded with a process known as forward error correction (or "FEC"). The FEC adds redundancy to the data, such that all of a message may be successfully recovered even if some bits are not received by the receiver. (The particular code used for JT65 is Reed-Solomon.) Because of this FEC process, messages are either decoded correctly or not decoded at all, with very high probability. After messages are encoded, they are transmitted using MFSK with 65 tones. Operators have also begun using the JT65 mode for contacts on the HF bands, often using QRP (very low transmit power usually less than 5 watts). While the mode was not originally intended for HF use, its popularity has resulted in several new programs being developed and enhancements to the original WSJT in order to facilitate HF operation.

Project 25 (P25 or APCO-25)

P25 is a suite of standards for digital radio communications for use by federal, state, province and local Public safety organizations in many countries enable them to communicate with other agencies and mutual aid response teams in emergencies. In this regard, P25 fills the same role as the European Terrestrial Trunked Radio (TETRA) protocol, but the two are not interoperable. The major difference between the two is P-25 is expected to work jointly with existing analogue systems. Therefore, it uses "simulcast" method for control. "Simulcast" refers to the use of the same set of control channels throughout a given area, which are "simultaneously broadcast, or simulcast" in the region. In contrast, TETRA uses "Multicast," which means the control channel is embedded. Therefore there is no need to use a separate channel to broadcast control signals. TETRA provides 4 slots per channel, which means four voice calls can be handled on one channel. This is similar to the GSM cellular system, which allows eight slots or eight users per channel.

DSTAR

D-STAR (Digital Smart Technologies for Amateur Radio) is a digital voice and data protocol specification for amateur radio. The system was developed in the late 1990s by the Japan Amateur Radio League and uses frequency-division multiple access and minimum-shift keying in its packet-based standard. There are newer digital modes (Codec2, for example) that have been adapted for use by amateurs, but D-STAR was the first that was designed specifically for amateur radio.

Several advantages of using digital voice modes are that it uses less bandwidth than older analogue voice modes such as amplitude modulation, frequency modulation, and single sideband. The quality of the data received is also better than an analogue signal at the same signal strength, as long as the signal is above a minimum threshold and as long as there is no multipath propagation.

D-STAR compatible radios are available for HF, VHF, UHF, and microwave amateur radio bands. In addition to the over-the-air protocol, D-STAR also provides specifications for network connectivity, enabling D-STAR radios to be connected to the Internet or other networks, allowing streams of voice or packet data to be routed via amateur radio.

D-STAR compatible radios are manufactured by both Icom and FlexRadio Systems.

The system today is capable of linking repeaters together locally and through the Internet utilizing callsigns for routing of traffic. Servers are linked via TCP/IP utilizing proprietary "gateway" software, available from Icom. This allows amateur radio operators to talk to any other amateurs participating in a particular gateway "trust" environment. The current master gateway in the United States is operated by the K5TIT group in Texas, who were the first to install a D-STAR repeater system in the U.S.

D-STAR transfers both voice and data via digital encoding over the 2M (VHF), 70cm (UHF), and 23cm (1.2 GHz) amateur radio bands. There is also an interlinking radio system for creating links between systems in a local area on 10GHz, which is valuable to allow emergency communications oriented networks to continue to link in the event of internet access failure or overload.

Within the D-STAR Digital Voice protocol standards (DV), voice audio is encoded as a 3600 bit/s data stream using proprietary AMBE encoding, with 1200 bit/s FEC, leaving 1200 bit/s for an additional data "path" between radios utilizing DV mode. On air bit rates for DV mode are 4800 bit/s over the 2M, 70cm and 23cm bands.

In addition to digital voice mode (DV), a Digital Data (DD) mode can be sent at 128kbit/s only on the 23cm band. A higher-rate proprietary data protocol, currently believed to be much like ATM, is used in the 10 GHz "link" radios for site-to-site links.

Radios providing DV data service within the low-speed voice protocol variant typically use an RS-232 or USB connection for low speed data (1200 bit/s), while the Icom ID-1 23 cm band radio offers a standard Ethernet connection for high speed (128kbit/s) connections, to allow easy interfacing with computer equipment.

DMR

Digital mobile radio (DMR) is an open digital mobile radio standard defined by the European Telecommunications Standards Institute (ETSI) Standard TS 102 361 parts 1–4. and used in commercial products around the world. DMR, along with P25 phase II and NXDN are the main competitor technologies in achieving 6.25 kHz equivalent bandwidth using the proprietary AMBE+2 vocoder. DMR and P25 II both use two-slot TDMA in a 12.5 kHz channel, while NXDN uses discrete 6.25 kHz channels using frequency division.

The primary goal of the standard is to specify a digital system with low complexity, low cost, and interoperability across brands, so radio communications purchasers are not locked into a proprietary solution. In practice, many brands have not adhered to this open standard and have introduced proprietary features that make their product offerings non-interoperable.

The DMR standard operates within the existing 12.5kHz channel spacing used in land mobile frequency bands globally but achieves two voice channels through two-slot TDMA technology built around a 30ms structure. The modulation is 4-state FSK, which creates four possible symbols over the air at a rate of 4,800 symbols/s, corresponding to 9,600 b/s. After overhead, forward error correction, and splitting into two channels, there is 2,450 b/s left for a single voice channel using DMR, compared to 4,400 b/s using P25 and 64,000 b/s with traditional telephone circuits.

DMR covers the RF range 30MHz to 1GHz.

The DMR Association and manufacturers often claim that DMR has superior coverage performance to analogue FM. Forward error correction can achieve a higher quality of voice when the received signal is still relatively high. In practice, however, digital modulation protocols are much more susceptible to multipath interference and fail to provide service in areas where analogue FM would otherwise provide degraded but audible voice service. At a higher quality of voice, DMR outperforms analogue FM by about 11 dB. But at a lower quality of voice, analogue FM outperforms DMR by about

5dB.

YAESU SYSTEM FUSION

Yaesu System Fusion integrated digital voice and analogue FM. The system switches the user automatically between C4FM digital voice and conventional analogue FM. Automatic modes select (AMS) switches the users radio, working into a Fusion Repeater, from digital to analogue based on what is detected by the repeaters receiver. If you were using digital and communicating with someone on analogue through a Fusion repeater, the repeater would convert your outgoing signal from the repeater to analogue.

System Fusion has four communications modes.

In the voice data or V/D mode half of the 12.5KHz channel bandwidth is used for digital voice and the other half for data.

Voice F/R mode uses the full 12.5kHz to transmit high-quality digital audio.

Data F/R mode uses the full 12.5kHz bandwidth for higher speed data. This can be used for transmitting images or pictures.

The fourth mode is Analogue FM. Yaesu claims greater distance a lower power consumption in Analogue mode.

C4FM digital group mode notifies you when members registered to a group are within communication range. Also so providing you with the distance and bearing of other group members. Group members and send text messages and images when they are within simplex range. In digital DV mode, your position and bearing are shared with group members in real time.

Fusion transceivers have an SD card and an optional microphone with an inbuilt in camera for taking and transmitting photographs. Some Fusion transceivers have a built in GPS. I suspect these features will become very much the standard for all transceivers even those not using the Fusion system.

A

A practical PLL	350
absorption limiting frequency	362
alkaline cell	159
am transmitters and receivers	383
ammeter shunt calculation	145
ammeter shunts	144
amplifier impedance	291
amplifier classes	310
amplifier efficiency	312
amplifiers	309
amplifiers class A operating curve	313
amplifiers class B operating curve	314
amplifiers class C - operating curve	315
amplifier class AB	316
amplitude modulation	368
Amtor	549
antenna 5/8 wave	477
antenna co-linear	490
antenna corner reflector	491
antenna doublet	485
antenna fan dipole	487
antenna feedpoint impedance	476
antenna folded dipole	483
antenna gain	474
antenna impedance analysers	546
antenna j-pole	496
antenna loading	479
antenna loss resistance	475
antenna off centre fed	488
antenna physical and electrical length	473
antenna quad loop	483
antenna radiation patterns	493
antenna radiation resistance	475
antenna trapped	484
antenna V-beam	493
antenna Yagi	481
antennas	469
Armstrong oscillator	328
ASCII	549
automatic gain control	405

B

balanced and unbalanced line	503
balanced ring modulator	397
bandwidth of an AM	372
Barkhausen effect	83
battery - testing	164
battery construction	25
Bessel functions of the first kind	441
biasing transistors	298
binary	548
BJT Hartley oscillator	335
BJT transistors	287
bridge rectifier	211
buffer amplifier	321

C

capacitance	99
capacitance stray	120
capacitance factors determining	107
capacitance unit	106
capacitive reactance	121
capacitive reactance factors determining	123
capacitor charge in coulombs	111
capacitor types of	112
capacitor voltage rating	111
capacitor charging and discharging	102
capacitor symbols	100
capacitors series; parallel	108
capacitors voltage across series	110
capacitive circuit power in	129
capacitive reactance example calculation	125
capture effect	445
carrier oscillator	403
carrier oscillator dual role	405
cathode ray tubes	533
cavity filters	420
cell/batteries - series and parallel	163
cells and batteries	153
characteristic Z determining factors	501
characteristic impedance	499
chirping	327
chirping	459
choke coils	137
circuit of an overtone oscillator	340
circulators	423
Clapp oscillator	336
Clover	552
coefficient of coupling	138
Colpitts oscillator	335
Colpitts quartz crystal oscillator	342
common base	295
common emitter	295
comparing series and parallel circuits	71
complementary symmetry amplifier	319
conductance	38
constant current source	223
conversion efficiency mixer	392
converting prefixes	33
converting prefixes	34
copper loss	180
Coulomb	21
covalent bonding	273
critical resistance	464
critical angle	364
critical frequency	362
crystal filter	415
current	18
current measurement of	143
current and voltage distribution on a halfwave antenna	471
cycles per second	86

D

damped oscillations	194
Darlington pair	225
decibel absolute measurement	254
decibel equation	251
decibels	247
decibel calculations with	259
dielectric constant	106

dielectric loss	129
digital SSTV	557
dip oscillator	542
direct digital syntheses (DDS)	354
displacement of charge	21
DMR	560
doping	274
Doppler shift	270
DRM	557
DSTAR	558

E

eddy currents	178
affects of electric current	24
electric cell local action	159
electric cell separation of charges	156
electric cell polarisation	158
electric shock	48
electromagnet	74
electromagnetic radiation	260
electron	9
electron tube Armstrong oscillator	331
electron tubes	231
electron tubes cathodes	232
electron tubes diodes	233
electron tubes feedback	238
electron tubes microphonics	244
electron tubes pentode	242
electron tubes soft tube	244
electron tubes tetrodes	237
electron tubes triodes	234
electroscope	16
element	13
EMF	21
emitter follower	296
EMR does not need a medium	262
EMR spectrum	266
EMR the inverse square law	263
EMR the velocity of	264
equalising resistors	219
external induction loss	181

F

fading	366
Faraday's law	84
filter curves	416
filters	409
filters identification	413
filters insertion loss	414
flat topping	465
flywheel effect	203
FM a complete transmitter	455
FM receiver block diagram	447
FM transmitters	452
Foster Seeley discriminator	449
frequency modulation	435
frequency and wavelength	267
frequency bands	268
frequency conversion	370
frequency multiplier	322
frequency quartz crystal	338

FSK and AFSK	554
full wave rectifier two diodes	212

G

gamma match	495
grid leak bias	333
ground wave	364
GTOR	551
Guanella balun	526

H

half wave dipole	469
half wave transformer	516
halfwave rectification	210
harmonic crystal oscillator	341
Hartley oscillator	334
Henry unit	137
high and low level modulation	372
high level amplitude modulated transmitter	388
homodyne receiver	345
hysteresis loss	180

I

image interference	391
impedance	188
impedance polar form	189
impedance new perspective	270
impedance matching	510
impedance rectangular form	190
inductive reactance series parallel	140
inductance formula	134
inductance lagging current	135
inductance schematic symbols	134
inductance self back emf	132
inductance series and parallel	139
inductance RL time constant	136
inductance the property of	131
inductive reactance	140
inductor losses	141
insertion loss	415
interference	425
interference digital services	432
intermediate frequency amplifier	321
intermodulation interference	467
ionisation	25
ionisation causes of	360
ionosphere	359
isotropic antenna	474

J

JEFT amplifier	304
JT65	557
junction field effect transistor	301

K

key clicks	460

L

laminated iron core	178
law of charges	12
lead-acid cell	159
lead acid cell safety precautions	161
lead acid cell specific gravity	160
lead acid cell - sulphation	160
leading current concept	119
Leclanche cell	157
left hand rule	73
light emitting diode	285
limiter	448
lithium cells	162
logarithms	249
low level amplitude modulated transmitter	386

M

magnet demagnetising	79
magnetism	71
magnetism domain theory	76
mains filter	419
mains supply colour code	228
master slave crystal	346
maximum usable frequency	362
meters	142
MFSK	556
microphone capacitor	171
microphone carbon	167
microphone crystal	171
microphone moving coil	168
microphones	166
microstrip stripline	519
mixer schematic diagram	381
modulation index	440
molecule	10
MOSFET	304
moving coil meter	142
MT63	556
multimeter	152
mutual inductance	138

N

negative	11
neutralisation	462
neutron	10
nickel cadmium cell	161
noise semiconductors	291
noise bridge	545

O

Ohmmeter	151
Ohm's law	40
oscillation requirements for	325

oscillators	325
oscilloscope	535
other operating modes	547
overtones	339

P

P25	558
packet radio	552
Pactor	551
parallel circuits	57
parasitic oscillations	463
parasitic stopper	464
peak inverse voltage	208
PEP measuring	531
periodic table of elements	15
permeability	75
permittivity	106
phased locked loop discriminator	451
phased locked loop frequency synthesis	345
pi-coupling	465
pierce crystal oscillator	341
piezoelectric effect	344
PN diodes	278
PN diodes characteristic curve	282
PN diodes reverse bias	279
PN diodes forward bias	280
point contact diode	292
power	44
power parallel circuits	65
power supplies	208
power supplies electronic regulation	221
power supplies pass transistors	225
power supplies capacitive filtering	213
power supplies transient protection	219
power supply filtering	216
power supply practical	214
power supply filters types	218
power supply with pass transistors	226
pre-emphasis and de-emphasis	446
primary and secondary cells	157
propagation	359
protected gate MOSFET	306
protons	10
PSK31	556
push pull amplifiers	317

Q

Q calculating	203
Q of a resonant circuit	199
quarter wave antenna	478
quarter wave notch filter	517
quarter wave transformer	515
quartz crystals	337

R

radio & TV broadcast filters	424
radio teletype	549
RC time constant	118
reactance modulator	438
reciprocal equation deriving	64

reflectometer	542
resistance in parallel	63
relays	82
repeater duplexer	421
resistor types	38
resistance	18
resistance	26
resistance in parallel	61
resistance of earth	50
resistance power dissipation in	48
resistance of a conductor calculating	29
resistivity	29
resonance	192
resonant circuits	192
resonant circuits impedance	199
resonant frequency	197
RF probe	539
RLC circuits	188
rms	94
Ruthroff balun	526

S

Schottky diode	292
selectivity curve describing	207
self oscillation	461
semiconductors	272
semiconductors holes	276
series circuits	51
series circuits power in	54
series parallel calculations	69
series regulation	221
series parallel circuits	66
shunt regulation	221
sidebands	396
significant attenuation	411
silicon controlled rectifier	307
simple electric circuit	19
simple FM transmitter	437
sinewave	92
single sideband suppressed carrier	376
single tone test	378
skip distance	365
slow scan television	555
speed of the electrons	20
sporadic-e	362
SSB measurements	530
SSB reception	403
SSB transceiver complete	406
SSB transmitter block diagram	399
stripline filters	424
sunspot cycle	361
superheterodyne receiver	389
swinging chokes	217
switching regulator	221

T

test equipment and measurements	530
the mains supply	227
the metric system	32
the ratio detector	451
the resistor colour code	35
thermionic emission	232
three terminal regulator	222
transformer auto	183
transformer impedance equation	184
transformer power construction	178
transformer using impedance equation	185
transformers	174
transformers impedance matching	183
transformers isolation	183
transformers voltage ratio	181
transistor configurations	294
transistor current gain	289
transmission line baluns	521
transmission lines	499
transmission lines reflected waves	505
transmission lines standing waves	506
transmitter faults	459
trapezoid patterns	375
transmission line stubs	514
triodes amplifier	238
tropospheric ducting	366
two-tone test	530
two-tone test signal	377
types of filters	411
types of interference	426
types of voltage and current	26

U

unit of charge	21

V

varactor diode	284
velocity of propagation	504
virtual height	363
voltage	18
voltage controlled oscillator	343
voltage divider	55
voltage doubler	219
voltaic cell	154
voltmeter	146
voltmeter multiplier	147
voltmeter multi-range	149
voltmeter resistance	148
voltmeter sensitivity	149
VSWR equations	509
VSWR meter	509
VSWR meter	540

W

work	47
WSPR	557

X

X-rays	243

Y

Yaesu system fusion	561

Z

zener diode voltage regulator	283
zener diodes	282

NOTES

Printed in Poland
by Amazon Fulfillment
Poland Sp. z o.o., Wrocław